AF614800

METHODS IN MOLECULAR BIOLOGY™

Series Editor
John M. Walker
School of Life Sciences
University of Hertfordshire
Hatfield, Hertfordshire, AL10 9AB, UK

For further volumes:
http://www.springer.com/series/7651

Genetic Toxicology

Principles and Methods

Edited by

James M. Parry and Elizabeth M. Parry

7, Cedar Mount, Lyndhurst, SO43 7ED, United Kingdom

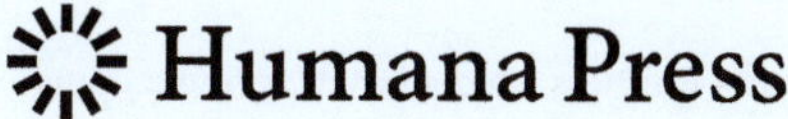
Humana Press

Editors
James M. Parry
7 Cedar Mount
Lyndhurst SO43 7ED
United Kingdom
j.m.parry@swansea.ac.uk

Elizabeth M. Parry
7 Cedar Mount
Lyndhurst SO43 7ED
United Kingdom
empsafechem@gmail.com

ISSN 1064-3745 e-ISSN 1940-6029
ISBN 978-1-61779-420-9 e-ISBN 978-1-61779-421-6
DOI 10.1007/978-1-61779-421-6
Springer New York Dordrecht Heidelberg London

Library of Congress Control Number: 2011942910

Printed on acid-free paper

Humana Press is part of Springer Science+Business Media (www.springer.com)

Preface I

Evaluation of potential mutagenic activity is a critical step in the assessment of the safety of both new and preexisting chemical types. Such assessments are critical in the development of new pharmaceuticals and consumer products. The UK Environmental Mutagen Society has identified an unsatisfied demand for education in the discipline of genetic toxicology, which provides the academic basis for the science behind mutagenicity testing.

To provide for education in genetic toxicology, the UKEMS is sponsoring postgraduate education and the production of both electronic and printed materials.

This book covers three basic areas: the scientific basis of the discipline, the methodologies of the main test assays, and the application of the methods.

The text is aimed primarily at workers in the safety departments of the industries working with both natural and synthetic chemicals. Such workers need to undertake continual updating in assay methods and their application. Changes in regulations for the assessment of chemical safety in areas such as the EU are resulting in substantial increases in the demand for mutagenicity testing. We aim to provide support to both laboratory workers in providing quality information on the appropriate application of techniques and to study directors in their assay selection and protocol design. The text will provide information for both individuals undertaking personal study programs and for those undertaking formal qualifications in genetic toxicology.

Jim Parry was a leading light and inspiration in this field and my lifelong collaborator. The project was devised, driven, and produced due to his unbounded enthusiasm, and he would have relished its completion. This book is dedicated to his memory.

Elizabeth M. Parry

Preface II

Mutation is a broad term covering a whole range of changes to the informational molecule, DNA (made up of the four nucleotides: the purines, adenine and guanine, and the pyrimidines, thymine and cytosine) packaged into chromosomes, of an organism from point/gene changes to modifications of the number and/or structure of chromosomes.

Point/gene mutations are changes to the sequence of nucleotides and may involve the substitution of individual bases (classified as base substitution mutations). When one purine nucleotide is replaced by another purine or a pyrimidine by a pyrimidine, it is called a transition mutation. When a purine is replaced by pyrimidine or vice versa, the mutation is called a transversion.

Since the genetic code is degenerate, not all base substitutions will result in coding changes leading to changes at the protein level. When mutations result in amino acid changes, they are classified as missense, and when they lead to a codon that terminates protein production, they are classified as nonsense mutations.

Mutations involving the loss or gain of DNA may range from a single base pair change, called a frameshift mutation, to many bases (often megabases), called deletion or duplications, or to whole chromosome changes called aneuploidy. All these types of change are produced spontaneously during the life cycle of living organisms and may also be produced by exposure to some chemicals which interact with the DNA (adduct formation) and to ionizing radiation and nonionizing radiation such as ultra violet light. Such chemical and physical agents are classified as mutagens and/or genotoxins.

Not all the DNA in a cell carries coded information for protein synthesis. Some of it is noncoding and is important for chromosome structure.

Although all cells of an organism contain the same DNA, somatic cells in different organs and tissues of the adult body become specialized to perform defined functions so that only some parts of the genome are expressed. A common feature of mutations in cancer causing genes, such as those controlling cell division and proliferation, is that this results in genes being expressed in the wrong tissue at the wrong time. The effect of a mutation will depend upon the position of the mutation within the DNA and the location and activity of the particular gene in which the mutation has been induced.

Mutations in the many genes that have been implicated in the multistage events leading to cancer can be produced by a variety of mechanisms and interactions and modifications of the genetic material, as is illustrated by the molecular changes in the DNA that occur in the progression of colorectal cancer as identified by Volgelstein and colleagues [1].Chemicals that induce mutations in cancer causing genes are classified as genotoxic carcinogens and the potential of mutagen test systems to detect such compounds has been a major stimulus to the development and application of the science of environmental mutagenesis and genetic toxicology.

There are a number of mechanisms by which chemicals can interact with DNA and lead to the induction of mutations. We have summarized some of these mechanisms below:

1. Direct interaction with the components of the DNA as illustrated by the reaction of alkylating agents such as methyl methanesulfonate with the components of the DNA helix [2].
2. The activation of a compound by cellular metabolism to produce compounds which are now capable of reacting with the DNA. An example of the production of active metabolites is the metabolism of benzo(a)pyrene by the arylhydrocarbon hydroxylases into the DNA reactive diol epoxide [3]. The metabolic activation of potential mutagens is a property of intact animals. However, such metabolically activated compounds may be detected using in vitro test systems by the incorporation into the test protocols of metabolic activation preparations, most frequently based upon the inclusion of microsomes prepared from rat liver (generally called S9 mix). A standard feature of in vitro protocols for the screening of chemicals for potential mutagenic and genotoxic activity is that chemicals are tested for activity in both the presence and absence of S9 mix.
3. The test compound may react with cellular components which may lead to the production of secondary active molecules which are themselves capable of reacting with the DNA. An example of such a mechanism is the production of reactive oxygen species (ROS). In the case of ROS, their level of production can be reduced by the cellular antioxidant mechanisms [4, 5].

Not all of the agent-induced changes in DNA lead to mutations, as all living cells have been shown to possess repair mechanisms which are capable of removing the damaged and modified DNA and reconstituting the original DNA structure (for review *see* ref. 5).

The potency of a compound can be modified by metabolic interactions prior to DNA reaction, which can prevent and/or increase the formation of mutagenic DNA changes, for example, the reduction of the levels of the benzopyrene diol epoxide produced by phase II conjugation reactions in the intact animal [6].

DNA modifications can be processed by the mammalian cell to produce mutations or cellular repair systems can "correct" the compound related modifications before they are processed to produce mutagenic changes (reviewed by Friedberg et al. [5]).

The effects of mutations upon an individual animal will depend upon the site of the mutation within the DNA and the location of the mutated cell within the body. Some mutations will have little or no effect upon protein production, whereas others may produce major changes. Mutations in somatic cells will depend for their effects upon whether or not the mutated cell is expressing that particular gene and if the cells are dividing. Thus, mutations in somatic cells that change normal growth controls are important in the development of cancer. If mutations occur in germ cells, the changes involved may be passed on to the next generation.

Mutagenic chemicals can potentially induce genetic changes in somatic cells in those genes (the oncogenes) whose modifications may be involved in cancer formation and in germ cells where gene modifications may lead to various types of birth defects. Because of the potential health hazard represented by exposure to mutagenic chemicals, it is important that all chemicals for which there is possible human exposure be screened for mutagenic activity. If mutagenic hazard is detected, then the risks of exposure can be assessed and the use of the chemical controlled and when appropriate eliminated from the market and the environment.

Over the past 20 years, more than 300 methods have been developed which were proposed for use as test systems suitable for use in the detection of mutagenic activity and assessment of risks. There have been a number of international collaborative research projects which have evaluated the various methods and determined their reliability, sensitivity, and cost-effectiveness. The aims of these collaborative projects have been to:

1. Provide in vitro methods capable of detecting most if not all of those chemicals with mutagenic potential at early stages of product development without the use of animal experiments.
2. Provide methods to determine whether the mutagenic activity detected by the in vitro methods is reproduced in experimental animals and thus potentially in humans.

Regulatory bodies for all chemical types and products at both the national and international level have produced guidance documents which recommend and/or require mutagenicity testing involving the use of specific types of assays. Although there are some minor differences between the requirements of the various regulators, there is now considerable international agreement on the use and application of the recommended methods. Basically, all international regulations require compounds to be initially evaluated using in vitro assays which measure their ability to induce DNA damage and chromosome damage, the induction of the repair of DNA and the induction of point/gene and chromosome mutations.

In this book, we have provided protocols for those methods which have been extensively validated and in most cases have received approval (or are currently undergoing the stages leading to approval) for usage by International regulatory bodies such as the Organization for Economic Cooperation and Development (OECD).

This collection of mutagenicity testing protocols has been organized on the basis of the testing strategy recommended by the UK Committee on Mutagenicity of Chemicals in Food, Consumer Products and the Environment (COM) published in 2000.

Stage 1. Initially those organizing testing programs should consider factors which may be relevant to the potential mutagenic activity of a compound such as possible metabolism, physiochemical properties, purity, and the nature of contaminants. Structure/activity models and software can be applied at this early stage.

Stage 2. The application of in vitro tests methods measuring the induction gene and chromosome mutation. In this volume, we have outlined protocols for:

1. The measurement of gene mutations in bacterial cultures.
2. The measurement of gene mutation in the Thymidine kinase gene of mouse lymphoma cells.
3. The measurement of gene mutations in the HPRT gene of cultured mammalian cells.
4. The measurement of the induction of chromosome aberrations in cultured mammalian cells.
5. The classification and analysis of chromosome aberrations.
6. The measurement of the induction of micronuclei in cultured mammalian cells.

   We have also reviewed methodologies which enable the identification of modifications of DNA (adducts) or the induction of the repair of DNA damage.

7. The measurement DNA adducts by the use of P32 postlabeling.
8. The measurement of damage to the chromatin material of the cell which when placed on an electrophoretic gel results in the production of "flares" of DNA from the cell, the

so-called Comets. In this volume, we describe the application of the Comet assay to assess chromatin damage in both cultured cells and in a range of tissues in intact animals. An additional protocol which can be used to identify those compounds which cross-link DNA molecules is also described.

9. The measurement of the activity of cell damage and repair related genes in the Green Screen assay is described here. This assay is yet to be extensively validated but is showing considerable potential for application in high-throughput programs where there is a need to screen substantial numbers of chemicals.
10. Although not a part of the standard mutagenicity testing packages, the measurement of the effects of test compounds upon gene activity is proving to be a valuable methodology when identifying the mechanisms of action of potential mutagens. In this volume, we outline a protocol for the use of real time reverse-transcription chain reaction for gene expression analysis. This methodology can be particularly informative when used to determine the effects of mutations induced at specific genes.

If a compound induces genetic damage and/or genetic changes in vitro, the next question to be asked is whether this activity is reproduced in vivo in intact animals and potentially in humans. In vivo assays can be divided into those undertaken in somatic cells, such as rodent bone marrow and peripheral blood, and those in germ cells. Although chromosome aberrations and the induction of sister chromatid exchange can be measured in blood, in this volume we have focussed upon the measurement of the induction of miconuclei in rodents. The micronucleus assay has the advantage of being able to detect and quantify the ability of a compound to induce both structural and numerical chromosome mutations.

In those situations where actual or suspected exposure to mutagenic chemicals has occurred in a human population, it may be necessary to monitor the population for genetic damage and to estimate the hazards and risks of mutagen exposure. In this volume, we review biomonitoring methods which can be applied to human populations. The methods described here are based upon chromosome endpoints, i.e., the quantification of sister-chromosome exchange, chromosome aberrations, and micronuclei.

The in vivo methods described thus far have been based upon the detection and quantification of chromosome mutations. The Comet assay can also be used to measure the induction of chromatid damage in vivo.

The in vivo genetic toxicology assays thus far described are utilized to detect and assess the potential of compounds to induce mutations in somatic tissues. When a compound produces positive results in somatic cells, then it can be considered to be a potential carcinogen and a possible germ cell mutagen.

We have described in this volume a range of cytogenetic methods which can be used to detect and assess the induction of structural and numerical chromosome mutations in germ cells:

(a) Metaphase analysis of mitotically dividing spermatozoa
(b) Metaphase analysis of meiotically dividing primary and secondary spermatocytes
(c) The spermatid micronucleus assay
(d) Sperm FISH assay
(e) Analysis of metaphase II oocytes
(f) Dominant lethal assay

Gene mutations can be monitored in human blood using modifications of the HPRT assay described in the in vitro section. However, considerable progress has been made in the development of genetically engineered rodents carrying genes that can be exposed to potential mutagens in vivo and analyzed in vitro. The basic principles and applications of rodent transgenic mutation models are described in this volume. An important feature of transgenic animals is that they can be used to analyze the induction of mutations in both somatic tissues and in germ cells.

References

1. Feron E R and Vogelstein B (1996) A genetic model for colorectal tumour genesis. *Cell* **62**: 759–767.
2. Singer B (1975) The chemical effects of nucleic acid alkylation and their relationship to mutagenesis and carcinogenesis. In *Progress in Nucleic Acid Research and Molecular Biology*. **15:** 218–184.
3. Levin W A Y, Luz H, Ryan D, Wood A W, Kpitulnik J, West S, Huang M T, Conney A H, Thakker D R, Holder G, Hogi G and Jerina D M. (1977). Properties of the liver microsomal monoxygenase and epoxide hydrolase: factors influencing the metabolism and mutagenicity of benzo(a)pyrene. P659–682. In H.H. Hiatt and J.A. Watson (Ed) Origins of Human Cancer. Cold Spring Harbour Laboratory. N.Y.
4. Riley PA (1994) Free Radical Biology: Oxidative Stress and the effects of ionising radiations. *International J. Radiation Biology* **65**: 27–33.
5. Freidberg E C, Walker G C, Siede W, Wood R D, Shultz R D and Ellenberger T (2006) DNA Repair and Mutagenesis 2nd Edition ASM Press Washington DC
6. Sies H and Ketterer B eds (1988) Glutathione Conjugation: Mechanisms and Biological Significance. Academic Press, London.

Contents

Contributors

Ilse-Dore Adler • *Marktplatz 15B85375, Neufahrn, Germany*
Diana Anderson • *Division of Biomedical Sciences, School of Life Sciences, University of Bradford, West Yorkshire, UK*
Adi Baumgartner • *Division of Biomedical Sciences, University of Bradford, West Yorkshire, UK*
Karen Brown • *Department of Cancer Studies and Molecular Medicine, University of Leicester, Leicester, UK*
Brian Burlinson • *Director of Safety Assessment, Huntingdon Life Sciences, Woolley Road, Alconbury Huntingdon, Cambridgeshire, PE28 4HS, UK*
Gill Clare • *15 The Knoll Framlingham, Suffolk IP13 9DH, UK*
Natalie Danford • *Microptic Cytogenetic Services, 2 Langland Close, Mumbles, Swansea, Wales SA3 4LY, UK*
Steve Dean • *Addleston, UK*
Ilse Decordier • *Vrije Universiteit Brussel, Brussels, Belgium*
Shareen H. Doak • *DNA Damage Research Group, Institute of Life Science, College of Medicine, Swansea University, Swansea, Wales, UK*
Ann T. Doherty • *Safety Assessment, AstraZeneca, Cheshire, UK*
David Gatehouse • *Old Barn, Cherry Orchard LaneWyddial, Hertfordshire SG9 0EN, UK*
George E. Johnson • *DNA Damage Group, School of Medicine, Swansea University, Swansea, Wales, UK*
Nigel J. Jones • *Department of Biochemistry and Cell Biology, Institute of Integrative Biology, University of Liverpool, Biosciences Building, Crown Street, Liverpool L69 7ZB, UK*
Philip Judson • *Lhasa Limited, 22-23 Blenheim Terrace, Woodhouse Lane, Leeds LS2 9HD, UK*
Darren Kidd • *Covance Laboratories Limited, Harrogate, UK*
Micheline Kirsch-Volders • *Vrije Universiteit Brussel, Brussels, Belgium*
Melvyn Lloyd • *Covance Laboratories Limited, Harrogate, UK*
Raluca A. Mateuca • *Vrije Universiteit Brussel, Brussels, Belgium*
Ian de G. Mitchell • *Chilfrome Cottage, Chilfrome, Dorchester, DT2 0HA, UK*
Francesca Pacchierotti • *Unit of Radiation Biology and Human Health, ENEA CR Casaccia, Rome, Italy*
Antonella Russo • *Dipartimento di Biologia, Università di Padova, Padova, Italy*
David O.F. Skibinski • *College of Medicine, Institute of Life Sciences, Medical School, Swansea University, Swansea, UK*
Matthew Tate • *Gentronix Ltd., University of Manchester, Manchester, UK*

RICHARD M. WALMSLEY • *Gentronix Ltd, Manchester, UK*
JIAN HONG WU • *Department of Biochemistry and Cell Biology, Institute of Integrative Biology, University of Liverpool, Liverpool, UK*
ZOULIKHA M. ZAÏR • *DNA Damage Research Group, Institute of Life Science, College of Medicine, Swansea University, Swansea, Wales, UK*

Chapter 1

The Application of Structure–Activity Relationships to the Prediction of the Mutagenic Activity of Chemicals

Philip Judson

Abstract

Prediction of mutagenicity by computer is now routinely used in research and by regulatory authorities. Broadly, two different approaches are in wide use. The first is based on statistical analysis of data to find patterns associated with mutagenic activity. The resultant models are generally termed quantitative structure–activity relationships (QSAR). The second is based on capturing human knowledge about the causes of mutagenicity and applying it in ways that mimic human reasoning. These systems are generally called knowledge-based system. Other methods for finding patterns in data, such as the application of neural networks, are in use but less widely so.

Key words: QSAR, Structure–activity relationships, Expert systems, Knowledge-based systems, Computer prediction, Toxicity prediction, Metabolism prediction

1. Introduction

Computer models are widely used to predict potential chemical toxicity. Such predictions are still far from certain but they have become meaningful enough to be useful and one of the areas where most progress has been made is in the prediction of mutagenicity. Because mutagenicity is of high concern, computer researchers may have given it special attention but that is probably not the main reason for the success. The extensive use of the Ames test and the relatively simple way in which results are recorded, compared with, say, the results of a long-term, repeated dose studies in mammals, have provided ideal data for computer analysis. The common mechanism of action of many mutagens makes collective statistical analysis of data about their activity possible, and having an understanding of mechanism supports reasoning-based methods as well.

James M. Parry and Elizabeth M. Parry (eds.), *Genetic Toxicology: Principles and Methods*, Methods in Molecular Biology, vol. 817, DOI 10.1007/978-1-61779-421-6_1,

1.1. Why Are Compounds Mutagenic?

Some very simple molecules are mutagenic and one such, acrolein (structure I in Fig. 1), conveniently illustrates one of the most common chemical mechanisms of action of mutagens. Acrolein is a so-called Michael acceptor. It can accept electrons from functional groups such as amino groups to form covalent derivatives (Fig. 1). Three of the four DNA bases contain amino groups capable of reacting like this, and the formation of a covalent substituent is very likely to disrupt base pairing and hence DNA replication. Michael acceptors are just one class of electrophiles capable of reacting with amines, and many other electrophilic functional groups are associated with mutagenicity – for example, saturated aldehydes, epoxides, and benzyl halides (Fig. 2). Functional groups, or sub-structural features, believed to confer potential mutagenicity on the compounds that contain them are generally termed "alerts", following the publication some years ago by Ashby and Tennant (1) of their famous composite alert molecule – an imaginary structure incorporating all the groups known at that time to be associated with mutagenicity.

This chemical mechanism of action is fairly easy to hit upon but it is not the only one. Even within the Ashby and Tennant set of alerts there are some that cannot be explained in simple, direct terms of electrophilicity. An alert that would not itself react with an amino group may be a precursor to more reactive species, as mentioned again in Subheading 1.2.3, and there are different mechanisms of action such as intercalation, in which a planar molecule

R-NH2 I R-NH-CH2CH2CHO

Fig. 1. Reaction of an amine with a Michael Acceptor.

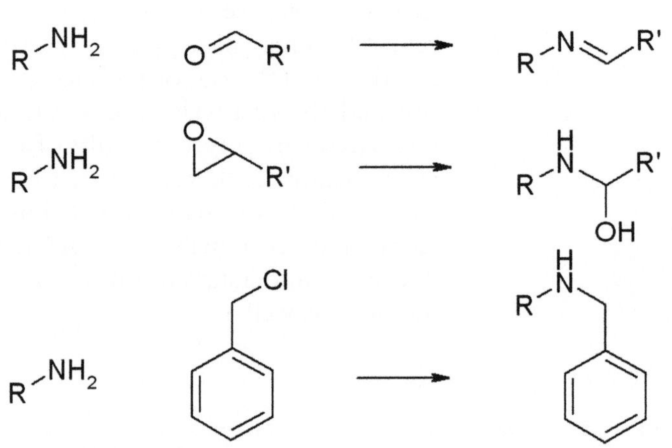

Fig. 2. Examples of electrophiles that react with amines.

becomes sandwiched between bases in the DNA helix, held there by non-covalent interactions, causing distortion and interfering with replication. Computer methods are used both to make predictions about the behaviour of structures containing known alerts and to help with the discovery of new ones.

1.2. Why Are Not All Electrophiles Mutagenic?

Compounds containing electrophilic groups are extremely common but many – probably the majority – are not mutagenic. Some may give positive results in assays such as the Ames test but have no activity in vivo. Others may show no activity in any test. Physicochemical properties, chemical reactivity, and susceptibility to metabolism can all be responsible for inactivity.

1.2.1. Physicochemical Properties

It is fairly obvious that if a compound is virtually insoluble in water, it will be unlikely to reach its site of action in a biological system but being water soluble is not enough. Indeed a compound might be too water soluble. A more informative property is the partition coefficient of a compound between lipids and water. To reach its site of action, even in a seemingly simple situation such as the Ames test where it has only to enter bacterial cells and find its way to DNA, a compound must cross lipid membranes. So to be active, it must have the right balance of solubility between water and lipids: if it has very low solubility in either it cannot progress; if it partitions too much in favour of water it will not enter and cross membranes; if it partitions too much in favour of lipids it will become trapped in the first membrane it encounters. Since laboratory measurement of partition between membranes and water is no easy matter, it is usually modelled by measuring partition between octanol and water and expressed logarithmically as "log$P$", or "K_{ow}". Methods for estimating physicochemical properties can be very unreliable but log$P$ is an exception and computer estimation is adequate for most purposes. Key confounding factors are understood and it is possible to recognise exceptions for which a model is likely to be untrustworthy.

Another important physicochemical property is pK_a – a measure of how acidic a molecule is – i.e. how easily it loses a proton to become a negatively charged ion. A corresponding property, pK_b, is a measure of how basic a molecule is – i.e. how readily it will accept a proton to become a positively charged ion – but some cunning juggling with mathematics allows both properties to be expressed in terms of pK_a – the pK_a of the ion in the case of a base. pK_a matters because, normally, ionised materials are much more soluble in water, and much less soluble in fats, than unionised ones. So partition between fat and water becomes dependent on what proportion of a compound is ionised. The situation is complicated by the fact that ionisation depends on the environment as well as the inherent pK_a of a compound – an acidic compound will be less ionised in, say, the acidic environment of the mammalian stomach than in the near-neutral internal environmental of a typical cell. A further issue is that a molecule may contain several acidic or basic groups (or both) and each will have its own pK_a. Methods for calculating, rather than measuring, pK_a are only moderately reliable. Happily, however, a great many compounds are neither acids nor bases, and the calculations for those that are, are generally good enough at least to support qualitative predictions.

1.2.2. Chemical Reactivity

There is a big range of chemical reactivity even for a single kind of reaction, and an even bigger one for a class of reactions such as all electrophilic reactions. The compounds shown in Fig. 3 are all chloro-compounds potentially capable of electrophilic reaction with amines, but out of them only benzyl chloride (II), chloroacetic acid (IV), chloropropane (VI), and benzoyl chloride (VII) are believed to be mutagenic. Aromatic halides are poor electrophiles and so chlorobenzene (III) would not be expected to react with amines in biological systems. 2-Chloro-3,3-dimethylbutane (V) is unreactive because of steric hindrance. Phenylsulphonyl chloride (VIII) would not be expected to be mutagenic because it is too reactive, rather than not reactive enough. Indeed, it reacts spontaneously with water and would not survive long enough to reach DNA in a biological system. This is an extreme example of a more general mechanism for removal of a potentially mutagenic compound before it can reach and react with DNA. Proteins contain amino groups and thiol groups, and a host of compounds present in biological systems contain hydroxyl groups, any of which may compete with the amino groups in DNA for reaction with an electrophile.

So having the right reactivity to be a mutagen is a fine balance between being reactive enough with the DNA bases and resistant enough to reaction with other molecules in a biological system.

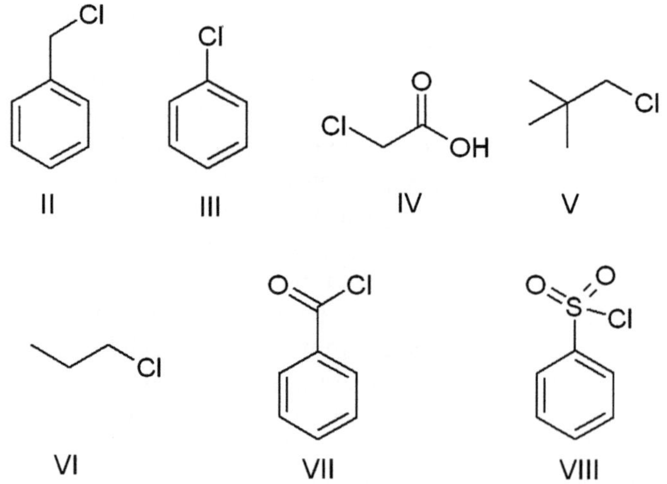

II III IV V

VI VII VIII

Fig. 3. Chloro-compounds with differing electrophilic reactivity.

1.2.3. Metabolic Activation and Deactivation

Removal of potential mutagens through competing reactions is not confined to spontaneous chemical reactions such as might be expected in a test tube. Enzymes promote reactions that would be a surprise to someone aware only of non-enzymatic chemistry, and in particular, there are enzymes to assist with the disposal of xenobiotic compounds – compounds that are foreign to biological systems. In higher animals, these enzymes are concentrated in the liver, where components of food are processed to render them safe to the organism and/or easy to dispose of (e.g. by converting them to readily water-soluble derivatives that can be excreted via the kidneys). The most studied enzymes of this kind are the P-450 enzymes which promote oxidative processes, but other enzymes convert the oxidation products into highly water-soluble derivates (e.g. gluconurides and sulphates), and there are enzymes to promote processes such as methylation and acetylation.

Aldehydes, for example, would be expected to react with amino groups in DNA bases but even if they escape prior capture by amino groups in proteins (a process that can lead to skin sensitisation) they are likely to be oxidised to carboxylic acids, which are much less reactive towards amines in the absence of enzymes to help the process, or reduced to alcohols, which are also unreactive towards amines. Only aldehydes with properties and reactivities within narrow margins are likely to reach and react with DNA in a living cell.

Modifying a xenobiotic compound does not necessarily convert it from a toxic to a non-toxic one. Mutagenicity is often the result of metabolic activation. For example, the aromatic polycyclic hydrocarbons present in vehicle exhausts are thought to be converted to epoxides, which react with DNA bases leading to mutagenicity and carcinogenicity. One would not expect aromatic amines or aromatic nitro compounds to react with DNA, but enzymatic processes lead to the generation of highly reactive radical ions, as illustrated in Fig. 4, which are believed to be responsible for mutagenic activity.

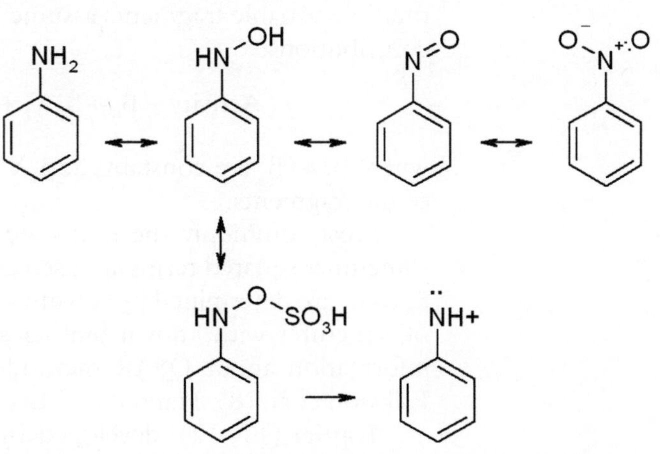

NH_2 HN OH N O O N O

HN O SO_3H NH+

Fig. 4. Suspected mechanism of metabolic activation of amines and nitro compounds.

2. Materials and Methods

2.1. Computer Methods

Broadly speaking two different approaches are taken to predicting toxicity by computer: automated methods and methods based on capturing and applying human knowledge. Automated methods include statistical ones, which are currently the most widely researched and applied, and the use of computer systems capable of learning such as neural networks and genetic algorithms. Methods that draw on human knowledge are nowadays usually called knowledge-based expert systems.

2.2. Statistical and Computer Learning Methods for Finding Alerts and Predicting Potency

Statistical quantitative structure–activity relationships (QSAR) modelling is based on applying a mathematical function to a set of numerical descriptors – values that describe key features or properties of chemical structures. One of the first to be described and still the most well known is the Hansch equation for calculating biological activity on the basis of the attributes of one variable substituent in a chemical structure (2):

$$\log(1/C) = a\pi + b\pi^2 + c\sigma + dE_S + k$$

where π is a hydrophobic term, σ is an electronic term, E_S is a steric term, and a, b, c, d, and k are constants.

The measure chosen for the hydrophobic term is usually the contribution to octanol–water partition coefficient, log*P*, associated with the fragment, and it was Corwin Hansch's group who pioneered the estimation of log*P* for whole structures by summing contributions from sub-structural fragments (3) and developed the widely used program, ClogP (4). The "electrotopological states" of Hall et al. (5) are widely used for σ values, and Taft values (6, 7) for E_S.

Analyses taking account of contributions to activity from multiple variable fragments assume that activity is the sum of their contributions:

$$\text{Activity} = \beta_0 + \beta_1 X_1 + \beta_2 X_2 + \ldots \beta_n X_n$$

where β_0 to β_n are constants and X_1 to X_n are numerical attributes of the fragments.

Most commonly the terms are all linear, as shown here, but sometimes squared terms are used as well. Values for the constants, β_0 to β_n are determined by solving simultaneous equations for a set of structures with known biological activities. For more detailed information about QSAR methods, see, for example, books by Eriksson et al. (8), Hansch and Leo (9), and Livingstone (10).

TopKat (11, 12), developed by Kurt Enslein and colleagues, and currently supplied by Accelrys Inc. (13), makes predictions on the basis of quantitative relationships between a pre-defined set of sub-structures and toxicological activity. There are several thousand sub-structures in the library that it uses. In addition to descriptors associated with sub-structural fragments, properties of the whole molecule can be included in the analysis, such as values encoding information about size and shape, and the estimated octanol–water partition coefficient (log*P*). TopKat includes modules for predicting mutagenicity as well as rat acute oral and inhalational toxicity, skin sensitisation, rodent carcinogenicity, developmental toxicity, and skin and eye irritation. It also covers the environmental end points of toxicity to fathead minnow and daphnia, and it estimates aerobic biodegradability. See Notes 1 and 2.

To reduce the risk of bias that using pre-defined sets of sub-structural fragments for modelling might introduce, the Multicase programs, Casetox, M-Case, and MC4PC, use all the linear fragments it is possible to generate automatically from the set of structures used to train a model, subject for practical reasons to constraints on minimum and maximum chain length (14–16). To improve discrimination, branching points are flagged in the linear fragments. Two kinds of fragment are recognised: the first are fragments primarily responsible for the observed biological activity; the second kind are fragments that do not cause activity in themselves but increase or decrease activity if there is any. The set of linear fragments associated with activity is called a "biophore" (see Note 3). Once a biophore has been identified, a QSAR can be constructed for structures that contain it. Partition coefficient, log*P*, is usually the variable most strongly correlated with activity in a series of compounds with a common biophore.

Prediction modules available from Multicase Inc. cover mutagenicity, acute mammalian toxicity, hepatotoxicity, renal toxicity, cardiac toxicity, carcinogenicity, developmental toxicity, skin and eye irritation, and more (see Note 2). Some environmental end points are covered, including fish toxicity, biodegradability, and bioaccumulation. The modules for some of these end points comprise sets of more specific ones, relating to one sex, a particular type of symptom or, in some cases, interaction with a particular enzyme.

Kühne et al. have shown that you can make predictions in the field of ecotoxicity by building statistical models based on atom-centred fragments (17, 18). The same approach could be applied to the prediction of mutagenicity. Atom-centred fragments can be based on many kinds of atom attributes but the method is illustrated here by how they might be based simply on elemental type. First all the atoms in a structure would be labelled according to their elemental type. Then a second list of labels would be attached to each atom, containing the types of its neighbouring atoms. This process could be repeated for bigger and bigger shells around each atom until the limits of the structure were reached. In practice, a constraint is placed on the number of shells built around an atom, and it is usually quite a low one. If the shells are too big, almost every atom is uniquely described, whereas model building depends on discovering similarities between atoms. In some systems, an algorithm converts the set of labels for each atom into a corresponding, single integer to facilitate fast computer processing.

The LeadScope Predictive Data Miner (19) is a toolbox to support data mining for chemistry-related problems. Graphs and bar charts help you to recognise trends and features in common between structures and the data associated with them. The software includes a large library of structural fragments, different kinds of descriptors and methods for generating them, and tools such as the ones used in TopKat and Multicase for building predictive models. In addition, you can enter and store your own structural alerts – some of which the other tools in the package may have helped you to discover.

Many groups have experimented with applying general machine learning techniques to the problem of toxicity prediction. For example, ID3 (20) has been used to build decision trees, and inductive logic programming (21), neural nets (22), and genetic algorithms (23) have been used. Some have been more successful than others, but none have so far matched the popularity of the statistical and knowledge-based methods. Research groups have compared and continue to compare, different approaches to find out how they might best be used in combination (24). See Notes 4 and 5.

2.3. Storing and Using Human Knowledge

Kaufman et al. were the first researchers to describe a computer system, TOX-MATCH (25, 26), to predict toxicity from structural alerts, or toxicophores, contained in a knowledge base written by human experts, but their research project came to an end without TOX-MATCH becoming commercially available. (Toxmatch (27–29) from the European Chemicals Bureau is entirely unconnected with TOX-MATCH. Toxmatch helps with the categorisation of chemicals and is not a toxicity prediction system. It provides methods for calculating a variety of physico-chemical descriptors such as log*P*, ionisation potential, and molecular surface area and for grouping chemicals according to their similarity).

Oncologic, developed by staff at the US Environmental Protection Agency (30–32), uses the concept of toxicophores but it is not strictly a knowledge-based system in the modern sense. It is driven by predefined decision trees. A question and a set of valid answers are associated with each node in a decision tree, one answer for each branch at that node. The computer follows a path through the tree directed by the answers a user gives to the questions. Modules in Oncologic can consider other factors as well as the presence or absence of structural alerts in organic chemicals. For example, there is a module for making predictions about the potential carcinogenicity of fibres, based upon particle size, shape, and surface properties. There is a module for making predictions about polymers, one for metals, and one for organic compounds. As its name implies, Oncologic gives advice about potential carcinogenicity but this end point may also be of concern to you if you are interested in mutagenicity. Oncologic is available for download free of charge from the EPA web site (33).

HazardExpert (34), from Compudrug (35), gives a numerical estimate of the probability that a compound will be toxic against the end point of interest to the user. Probabilities are based on expert assessments, supported by statistical analysis, of the proportion of compounds containing each alert that show the associated toxicity. See Note 6.

ToxTree was developed for the European Chemicals Bureau by IdeaConsult (29) and it can be downloaded free of charge from the European Chemicals Bureau (36). It is a simple decision tree system and is not primarily intended for toxicity prediction. It incorporates several classification schemes to help with deciding, for example, which QSAR (Quantitative Structure-Activity Relationship) models are most suitable for a given compound, and that is its primary purpose. However, there are some toxicity prediction rules in it, including some relating to mutagenicity and carcinogenicity and some about skin and eye irritancy and corrosivity.

Derek for Windows (37–39), a knowledge-based system for predicting toxicity on the basis of alerts present in query structures, superseded DEREK (40–43). It uses the Logic of Argumentation (44, 45) to reason about evidence for and against toxicity. In essence, this mimics the way that humans reach a conclusion by considering the arguments for and against a proposition – in a court of law, for example.

By way of illustration, Derek for Windows considers it plausible that 2-methyliodopropane will be mutagenic in bacteria (Fig. 5) but does not predict 2,2,-dimethyliodopropane (neopentyl iodide) to be mutagenic (Fig. 6): comments associated with the alert for alkylating agents in the knowledge base explain that compounds of the latter kind are not normally mutagenic (Fig. 7). This is presumably because steric hindrance reduces the reactivity of the halogen substituent, as mentioned in Subheading 1.2.2. Iodopropan-2-one is not a potential mutagen according to Derek for Windows (Fig. 8), and this kind of compound is also mentioned in the section of the comments about the alert shown in Fig. 7. In this case, the likely reason for inactivity is that the halogen is too reactive for the compound to reach DNA inside a cell. Indeed, alpha-iodoketones are sufficiently reactive to hydrolyse spontaneously in tear fluid, and it is believed that the liberated hydriodic acid is the cause of eye irritation and lachrymation about which Derek for Windows does give warnings in Fig. 8.

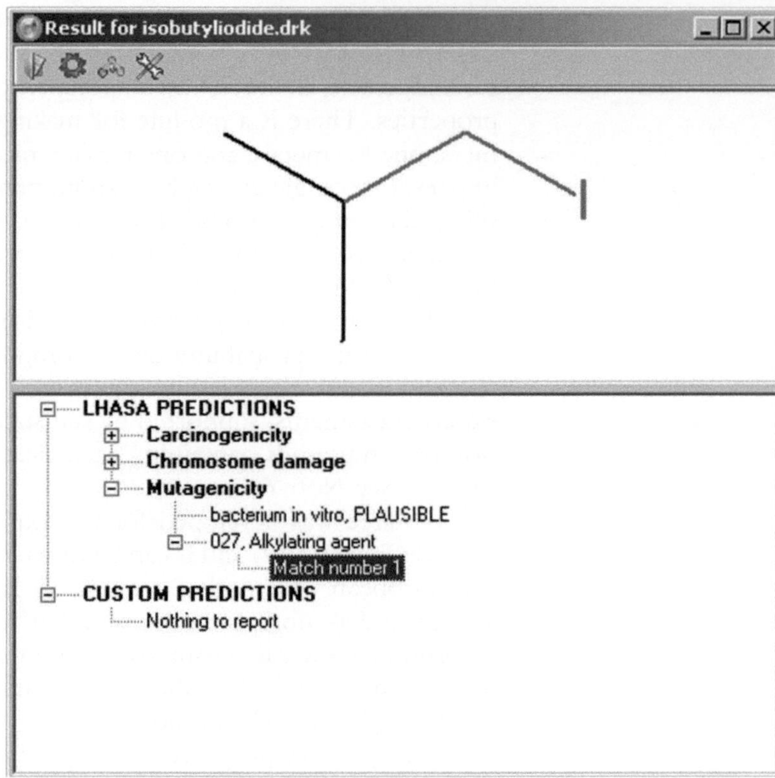

Fig. 5. A Derek for Windows report for *iso*-butyl iodide.

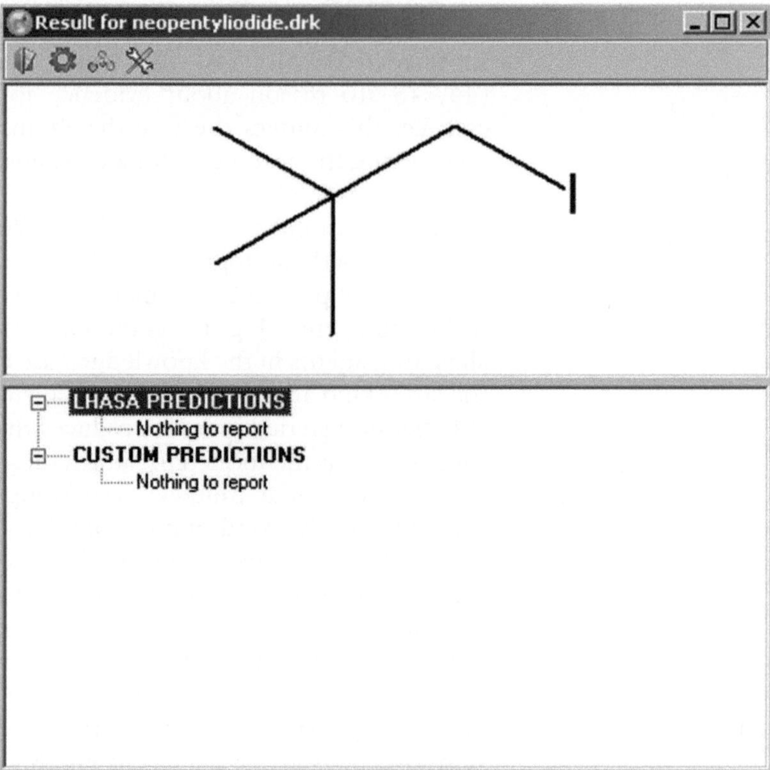

Fig. 6. A Derek for Windows report for *neo*-pentyl iodide.

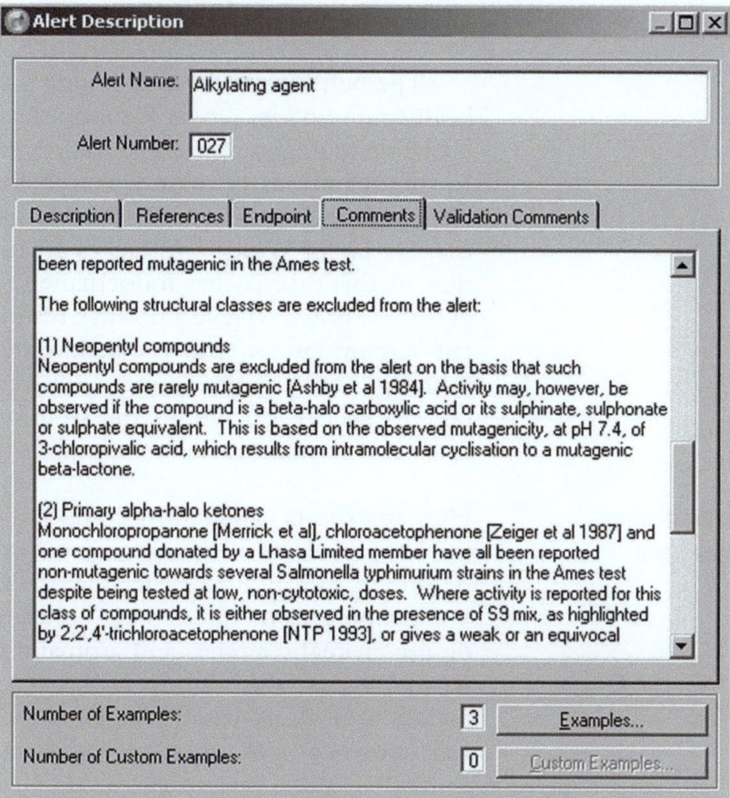

Fig. 7. Comments about the alkylating agent alert in Derek for Windows.

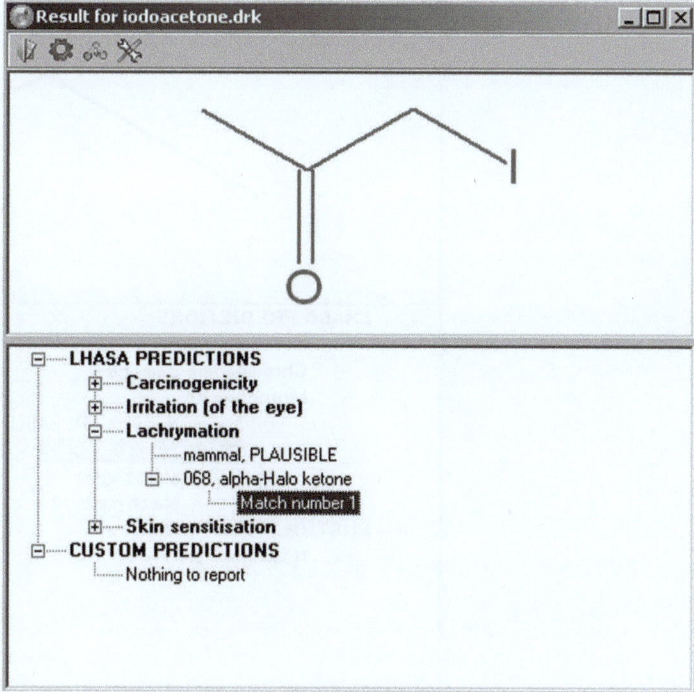

Fig. 8. A Derek for Windows report for Iodopropan-2-one mentions several potential toxic effects, including lachrymation, but not mutagenicity.

The behaviour of the reasoning system in Derek for Windows is illustrated by its report for iodoethane, which it designates as a probable mutagen in bacteria, rather than a plausible one (Fig. 9). In the definitions of these words in Derek for Windows, "probable" indicates a greater likelihood than "plausible". The reason that Derek for Windows attaches greater confidence to the prediction in this case is that iodoethane is listed in its database as an example known to give a positive result in the Ames test. In different circumstances, the prediction might have been classed as "certain", but the prediction is for mutagenic activity in bacteria, not the specific bacterium and strains used in the reported Ames tests, and so there remains room for some doubt.

2.4. Predicting Metabolism

Human experts can identify compounds as potential mutagens on the basis of alerts in their structures even though the activity may be due to metabolites rather than to the compounds themselves. For example, as mentioned in Subheading 1.2.3, toxicologists recognise aromatic amines and aromatic nitro-compounds as potential mutagens, although in both cases the active compounds are metabolites. Alert-based programs make predictions in the same way about compounds requiring metabolic action and Derek for

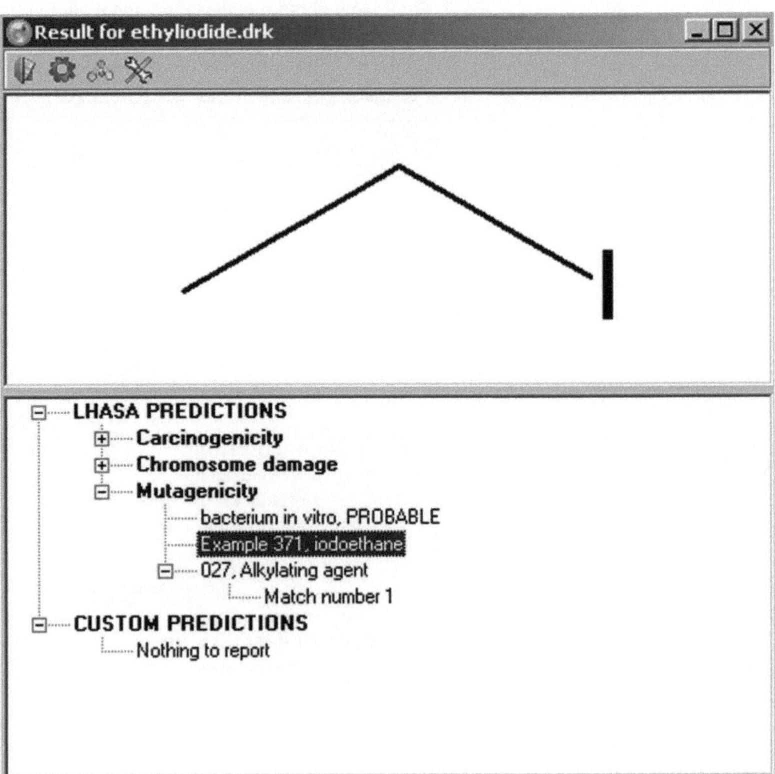

Fig. 9. A Derek for Windows report for ethyl iodide (iodoethane).

Windows, for example, issues warnings for aromatic amines and nitro compounds. However, prediction of mutagenicity via metabolic activation does not have to depend on alerts that anticipate metabolism; there are programs for predicting metabolism and they can be used in conjunction with toxicity prediction programs.

Derek for Windows is linked to Meteor (39, 46, 47). As an example, with default settings to see the more likely biotransformations, Meteor predicts two probable metabolism pathways for 2-cyclohexyl-2-hydroxypropene, leading to a glucuronide and a glutathione conjugate. The route to the latter has been selected for display in Fig. 10, and it shows an isolable, intermediate enone (the "][" symbols in the display indicate reaction steps which have been suppressed by choosing an option not to view less stable intermediates, to keep the figure simple). Derek for Windows reports the enone to be a potential mutagen (Fig. 11).

Predicting the metabolism of xenobiotic chemicals with a knowledge-based system was pioneered by Todd Wipke et al., who developed XENO (48), but the program was not developed much beyond the original prototype. MetabolExpert, a knowledge-based system for predicting metabolism developed by Ferenc Darvas (49, 50), is a sister program to HazardExpert, mentioned in Subheading 2.2.

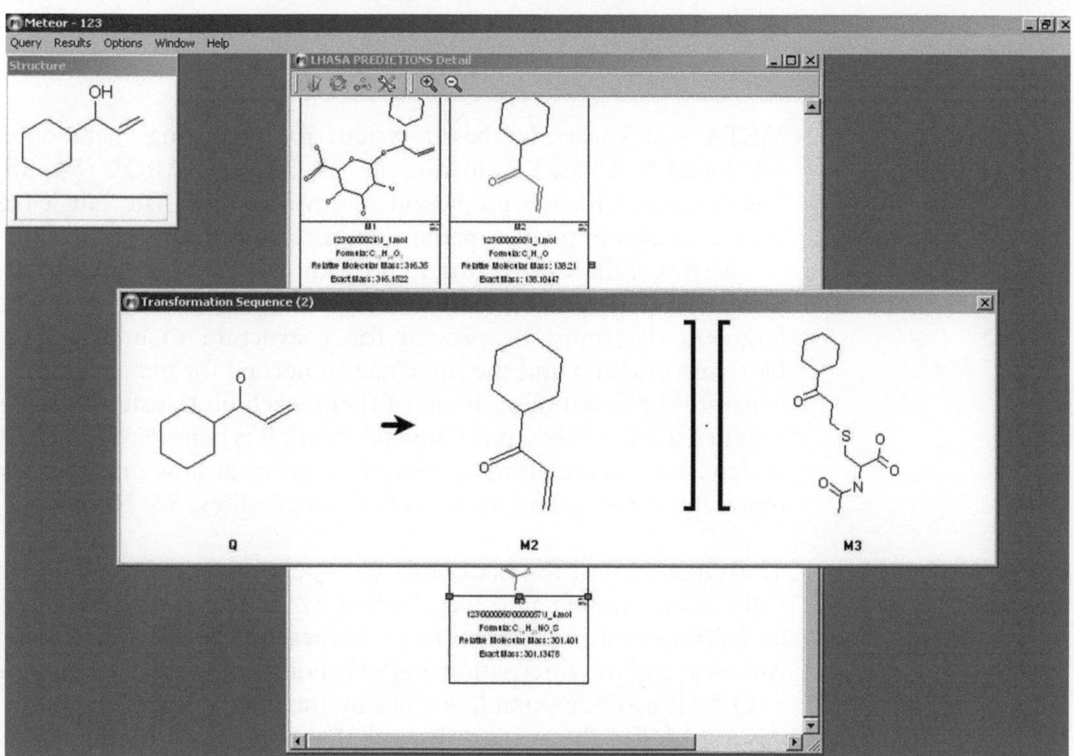

Fig. 10. One potential metabolic pathway for 2-cyclohexyl-2-hydroxypropene suggested by Derek for Windows.

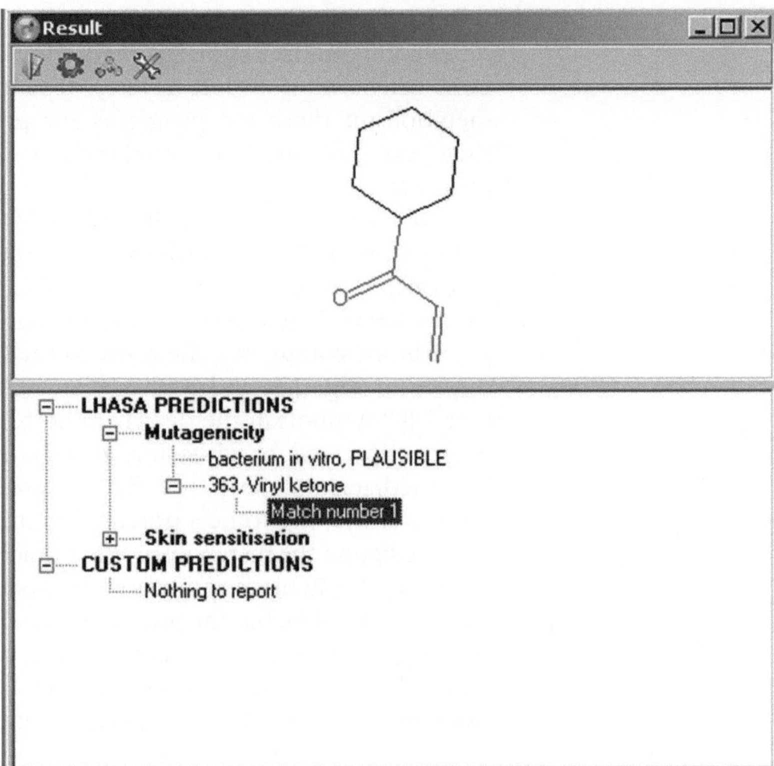

Fig. 11. A Derek for Windows report for the intermediate enone shown in Fig. 10.

META is a knowledge-based system for predicting metabolism developed by Gilles Klopman et al. (51–53). CATABOL (54, 55) was developed for the prediction of environmental fate but it has been extended to predict mammalian metabolism.

All five of these programs draw on knowledge bases of biotransformation descriptions, each of which includes the sub-structural fragment that must be present for a structure to undergo the biotransformation and the information needed for the program to generate the metabolite. In all of them, each biotransformation is assigned a priority based on how prevalent it is considered to be by metabolism experts, but the programs differ in how priorities are represented and manipulated to rank metabolites. See Note 7.

2.5. The OECD QSAR Toolbox

The Organisation for Economic Co-operation and Development (OECD) is running a project, funded by the European Union and in collaboration with regulatory bodies in Europe, Japan, and America, and the International QSAR Foundation (56), to provide a (Q)SAR toolbox suitable for use by the regulators. A prototype, developed for the project by the Laboratory of Mathematical Chemistry (57) under the leadership of Ovanes Mekenyan can be downloaded free of charge (58).

The toolbox is not intended to make predictions itself about toxicity but to help a user to find information and/or to develop and use suitable prediction models. It puts together a profile for it a query, giving information on whether it is included in one or more regulatory databases about which the Toolbox has information (rarely will full toxicological details for the query be found in the databases but if they are, of course, the search may be over at this stage); a broad classification of the chemical such as whether it is defined as organic or inorganic for regulatory purposes, whether it is a single compound or a mixture, whether it is a polymer, etc. The Toolbox lists how the chemical fits into systems for classifying chemicals that are used for some regulatory purposes, based on physico-chemical properties or the presence of sub-structural features in the chemical. Although these sub-structural features are like alerts, as described earlier, in the Toolbox they are not used to make predictions but to group chemicals into classes likely to have the same mechanism of toxicity. Thus other members of the class for which toxicity is known can be used to make judgements by comparison or as a training set to build a quantitative structure–activity relationship model that may be suitable for estimating the toxicity of the query.

3. Notes

1. Best practice in statistics is important in quantitative structure–activity relationship modelling and the systems described provide accepted statistical performance measures. Be aware in addition that, no matter how rigorous a statistical analysis is, the validity of the resultant model depends on the appropriateness of the descriptors used in the analysis. Keep in mind also that when a statistics-based system presents indications of reliability, such as ranges in reported values or error bars in bar charts and measures of statistical performance, the models have been judged in statistical terms, not on the basis of the behaviour of the biological processes being modelled.
2. It is increasingly considered that for a model to be reliable it must predict for a clearly defined toxicological end point and for compounds that act by the same mechanism. Models for properties such as acute toxicity and developmental toxicity may not fit well with these criteria since both comprise a host of different, more specific end points involving very different mechanisms of action. The models have been shown to work in the statistical tests that have been done on them, but caution is needed when using them and basing decisions on their predictions.

3. A perceived weakness with systems like M-Case is that a biophore is a collection of fragments with no specified relationship between them. A real toxicophore is typically a single, branched fragment – comprising the components of the biophore all joined together. In many cases, more than one toxicophore could be constructed from the set of fragments in a biophore. In addition, structures may exist that contain the biophore as separated fragments, not joined together to make the toxicophore actually responsible for activity in compounds that are active. In practice, mis-prediction arising from these causes does not seem to be a problem; probably because the chance is so low of submitting a structure containing all the components of a biophore but distributed differently from ways they were found in compounds in the training set.
4. Approaching the problem from a mathematical, or information science, point of view rather than from the point of view of a chemist carries risks. For example, many mining tools automatically apply Occam's razor. That is, if several solutions are available, they select the simplest one. Imagine a training set in which all of the active molecules are acid chlorides and there are no examples of other chlorine-containing compounds. A system seeking out the simplest explanation will associate the presence of the chlorine atom with activity. It will not associate oxygen or the carbonyl group also present in an acid chloride with activity if they are present in other functional groups in lots of inactive molecules in the training set (as would typically be the case). So, it will flag any compound containing chlorine as a potential skin sensitiser. Statistically that is correct for the structures in the training set, but a chemist would recognise the likely significance of finding the chlorine atom always in an acid chloride group in the training set.
5. A difficulty with models built by neural networks is that they cannot explain the patterns they find. This can be an issue with the other automated methods for modelling toxicity, too, included the statistical methods. Human users are more comfortable with predictions that are delivered with explanations that relate to mechanistic understanding. So, an ideal prediction system based on statistics or automatic learning should provide tools to allow the user to explore the underlying data, so that he/she can seek out mechanistic information to support the prediction.
6. Conventional, numerical probability theory is based on the laws of chance. The biological activity of a compound is not a chance event – there are mechanistic reasons why a structure containing a particular sub-structure is or is not active. If a structure contains more than one toxicophore, there is no apparent mechanistic reason for applying the laws of chance.

The numerical approach to probability taken in HazardExpert and other, similar, systems has a pragmatic basis – it appears to work well enough to be useful in practice – but too much reliance should not be placed on the precise numbers.

7. The assumption that observed probabilities, reaction rates, and hence product quantities are related is not necessarily true, but it is pragmatic. What catabolites or metabolites are seen in practice in an experiment depends very much, though not exclusively, on the competitive success of different reactions which in turn can be expected to depend on their relative rates. So one can hope that, when averaged out over many studies, the probabilities of observing reactions (i.e. for each reaction, the ratio of the number of studies in which a reaction is observed to the number of studies in which the keying substructure is present in a test structure or a degradant) will reflect their relative rates. Given the relative reaction rates, you can estimate the relative quantities of degradants which will be formed.

References

1. Ashby J and Tennant R W (1988) Chemical structure, *Salmonella* Mutagenicity and Extent of Carcinogenicity as Indicators of Genotoxic Carcinogenesis Among 222 Chemicals Tested in Rodents by the US NCI/NTP. *Mutagenesis,* **204**, 17–115.
2. Hansch C and Fujita T (1964) A Method for the Correlation of Biological Activity and Chemical Structure. *J. Amer. Chem. Soc.* **86**, 1616–1626.
3. Fujita T, Isawa J, and Hansch C (1964) A New Substituent Constant, π, Derived from Partition Coefficient. *J. Amer. Chem. Soc.*, **86**, 5175–5180.
4. ClogP is supplied by Biobyte Corporation, 201 West 4th. Street, #204, Claremont, CA 91711–4707, USA.
5. Hall L H, Mohney B, and Kier L B (1991) The Electrotopological State: Structure Information at the Atomic Level for Molecular Graphs. *J. Amer. Chem. Soc.*, **31**, 76–82.
6. Taft R W, (1956) Separation of Polar, Steric, and Resonance Effects. In: Newmann MS (ed.). Steric Effects in Organic Chemistry, John Wiley, New York, pp. 559–675.
7. Kamlet MJ, Abboud JML, Abraham MH, and Taft RW (1983) Linear Solvation Energy Relationships. 23. A Comprehensive Collection of the Solvatochromic Parameters, π*, α, and β, and Some Methods for Simplifying the Generalised Solvatochromic Equation, *J. Org. Chem.*, **48**, 2877–2887.
8. Eriksson L, Johansson E, Kettaneh-Wold N, and Wold S (2001) Multi- and Megavariate Data Analysis, Umetrics AB, Umeå, Sweden.
9. Hansch C and Leo A (1995) Exploring QSAR: Fundamentals and Applications in Chemistry and Biology, American Chemical Society, Washington DC, USA.
10. Livingstone D (1995) Data Analysis for Chemists: Applications to QSAR and Chemical Product Design, Oxford University Press, England.
11. Enslein K, and Craig P N (1982) Carcinogenesis: a Predictive Structure-Activity Model, *J. Toxicol. Environ. Health*, **10**, 521–530.
12. Enslein K, Gombar V K, and Blake B W (1994) Use of SAR in Computer-Assisted Prediction of Carcinogenicity and Mutagenicity of Chemicals by the TOPKAT Program, *Mutation Res.*, **305**, 47–62.
13. Accelrys Inc., 10188 Telesis Court, Suite 100, San Diego, CA 92121, USA. http//accelrys.com/.
14. Klopman G (1984) Artificial Intelligence Approach to Structure-Activity Studies: Computer Automated Structure Evaluation of Biological Activity of Organic Molecules, *J. Amer. Chem. Soc.*, **106**, 7315–7321.
15. Klopman G, Ivanov J, Saiakhov R and Chakravarti S (2005) MC4PC – An Artificial Intelligence Approach to the Discovery of Structure Toxic Activity Relationships (STAR),

in Predictive Toxicology, ed. Christoph Helma, CRC Press, Boca Raton, pp. 423–457.

16. Klopman G, Chakravarti S K, Zhu H, Ivanov J M, and Saiakhov R D (2004) ESP: a Method to Predict Toxicity and Pharmacological Properties of Chemicals Using Multiple MCASE Databases. *J. Chem. Inf. Comput. Sci.* **44**, 704–715.
17. Kühne R, Kleint F, Ebert R U, and Schüürmann G (1996) Calculation of Compound Properties Using Experimental Data from Sufficiently Similar Chemicals, in Software Development in Chemistry *10*, ed. Johann Gasteiger, Gesellschaft Deutscher Chemiker, Frankfurt, Germany, pp. 125–134.
18. Kühne R, Ebert R-U, and Schüürmann G. (2007) Estimation of Compartmental Half-Lives of Organic Compounds – Structural Similarity vs. EPI-Suite, *QSAR Comb. Sci.* **26**, 542–549.
19. LeadScope Predictive Data Miner comes from Leadscope Inc., 1393 Dublin Road, Columbus, Ohio 43215, USA.
20. Quinlan J R (1986) Induction of Decision Trees, *Machine Learning*, **1**, 81–106.
21. King R D and Srinivasan A (1996) Prediction of Rodent Carcinogenicity Bioassays from Molecular Structure Using Inductive Logic Programming, *Environ. Health, Persp.*, **104**, 1031–1040.
22. Vracko M, Bandelj V, Barbieri P, Benfenati E, Chaudhry Q, Cronin M, Devillers J, Gallegos A, Gini G, Gramatica P, Helma C, Neagu D, Netzeva T, Pavan M, Patlevicz G, Randic M, Tsakovska I, and Worth A (2006) Validation of Counter Propagation Neural Network Models for Predictive Toxicology According to the OECD Principles. A Case Study, *SAR QSAR Environ. Res.,* **17**, 265–284.
23. Buontempo F V, Zhong Wang X, Mwense M, Horan N, Young A, and Osborn D (2005) Genetic Programming for the Induction of Decision Trees to Model Ecotoxicity Data, *J. Chem. Inf. Model.*, **45**, 904–912.
24. Helma C, Cramer T, Cramer S, and de Raedt L (2004) Data Mining and Machine Learning Techniques for the Identification of Mutagenicity Inducing Substructures and Structure Activity Relationships of Noncongeneric Compounds, *J. Chem. Inf. Comput. Sci.,* **44**, 1402–1411.
25. Kaufman J J, Koski W S, Harihan P, Crawford J, Garmer D M and Chan-Lizardo L. (1983) Prediction of Toxicology and Pharmacology Based on Model Toxicophores and Pharmacophores using the New TOX-MATCH-PHARM-MATCH Program, *Int. J. Quantum Chem., Quantum Biol. Symp.*, **10**, 375–416.
26. Koski W S and Kaufman J J (1988) TOX-MATCH/PHARM-MATCH Prediction of Toxicological and Pharmacological Features by Using Optimal Substructure Coding and Retrieval Systems, *Anal. Chim. Acta*, **210**, 203–7.
27. Patlewicz G, Jeliazkova N, Saliner A G, and Worth A P (2008) Toxmatch – a New Software Tool to Aid in the Development and Evaluation of Chemically Similar Groups, *SAR and QSAR in Environ. Res.,* **19**, 397–412.
28. http://ecb.jrc.it/qsar/qsar-tools/index.php?c=TOXMATCH
29. Ideaconsult Limited., 4 Angel Kanchev Street, 1000 Sofia, Bulgaria.
30. Woo Y T, Lai D Y, Argus M F, and Arcos J C, (1995) Development of Structure Activity Relationship Rules for Predicting Carcinogenic Potential of Chemicals, *Toxicol. Lett.,* **79**, 219–28.
31. Lai D Y, Woo Y T, Argus M F, and Arcos J C, (1996) Cancer Risk Reduction Through Mechanism-Based Molecular Design of Chemicals, in Designing Safer Chemicals, eds. S. De Vito and R. Garrett, ACS Symposium Series Vol. 640, American Chemical Society, Washington DC, pp. 62–73.
32. Woo Y T and Lai D Y (2005) OncoLogic: a Mechanism-Based Expert System for Predicting the Carcinogenic Potential of Chemicals, in Predictive Toxicology, ed. Christoph Helma, Marcel Dekker, New York, 385–413.
33. http://www.epa.gov/oppt/newchems/tools/oncologic.htm
34. Smithing M P and Darvas F (1992) HazardExpert – an Expert System for Predicting Chemical Toxicity, in Food Safety Assessment, eds. J. W. Finley, S. F. Robinson, and D. J. Armstrong, ACS Symposium Series Vol. 484, American Chemical Society, Washington DC, pp. 191–200.
35. CompuDrug International Inc., 115 Morgan Drive, Sedona, AZ 86351, USA.
36. http://ecb.jrc.ec.europa.eu/
37. Judson P N, Marchant C A, and Vessey J D (2003) Using Argumentation for Absolute Reasoning About the Potential Toxicity of Chemicals, *J. Chem. Inf. Comput. Sci.,* **43**, 1364–1370.
38. Langton K and Marchant C A (2005) Improvements to the Derek for Windows Prediction of Chromosome Damage, *Toxicology Letters,* **158**, S36–S37.
39. Derek for Windows and Meteor are developed by, and available from, Lhasa Limited, 22–23

Blenheim Terrace, Woodhouse Lane, Leeds LS2 9HD, United Kingdom.

40. Sanderson D M, (1989) Computer Prediction of Possible Toxicity from Chemical Structure. Paper presented at the autumn meeting of the British Toxicological Society.
41. Judson P N (1989) The Use of Expert Systems for Detecting Potential Toxicity. Paper presented at the autumn meeting of the British Toxicological Society.
42. Sanderson D M, Earnshaw C G, and Judson P N (1991) Computer Prediction of Possible Toxic Action from Chemical Structure; the DEREK System. *Human. Exp. Toxicol.,* **10**, 261–273.
43. Ridings J E, Barratt M D, Cary R, Earnshaw C G, Eggington E, Ellis M K, Judson P N, . Langowski J J, Marchant C A, Payne M P, Watson W P and Yih T D (1996) Computer Prediction of Possible Toxic Action from Chemical Structure: an Update on the DEREK System. *Toxicology,* **106**, 267–279.
44. Krause P J, Ambler S, Elvang-Gøransson M, and Fox J (1995) A Logic of Argumentation for Reasoning Under Uncertainty, *Computational Intelligence*, **11**(1), 113–131.
45. Judson P N and Vessey J D (2003) A Comprehensive Approach to Argumentation, *J. Chem. Inf. Comput. Sci.,* **43**, 1356–1363.
46. Greene N, Judson P N, Langowski J J, and Marchant C A (1999) Knowledge-Based Expert Systems for Toxicity and Metabolism Prediction: DEREK, StAR and METEOR, *SAR and QSAR in Environ. Res.*, **10**, 299–314.
47. Testa B, Balmat A L, Long A, and Judson P (2005) Predicting Drug Metabolism – an Evaluation of the Expert System METEOR, *Chemistry and Biodiversity*, **2**, 872–885.
48. Wipke W T, Ouchi G I, and Chou J T (1983) Computer-Assisted Prediction of Metabolism, in Structure-Activity Correlation as a Predictive Tool in Toxicology: Fundamentals, Methods, and Applications, ed. Leon Golberg, Hemisphere, Washington, DC, pp 151–169.
49. Darvas F (1987) METABOLEXPERT: an Expert System for Predicting Metabolism of Substances, in QSAR. Environmental Toxicology, proceedings of an international workshop 1986, ed. K .L. E. Kaisler, Reidel, Dordrecht, pp. 71–81.
50. MetabolExpert is supplied by CompuDrug International Inc., 115 Morgan Drive, Sedona, AZ 86351, USA.
51. Klopman G (1994) Mario Dimayagu, and Joseph Talafous, META. 1. A Program for the Evaluation of Metabolic Transformations of Chemicals, *J. Chem. Inf. Comput. Sci.*, **34** 1320–1325.
52. Talafous J, Sayre L M, Mieyal J J, and Klopman G (1994) META. 2. A Dictionary Model of Mammalian Xenobiotic Metabolism, *J. Chem. Inf. Comput. Sci.,* **34**, 1326–1333.
53. META is supplied by Multicase Inc, 23811 Chagrin Blvd Ste 305, Beachwood, OH, 44122, USA.
54. Jaworska J, Dimitrov S, Nikolova N, and Mekenyan O (2002) Probabilistic Assessment of Biodegradability Based on Metabolic Pathways: CATABOL System, *SAR and QSAR in Environ. Res.*, **13**, 307–323.
55. CATABOL is supplied by the Laboratory of Mathematical Chemistry, University “Prof. Assen Zlatarov”, 1 Yakimov Street, Bourgas, 8010 Bulgaria.
56. International QSAR Foundation, 1501 W. Knife River Road, Two Harbors, Minnesota 55616. Their website address is http://www.qsari.org/.
57. Laboratory of Mathematical Chemistry, University “Prof. Assen Zlatarov”, 1 Yakimov Street, Bourgas, 8010 Bulgaria.
58. Information about the (Q)SAR Toolbox Project can be found at http://www.oecd.org/document/23/0,3343,en_2649_34379_33957015_1_1_1_1,00.html and the toolbox can be downloaded from there. The address of the OECD is 2, rue André Pascal, F-75775 Paris, Cedex 16, France.

Chapter 2

Bacterial Mutagenicity Assays: Test Methods

David Gatehouse

Abstract

The most widely used assays for detecting chemically induced gene mutations are those employing bacteria. The plate incorporation assay using various *Salmonella typhimurium* LT2 and *E. coli* WP2 strains is a short-term bacterial reverse mutation assay specifically designed to detect a wide range of chemical substances capable of causing DNA damage leading to gene mutations. The test is used worldwide as an initial screen to determine the mutagenic potential of new chemicals and drugs.

The test uses several strains of *S. typhimurium* which carry different mutations in various genes of the histidine operon, and *E. coli* which carry the same AT base pair at the critical mutation site within the trpE gene. These mutations act as hot spots for mutagens that cause DNA damage via different mechanisms. When these auxotrophic bacterial strains are grown on a minimal media agar plates containing a trace of the required amino-acid (histidine or tryptophan), only those bacteria that revert to amino-acid independence (His^+ or $Tryp^+$) will grow to form visible colonies. The number of spontaneously induced revertant colonies per plate is relatively constant. However, when a mutagen is added to the plate, the number of revertant colonies per plate is increased, usually in a dose-related manner.

This chapter provides detailed procedures for performing the test in the presence and absence of a metabolic activation system (S9-mix), including advice on specific assay variations and any technical problems.

Key words: Plate-incorporation test, Pre-incubation test, *Salmonella typhimurium*, *E. coli*, S9-mix, Metabolic activation, Histidine auxotroph, Tryptophan auxotroph

1. Introduction

The most widely used assays for detecting chemically induced gene mutations are those employing bacteria. These assays feature in all test batteries for genotoxicity as it is relatively straight forward to use them as a sensitive indirect indicator of DNA damage. Bacteria can be grown in large numbers overnight, permitting the detection of rare mutational events. The extensive knowledge of bacterial genetics that was obtained during the twentieth century allowed the construction of special strains of bacteria with exquisite sensitivity

James M. Parry and Elizabeth M. Parry (eds.), *Genetic Toxicology: Principles and Methods*, Methods in Molecular Biology, vol. 817, DOI 10.1007/978-1-61779-421-6_2,

to a variety of genotoxins. An offshoot of the studies on genes concerned with amino acid biosynthesis led to the development of *Escherichia coli* and *Salmonella typhimurium* strains with relatively well defined mutations in known genes. The most commonly used bacteria are the *S. typhimurium* strains which contain defined mutations in the histidine operon. These were developed by Bruce Ames and form the basis of the "reverse" mutation assays (1). In these assays, bacteria which are already mutant at the histidine locus are treated with a range of concentrations of test chemical to determine whether the compound can induce a second mutation that directly reverses or suppresses the original mutations. Thus, for the *S. typhimurium* strains which are histidine auxotrophs, the original mutation resulted in the loss of ability to grow in the absence of histidine. The second mutation (induced by the chemical) restores prototrophy, i.e. the affected cell is now able to grow in the absence of histidine, if provided with inorganic salts and a carbon source. This simple concept underlines the great strength of these assays for it provides enormous selective power which can identify a small number of the chosen mutants from a population of millions of unmutated cells and cells mutated in other genes. Each of the *S. typhimurium* strains contains one of a number of possible mutations in the histidine operon, and each can be reverted by either base-change or frameshift mutations. The genotype of the most commonly used strains is shown in Table 1, together with the types of reversion events that each strain detects.

In order to make the bacteria more sensitive to mutation by chemical agents, several additional traits have been introduced. Ames and colleagues realised that many carcinogens (or their metabolites) are large molecules that are often unable to cross the protective cell wall of the bacteria. Wild-type cells produce a lipopolysaccharide that acts as a barrier to bulky hydrophobic molecules. Consequently, an *rfa* mutation was introduced into the *Salmonella* strains, which resulted in defective lipopolysaccharide and increased permeability.

Bacteria possess several major DNA repair pathways that appear to be error-free. The test strains were constructed, therefore, with a deletion removing the *uvrB* gene. This codes for the first enzyme in the error-free excision repair pathway, and so gene deletion renders the strains excision repair deficient, thus increasing their sensitivity to many genotoxins by several orders of magnitude. Lastly, some of the bacterial strains do not appear to possess classical error-prone repair as found in other members of the Enterobacteria such as *E. coli*. This results from a deficiency in *umuD* activity. This deficiency is overcome by insertion of a plasmid containing *umuDC* genes. Plasmid pKM101 is the most useful (2) conferring on the bacteria sensitivity to mutation without a concomitant increase in sensitivity to the lethal effects of test compounds. Further sensitivity is gained by the fact that the initial mutation responsible for the

Table 1
Genotype of commonly used *S. typhimurium* LT2 and *E. coli* WP2 strains

Bacterial strains	Histidine or tryptophan mutation	Full genotype[a]	Reversion events
S. typhimurium			
TA1535	hisG46	Dgal chlD bio *uvr*B rfa	Sub-set of base pair substitution events, extragenic suppressors
TA100	hisG46	Dgal chlD bio *uvr*B rfa (pKM101)	
TA1538	hisD3052	rfa Dgal chlD bio *uvr*B	Frameshifts
TA98	hisD3052	rfa Dgal chlD bio *uvr*B (pKM101)	Frameshifts
TA1537	hisC3076	rfa Dgal chlD bio *uvr*B	Frameshifts
TA97	hisD6610	his O_{1242}rfa Dgal chlD bio *uvr*B (pKM101)	Frameshifts
TA102	hisG428	his D (G) $_{8476}$rfa galE (pAQ1) (pKM101)	All possible transitions and transversions, small deletions
E. coli			
WP2 *uvr*A	trpE	*uvr*A	All possible transitions and transversions, small deletions
WP2 uvrA (pKM101)	trp E	uvrA (pKM101)	

[a]*rfa* Deep rough, *galE* UDP galactose 4-epimerase, *ChlD* nitrate reductase (resistance to chlorate), *bio* biotin, *uvrB* UV endonuclease component B, *D* deletion of genes following this symbol, *pAQ1* a plasmid containing the his G_{428} gene, *pKM101* a plasmid carrying the *uvr*A and B genes that enhance error-prone repair

histidine growth requirement is situated at a site within the gene that is particularly sensitive to reversion by specific classes of genotoxin (i.e. hotspots). The incorporation of strain TA102 into the test battery was subsequently proposed, as the target mutation has an AT base pair at the critical site. This allows the detection of genotoxins not detected by the usual battery of *S. typhimurium* strains that possess mutations exclusively at GC base pairs. As an alternative many guidelines recommend the use of the *E. coli* WP2 trpE strains which contain a terminating ochre mutation in the *trpE* gene. The ochre mutation involves an AT base pair, and so reverse mutation can take place at the original site of mutation or in the relevant tRNA loci. A combination of *E. coli* WP2 *trpE* (pKM101) and *E. coli* WP2 *trpE uvrA* (pKM101) can be used as alternatives to *S. typhimurium* TA102 for the detection of point mutations at AT sites.

Consequently the following base set of bacterial test strains have been recommended in several guidelines (3–5):

S. typhimurium: TA98, TA100, TA1535

S. typhimurium: TA1537, or TA97, or TA97a

S. typhimurium: TA102, or E. coli WP2 *uvrA* or E. coli WP2 *uvrA* (pKM101).

The use of the repair-proficient *E. coli* strain WP2 (pKM101) allows the detection of cross-linking agents that require an intact excision repair pathway to generate mutations and this strain may also be selected.

In general, the most widely used protocol is the "plate incorporation assay". In this method, the bacterial strain, test material and an in vitro metabolic activation system (S9 mix) are added to a small volume of molten agar containing a trace of histidine and biotin (or tryptophan alone in the case of the *E. coli* strains). The mixture is poured across the surface of a basal (minimal glucose) agar plate and allowed to set prior to incubation at 37°C for 48–72 h. The trace of histidine allows the growth of the auxotrophic bacteria in the presence of the test compound and/or any in vitro metabolites. The period of several cell divisions is essential to allow fixation of any premutagenic lesions that have been induced in the bacterial DNA. Exhaustion of the histidine halts growth of the auxotrophic cells. Only those cells that have been reverted to histidine independence will continue to grow and form discrete visible colonies (Fig. 1). The growth of non-reverted cells forms a visible background lawn, the thinning of which can be used as a non-quantitative indication of chemical related toxicity. Revertant colonies can be counted manually or by use of an Image analyser. Untreated (vehicle) and suitable positive controls are included. The test concentration range is determined by performance of a preliminary toxicity test, whilst for non-toxic compounds a maximum concentration of 5 mg per plate is generally recommended.

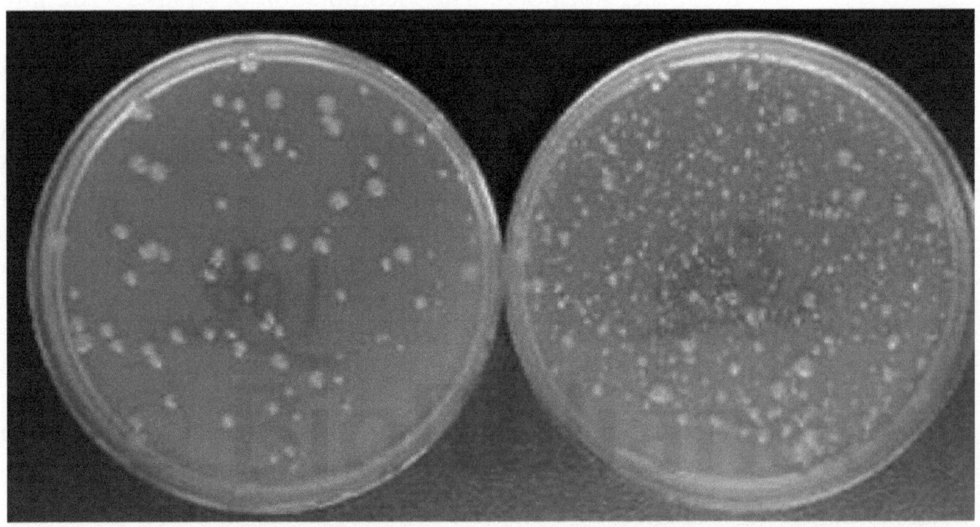

Fig. 1. Examples of a solvent control plate and a positive mutagen plate.

Bacterial mutation tests have been subjected to several large-scale trials over the years (6). These studies were primarily concerned with assessing the correlation between results obtained in the assays and the carcinogenic activity of chemicals. Most of the studies suggest that there is a good qualitative relationship between genotoxicity in the *Salmonella* assay and carcinogenicity for many, although not all, chemical classes. This figure varies between a sensitivity of 60 and 90% dependent upon chemical class. The bacterial assays seem to be particularly efficient in detecting trans-species, multi-organ animal carcinogens (7).

2. Materials

2.1. Vogel-Bonner Salts Medium E (50×)

Use: Salts for GM plates.

Ingredients	Per litre
Warm distilled water (45–50°C)	670 ml
Magnesium sulphate ($MgSO_4 \cdot 7H_2O$)	10 g
Citric acid monohydrate	100 g
Potassium phosphate dibasic anhydrous (K_2HPO_4)	500 g
Sodium ammonium phosphate ($NaHNH_4PO_4 \cdot 4H_2O$)	175 g

Add salts in the order indicated to warm water in a 2-L beaker with stirring (magnetic stirrer) making sure each one is dissolved before adding next salt. Adjust volume to 1 L Distribute into 20 ml aliquots and autoclave, loosely capped for 30 min at 121°C. When cooled tighten caps and store at room temperature in dark.

2.2. Glucose Solution (10% v/v)

Use: Carbon source for GM plates.

Ingredients	Per litre
Distilled water	700 ml
Dextrose	100 g

Add dextrose to water in 3 L flask. Stir until mixture is clear. Add more water to bring to 1,000 ml and then dispense in 50 ml aliquots and autoclave at 121°C for 20 min. When cooled tighten caps and store at 4°C.

2.3. Glucose Minimal Agar Plates

Use: Basal agar for mutagenicity assay.

Ingredient	Per litre
Agar	15 g
Distilled water	930 ml
VB Salts solution (50×)	20 ml
Glucose solution (10%v/v)	50 ml

Add agar to water in 3 L flask and autoclave for 30 min at 121°C. After cooling (to approx 65°C) add VB salts, mix thoroughly, then add glucose solution and swirl thoroughly. Dispense the medium into Petri dishes (approx 25 ml per dish). When solidified, plates can be stored at 4°C for several weeks in sealed plastic bags.

2.4. Histidine–Biotin Solution 0.5 mM (When Using *S. typhimurium* Strains)

Use: To supplement top agar with excess biotin and a trace amount of histidine.

Ingredient	Per litre
Distilled water	1,000 ml
d-Biotin (F.W. 247.3)	124 mg
l-Histidine. HCl (F.W. 191.7)	96 mg

Add biotin and histidine to boiling water. After cooling sterilise by filtration (0.45 μm) or by autoclaving at 121°C for 20 min. Dispense in 20 ml aliquots and store in glass bottle at 4°C.

2.5. Tryptophan Solution 0.25 mM (When Using *E. coli* Strains)

Use: To supplement top agar with a trace amount of tryptophan.

Ingredient	Per litre
Distilled water	1,000 ml
L-tryptophan HCl (F.W. 204.2)	51 mg

Dissolve tryptophan in water. Sterilise by filtration (0.45 μm) or by autoclaving at 121°C for 20 min. Dispense in 200 ml aliquots and store at 4°C.

2.6. Top Agar Supplemented with Histidine–Biotin or Tryptophan (Dependent Upon Test Strains)

Use: To deliver bacteria, chemical and buffer or S9-mix to the bottom agar.

Ingredient	Per litre
Distilled water	1,000 ml
Agar	6 g
Sodium chloride	5 g

Add agar and sodium chloride to water in 3 L flask. Agar may be dissolved in steam bath, microwave or by autoclaving briefly (10 min, liquid cycle). Dispense in 180 ml aliquots and autoclave at 121°C for 20 min. When ready to use, melt top agar in boiling water and add following to each 180 ml aliquot:

For *S.typhimurium* strains: Sterile histidine–biotin solution (0.5 mM)	20 ml
For *E. coli* strains: Sterile tryptophan solution (0.25 mM)	20 ml

2.7. Sodium Phosphate Buffer (0.1 mM, pH 7.4)

Use: For testing chemicals in the absence of metabolic activation (S9-mix).

Ingredients	Per litre
Sodium phosphate, monobasic (0.1 M): To 1 L water add 13.8 g $NaH_2PO_4{\cdot}H_2O$	120 ml
Sodium phosphate dibasic (0.1 M): To 1 L water add 14.2 g $Na_2HPO4{\cdot}H_2O$	880 ml

After mixing, adjust pH to 7.4 using 0.1 M dibasic sodium phosphate solution. Dispense in 100 ml aliquots and autoclave with loose caps at 121°C for 30 min. When cooled tighten caps and store at room temperature in the dark.

2.8. S9-mix

A factor of critical importance in bacterial mutagenicity screening is the need to include some form of in vitro metabolising system. This is because the bacterial indicator cells possess a very limited capacity for endogenous metabolism of xenobiotics. Many carcinogens and mutagens are unable to interact with DNA unless they have undergone some degree of metabolism. To improve the ability of the bacterial test systems to detect as many authentic in vivo mutagens and carcinogens as possible, extracts of mammalian liver (usually rat) are incorporated. The liver is a rich source of mixed-function oxygenases capable of converting carcinogens to reactive electrophiles. Crude homogenate such as the 9,000 × *g* supernatant (S9 fraction) is used, which is composed of free endoplasmic reticulum, microsomes, soluble enzymes, and some cofactors. The oxygenases require the reduced form of nicotinamide adenine dinucleotide phosphate (NADPH), which is normally generated in situ by the action of glucose-6-phosphate dehydrogenase on glucose-6-phosphate and reducing NADP, both of which are normally supplied as cofactors. Normal uninduced S9 preparations are of limited value for screening as they are deficient in particular enzyme activities. In addition, species and tissue differences are most divergent in such preparations. These problems are reduced when enzyme inducers are used, and most commonly preparations are made from rat livers after enzyme induction with Aroclor 1254, which is a mixture of polychlorinated biphenyls. Concern about the toxicity, carcinogenicity and persistence of this material in the environment has led to the introduction of alternatives, such as a combination of phenobarbitone and β-naphthoflavone. This combination induces a similar range of mono-oxygenases and has been recommended as a safer alternative to Aroclor (8).

It should be noted that this system is only a first approximation to the complex metabolic processes that occur in vivo, and in particular there is little account taken of the phase II detoxification reactions.

The S9-mix has the following composition:

Cofactors for S9-mix

Ingredient	Per litre
Distilled water	900 ml
D-Glucose-6-phosphate	1.6 g
Nicotinamide adenine dinucleotide phosphate (NADP)	3.5 g
Magnesium chloride (MgCl)	1.8 g
Potassium chloride (KCl)	2.7 g
Sodium phosphate dibasic ($Na_2HPO_4 \cdot H_2O$)	12.8 g
Sodium phosphate monobasic ($NaH_2PO_4 \cdot H_2O$)	2.8 g

To 900 ml of water add each ingredient sequentially making sure all dissolve thoroughly (may take up to an hour). Filter sterilise (0.45 μm filter) and dispense into sterile glass bottles in aliquots of 7, 9 or 9.5 ml (or multiples of these volumes). This allows the convenient preparation of 30, 10, and 5% v/v S9, by addition of 3, 1, or 0.5 ml of S9 fraction, respectively, to produce the final S9-mix (10 ml volumes). Store the cofactor solution at −20°C.

3. Methods

3.1. Plate Incorporation Assay

The most widely used protocol is the "Plate Incorporation Assay" which is carried out as follows (see Fig. 2, Notes 1 and 2):

1. Each selected strain is grown for 10–15 h at 37°C in nutrient broth (Oxoid No 2) or supplemented media (Vogel-Bonner) on an orbital shaker, to a density of $1–2 \times 10^9$ colony forming units/ml. A timing device can be used to ensure that cultures are ready at the beginning of the working day.

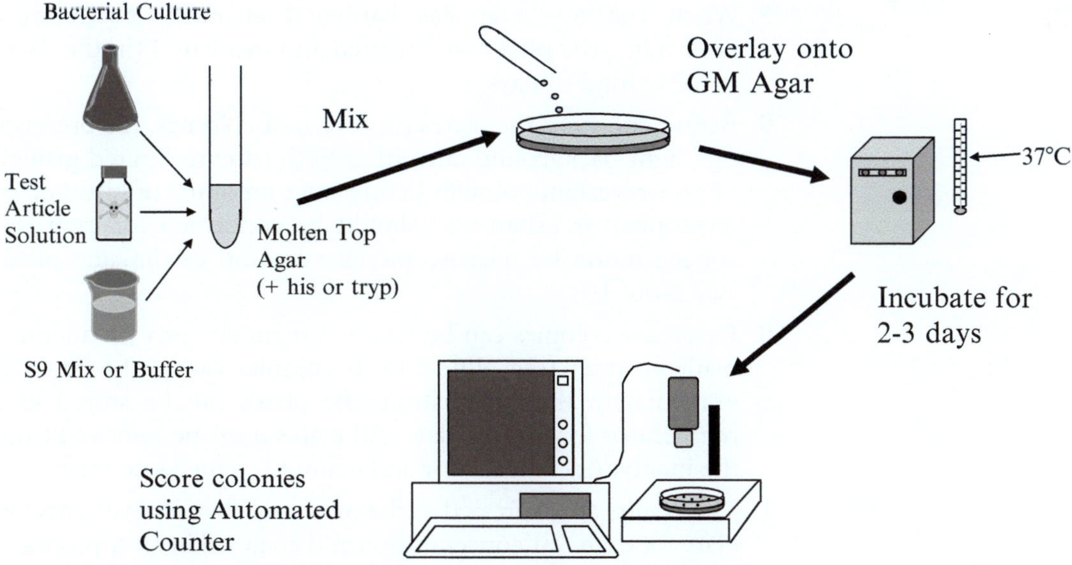

Fig. 2. Schematic representation of the conduct of the plate incorporation method.

2. Label an appropriate number of pre-dried GM agar plates and sterile test tubes for each test chemical and bacterial strain.
3. Prepare S9-mix and keep on ice until use.
4. Prepare chemical dilutions in a suitable solvent (see Note 3).
5. Melt top agar supplemented with either 9.6 μg/ml (0.05 mM) histidine and 12.4 μg/ml (0.05 mM) biotin (for *S. typhimurium* strains), or 5.1 μg/ml (0.025 mM) tryptophan alone (for *E. coli* WP2 test strains) (see Note 4). Dispense into tubes in 2 ml aliquots and keep semi-molten (43–48°C) by holding tubes in a thermostatically controlled aluminium block (see Note 5).
6. To each 2 ml volume of top agar, add in the following order (with thorough mixing/vortexing after each addition):
   (a) Chemical dilution (10–200 μl) or equivalent volume of solvent (see Note 6) or positive control chemical for each test strain (see Note 7).
   (b) S9-mix (500 μl) or equivalent volume of phosphate buffer.
   (c) Overnight culture of required test strain (100 μl) to give approx 1×10^8 viable bacteria per tube.
7. The contents of the tubes are then quickly mixed and poured onto the surface of dried GM agar plates (see Note 8). There should be at least three replicate plates per treatment with at least five test concentrations (covering a range of three logs) and solvent/untreated controls. Duplicate plates are sufficient for positive controls and sterility controls (see Note 9).

8. When the top agar has hardened at room temperature (2–4 min), the plates are inverted and incubated (in the dark) at 37°C for 2–3 days.
9. Before counting the plates for revertant colonies, the presence of a light background lawn of growth (due to limited growth of non-revertant colonies before trace amounts of histidine or tryptophan is exhausted) should be confirmed for each test concentration by microscopic examination of the agar plates (see Note 10).
10. Revertant colonies can be counted manually or with an automatic counter (see Note 11). If colonies cannot be counted immediately after incubation, the plates can be stored in a refrigerator for up to 2 days. All plates must be removed from the incubator/refrigerator and counted at the same time.
11. The results are expressed as the mean number of revertants per plate for each test concentration and analysed in an appropriate manner (see Notes 12 and 13).
12. For any technical assay problems, see Notes 14–20.

4. Notes

1. Some genotoxins are poorly detected using the plate incorporation procedure, particularly those that are metabolised to short-lived reactive electrophiles (e.g. aliphatic nitrosamines). In these cases, a preincubation procedure should be used in which bacteria; test compound and S9-mix are incubated together in a small volume at 37°C for 30–60 min prior to agar addition (9). This maximises exposure to the reactive species and limits non-specific binding to agar. In addition, in the plate-incorporation method soluble enzymes in the S9-mix, cofactors, and the test chemical can diffuse into the basal agar. This can interfere with the detection of some mutagens.
2. Neither the plate incorporation assay nor pre-incubation assay is suitable for testing highly volatile substances. The use of a closed chamber is recommended for testing such chemicals in a vapour phase, as well as for gases (10). Procedures using plastic bags in lieu of desiccators have also been described (11).
3. Solvent of choice is sterile distilled water. Dimethylsulphoxide is often used for hydrophobic chemicals, although other solvents have been used (e.g. acetone, ethyl alcohol, dimethyl formamide, and tetrahydrofuran). It is essential to use the minimum amount of organic solvent (e.g. <2% w/w) compatible with adequate testing of the chemical, and to use fresh batches of solvent of the highest grade possible.

4. Because the bacteria stop growing when the amino-acid (histidine or tryptophan) is depleted, the final population of auxotrophic bacteria on the plate is dependant upon this concentration, which in turn affects the number of spontaneous revertant (prototrophic) colonies. It is therefore important that the utmost care is taken to accurately supplement the top agar with the correct amount of amino acid.
5. It is best to avoid water baths, as microbial contamination can cause problems.
6. The exact volume of test chemical or solvent may depend upon toxicity or solubility.
7. The diagnostic positive control chemicals validate each test run and help to confirm the nature of the bacterial test strain. Table 2 lists representative positive controls.
8. It is important to quickly swirl the plates after addition of the top agar to the surface of the GM agar plates to ensure an even distribution of the top agar which contains the bacteria, test chemical, and S9-mix or buffer.
9. Checks should be made on the sterility of S9-mix, media and test chemicals to prevent unnecessary loss of valuable time and resources.
10. When plates are inspected after 48 h incubation and growth retardation is seen, as evidenced by smaller than anticipated colony sizes, the plates should be incubated for an additional 24 h.

Table 2
Representative positive control chemicals

Bacterial strain	Chemical (μg/plate[a])	
	Without S9-mix	**With S9-mix**
S. typhimurium		
TA97/TA97a	9-Aminoacridine (50)	2-Aminoanthracene(1–5)[b]
TA98	4-Nitro-o-phenylenediamine (2.5)	2-Aminoanthracene(1–5)
TA100	Sodium azide (5)	2-Aminoanthracene(1–5)
TA102	Mitomycin C (0.5)	2-Aminoanthracene(5–10)
TA1535	Sodium azide (5)	2-Aminoanthracene(2–10)
TA1537	Neutral red (10)	2-Aminoanthracene(2–10)
TA1538	4-Nitro-o-phenylenediamine (2.5)	2-Aminoanthracene(2–10)
E. coli		
WP2 *uvr*A	Nifuroxime(5–15)	2-Aminoanthracene (1–10)
WP2 *uvr*A (pKM101)	Cumene hydroperoxide (75–200)	2-Aminoanthracene (1–10)

[a] Concentration based upon 100× 15 mm Petri plate containing 20–25 ml GM agar

[b] Some researchers have suggested that 2-aminoanthracene should not be used as the only positive control to evaluate S9-mix activity as it has been shown that the chemical may be activated by enzymes other than the microsomal cytochrome P_{450} family (15)

This should be done for all plates from an experiment even if growth retardation is only seen at the higher test concentrations. At concentrations that are toxic to the test strains the lawn will be depleted and colonies may appear that are not true revertants but surviving, non-prototrophic cells. If necessary the phenotype of any questionable colonies (pseudo-revertants) should be checked by plating on histidine or tryptophan-free medium.

11. Automatic colony counters are relatively accurate in the range of colonies normally observed (although calibration against manual counts is a wise precaution). Manual counting may be needed where a potent mutagen is being tested and accurate quantitative colony counts are required. In addition, manual counting could be required if heavy precipitate and lack of contrast between the colonies is observed on the plates (especially at higher dose levels).

12. If the test chemical induces increases in revertant counts, these values are compared with the spontaneous revertant counts for each bacterial test strain. Each test strain has a characteristic spontaneous revertant frequency. There is usually some day-to-day and laboratory-to-laboratory variation in the number of spontaneous revertant colonies. Choice of solvent may also affect this value. It is recommended that each laboratory establish its own historical data base on which to base an acceptable spontaneous revertant rate. Table 3 presents a range of spontaneous revertant values per plate considered valid by many experienced test laboratories.

Table 3
Representative ranges for spontaneous revertant values

	Number of revertants per plate	
Bacterial strain	**Without S9-mix**	**With S9-mix**
S. typhimurium		
TA97/TA97a	75–200	100–200
TA98	20–50	20–50
TA100	75–200	75–200
TA102	100–300	200–400
TA1535	5–20	5–20
TA1537	5–20	5–20
TA1538	5–20	5–20
E. coli		
WP2 *uvr*A	15–50	15–50
WP2 *uvr*A (pKM101)	45–151	45–151

13. Several statistical approaches have been applied to the results of these assays and all have their strengths and weaknesses (12, 13). Another approach that has been widely used is to set a minimum fold increase, usually two- to threefold, in revertants (over the solvent control) as the cut-off between a mutagenic and non-mutagenic response (14). In general, positive results should be statistically significant, dose related and reproducible.
14. If the spontaneous revertant count is *too low* relative to the historical control range for an individual laboratory, this may be due to:
    (a) Toxicity associated with new batch of agar.
    (b) Too little histidine added to top agar.
15. If the spontaneous reverant count is *too high* relative to the historical control range for an individual laboratory, it may be due to the following:
    (a) Too much histidine added to top agar.
    (b) Initial inoculums from working culture contained unusually high number of His^{+} bacteria (jackpot effect).
    (c) Petri plates have been sterilised with ethylene oxide with residual levels inducing mutations in *S. typhimurium* strains TA1535, TA100 and the *E. coli* strains.
    (d) Contamination might be present.
16. If the positive control does not work, it may be due to the following:
    (a) Use of incorrect chemical.
    (b) Chemical may have deteriorated during storage.
    (c) Omission of S9-mix if required.
    (d) S9-mix has lost its activity.
    (e) Incorrect bacterial test strain used.
17. If there are few, if any, revertant colonies on any plates, it may be due to the following:
    (a) Test chemical is highly toxic.
    (b) Temperature of top agar was too high.
    (c) Solvent is toxic.
18. If most of colonies are concentrated on one half of plate, it is because the basal and/or top agar were allowed to solidify on a slant.
19. If the top agar does not solidify, it may be due to the following:
    (a) Agar concentration was too low.
    (b) Gelling is retarded by acidic pH of test material.

20. If top agar slips out of place, it is because surface of basal agar was too wet. If too much moisture is present, the GM plates should be dried by incubating plates at 37°C overnight before use.

References

1. Ames, B. N. (1971) The detection of chemical mutagens with enteric bacteria, in *Chemical Mutagens, Principles and Methods for Their Detection.* Volume **1** (Hollander, A; ed.), Plenum Press, New York, pp. 267–282
2. Mortelmans, K. E., and Dousman, L. (1986) Mutagenesis and plasmids, in *Chemical Mutagens, Principles and Methods for their Detection*, Volume **10**. (de Serres, F. J; ed.), Plenum Press, New York, pp. 469–508.
3. *ICH (International Conference on Harmonisation), Guideline on Specific Aspects of Regulatory Genotoxicity Tests,* Fed.Reg. 61: 18199, April 24, 1996 (ICH S) and 62:62472, November 21, 1997 (ICH 2B).2A, 1995.
4. OECD *Guideline for Testing of Chemicals, Genetic Toxicology* No. 471, Organisation for Economic Co-Operation and Development, Paris, 21 July 1997.
5. USEPA, Part 712-C-98-247, *Health Effects Test Guidelines* OPPTS 870.5100, Bacterial Reverse Mutation Test, United States Environmental Protection Agency, Prevention and Toxic Substances (7101), August 1998.
6. Tennant, R. W., Margolin, B., Shelby, M., Zeiger, E., Haseman, J., Spalding, J., Resnick, W., Stasiewicz, M., Anderson, B., and Minor, R. (1987) Prediction of chemical carcinogenicity in rodents from *in vitro* genetic toxicity assays. *Science,* **236**, 933–941.
7. Ashby, J. and Tennant, R. W. (1988) Chemical structure, *Salmonella* mutagenicity and extent of carcinogenicity as indices of genotoxic carcinogens among 222 chemicals tested in rodents by the US NCI/NTP. *Mutat. Res.* **204**, 17–115.
8. Eliott, B., Coombs, R., Elcombe, C., Gatehouse, D., Gibson, G., MacKay, J., and Wolf, R. (1992) Report of UKEMS Working Party: Alternatives to Aroclor1254 induced S9 in *in vitro* genotoxicity assays. *Mutagenesis,* 7, 175–177.
9. Yahagi, T., Degawa, W., Seino, Y., Matsushima, T., Nagao, M., Sugimura, T., and Hashimoto, Y. (1975) Mutagenicity of carcinogenic azo-dyes and their derivatives. *Cancer Lett.* **1,** 91–97.
10. Simmon, V. F., Kauhanen, K., Tardiff, R. (1977) Mutagenic activities of chemicals identified in drinking water, in *Progress in Genetic Toxicology* (Scott, D., Bridges, B., and Sobels, F., eds.), Elsevier/North Holland, Amsterdam, pp.249–258.
11. Araki, A., Noguchi, T., Kato, F., and Matsushima, T. (1994) Improved method for mutagenicity testing of gaseous compounds by using a gas sampling bag. *Mutat. Res.* **307**, 335–344.
12. Margolin, B., Kaplan, N., and Zeiger, E . (1981) Statistical analysis of the Ames *Salmonella/*microsome test. *Proc. Nat. Acad. Sci. USA.* **78**, 3779–3783.
13. Stead, A., Hasselblad, J., and Claxton, L. (1981) Modeling the Ames test. *Mutat. Res.* **85**, 13–27.
14. Mahon, G., Green, M., Middleton, B., Mitchell, I., Robinson, W., and Tweats, D. (1989) *Analysis of data from microbial colony assays, in UKEMS Sub-Committee on Guidelines for Mutagenicity Testing. Report: Part III. Statistical Evaluation of Mutagenicity Test Data* (Kirkland, D. J., ed.), Cambridge Univ. Press, Cambridge. pp.28–65.
15. Gatehouse, D., Haworth, S., Cebula, T., Gocke, E., Kier, L., Matsushima, T., Melcion, C., Nohmi, T., Ohta, T., Venitt, S., and Zeiger, E. (1994) Recommendations for the performance of bacterial mutation assays. *Mutat. Res.* **312**, 217–233.

Chapter 3

The Mouse Lymphoma Assay

Melvyn Lloyd and Darren Kidd

Abstract

The mouse lymphoma TK assay (MLA) is part of an in vitro battery of tests designed to predict risk assessment prior to in vivo testing. The test has the potential to detect mutagenic and clastogenic events at the thymidine kinase (*tk*) locus of L5178Y mouse lymphoma $tk^{+/-}$ cells by measuring resistance to the lethal nucleoside analogue triflurothymidine (TFT). Cells may be plated for viability and mutation in semi-solid agar (agar assay) or in 96-well microtitre plates (microwell assay). When added to selective medium containing TFT, wild-type $tk^{+/-}$ cells die, but TFT cannot be incorporated into the DNA of mutant $tk^{-/-}$ cells, which survive to form colonies that may be large (indicative of gene mutation) or small (indicative of chromosomal mutation) in nature. Mutant frequency is expressed as the number of mutants per 10^6 viable cells.

Key words: Mouse lymphoma, Thymidine kinase, TFT, Viability, Mutation, Large and small colonies

1. Introduction

The Mouse Lymphoma Assay (MLA) at the thymidine kinase (*tk*) locus is a mammalian gene mutation assay used (but not solely) within a regulatory framework for genetic toxicology testing as laid out by the Organisation for Economic Co-Operation and Development (OECD) and the International Conference on Harmonisation (ICH) guidance documents (1, 2). It is an assay designed to assist in the elucidation and evaluation of the potential risk a novel compound may pose to humans. It is currently one component in a battery of in vitro tests that also includes assays for bacterial mutation (Ames test) and chromosomal damage (chromosome aberration or in vitro micronucleus test). Each assay within the battery plays a part in the overall assessment of potential risk to humans. Much work has been undertaken to improve the predictive ability of the battery.

James M. Parry and Elizabeth M. Parry (eds.), *Genetic Toxicology: Principles and Methods*, Methods in Molecular Biology, vol. 817, DOI 10.1007/978-1-61779-421-6_3, © Springer Science+Business Media, LLC 2012

1.1. The Cell Line

The cell line used in the MLA is the L5178Y $tk^{+/-}$ clone 3.7.2C and is a derivative of the L5178 cell line. This parent cell line began as a transplantable mouse leukaemia from a thymic tumour induced in the DBA/2 strain of mouse using 3-methylcholanthrene and was designated L5178 (3). Six years later, the tumour was adapted to create an established suspension cell line and was designated L5178Y (4). From this cell line, a $tk^{-/-}$ derivative was established following mutagenesis by ethyl methanesulphonate (5). The MLA cell line, L5178Y $tk^{+/-}$ 3.7.2C, was isolated as a spontaneous THAG (thymidine, hypoxanthine, amethopterin, glycine) resistant revertant (6) of the $tk^{-/-}$ cell line (see Note 1) and the assay using the soft agar method was subsequently described (7).

1.2. Thymidine Kinase Gene

Thymidine kinase (TK) in this cell line is encoded by a heterozygous gene located on chromosome 11 and is part of the salvage pathway for pyrimidine nucleic acid breakdown products. As a gene, it is not essential for survival therefore if it becomes inactivated in some way, it does not lead to mortality of the cell. It catalyses the phosphorylation of thymidine deoxyriboside (dThd) to form deoxythymidylate (dTMP). Two further phosphate groups are added forming dTTP (the deoxyribonucleotide, thymidylic acid). This base–sugar complex is incorporated into the DNA via the binding of the 3′-hydroxyl of the deoxyribose moiety to the 5′-hydroxyl of the adjacent sugar creating a phosphodiester bridge. If *tk* has been damaged or disrupted in some manner as a result of mutation (internal or chromosomal), transcription (if successful) will lead to a non-functional enzyme.

The original selective agent used for the recovery of $tk^{-/-}$ cells was 5′-bromodeoxyuridine (BUdR) but 5-trifluorothymidine (TFT: Note 2) was found to be a more effective selective agent, more versatile in its use and 50 times more potent (8). If the TK enzyme is normal, it incorporates the TFT into the DNA during repair/cell division, thus killing the cell. Only cells carrying the forward mutation in the *tk* gene ($tk^{-/-}$) survive in its presence due to breakdown of the salvage pathway. After addition of TFT, the cells are typically cultured for 10–14 days during which any $tk^{-/-}$ cell forms a clone. The clone forms one of two phenotypes, small or large, dependent on the type of mutation that has occurred. Visually, they are distinct as the small clone is dense and has uneven edges (Fig. 1), whereas the large clone is semi-opaque and has smooth edges (Fig. 2). Small clones tend to contain inter-gene mutations (chromosomal rearrangement, translocation, etc.) and have an extended doubling time whereas the large clones tend to contain intra-gene mutations (point mutations, base deletions, etc.) and generally have a doubling time closer to the parent L5178Y $tk^{+/-}$ (clone 3.7.2C) cell line (9, 10).

Fig. 1. Photograph of a single well in a 96-well plate showing a large clone. Note the smooth edge and relative opaqueness of the cell growth pattern.

Fig. 2. Photograph of a single well in a 96-well plate showing a small clone. Note the uneven edge and relative density of the cell growth pattern.

1.3. Assay Development

The MLA assay development continued during the 1970s (11–13) but technical challenges in cloning the mutant cells raised questions over the operation of the test system (see Note 3).

In response to this, a microwell version of the assay was developed (14) using liquid media at all stages and cloned the cells in

96-well plates after the addition of TFT. Both the microwell and agar methodologies, when performed to a good standard, are equally acceptable and internationally accepted guidelines reflect this.

In 1988, five publications resulted from a small inter-laboratory trial testing 63 chemicals in the MLA to validate its potential as a mutagenesis assay (15–19). The trial concluded that the assay was effective and suitable for detecting a range of mutations including point, base deletion, nonsense, mis-sense and larger events affecting chromosomes such as deletions and potentially aneuploidy in a mammalian cell line. Cytogenetic characterisation was also published around this time using standard G-banding (20) and as technology advanced, the results were improved upon by using multicolour spectral karyotyping (21). These two publications are the benchmark for laboratories wanting to verify that L5178Y cells have not altered karyotypically from the original cells.

As the understanding of the MLA cell line increased, its potential to be used instead of the chromosome aberration assay was proposed. It was first suggested that this was possible due to the nature of the small clone being indicative of chromosomal mutation (22) and the MLA and chromosomal aberration assays are both considered acceptable in vitro mammalian genetic toxicology assays.

In the early 1990s, ICH was formed to bring together labs from across the globe with the goal of harmonising practised procedures. ICH published documents in 1995 (23) and 1997 (2) detailing genotoxicity testing regimes and an update to OECD Test Guideline 476 for performing the MLA assay was also published around this time (1). In ICH guideline S2B (2) it was stated that the extended continuous treatment of the cells for 24-h in the presence of the test chemical in the MLA was required for the detection of certain groups of chemicals (nucleoside and base analogues) as these were commonly non-mutagenic following a 3–4-h treatment. The committee also expected the 24-h treatment to increase the detection of aneugens, as there were few examples available at the time of this chemical class being positive in the MLA following short treatment times (up to 6-h).

Around the same time as ICH S2B was published, it was reported that the loss of the functional *tk* allele (designated $tk1_b$) was common in both large and small clones (24). This confirmed that the cells could survive in the absence of a functional *tk* gene, increasing confidence in the opinion that aneuploidy was detectable using this assay. The following year, a Japanese group (25) concluded that aneugens were detectable following the 24-h treatment incubation protocol, however some of the results were observed at highly toxic levels (<10% relative survival). The maximum recommended toxicity for the MLA was and still is 10% survival (now measured by RTG).

The toxicity measure and other assay-based issues have been debated at the International Workgroup on Genotoxicity Tests for the MLA (IWGT-*ML*) who have steered the assay's development since the late 1990s (26–30). In 2002, an update on the decision made in Portland, USA (31) was that the spontaneous mutation frequency for the negative controls should be at least 50 or 35 mutants per 10^6 cells for the microwell or agar version of the assay, respectively (28). In 2003, relative total growth (RTG), which incorporates culture growth in suspension and cloning efficiency, became the recommended measure of toxicity (29) – prior to this, RTG and/or relative survival (RS) had been used to assess toxicity. When the IWGT-*ML* met in 2003 and 2005 (29, 30), a standardised, empirically derived numerical factor, calculated as a result of analysing large volumes of test data submitted by laboratories worldwide, was recommended for assessing a positive response (taken in conjunction with a dose-related linear trend observed in mutation frequency). This factor, termed the Global Evaluation Factor (GEF) defines the numerical cut-off value to determine a positive response. The workgroup's recommendation was that the test compound should induce a mutation frequency greater than 90 (agar method) or 126 (microwell method) mutants per 10^6 viable cells above the concurrent negative control frequency to be classified as a positive response. It also became a requirement that the positive control chemicals must induce increases in the number of small colonies, thus showing generation of significant genetic damage and proving the assay's functionality. The latter (2005) meeting reinforced the continued need for the 24-h treatment, but acknowledged that few compounds are uniquely positive following this protocol (30).

Research into the assay is ongoing, with the current focus being the genetic and molecular determination of clones harvested from the assay. Microarray techniques have been employed to determine the genetic make-up of the two clone-phenotypes the assay produces (large and small), with the aim of further understanding their generation. One group (32) analysed a large number of genes involved in cell growth, all of which were located close to the *tk* gene itself on chromosome 11 and found that the large and small phenotypes displayed suitably different profiles for the analysed genes for the authors to deem them significantly different. These authors also concluded that this pocket of genes may play a part in pathways such as apoptosis.

2. Materials

2.1. Safety

When testing compounds where the hazards are unknown, it is always prudent to consider them as potential mutagens (see Note 4).

2.2. Cell Line

1. L5178Y *tk*$^{+/-}$ (3.7.2C) mouse lymphoma cells should be used for the assay. The cultures originated from Dr Donald Clive, Burroughs Wellcome Co, NC, USA and an assurance should be sought that the cell line has been originally derived from this source (see Note 5).
2. The cells are stored as frozen stocks in liquid nitrogen.
3. Each batch of frozen cells should be purged of *tk*$^{-/-}$ mutants, checked for spontaneous mutant frequency and that they are mycoplasma free.
4. The population doubling time of the cell cultures should also be ascertained: typically, this cell line has a doubling time of approximately 8–10 h.

2.3. Growth Media

1. RPMI 1640 medium is most commonly used for culturing mouse lymphoma cells (see Note 6).
2. The medium is supplemented with antibiotics (e.g. penicillin/streptomycin and amphotericin B) to prevent contamination.
3. Horse serum is added to the medium to supplement cell growth and should be checked prior to use to check effects on cell growth, cloning efficiency, and spontaneous and induced mutant frequencies. Typically, media containing 10% (v/v) horse serum (RPMI 10) should be used for normal sub-culturing procedures and media containing 20% (v/v) horse serum (RPMI 20) is often recommended to be used when cells are plated for mutant selection.
4. Heat-inactivated horse serum is used to eliminate a factor which degrades TFT, the selective agent (33). Serum is heat inactivated by warming at 56°C for 30 min.

2.4. Cell Culture

1. For each experiment, one or more vials are thawed rapidly, the cells diluted in RPMI 10 and incubated at 37 ± 1°C in a humidified atmosphere of 5% (v/v) CO_2 in air.
2. When cells are in exponential growth, sub-cultures are established in an appropriate number of flasks.
3. The cells grow in suspension and do not form monolayers.

2.5. Metabolic Activation

1. Direct-acting effects can be assessed in the absence of metabolic activation but OECD Guideline 476 (1) recommends that the assay should also be performed in the presence of an exogenous metabolising system.
2. The most commonly used form of metabolic activation is a mammalian liver post-mitochondrial fraction (S-9: Molecular

Toxicology Inc., Boone, NC, USA). The fraction is prepared from male Sprague Dawley rats induced with Aroclor 1254 or phenobarbital/5,6-benzoflavone and is stored frozen (at –80°C nominal) prior to use (see Note 7).

3. A typical S-9 mix for use in this assay would contain glucose-6-phosphate (G6P: 180 mg/mL), β-Nicotinamide adenine dinucleotide phosphate (NADP: 25 mg/mL), potassium chloride (KCl: 150 mM), and rat liver S-9, mixed in the ratio 1:1:1:2 (other co-factor mixtures may also be used).
4. The final concentration of the liver homogenate in the test system is 1–10% (typically 2%).
5. Cultures treated in the absence of S-9 receive an equivalent volume of KCl (150 mM).

2.6. Test Item

1. The test item is formulated in a suitable vehicle prior to administration to the test system.
2. Aqueous or organic vehicles may be used but it is important to achieve full solubility in the primary vehicle.
3. Stock test solutions prepared in aqueous vehicles are normally filter-sterilised either before further dilution or before use.

2.7. Choice of Vehicle

1. Aqueous vehicles (such as water or physiological saline) may be added to the test system at a final concentration up to 10% v/v. Poorly soluble test items may be diluted directly in tissue culture medium.
2. Organic vehicles (such as dimethyl sulphoxide [DMSO], acetone, or dimethyl formamide [DMF]) may be added at a final concentration of up to 1% v/v.
3. If other vehicles (or volumes exceeding those described) need to be used, the effects on mutant frequencies may need to be checked and untreated controls should also be included in the experimental design (see Note 8).

2.8. Choice of Positive Controls

1. Positive controls are included in each experiment as a quality control measure, to ensure that the test system is capable of responding to known mutagens.
2. In the absence of S-9, 4-nitroquinoline-1-oxide (4 NQO) or methyl methane sulphonate (MMS) are most commonly used. Ethyl methane sulphonate (EMS) may also be used, but results in a lower recovery of small colony mutants.
3. In the presence of S-9, benzo[*a*]pyrene (BP), cyclophosphamide (CPA), or 3-methyl cholanthrene (3-MC) are recommended.

3. Methods

3.1. Study Design

Treat cells
(Treatment period)
↓
Subculture cells for 2 days
(Expression period)
↓
Plate for viability / TFT resistance
(Plating)
↓
Score viability plates and TFT plates
(Scoring)

Initially, a preliminary cytotoxicity Range-Finder Experiment may be performed to establish an appropriate concentration range for the Mutation Experiments (see Note 9). The Range-Finder is performed in the absence and presence of S-9, as toxicity is often observed at different concentrations under these two test conditions. This experiment will typically contain short (3–6 h) treatments in the absence and presence of S-9 and may also contain an extended (24-h) treatment in the absence of S-9 if required (see Note 10 e.g. it is recommended in the testing of pharmaceuticals).

The mouse lymphoma cells grow as a suspension culture and heavy precipitates can interfere with the assay. At the end of the treatment incubation, the cells are pelleted by centrifugation and the precipitate may pellet with the cells, making the control of exposure impossible. Thus, normally the lowest precipitating concentration will be the highest test item treatment analysed.

When the toxic range has been determined, a minimum of five concentrations, ranging from non-toxic to toxic (approximately 10–20% RTG) where possible, are selected for the first Mutation Experiment, in which short treatments are performed in the absence and presence of S-9 (see Note 11). Depending on the results of Experiment 1, a second confirmatory experiment may be performed. Clearly positive results in Experiment 1 need not be repeated but if the results in the absence of S-9 are clearly negative, Experiment 2 in the absence of S-9 may be tested using the 24-h treatment incubation period. A confirmatory 3-h treatment in the presence of S-9 may also be included.

Changes in osmolality of more than 50 mOsm/kg and fluctuations in pH of more than one unit may be responsible for an increase in mutant frequency (34, 35). Osmolality and pH of the culture medium should be assessed during the study.

Toxicity is measured by assessment of RTG (Sub-heading "Microwell Assay"). Other data relevant to toxicity generated during the conduct of the study (suspension growth and Day 2 plating efficiency) may also contribute to the interpretation of the data.

In this section, it is assumed that short treatments are for 3-h and that the final culture volume at the time of treatment (following addition of the cell suspension, test item or vehicle/positive control, and S-9 or KCl) is 20 mL. It is critical that sufficient numbers of cells (see Note 12) are treated, sub-cultured, and plated for mutation (taking into account toxicity and spontaneous mutant frequency) to demonstrate a sufficient increase in mutant frequency by comparison to the concurrent vehicle control values (36).

3.2. Cytotoxicity Range-Finder Experiment

For 3-h treatments in the absence and presence of S-9, at least 10^7 cells/culture resuspended in RPMI 1640 medium containing 5% v/v horse serum (RPMI 5) are seeded into sterile, disposable centrifuge tubes. For 24-h treatments in the absence of S-9, at least 4×10^6 cells/culture resuspended in RPMI 10 are seeded into 75-cm^2 tissue culture flasks. Vehicle or test item solutions are added, together with S-9 mix or 150 mM KCl, prepared as in Sub-heading 2.5. The cultures are gassed with 5% v/v CO_2 in air and incubated at 37 ± 1°C for the appropriate treatment period in a humidified incubator gassed with 5% (v/v) CO_2 in air (see Note 13).

3.3. Mutation Experiments

Mutation Experiments are performed as required (see Sub-heading 3.1), using the same numbers of cells/culture as the Cytotoxicity Range-Finder Experiment. Treatment of cultures with the test item, controls, and S-9 or KCl are also as per the Range-Finder, with positive control compounds (in the absence and presence of S-9) included to demonstrate the effective functioning of the test system (see Note 14).

3.4. Post-treatment Procedures

Following 3-h treatment, the cultures are centrifuged ($200 \times g$) for 5 min, washed with medium and resuspended in 50 mL RPMI 10. This assumes that if all cells had survived treatment, the cell concentration is 2×10^5 cells/mL (based on the pre-treatment cell number of 10^7 cells). This gives a true indication of cell growth and also accounts for cell loss during the treatment incubation period.

Following 24-h treatment, the cultures are centrifuged ($200 \times g$) for 5 min, washed and resuspended in 20 mL of RPMI 10. Cell counts are determined for each culture using a Coulter counter or haemocytometer and adjusted (where sufficient cells survive) to 2×10^5 cells/mL (see Note 15).

3.5. Expression Period

The expression period is defined as the time in which any mutations induced by test item treatment will be expressed in the cell population.

In mouse lymphoma cells, the period for expression of mutation at the *tk* locus is 2 days (but may extend to 3 days if the test item induces cell cycle delay).

3.5.1. Expression Period: Day 1

Approximately 24-h after the end of the respective treatment periods, all cultures are counted and the cell concentrations readjusted (where possible) by adding RPMI 10 to give 2×10^5 cells/mL in each culture (see Note 16).

3.5.2. Expression Period – Day 2: Microwell Plate Method

Approximately 48-h after the end of the respective treatment periods, all cultures are counted and the cell concentrations readjusted (where possible) by adding RPMI 20. The higher serum content enables effective growth during the following incubation period.

Mutant Selection and Viability

For mutant selection plating, 0.2 mL aliquots of a cell suspension at 10^4 cells/mL (i.e. 2,000 cells/well) in RPMI 20 containing TFT (see Note 17) are plated into each well of at least two (normally four) 96-well plates. TFT should be at a sufficiently high concentration to kill all non-mutant cells (a final concentration of 3 µg/mL should be sufficient).

For viability plating, 0.2 mL aliquots of a cell suspension at 8 cells/mL (i.e. 1.6 cells/well) in RPMI 20 are plated into each well of two 96-well plates.

Colony Counting

Mutant plates are typically incubated for 10–14 days and viability plates are typically incubated for 7–10 days (see Note 18) in a humidified incubator at 37 ± 1°C gassed with 5% v/v CO_2 in air. Some automated plate scoring systems are available but the plates may also be counted by eye, using background illumination.

3.5.3. Expression Period – Day 2: Agar Plate Method

Mutant Selection and Viability

Cultures with low cell densities (nominally less than approximately 3×10^5 cells/mL) on Day 2 will not be considered for mutant selection.

A total of 3×10^6 cells/culture will be suspended in selection medium containing TFT and distributed into three 100-mm dishes. The absolute cloning efficiency at the time of selection will be determined by seeding a total of 600 cells into three 100-mm dishes in cloning medium. All dishes will be incubated in a humidified incubator at 37 ± 1°C gassed with 5% v/v CO_2 in air.

Colony Counting

After 10–14 days, colonies may be counted by eye or by using an automated colony counter, e.g. Loats Associates, Inc. High Resolution Colony Counter System for the Mouse Lymphoma Assay (see Note 19).

3.6. Analysis of Data

Calculations may be performed manually, but validated computer software systems are also available.

3.6.1. Cell Growth Characteristics

Suspension Growth (SG) is a measure of the growth in suspension during treatment and the expression period.

$$\mathrm{SG} = a \times b \times c$$

$$\text{where } a = \left(\frac{D_0 \text{post - treatment cell count}}{\text{pre - treatment cell density}} \right)$$

$$\text{where } b = \left(\frac{D_1 \text{cell count}}{\text{Cell count set up on } D_0 \text{post - treatment}} \right)$$

$$\text{where } c = \left(\frac{D_2 \text{ cell count}}{\text{Cell count set up on } D_1} \right)$$

For 3-h treatments, "*a*" is assumed to equal 1.

Usually, the denominators for "*b*" and "*c*" are 2×10^5 cells/mL. However, if cytotoxicity causes the cell count to be lower than 2×10^5 cells/mL following treatment and/or if the cells do not grow satisfactorily during the expression period, it can be lower. In these cases, the respective cell count values are entered into the calculation above.

Relative suspension growth (RSG) is a measure of the growth in suspension during treatment and the expression period relative to the mean control.

$$\mathrm{RSG}(\%) = \left(\frac{\text{Individual SG value}}{\text{Mean control SG value}} \right) \times 100$$

3.6.2. Viability and Mutant Selection Calculations

Microwell Assay

Cloning efficiency (CE), also known as viability, is the measure of the ability of the cells to clone.

From the zero term of the Poisson distribution the probable number of clones/well, $P(0)$, is given by:

$$P(0) = \left(\frac{\text{Number of wells with no colony}}{\text{Total number of wells}} \right)$$

The CE for each culture is therefore calculated as follows:

$$\mathrm{CE} = \left(\frac{-\ln P(0)}{\text{Number of cells per well *}} \right) \times 100$$

*Number of cells per well is 1.6 cells per well on average on all viability plates

Relative Total Growth (RTG) is the primary measure of cytotoxicity and is relative to the control (vehicle control RTG = 100%). RTG takes into account all cell growth and cell loss during the treatment period and the 2 day expression period (RSG), and the cells' ability to clone 2 days after treatment (viability).

$$\text{RTG} = \text{RSG} \times \left(\frac{\text{Individual Viability Value}}{\text{Mean Control Viability Value}} \right)$$

Mutant frequency (MF) is calculated as follows:

$$\text{MF} = \frac{-\ln P(0) \text{for mutant plates}}{\text{Number of cells per well}^* \times (\text{viability} / 100)}$$

*Number of cells per well is 2,000 cells per well on average on all mutant plates.

Small and large colony mutant frequencies will be calculated in an identical manner, using the relevant number of empty wells for small and large colonies, as appropriate.

Agar Assay

Mutant frequency is calculated as the number of mutant colonies (total of three dishes), divided by the number of cells seeded, adjusted by the absolute CE at the time of selection, and is reported as TFT^r mutants/10^6 clonable cells. Absolute CE is calculated as the ratio of the total number of viable colonies to the number of cells seeded.

Mutant frequencies are normally derived from sets of three dishes for both mutant colony count and viable colony count. In order to allow for losses due to contamination or other reasons, an acceptable mutant frequency can be calculated from a minimum of two dishes per set.

3.7. Data Interpretation

The significance of increases in mutant frequencies (total wells with clones) is assessed according to the recommendations of the Mouse Lymphoma Workgroup, Aberdeen, 2003 (29). In accordance with the recommendations of the test guidelines, the biological relevance of the result is the most critical issue, but statistical analysis may also be used as an aid to data interpretation.

3.8. Assay Acceptance Criteria

The assay will be considered valid if the following criteria, agreed in consensus documents published by the MLA workgroup (28–30), are met:

1. The mean mutant frequencies in the vehicle control cultures should fall within the normal ranges (50–170 mutants per 10^6 viable cells for the microwell assay, 35–140 mutants per 10^6 viable cells for the agar assay).
2. At least one positive control should show either an absolute increase in mean total MF of at least 300×10^{-6} (at least 40% of this should be in the small colony MF), or an increase in small colony mutant frequency of at least 150×10^{-6} above the concurrent vehicle control.
3. The mean RTG for the positive controls should be greater than 10%.

4. The mean cloning efficiencies of the negative controls from the Mutation Experiments should be within the range 65–120% on Day 2.
5. The mean suspension growth of the negative controls from the Mutation Experiments should be between the range 8 and 32 following 3-h treatments or between 32 and 180 following 24-h treatments.

3.9. Assay Evaluation Criteria

If the acceptability criteria are fulfilled, the test item will be considered as mutagenic in this assay if:

1. The MF of any test concentration exceeds the sum of the mean control mutant frequency plus GEF (see Note 20), as defined by the IWGT MLA Workgroup in 2005 (30).
2. The increases are concentration-related, as defined by a statistically significant linear trend test.

The test item will be considered positive in this assay if the above criteria are met and negative if neither of the criteria are met. Results that only partially satisfy the assessment criteria described above are considered on a case-by-case basis (see Note 21).

Colony sizing will also be taken into consideration: increases in large colony MF are indicative of small genetic events (such as point mutations), whereas increases in small colony MF are indicative of potential clastogenic events.

3.10. Further Areas of Research

3.10.1. Microsatellite Analysis

Microsatellite analysis is a technique that exploits the characteristic of polymorphic loci on a chromosome. There are several examples of polymorphic loci such as single nucleotide polymorphisms (SNPs) and restriction fragment length polymorphisms (RFLPs) and microsatellites also fall into this category. Whereas SNPs and RFLPs are short sequences that vary from person to person, microsatellites (simple sequence repeats) are longer regions of DNA (typically a few hundred base pairs). They have been used in the past as DNA linkage markers enabling researchers to generate genome-scale linkage maps in both human and mouse (37, 38). On a different scale, they have also been used in molecular analysis of clones harvested post-treatment by genetic toxicologists (5). The technique is a very useful molecular-level analysis technique with the MLA cell line due to the known differences between the two copies of chromosome 11. The benefit of this is that each microsatellite marker will yield a different length sequence depending on which copy of chromosome 11 it is located on.

The method involves two stages. The first is a PCR stage to amplify the region of interest by using a primer pair designed to flank the microsatellite region. As with any PCR reaction, optimisation of the conditions is paramount to success with annealing and extension temperatures/times all affecting the quality of the product at the end. The second stage is visualisation of the PCR product, using gel electrophoresis. As mentioned earlier, the unusual trait of chromosome 11 in these cells allows for two bands to appear on the gel if both copies of chromosome 11 are present. As the technique uses lysed cells, the whole genome is present, therefore if the locus of interest is present anywhere in the genome (i.e. by translocation etc.), the PCR product will still be produced and the band appears on the subsequent gel. Likewise, if the locus of interest has been deleted from the genome by (for example) some chromosomal event, the gel will reveal either one band (one copy of the locus present) or no bands (both copies of the locus deleted).

Microsatellite loci are common in the mouse genome with almost 400 loci known on chromosome 11 alone, but when the technique first became popular in the early 1990s, this figure was far lower. The benchmark publication using the technique on MLA cells described just 21 loci on chromosome 11 (5). Two years later, the same author described a microsatellite locus within the *tk* gene. Designated *Agl2*, this locus also produced two bands following PCR amplification and PCR product sequence analysis enabled the smaller band to be attributed to the non-functional $tk1_b$ and the larger band to $tk1_a$ (39). This method complemented the previously employed methods of Southern blotting and cytogenetic analysis (9, 13, 40–47) for mutant allele detection but also offered a simpler technique for the same result.

More recently, there have been a limited number of publications using this technique in MLA cells (e.g. (48)). Authors have selected loci for analysis referencing Liechty et al. (5) and have demonstrated the technique to be useful when looking at a degree of complexity between cytogenetic detection of mutants and the use of blots. It adds valuable contributions to the information data base and can be a useful molecular-level tool in the understanding of the mutations detected by the MLA.

3.10.2. p53 Status

Transformation of cells by Simian Vacuolation Virus 40 (SV40) led to the discovery of a protein that was produced in response to the viral infection. Co-precipitation experiments showed the new protein had a molecular weight around 54 kDa. This protein is known today as p53 (49–55). Extensive investigative work has demonstrated the role of p53 in DNA damage (56), cell cycle (51), and subsequently tumour development (57–59) and today, many of p53's roles in the responses of cells to various stress signals are known.

Activation of p53 occurs through post-translational modifications such as phosphorylation and acetylation, but also by methylation, ubiquitination or sumolation (60). The extent of these post-translation modifications forms a code that transmits the nature of the stress to p53 (61). Post-translation modification stabilises and activates p53 in a manner dependent on the position(s) modified leading to the protein's involvement in cell cycle arrest, cellular senescence, or cellular apoptosis (62, 63).

Sequence conservation across species for p53 is high and the difference between mouse and human is only a few amino acids (causing a slight reading frameshift). Human p53 has 393 amino acids (aa) whereas mouse is a little shorter at 387 aa. The genetic location also differs, with the human p53 locus on chromosome 17, whereas it is on chromosome 11 in the mouse.

In L5178Y $TK^{+/-}$ clone 3.7.2C cells, the p53 gene is located at cytogenetic band B2-C (39.0 centiMorgans [cM]) with TK further toward the distal end at cytogenetic band E1-E2 (78.0 cM). Owing to the known slight difference between the two copies of chromosome 11, it has been possible to assign the TK^+ and TK^- genes to each copy of the chromosome (42). In 1997, a publication followed up the possibility of the MLA cell line having mutant p53 (64). The MLA cell line is a derivative of the L5178Y-R cell line, which had been proven to react with antibodies against mutant p53. Storer and co-workers described a mutation at codon 170 which had a structural effect on the protein at the L2-L3 loops and concluded that the cell line was p53-mutant. Clark et al. (65) then identified two heterozygous mutations in p53 and also assigned them to specific copies of chromosome 11. Using a clone of MLA cells that was known to be $TK^{-/-}$, they were able to assign a nonsense mutation (causing a premature stop codon) to the TK^+ copy and the previously reported mutation to the TK^- copy. Both groups commented that this feature of the cell line removes an element in the natural protection of the genome. The potential effects of this allow MLA cells to harbour lethal mutations that are subsequently detected by the assay that would not be found if using a p53 wild-type cell line. Since these publications, it has been shown that the MLA cell line has mutant p53 and has long been thought to fail to undergo apoptosis and cell cycle arrest as a result. Investigations into the functionality of the existing protein are now in progress.

3.10.3. Future Challenges

Several recent publications (66–68) have specifically examined the effectiveness of the in vitro test battery to accurately predict rodent carcinogenesis and a revision of the guidance for genetic toxicology testing has been initiated by ICH. The current documents (ICH S2A and S2B) recommended and accepted strategies for genotoxicity from in vitro through to in vivo testing. Within the revision (ICH S2), two prospective strategies will become available. One of these will use a range of in vitro genotoxicity assays in combination with an in vivo strategy, as at present, but a second option has been proposed in which the Ames test is the sole in vitro assay, followed by two in vivo genetic toxicology tests. It remains to be seen which option will be favoured: on the one hand, publications regarding false-positive results would favour the predominance of in vivo testing, but on the other hand the initiative to reduce animal usage in toxicology testing may favour the continued use of in vitro tests. With current research adding further understanding to the MLA, it can only strengthen the assay's suitability as a mammalian gene mutation assay.

4. Notes

1. Amethopterin (methotrexate) resistance is important as it indicates that formation of deoxythymidylate (dTMP) is not possible. Amethopterin is an analogue of dihydrofolate, the precursor of tetrahydrofolate. Tetrahydrofolate is the precursor of *N*5,*N*10-methylenetetrahydrofolate which is the one-carbon donor and electron donor in the methylation reaction causing reduction of tetrahydrofolate to dihydrofolate and subsequently conversion of dUMP (deoxyuridylate) to dTMP (deoxythymidylate). As this pathway for production of dTMP is non-functional in the MLA cell line, then the primary method of thymidine base repair operates via the salvage pathway.
2. TFT was originally developed as an anti-viral treatment for herpes simplex virus infections but when added to the MLA test system was found to generate a selective pressure on cells that have a disfunctional/absent enzyme, as it is a toxic analogue of thymidine.
3. Some batches of the soft agar, as used in the original assay design, contained impurities and had the potential to be used at too high (or too low) a temperature, compromising the assay results.
4. Where available, safety data should be consulted and appropriate precautions taken. Tissue culture laboratory facilities must be available to facilitate the performance of assays of this type, particularly safety cabinets where the air flow protects the materials under test and the operator. The use of sterile equipment and reagents is essential and aseptic techniques must be used to prevent contamination. Appropriate laboratory clothing and/or apparatus should be worn.
5. Karyotyping of the L5178Y $tk^{+/-}$ (3.7.2C) cell line is recommended when cryopreserving a master stock and chromosome painting is considered useful to confirm that the cells have two normal copies of chromosome 11.
6. Fischer's medium may also be used.
7. Each batch of S-9 fraction is checked for sterility, protein content, ability to convert known promutagens to bacterial mutagens, and cytochrome P-450-catalysed enzyme activities (alkoxyresorufin-*O*-dealkylase activities).
8. It is considered unwise to exceed 10% v/v (aqueous) or 1% v/v (organic) vehicle additions due to risks associated with increased osmolality, reduced cell growth, and/or increased mutant frequency.

9. Treatment in the absence and presence of S-9 is recommended in the Range-Finder, as toxicity is often observed at different concentrations under these two test conditions.
10. The 24-h treatment is performed in the absence of S-9 only because the S-9 fraction becomes toxic if exposed to the test system for over 6 h.
11. Where duplicate cultures are used, the guidelines recommend that a minimum of four test item concentrations should be carried through all stages of the assay, but if single cultures are used, a minimum of eight concentrations is required.
12. The recommendation is that sufficient cells should be treated such that at least 100 mutants can survive treatment, based on a theoretical mutation rate of 10^{-4} (36).
13. Single cultures at each concentration are normally considered sufficient for the Range-Finder and positive control treatments are not included.
14. Duplicate cultures are recommended for the Mutation Experiments, as sub-culturing may induce variation that cannot be estimated accurately from a single culture. If single cultures are used, more concentrations of the test item should be tested.
15. Cultures may be plated immediately post-treatment to determine relative survival (RS), but this procedure is not performed routinely.
16. Readjusting of cell concentrations is important because it improves the quantitative measure of cell growth and prevents overgrowth of cell cultures. It is recommended that cultures should be maintained at below 10^6 cells/mL, but at least 10^7 cells/culture, if possible, throughout the expression period.
17. TFT has a short half-life and is light-sensitive, therefore appropriate precautions must be taken.
18. Under exceptional cicumstances, viability plates may be removed from the incubator after <7 days. If so, it is recommended that they are scored by microscope.
19. If one agar dish is lost due to contamination or other cause, the colony count of the missing dish is determined by the relative weights of the three dishes in the set and the colony counts in the two acceptable dishes. If a lost plate is not available for weighing, the colony count of the lost plate will be determined from the average of the two remaining acceptable plates.
20. The GEF is defined as increases in MF of ≥126 (microwell assay) or ≥90 (agar assay) over the concurrent vehicle control MF and are required for the result to be considered biologically relevant.
21. Positive responses seen only at high levels of cytotoxicity (RTG <10%) are interpreted with extreme caution. Further experiments are usually required to clarify such data.

References

1. OECD (1997) "*In Vitro* Mammalian Cell Gene Mutation Test", in: OECD *Guidelines for the Testing of Chemicals,* Test Guideline 476. OECD Paris (http://titania.sourceoecd.org/vl=1382924/cl=17/nw=1/rpsv/cw/vhosts/oecdjournals/1607310x/v1n4/contp1-1.htm)
2. ICH, Topic S2B Genotoxicity*: A Standard Battery for Genotoxocity Testing of Pharmaceuticals, International Conference on Harmonisation of Technical Requirements for Registration of Pharmaceuticals for Human Use.* Step 4 Guideline, Brussels, July, 1997 (http://www.ich.org/cache/compo/502-272-1.html)
3. Law LW (1952) Increase in incidence of leukemia in hybrid mice bearing thymic transplants from a high leukemic strain. *Journal of the National Cancer Institute* **12**(4):789–805
4. Fischer GA (1958) Studies of the culture of leukemic cells *in vitro. Annals of the New York Academy of Sciences* **76**(3):673–680
5. Liechty MC, Hassanpour Z, Hozier JC and Clive D (1994) Use of microsatellite DNA polymorphisms on mouse chromosome 11 for in vitro analysis of thymidine kinase gene mutations. *Mutagenesis* **9**:423–427
6. Clive D, Flamm WG, Machesko MR and Bernheim NJ (1972) A mutational assay system using the thymidine kinase locus in mouse lymphoma cells. *Mutation Research* **16**:77–87
7. Clive D (1973) Recent Developments with the L5178Y TK Heterozygote Mutagen Assay System. *Environmental Health Perspectives* **6:**119–125.
8. Moore-Brown MM, Clive D, Howard BE *et al.* (1981) The utilization of trifluorothymidine (TFT) to select for thymidine kinase-deficient (TK-/-) mutants from L5178Y/TK+/- mouse lymphoma cells. *Mutation Research* **85**:363–378
9. Applegate ML, Moore MM, Broder CC, Burrell A., Juhn G., Kasweck KL, Lin P-F, Wadhams A and Hozier JC (1990) Molecular dissection of mutations at the heterozygous thymidine kinase locus in mouse lymphoma cells. *Proceedings of the National Academy of Sciences USA* **87**:51–55
10. Moore MM, Clive D, Hozier JC *et al.* (1985) Analysis of trifluorothymidine-resistant (TFTr) mutants of L5178Y/TK+/- mouse lymphoma cells. *Mutation Research* **151**:161–174
11. Clive D and Spector JFS (1975) Laboratory procedure for assessing specific locus mutations at the TK locus in cultured L5178Y mouse lymphoma cells. *Mutation Research* **31**:17–29.
12. Clive D and Voytek P (1977) Evidence for chemically-induced structural gene mutations at the thymidine kinase locus in cultured L5178Y mouse lymphoma cells. *Mutation Research* **44**:269–278.
13. Clive D, Johnson KO, Spector JFS *et al.* (1979) Validation and characterization of the L5178Y/TK+/- mouse lymphoma mutagen assay system. *Mutation Research* **59**:61–108
14. Cole J, Arlett CF, Green MHL *et al.* (1983) A comparison of the agar cloning and microtitration techniques for assaying cell survival and mutation frequency in L5178Y mouse lymphoma cells. *Mutation Research* **111**:371–386
15. Mitchell AD, Myhr BC, Rudd CJ *et al.* (1988) Evaluation of the L5178Y mouse lymphoma cell mutagenesis assay: Methods used and chemicals evaluated. *Environmental Mutagenesis* **12**:1–18
16. Mitchell AD, Rudd CJ and Caspary WJ (1988) Evaluation of the L5178Y mouse lymphoma cell mutagenesis assay: Intralaboratory results for sixty-three coded chemicals tested at SRI international. *Environmental Mutagenesis* **12**:37–101
17. Caspary WJ, Daston DS, Myhr BC, *et al.* (1988) Evaluation of the L5178Y mouse lymphoma cell mutagenesis assay: Interlaboratory reproducibility and assessment. *Environmental Mutagenesis* **12**:195–229
18. Caspary WJ, Lee YJ, Poulton S, *et al.* (1988) Evaluation of the L5178y mouse lymphoma cell mutagenesis assay: Quality-control guidelines and response categories. *Environmental Mutagenesis* **12**:19–36
19. Myhr BC and Caspary WJ (1988) Evaluation of the L5178Y mouse lymphoma cell mutagenesis assay: Intralaboratory results for sixty-three coded chemicals tested at Litton Bionetics, Inc. *Environmental Mutagenesis* 12:103–194
20. Sawyer J, Moore MM, Clive D and Hozier J (1985) Cytogenetic characterization of the L5178Y TK+/- 3.7.2C mouse lymphoma cell line. *Mutation Research* **147**:243–253
21. Sawyer JR, Binz RL, Wang J and Moore MM (2006) Multicolor spectral karyotyping of the L5178Y $TK^{+/-}$ 3.7.2C mouse lymphoma cell line. *Environmental and Molecular Mutagenesis* **47**:127–131
22. Clive D, Turner NT, Krehl R and Eyre J (1985) The mouse lymphoma assay may be used as a chromosome aberration assay. *Environmental Mutagenesis* 7 (suppl. 3):33
23. ICH, Topic S2A Genotoxicity: *Guidance on Specific Aspects of Regulatory Genotoxicity Tests*

for Pharmaceuticals, International Conference on Harmonisation of Technical Requirements for Registration of Pharmaceuticals for Human Use, Harmonised Tripartite Guideline CPMP/ICH/141/95, Approved September 1995 (http://www.ich.org/cache/compo/502-272-1.html)

24. Liechty MC, Scalzi JM, Sims KR *et al.* (1998) Analysis of large and small clone L5178Y *tk*$^{-/-}$ mouse lymphoma mutants by loss of heterozygosity (LOH) and by whole chromosome 11 painting: detection or recombination. *Mutagenesis* **13** (5): 461–474
25. Honma M, Hayashi M, Shimada H *et al.* (1999) Evaluation of the mouse lymphoma tk assay (microwell method) as an alternative to the in vitro chromosomal aberration test. *Mutagenesis* **14**:5–22
26. Moore MM, Honma M, Clements J *et al.* (2000). Mouse Lymphoma Thymidine Kinase Locus Gene Mutation Assay: International Workshop on Genotoxicity Test Procedures Workgroup Report. *Environmental and Molecular Mutagenesis* **35**:185–190
27. Moore MM, Honma M, Clements J *et al.* (2002). Mouse lymphoma thymidine kinase gene mutation assay: follow-up International Workshop on Genotoxicity Test Procedures, New Orleans, Louisiana, April 2000. *Environmental & Molecular Mutagenesis* **40**:292–9
28. Moore MM, Honma M, Clements J *et al.* (2003). Mouse Lymphoma Thymidine Kinase Locus Gene Mutation Assay: *International Workshop on Genotoxicity Test Procedures Workgroup Report-Plymouth. UK* 2002. Mutation Research **540**:127–140
29. Moore MM, Honma M, Clements J *et al.* (2006). Mouse lymphoma thymidine kinase gene mutation assay: Follow-up meeting of the international workshop on Genotoxicity testing-Aberdeen, Scotland, 2003-Assay acceptance criteria, positive controls, and data evaluation. *Environmental & Molecular Mutagenesis* **47**:1–5
30. Moore MM, Honma M, Clements J *et al.* (2007). Mouse lymphoma thymidine kinase gene mutation assay: Meeting of the International Workshop on Genotoxicity Testing, San Francisco, 2005, recommendations for 24-h treatment. *Mutation Research* **627**:36–40
31. Clive D, Bolcsfoldi G, Clements J, Cole J, Honma M, Majeska J, Moore M, Müller L, Myhr B, Oberly T, Ouldelhkim M-C, Rudd C, Shimada H, Sofuni T, Thybaud V and Wilcox P (1995), Consensus agreement regarding protocol issues discussed during the mouse lymphoma workshop: Portland, Oregon, May 7, 1994, *Environmental and Molecular Mutagenesis* **25:**165–168
32. Han T, Wang J, Tong W, *et al.* (2006) Microarray analysis distinguished differential gene expression patterns from large and small colony *Thymidine kinase* mutants of L5178Y mouse lymphoma cells. *BMC Bioinformatics* 7 (Suppl 2):S9
33. Moore MM and Howard BE (1982) Quantitation of small colony trifluorothymidine-resistant mutants of L5178Y/TK+/− mouse lymphoma cells in RPMI-1640 medium. *Mutation Research* **104**:287–294
34. Brusick D (1986) Genotoxic effects in cultured mammalian cells produced by low pH treatment conditions and increased ion concentrations. *Environmental Mutagenesis* **8**:879-886
35. Scott D, Galloway SM, Marshall RR, Ishidate M Jr, Brusick D, Ashby J and Myhr BC (1991) Genotoxicity under extreme culture conditions. A report from ICPEMC Task Group 9. *Mutation Research* **257**:147–204
36. Robinson WD, Green MHL, Cole J, Garner RC, Healy MJR and Gatehouse D (1990) Statistical evaluation of bacterial/mammalian fluctuation tests. In *Statistical Evaluation of Mutagenicity Test Data* (Ed Kirkland DJ) Cambridge University Press, pp 102–140
37. Gyapay G, Morissette J, Vignal A, Dib C, Fizames C, Millasseau P, Marc S, Bernardi G, Lathrop M and Weissenbach J (1994) The 1993-94 Genethon human genetic linkage map. *Nat Genet* 7:246–339
38. Love JM, Knight AM, McAleer MA and Todd JA (1990) Towards construction of a high resolution map of the mouse genome using PCR-analysed microsatellites. *Nucl. Acids Res.* **18**:4123–4130
39. Liechty MC, Crosby H, Murthy A, Davis LM, Caspary WJ and Hozier JC (1996) Identification of a heteromorphic microsatellite within the thymidine kinase gene in L5178Y mouse lymphoma cells. *Mutation Research* **371**:265–271
40. Clive D, Glover P, Applegate M and Hozier J (1990) Molecular aspects of chemical mutagenesis in L5178Y/tk+/− mouse lymphoma cells. *Mutagenesis* **5**:191–197
41. Hozier J, Sawyer J, Moore M, Howard .B and Clive D (1981) Cytogenetic analysis of the L5178Y/TK$^{+/-}$ mouse lymphoma mutagenesis assay system. *Mutation Research* **84**:169–181
42. Hozier J, Sawyer J, Clive D and Moore M (1982) Cytogenetic distinction between the TK+ and TK− chromosomes in the L5178Y TK+/− 3.7.2C mouse-lymphoma cell line. *Mutation Research* **105**:451–456
43. Hozier J, Sawyer J, Clive D and Moore M (1985) Chromosome 11 aberrations in small colony L5178Y TK$^{+/-}$ mutants early in their clonal history. *Mutation Research* **147**:237–242

44. Moore MM, Clive D, Howard .BE, Batson AG and Tumer NT (1985) In situ analysis of trifluorothymidine-resistant (TFT[r]) mutants of L5178Y/TK+/- mouse-lymphoma cells. *Mutation Research* **151** 147–159
45. Moore MM, Clive D, Hozier JC, Howard .BE, Batson AG, Turner NT and Sawyer J. (1985) Analysis of trifluorothymidine-resistant (TFT′) mutants of L5178Y TK+/- mouse lymphoma cells. *Mutation Research* **151**:161–174
46. Blazak WF, Stewart BE, Galperin I, Allen KL, Rudd CJ, Mitchell AD and Caspary WJ (1986) Chromosome analysis of trifluorothymidine-resistant L5178Y mouse lymphoma cell colonies. *Environmental Mutagenesis* **8**:229–240
47. Liechty MC, Rauchfuss HS, Lugo MH and Hozier JC (1993) Sequence analysis of tka(-)-1 and tkb(+)-1 alleles in L5178Y tk+/- mouse-lymphoma cells and spontaneous tk-/- mutants *Mutation Research* **286**:299–307
48. Soriano C, Creus A and Marcos R (2008) Arsenic trioxide mutational spectrum analysis in the mouse lymphoma assay. *Mutation Research* **646**:1–7
49. Chang C, Simmons DT, Martin MA and Mora PT (1979) Identification and partial characterization of new antigens from simian virus 40-transformed mouse cells. *J Virol* **31**:463–471
50. Kress M, May E, Cassingena R and May P (1979) Simian virus 40-transformed cells express new species of proteins precipitable by anti-simian virus 40 tumor serum. *J. Virol.* **31**:472–483
51. Lane DP and Crawford LV (1979) T antigen is bound to a host protein in SV40-transformed cells. *Nature* **278**, 261–263
52. LinzerDIHandLevineAJ(1979)Characterization of a 54 K dalton cellular SV40 tumor antigen resent in SV40-transformed cells and in infected embryonal carcinoma cells. *Cell* **17**: 43–52.
53. Melero JA, Stitt DT, Mangel WF and Carroll RB (1979) Identification of new polypeptide species (48-55 K) immunoprecipitable by antiserum to purified large T antigen and present in simian virus 40-infected and transformed cells. *J. Virol.* **93**:466–480.
54. DeLeo A, Jay G, Appella E, Dubois GC, Law LW and Old LJ (1979) Detection of a transformation-related antigen in chemically induced sarcomas and other transformed cells of the mouse. *Proceedings of the National Academy of Sciences, USA* **76**:2420–2424
55. Crawford LV, Pim DC and Bulbrook RD (1982) Detection of antibodies against the cellular protein p53 in sera from patients with breast cancer. *Int. J. Cancer* **30**: 403–408.
56. Maltzman W and Czyzyk L (1984) UV irradiation stimulates levels of p53 cellular tumor antigen in nontransformed mouse cells. *Mol. Cell. Biol.* **4**:1689–1694
57. Benchimol S., Matlashewski G. and Crawford L. (1984) The use of monoclonal antibodies for selection of a low-abundance mRNA: p53. *Biochem Soc Trans* **12(4)**:708–711
58. Wolf D and Rotter V (1985) Major deletions in the gene encoding the p53 tumor antigen cause lack of p53 expression in HL-60 cells. *Proceedings of the National Academy of Sciences USA* **82**:790–794.
59. Caron de Fromentel C, May-Levin F, Mouriesse H, Lemerle J, Chandrasekaran K and May P (1987) Presence of circulating antibodies against cellular protein p53 in a notable proportion of children with B-cell lymphoma. *Int. J. Cancer* **39**:185–189
60. Harris SL and Levine AJ (2005) The p53 pathway: positive and negative feedback loops. *Oncogene* **24**:2899–2908
61. Colman MS, Afshar, CA and Barrett JC (2000) Regulation of p53 stability and activity in response to genotoxic stress. *Mutation Research/ Reviews in Mutation Research* **462**:179–188.
62. Jin S and Levine AJ (2001) The p53 functional circuit. *J Cell Sci* **114**:4139–4140.
63. Levine AJ, Hu W and Feng Z (2006) The P53 pathway: what questions remain to be explored? *Cell Death Differ* **13**:1027–1036.
64. Storer RD, Kraynak AR, McKelvey TW, Elia MC, Goodrow TL and DeLuca JG (1997) The mouse lymphoma L5178Y TK+/- cell line is heterozygous for a codon 170 mutation in the p53 tumor suppressor gene. *Mutation Research* **373**:157–165
65. Clark LS, Hart DW, Vojta PJ, Harrington-Brock K, Barrett JC, Moore MM and TIndall KR (1998) Identification and chromosomal assignment of two heterozygous mutations in the Trp53 gene in L5178Y Tk(+/-) -3.7.2C mouse lymphoma cells. *Mutagenesis* **13**:427–434
66. Kirkland D, Aardema M, Henderson L and Müller L (2005). Evaluation of the ability of a battery of three in vitro genotoxicity tests to discriminate rodent carcinogens and non-carcinogens: I. Sensitivity, specificity and relative predictivity. *Mutation Research* **584**:1–256
67. Kirkland D, Aardema M, Müller L and Hayashi M (2006). Evaluation of the ability of a battery of three in vitro genotoxicity tests to discriminate rodent carcinogens and non-carcinogens: II. Further analysis of mammalian cell results, relative predictivity and tumour profiles. *Mutation Research* **608**:29–42
68. Kirkland D, Pfuhler S, Tweats D *et al.* (2007). How to reduce false positive results when undertaking in vitro genotoxicity testing and thus avoid unnecessary follow-up animal tests: Report of an ECVAM Workshop. *Mutation Research* **628**:31–55

Chapter 4

Mammalian Cell *HPRT* Gene Mutation Assay: Test Methods

George E. Johnson

Abstract

Using the combination of bacterial gene mutation assay and chromosomal aberrations test in mammalian cells may not detect a small proportion of mammalian specific mutagenic agents. Therefore, at the current time a third assay should be used, except for compounds for which there is little or no exposure (DOH (2000) Department of Health Guidance for the testing of chemicals for Mutagenicity. Committee on Mutagenicity of Chemicals in Food, Consumer Products and the Environment). The hypoxanthine phosphorybosyl transferase (*HPRT*) gene is on the X chromosome of mammalian cells, and it is used as a model gene to investigate gene mutations in mammalian cell lines. The assay can detect a wide range of chemicals capable of causing DNA damage that leads to gene mutation. The test follows a very similar methodology to the thymidine kinase (TK) mouse lymphoma assay (MLA), and both are included in the guidelines for mammalian gene mutation tests (OECD (1997) Organisation for Economic Co-operation and Development. Ninth addendum to the OECD Guidelines for the Testing of Chemicals. In Vitro Mammalian Cell Gene Mutation Test: 476). The *HPRT* methodology is such that mutations which destroy the functionality of the *HPRT* gene and or/protein are detected by positive selection using a toxic analogue, and *HPRT*$^-$ mutants are seen as viable colonies. Unlike bacterial reverse mutation assays, mammalian gene mutation assays respond to a broad spectrum of mutagens, since any mutation resulting in the ablation of gene expression/function produces a *HPRT*$^-$ mutant. Human cells are readily used, and mechanistic studies using the *HPRT* test methodology with modifications, such as knock-out cell lines for DNA repair, can provide details of the mode of action (MOA) of the test compound (24).

This chapter provides the methodology for carrying out the assay in different cell lines in the presence and absence of metabolism with technical information and general advice on how to carry out the test.

Key words: *HPRT* gene mutation, Mammalian cell mutation, Mutation testing, Human cells

1. Introduction

Mutations are commonly detected using mammalian cell test systems which can be used with great efficiency. Cultured mammalian cells in mutagenicity testing have some advantages over microbial tests and use similar methods. Two main advantages for testing on human-derived cells are: (1) The genomic organisation (DNA into chromosomes and nuclei) is absent in bacteria as is the cells division

James M. Parry and Elizabeth M. Parry (eds.), *Genetic Toxicology: Principles and Methods*, Methods in Molecular Biology, vol. 817, DOI 10.1007/978-1-61779-421-6_4,

apparatus, and (2) mammalian specific cell metabolism can not be replicated in bacteria. However, there are also disadvantages of testing on human-derived cells alone, for example, the high sensitivity but low specificity associated with mammalian genotoxicity assays in vitro, and thus a battery of assays is generally used to provide greater knowledge of the mode of action (MOA) and possible extrapolation to in vivo. The most common assays for detection of gene mutations in mammalian cells rely upon forward mutations that confer resistance to a toxic chemical, such as the thymidine kinase (TK) and hypoxanthine phosphorybosyl transferase (*HPRT*) gene mutation assays (3). In these assays, forward mutations cause inactivation of a wild-type gene at heterozygous autosomal loci (e.g. $TK^{+/-}$) or sex linked loci (e.g. $HPRT^{+}$).

The *TK* assay is the most commonly used mammalian cell gene mutation test to date. When this test is carried out using the L5178Y mouse lymphoma cell line, the test is referred to as the mouse lymphoma assay (MLA) (4). The *HPRT* assay follows very similar methodology to the *TK* assay, and both tests have been used for many years. The *HPRT* assay was first used in microtitre plates (microplates) by Furth et al. who used human lymphoblastoid cell lines (5). Chinese hamster ovary (CHO) cells then became the main cell line (6) and these adherent cells have been the most widely used cells to date. However, the test is relatively flexible and the human lymphoblastoid AHH-1 and MCL-5 suspension cell lines can also be used (7, 8). Our group in Swansea University has recently used the AHH-1 cell line to investigate the dose response relationships of DNA reactive genotoxic agents in the low dose region of exposure (9, 10). The AHH-1 cell line was very suitable for these experiments, and because it is heterozygous at the $TK^{+/-}$ locus we were able to validate our *HPRT* results by carrying out both the *HPRT* and *TK* gene mutation assays alongside one another.

There are three main features of the mammalian *HPRT* gene mutation assay which have led to it being widely used:

(a) The target gene is encoded on the mammalian X chromosome and consequently it is easy to select for loss of function mutants in cells derived from males, which in mammals are heterogametic for sex chromosomes.

(b) The biochemical selection systems for loss of function with cells that survive in presence of 6-thioguanine (6-TG) and/or 8-azoguanine are simple and effective (11).

(c) Also an advantage of the *HPRT* gene is that mutations in the same gene can be compared between cell lines, experimental animals, and with humans (12).

As large losses in the X chromosome lead to cell lethality, only small changes, such as point mutations and exon deletions are detected in the *HPRT* gene. The spontaneous mutation frequency (MF) is also lower at the *HPRT* loci than the *TK* loci, as high proportions of mutations at *HPRT* are lethal (but not *TK*). These events include non-disjunction and translocation which are known to lead to viable *TK* mutants but not viable *HPRT* mutants. After a chemical insult further multiple events are required to transform DNA changes (promutagenic DNA lesions) into selectable phenotypes:

(a) Fixation of the mutation.

(b) Reduction of the pre-existing enzyme to a level with no biological activity.

Fixation of a mutation requires the initial lesion in the DNA (i.e. adduct, strand break, or damage to DNA dependent proteins) being translated into a DNA sequence change, such as point mutation, deletion, or loss. For a point mutation, the mutant strand must be separated from the wild-type strand by cell division, thus one of the progeny cells no longer produces active mRNA and/or protein. Point mutations can only occur after the cells have undergone cell division, as the lesion affecting the base may be removed or the base may be repaired by DNA repair. The mutation is therefore only relevant in these assays once it has been incorporated into both strands. For selection, the existing enzyme or mRNA must be reduced either by cell division or degradation to non-functional levels, so that the original phenotype can no longer be identified. Furthermore, it is difficult to treat a sufficiently large number of cells ($>10^5$ cells per petri dish) to produce statistically robust assays. Therefore, we make use of a microplate protocol to improve the sensitivity of assay.

2. Materials

2.1. Cell Lines

Suitable cell lines for the *HPRT* assay include L5178Y mouse lymphoma cells, the CHO, AS52, and V79 lines from Chinese hamsters (2, 13, 14), and AHH-1, MCL-5, and TK6 human lymphoblastoid cells (7, 8, 10, 15). The *HPRT* loci are on the X chromosome and therefore primary male cell lines can also be used to study mutagenic effects in mice and rats (16, 17) and this methodology can also be used for human bio-monitoring (10, 18). In this chapter, we will focus on the in vitro *HPRT* assay mainly in human lymphoblastoid cell lines.

Each cell line requires specific culture medium and this along with the cell culture conditions are stated in the batch details, provided upon purchase of the cell line. Table 1 also shows detailed tissue culture instructions of the pre-mentioned cell lines.

Table 1
Cell culture conditions for a selection of recommended cell lines used in the *HPRT* gene mutation assay

Cell line	L5178Y	CHO	AS52	V79	AHH-1	MCL-5	TK6
Culture medium: Media	DMEM	Ham's F12	Ham's F12	DMEM	RPMI 1640	RPMI 1640 without L-histidine	RPMI 1640
+L-Glutamine	4 mM	2 mM	2 mM	2 mM	2 mM	2 mM	2 mM
+Serum	10% FBS	10% FBS	10% FBS	10% FBS	10% HS	10% HS	10% FBS
+ Other supplements	0.1% Pluronic; 1.5 g/L Sodium bicarbonate; 4.5 g/L Glucose				10 mM HEPES; 1 mM sodium pyruvate; 1.5 g/L Sodium bicarbonate; 4.5 g/L Glucose	0.1 mg/ml Hygromycin B[a]; 0.03 mg/ml 50 Aminolevulinic acid; 2 mM L-histidinol	1 mM sodium pyruvate
Growth conditions	Suspension	Adherent	Adherent	Adherent	Suspension	Suspension	Suspension
Maintain between "Re-suspension density→confluence" (cells/ml)	$2\times10^5\rightarrow 1\times10^6$	$1\times10^5\rightarrow 4\times10^5$	$1\times10^5\rightarrow 4\times10^5$	$1\times10^5\rightarrow 4\times10^5$	$1\times10^5\rightarrow 1\times10^6$	$1\times10^5\rightarrow 1\times10^6$	$2\times10^5\rightarrow 1\times10^6$
Trypsin EDTA or 0.25% trypsin used to subculture	No	Yes	Yes	Yes	No	No	No
Medium renewal	Two to three times a week	Two to three times a week	Two to three times a week	Two to three times a week	Two to three times a week	Two to three times a week	Two to three times a week
Preservation: Store in liquid nitrogen vapour phase	88% FBS and 12% DMSO	88% FBS and 12% DMSO	88% FBS and 12% DMSO	88% FBS and 12% DMSO	88% HS and 12% DMSO	88% HS and 12% DMSO	88% HS and 12% DMSO
References	http://www.atcc.org	HPAcutures.org.uk	Tindall and Stankowski Jr (13), Tindall et al. (14)	HPAcutures.org.uk	Crespi and Thilly (8) (http://www.atcc.org)	Crespi et al. (7), Crespi and Thilly (8) (http://www.atcc.org)	(http://www.atcc.org)

Penicillin/streptomycin mix can also be added to culture medium if required. *FBS* Fetal Bovine Serum, *HS* Horse Serum
[a]Essential in maintaining the plasmids

2.2. Compounds Required for the HPRT Gene Mutation Assay

Hypoxanthine–aminopterin–thymidine (HAT) supplement is added to culture medium at the mutant cleansing stage (Fig. 1 and Subheading 3.2). The aminopterin in HAT medium blocks the salvage pathway, leaving cells reliant on the endogenous pathway, that is, *HPRT* and *TK*, and therefore $HPRT^{-}$ and $TK^{-/-}$ mutants are killed (Fig. 2). This reduces the spontaneous (background) MF values and is a crucial step. Following HAT treatment, hypoxanthine–thymidine (HT) supplement is added to culture medium and both the de novo nucleotide biosynthesis pathway and the

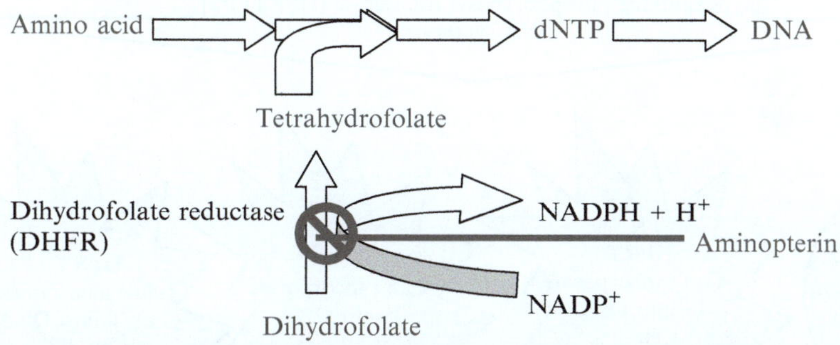

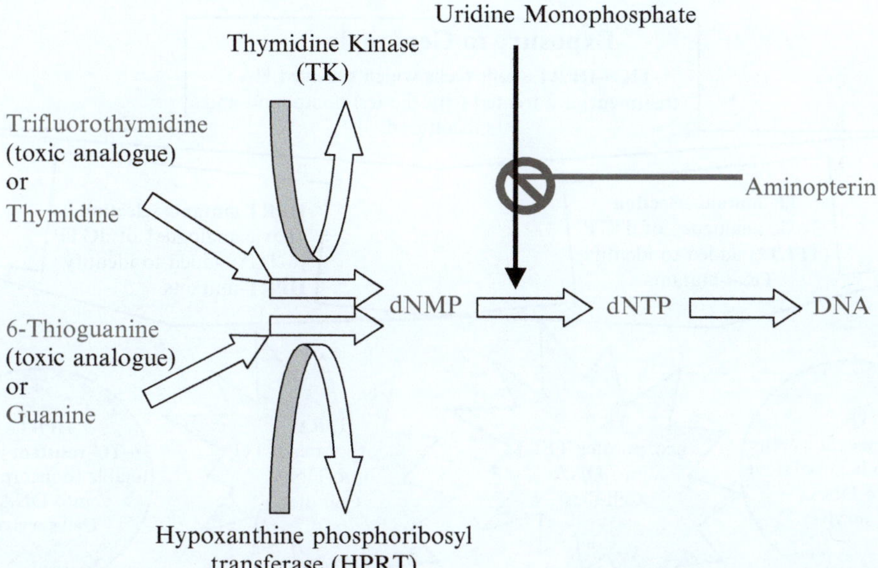

Fig. 1. Illustration of the (**a**) endogenous and (**b**) salvage pathways in production of dNTPs for DNA synthesis, showing the effect of aminopterin (present in HAT medium) in stopping enzymatic pathways to sequentially force cells into other pathways or cell death and this is used in mutant cleansing. 6-thioguanine (6-TG) and trifluorothymidine (TFLT) are also represented here to show that they are toxic analogues of guanine and thymidine, respectively, and can be incorporated into DNA by the hprt and tk enzymes, respectively.

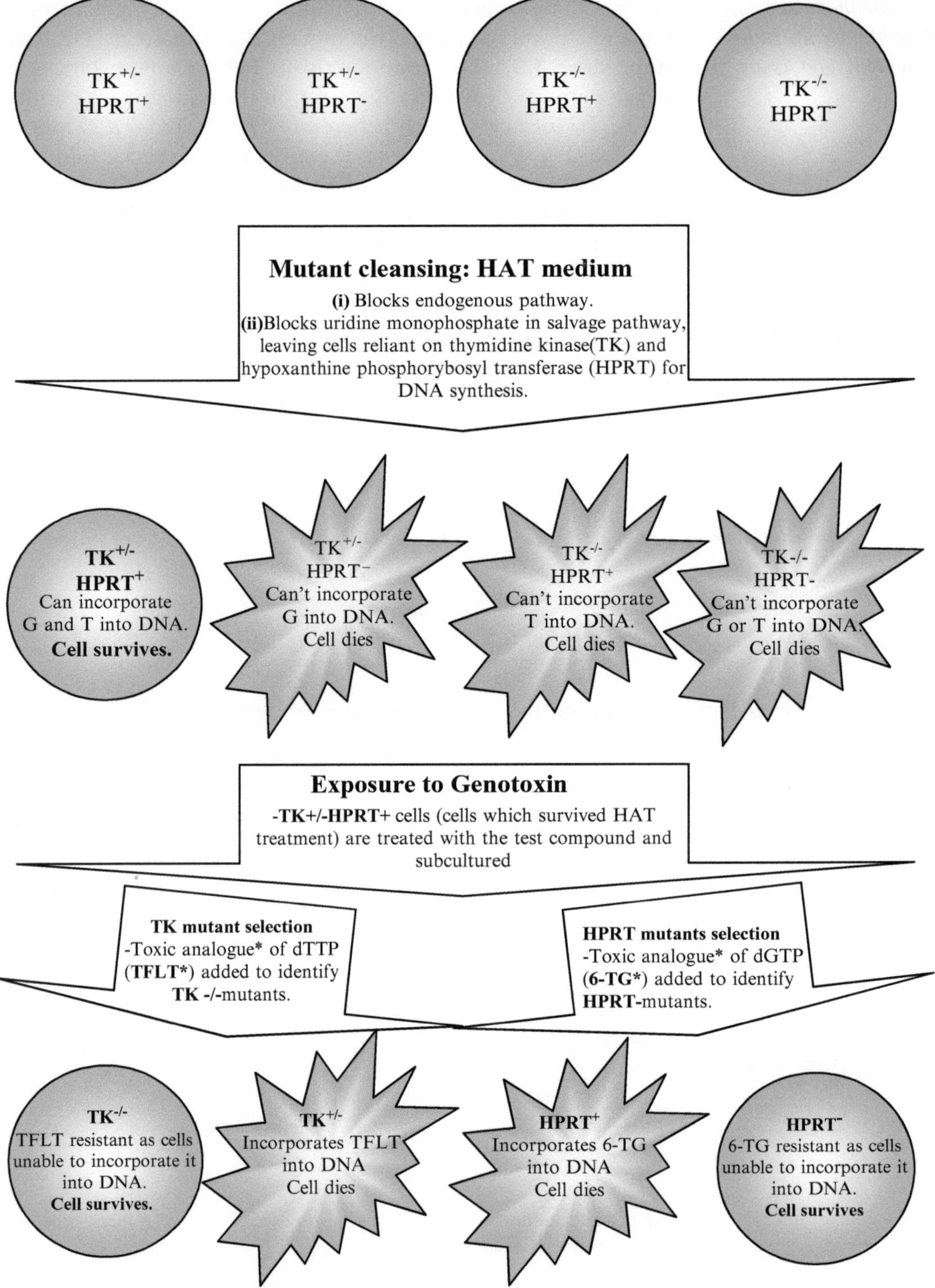

Fig. 2. Schematic representation of the HPRT and TK forward mutation assay methodology. Hypoxanthine–aminopterin–thymidine (HAT) medium.

salvage pathway are able to function from this stage onwards. After treatment with the test compound, the *HPRT*⁻ mutant cells are selected for using 6-TG. *HPRT*⁺ cells incorporate 6-TG into the DNA and die, and *HPRT*⁻ cells do not incorporate this toxic analogue into their DNA and they survive (Fig. 2).

Ethyl-methanesulphonate (EMS) and Ethyl-nitrosourea (ENU) can be used as the positive controls in the absence of exogenous metabolic activation (2). 3-Methylcholanthrene, *N*-nitrosodimethylamine, or 7,12-dimethylbenzanthracene can be used as a positive control in the presence of exogenous metabolic activation (2).

3. Methods

3.1. Metabolism

Compounds that require metabolic activation require either an exogenous source such as S9 (treatment time of 3–6 h with the test compound), or the cell line can be genetically modified. For example MCL-5 cells are derived from L3 cells, a subpopulation of AHH-1 cells that express a particularly high level of CYP1A1 activity (7, 19). The MCL-5 cell line has also been transfected with two plasmids: one containing two copies of CYP3A4 cDNA and one copy of CYP2E1 cDNA, and a second containing one copy of each CYP1A2, CYP2A6, and microsomal epoxide hydrolase cDNA (7, 19). Therefore MCL-5 cells stably express all five cDNAs and also have increased levels of CYP1A1 compared to AHH-1 cells, and test compounds that are known to be metabolised by these enzymes can be tested using MCL-5 cells. Genetically modified cell lines that stably express metabolic enzymes can promote more stable and reliable results as there are fewer variables such as pH, osmolality, or high levels of cytotoxicity than when adding crude cell extracts of liver, S9.

3.2. HPRT Mutant Cleansing

Mammalian gene mutation assays depend upon the ability to quantify mutant cells using selective media. The MF at each test concentration is compared to the control MF, and the control is a measure of the spontaneous MF. Therefore, the spontaneous MF should be maintained at a low and stable level within each laboratory. To decrease the spontaneous MF of cultures the number of *HPRT*⁻ mutants are reduced using HAT medium which inhibits the endogenous de novo nucleotide biosynthesis pathway such that while the salvage pathway is the required dNTP for DNA replication (Fig.1). Cells that are incapable of using the salvage pathway (i.e. *HPRT*⁻ mutants) can no longer divide and undergo cell death (Fig. 2).

3.3. Treatment Protocol (Fig. 3)

Following "mutant cleansing" the cells are sub-cultured and grown for 24 h in HT medium. The cells are washed and grown in normal culture medium for 3–4 days to attain sufficient numbers for treatment. This can be a good time to cryogenically freeze down the cells for storage, if for any reason the experiment needs to be carried out at a later date. However, one should be careful that there are no extra days added to the assay as these can cause clonal expansion and the spontaneous MF can be increased. 10 ml treatment

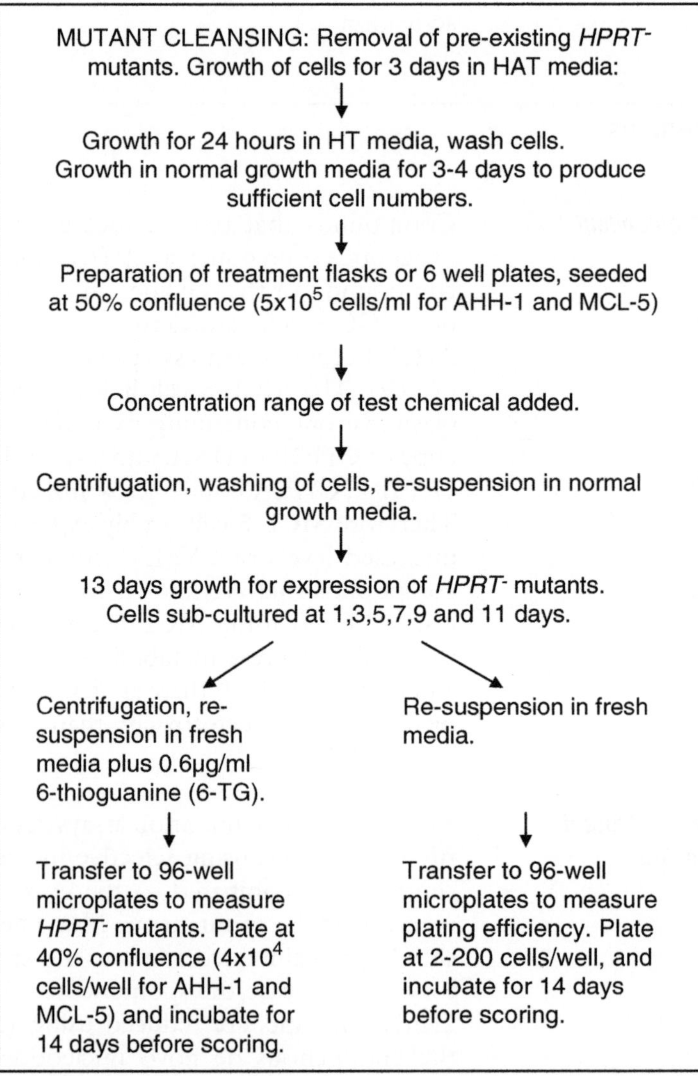

Fig. 3. Flow chart of *HPRT* gene mutation assay. 96-well microplates are maintained in a CO_2 incubator (37°C) for 14 days. *= cells can be cryogenically frozen down on days 3 and 4 after growth in HT medium. However, this can increase the time length of the assay which leads to clonal expansion and increased spontaneous MF. ** = cells can be frozen down at this stage by day 9 and the day should be recorded so that the time length of the assay is not altered.

flasks or 6-well plates are set up and the test chemical is added to the cultures at pre-defined concentrations, which should usually be separated by no more than a factor of between 2 and $\sqrt{10}$ and should also cover a range of high toxicity to little or no toxicity (2). Treated cultures are then incubated for 4 h or 24 h depending on the half-life of the compound and whether S9 is added, and each test compound should be prepared and dissolved in the correct solvent using sources of information such as the batch guidelines or the Merck Index. Each experiment is then carried out in at least duplicate, and each duplicate treatment flask is treated using a stock solution that is prepared each time (i.e. two times for a duplicate experiment) from the purchased product, to allow for variation in preparation procedures. Negative controls must be used, and the solvent should not have a significantly different MF than the spontaneous MF. Positive controls must also be used and these are specified in Subheading 2.2. Following chemical exposure the cells are centrifuged to remove the test medium, washed and re-suspended in fresh culture medium for mutant expression. For the *HPRT* mutation assay this involves incubation for 13 days. This allows any mutations to become fixed and any existing hprt proteins/RNA to become degraded. During the phenotypic expression period, the cells are sub-cultured every other day, on days 1, 3, 5, 7, 9, and 11, by centrifugation and re-suspension of the cells in fresh culture medium. Cells can be cryogenically frozen down at this stage. The day of freezing down should be recorded so that the number of expression period days is not altered due to this process. It is not advisable to freeze them down after day 9. After the phenotypic expression period, the cells are added to 96-well microplates at 40% confluence (4×10^4 cells/well for AHH-1 and MCL-5 cells) in culture medium with selection using the toxic analogue 6-TG (from Sigma UK) at 0.6 μg/ml. For the plating efficiency (PE) calculation, 2–200 cells/well can be plated with no selection at each dose. Plates are scored for colony formation after 14 days of incubation at 37°C in humidified incubator with 5% CO_2.

3.4. Scoring Method

The criteria for colony counting include only scoring colonies of 20+ cells in diameter and ensuring separate colonies are clearly apart, thereby accounting for clonal expansion. This value was defined by us in Swansea and other laboratories may wish to define other scoring criteria. There will also be a large number of dead *HPRT*$^+$ colonies due to 6-TG selection, and these must not be scored as viable colonies.

3.5. Clonal Expansion of Mutants and DNA Extraction

Mutant *HPRT*$^-$ colonies are removed from the 96-well microplates using Pasteur pipettes and re-suspended in fresh culture medium. Cultures are maintained and grown to confluence for sufficient numbers of cells for DNA/RNA extraction.

3.6. HPRT Mutation Spectrum Analysis

The human *HPRT* gene, when disrupted, results in the human disorders gouty arthritis and Lesch–Nyhan syndrome. Subsequently the *HPRT* gene has been characterised largely using the polymerase chain reaction (PCR), with multiplex PCR being used for exon deletion detection, that is, many DNA fragments (i.e. different exons) can be amplified simultaneously, which alongside sequencing can detect smaller mutations (20). The coding region of the human *HPRT* gene is distributed over 39.8 kb DNA and contains nine coding exons. The RNA transcript can also be sequenced using reverse transcriptase PCR, and the resulting mutation spectrum from this technique provides sequence information of mRNA (cDNA) and this is less costly, with regards to both time and money, than multiplex PCR.

3.7. MF and PE Equations (5)

PE

$$\text{Plating Efficiency \% (PE)} = -\text{Ln}\,(X_o / N_o) \times 100$$

Cell Viability (Relative PE)

$$\text{Cell Viability \%} = \frac{\text{PE}}{\text{PE of control}} \times 100$$

Mutant Fraction

$$\text{Mutation frequency (MF)} = \frac{-\text{Ln}(X_s / N_s)}{-\text{Ln}\left(X_o / N_o\right)} \times \text{DF}$$

$$\text{Dilution factor (DF)} =$$

$$\left[\frac{\text{(No. of initial cells per well) Non - selective conditions}}{\text{(No. of initial cells per well) Selective conditions}}\right]$$

$$\left.\begin{aligned} X_S &= \text{No. of wells without colonies} \\ N_S &= \text{Total no. of wells} \end{aligned}\right\}\text{Selective conditions}$$

$$\left.\begin{aligned} X_0 &= \text{No. of wells without colonies} \\ N_0 &= \text{Total no. of wells} \end{aligned}\right\}\text{Non - selective conditions}$$

Worked examples:

HPRT[-] MF calculation for replicate B at 0 μg/ml MMS

$$\text{MF} = \left(\frac{-\text{Ln}\left(1{,}699 / 1{,}800\right)}{-\text{Ln}\left(1{,}218 / 1{,}920\right)}\right) \times \frac{20}{40{,}000}$$

$$= 0.0577468 / 0.455115 \times 0.0005 = 6.344 \times 10^{-5}$$

If cells have a low PE, it can be due to (21):

1. A bad batch of 96-well microplates (relatively rare)
2. A bad batch of horse serum (relatively rare). Make sure you use the same batch of horse serum throughout your experiment.

3. High pH on the 96-well microplates (common) due to:
   (a) Opening the incubator too much during the first 4–5 days.
   (b) Low CO_2 setting.
   (c) Incorrect medium pH (should be 6.8–7.0 before adding serum).
4. Poorly growing cells.

If cells have a high negative control mutant fraction, it can be due to (21):

1. An artefact due to low PE.
2. Improper HAT/HT treatment due to:
   (a) Thymidine starvation.
   (b) Inadequate aminopterin.
3. Exposure to a mutagen (sunlight).
4. Inadequate selective agent.

4. Notes

1. L5178Y mouse lymphoma cells and CHO cells have a published spontaneous MF of $2–50 \times 10^6$ (22) (Subheading 2.1).
2. AHH-1 and MCL-5 human lymphoblastoid cells have a published spontaneous MF of $6–80 \times 10^6$ (21, 23, 25) (Subheading 2.1).
3. A typical design of the *HPRT* assay would be testing up to 10 mM or 5,000 μg/ml (2) with a maximum of one insoluble concentration, because cells grow in suspension and precipitate cannot be removed (Subheading 3.3).
4. A typical design of the *HPRT* assay would be with and without S9 (4 h treatments) plus a 24-h treatment without S9 (Subheading 3.3).
5. A typical design of the *HPRT* assay would usually test four concentrations with duplicate treatments per concentration, or in triplicate if more advanced statistical analysis is required. More concentrations can also be used (Subheading 3.3).
6. Each replicate should have a minimum of 4 × 96-well microplates for 6-TG selection and 2 × 96-well microplates for PE (viability). This can be increased if the experiment is designed to detect smaller changes in MF increase. For example, 100 × 96-well microplates for 6-TG selection and 50 × 96-well microplates for PE are required in total for threshold analysis (Subheading 3.3), for this example the experiments were carried out in triplicate (23, 26).

7. PE is a crucial step in the *HPRT* gene mutation assay. It is important to optimise the number of cells plated for this step (2–200 cells/well) for the particular laboratory and particular cell line (Subheading 3.3).
8. The outside wells of the 96-well microplate can dry up due to the long growth period in the incubator to give false negative colony growth. Special plates can be purchased that have a channel for water which keeps the cells in a humid environment, or other methods can be designed to keep the cells in a humid environment (Subheading 3.3).
9. Scoring colonies of attached cell lines is different from scoring colonies of suspension cell lines, and this should be considered when deciding whether they are defined as viable colonies (e.g. >20 cells diameter) or not (e.g. <20 cells diameter or the cells are dead) (Subheading 3.4).
10. Adding 6-TG can kill the colonies at a late stage of growth, and therefore you should be careful that you are scoring viable colonies and not dead cells. This can be determined by observing the morphology of the cells.Trypan blue can be used to stain for cell viability, which can also be determined by observing the morphology of the cells. For example, dead AHH-1 lymphoblastoid cells are darker and their cell walls are also less circular than viable cells (Subheading 3.4).

References

1. DOH (2000) Department of Health Guidance for the testing of chemicals for Mutagenicity. Committee on Mutagenicity of Chemicals in Food, Consumer Products and the Environment.
2. OECD (1997) Organisation for Economic Co-operation and Development. Ninth addendum to the OECD Guidelines for the Testing of Chemicals. *In Vitro* Mammalian Cell Gene Mutation Test: 476.
3. DeMarini D M, Brockman H E, de Serres F J, Evans H H, Stankowski, Jr L F and Hsie AW (1989) Specific-locus mutations induced in eukaryotes (especially mammalian cells) by radiation and chemicals: a perspective, *Mutat Res* **220:** 11–29.
4. Moore M M, Honma M, Clements J, Awogi T, Bolcsfoldi G, Cole J, Gollapudi B, Harrington-Brock K, Mitchell A, Muster W, Myhr B, O'Donovan M, Ouldelhkim M C, San R, Shimada H and Stankowski, Jr L F (2000) Mouse lymphoma thymidine kinase locus gene mutation assay: International Workshop on Genotoxicity Test Procedures Workgroup Report, *Environ Mol Mutagen* **35:** 185–190.
5. Furth E E, Thilly W G, Penman B W, Liber W L and Rand W M (1981) Quantitative assay for mutation in diploid human lymphoblasts using microtiter plates, *Anal Biochem* **110:** 1–8.
6. Moore M M, Parker L, Huston J, Harrington-Brock K and Dearfield K L (1991) Comparison of mutagenicity results for nine compounds evaluated at the hgprt locus in the standard and suspension CHO assays, *Mutagenesis* **6:** 77–85.
7. Crespi C L, Gonzalez F J, Steimel D T, Turner T R, Gelboin H V, Penman B W and Langenbach R (1991) A metabolically competent human cell line expressing five cDNAs encoding procarcinogen-activating enzymes: application to mutagenicity testing, *Chem Res Toxicol* **4**: 566–572.
8. Crespi C L and Thilly W G (1984) Assay for gene mutation in a human lymphoblast line, AHH-1, competent for xenobiotic metabolism, *Mutat Res* **128**: 221–230.
9. Doak S H (2008) Aneuploidy in upper gastrointestinal tract cancers – a potential prognostic marker? *Mutat Res* **651**: 93–104.

10. Parry J M, Parry E M, Johnson G, Quick E and Waters E M (2005) The detection of genotoxic activity and the quantitative and qualitative assessment of the consequences of exposures, *Exp Toxicol Pathol* **57:** Suppl 1 205–212.

11. Caskey C T and Kruh G D (1979) The HPRT locus, *Cell* **16**: 1–9.

12. Chen T, Harrington-Brock K and Moore M M (2002) Mutant frequency and mutational spectra in the Tk and Hprt genes of N-ethyl-N-nitrosourea-treated mouse lymphoma cells, *Environ Mol Mutagen* **39**: 296–305.

13. Tindall K R and Stankowski Jr L F (1989) Molecular analysis of spontaneous mutations at the gpt locus in Chinese hamster ovary (AS52) cells, *Mutation Research/Reviews in Genetic Toxicology* **220**: 241–253.

14. Tindall K R, Stankowski Jr L F, Machanoff R and Hsie A W (1986) Analyses of mutation in pSV2gpt-transformed CHO cells, *Mutation Research/Fundamental and Molecular Mechanisms of Mutagenesis* **160**: 121–131.

15. Moore M M, DeMarini D M, DeSerres F J and Tindall K R (1987) Banbury Report **28**: *Mammalian Cell Mutagenesis,* in: N.Y. Cold Sprind Harbor Laboratory, New York (Ed.).

16. Tates A D, van Dam F J, de Zwart F A,van Teylingen C M and Natarajan A T (1994) Development of a cloning assay with high cloning efficiency to detect induction of 6-thioguanine-resistant lymphocytes in spleen of adult mice following in vivo inhalation exposure to 1,3-butadiene, *Mutat Res* **309**: 299–306.

17. van Dam F J, Natarajan A T and Tates A D (1992) Use of a T-lymphocyte clonal assay for determining HPRT mutant frequencies in individual rats, *Mutat Res* **271:** 231–242.

18. Albertini R J (2001) HPRT mutations in humans: biomarkers for mechanistic studies, *Mutation Research/Reviews in Mutation Research* **489**: 1–16.

19. Woodruff N W, Durant J L, Donhoffner L L, Penman B W and Crespi C L (2001) Human cell mutagenicity of chlorinated and unchlorinated water and the disinfection byproduct 3-chloro-4-(dichloromethyl)-5-hydroxy-2(5 H)-furanone (MX), *Mutat Res* **495**: 157–168.

20. McPherson M, Taylor G and Quirke P (1991) *PCR 1: A Practical Approach* (Practical Approach S.), Oxford University Press, Oxford.

21. Gentest (1994) AHH-1 TK+/- *Human Lymphoblastoid Cells. Routine cell culture, metabolite production and gene-locus mutation assay. Procedures for use.* Gentest corporation, Woburn, MA 01801, USA.

22. DeMarini D M, Brock K H, Doerr C L and Moore M M (1988) Mutagenicity and clastogenicity of proflavin in L5178Y/TK +/- -3.7.2.C cells, *Mutat Res* **204**: 323–328.

23. Doak S H, Jenkins G L, Johnson G E, Quick E, Parry E M and Parry J M (2007) Mechanistic influences for mutation induction curves after exposure to DNA-reactive carcinogens, *Cancer Res* **67**: 3904–3911.

24. Zair Z M, Jenkins G J, Doak S H, Singh R, Brown K and Johnson G E (2011) N-Methylpurine DNA Glycosylase Plays a Pivotal Role in the Threshold Response of Ethyl Methanesulfonate-Induced Chromosome Damage. *Toxicological Sciences,* **119**: 346–358.

25. Johnson G E, Quick E L, Parry E M and Parry J M (2010) Metabolic influences for mutation induction curves after exposure to Sudan-1 and para red. *Mutagenesis,* **25**: 327–333.

26. Johnson G E, Doak S H, Griffiths S M, Quick E L, Skibinski D O F, Zaïr Z M and Jenkins G J (2009) Non-linear dose-response of DNA-reactive genotoxins: Recommendations for data analysis. *Mutation Research/Genetic Toxicology and Environmental Mutagenesis,* **678**: 95–100.

Chapter 5

The In Vitro Mammalian Chromosome Aberration Test

Gill Clare

Abstract

The short-term in vitro mammalian cell chromosome aberration test is used to assess potential genotoxic hazard of test substances. Mammalian cells are cultured in vitro, exposed to a test substance, harvested, and the frequency of asymmetrical structural chromosome aberrations is measured. Human peripheral blood lymphocytes do not normally divide. The assessment of the effects of cyclophosphamide on lymphocytes, stimulated to divide in whole blood cultures in vitro, is described. Procedures that are important in generating accurate results are emphasised, to avoid false positive results. The study design for a regulatory assay, the use of established cell lines, alternative methods of measuring cytotoxicity and analysis of results are included.

Key words: Whole blood cultures, Human lymphocytes, Structural chromosome aberrations, Cytogenetics, In vitro mammalian chromosome aberration test, Cyclophosphamide, Regulatory assay, Established cell lines, S-9 mix, Cytotoxicity, Accurate test result

1. Introduction

The adverse health effects caused by genotoxins in vivo are of concern. The effects of exposure to relatively low concentrations of genotoxic substances may not be immediately apparent. This means that there are no warning signs so that exposure can be avoided. Heritable chromosomal mutations in somatic cells are implicated in the development of cancer and those in germinal cells may lead to inherited disorders (1). Since chromosomal mutations are more likely to be deleterious than beneficial, they are associated with potential risk.

The reliance on the ability of results from in vitro assays to accurately predict human risk is increasing. This is mainly owing to the drive to reduce animal use in the safety assessment of test substances. Current legislation means that cosmetic products may no longer be tested on animals. REACH (Registration, Evaluation,

James M. Parry and Elizabeth M. Parry (eds.), *Genetic Toxicology: Principles and Methods*, Methods in Molecular Biology, vol. 817, DOI 10.1007/978-1-61779-421-6_5, © Springer Science+Business Media, LLC 2012

Authorisation and restriction of CHemicals) legislation (EC No. 1907/2006) means that thousands of chemicals, many of which have had little safety testing, will need to be shown to be safe or removed from the market. The size of this task mitigates against using animals.

The short-term in vitro mammalian cell chromosome aberration test is used to assess potential genotoxic hazard of test substances and guidance is provided by regulatory authorities (2–5). Mammalian cells are cultured in vitro, exposed to a test substance, harvested, then microscope slides are prepared and the frequency of asymmetrical structural chromosome aberrations is measured. The results from such analyses inform regulatory decisions about substances.

An ideal cell line for use in genotoxicity testing is defined by certain characteristics (6) and primary human peripheral lymphocytes possess some of these. Lymphocytes may give fewer false positive results than currently used rodent cell lines (7). For some advantages, disadvantages, and tips on using cell lines see Notes 1–3, respectively.

Human peripheral blood lymphocytes do not normally divide in vivo but can be stimulated to do so in vitro by addition of a mitogen such as phytohaemagglutinin (PHA) (8, 9). Cyclophosphamide (CPA) is commonly used as a positive control that requires metabolic activation, by an exogenous system (S-9 mix) in the case of lymphocytes. The methodology for the assessment of the effects of CPA on lymphocytes is described because it employs many of the procedures used in an assay for submission to regulatory authorities. The study timings rely on estimating the cell cycle time (3–5). A method for estimating the cell cycle time (average generation time, AGT) is provided, because it is important even though used only periodically in laboratories for this purpose.

The methodology is prescriptive, to help achieve a successful outcome, but this should not be taken to imply that it is best. The test is regularly used in commercial laboratories, and each laboratory uses variations in methodology that will produce equally acceptable outcomes.

The specificity for mammalian cell genotoxicity tests to predict human risk is considered to be low (e.g. ref. 7). The chapter is written with the intention of advocating procedures and developments that are important in generating accurate results. The study design for a regulatory assay, the use of established cell lines, alternative methods of measuring cytotoxicity, and analysis of results are included. The use of human lymphocytes for the analysis of chromosome aberrations (10), the rationale underlying the in vitro cytogenetic test (11), and the principles of cell culture (12) are helpful background reading.

2. Materials

2.1. In Vitro Chromosome Aberration Test

2.1.1. Whole Blood Culture

1. Unless otherwise specified, water should be distilled or de-ionized.
2. Plan the experiment in advance to determine the quantities of materials, blood, cultures, etc., required.
3. Heparinised tubes, to avoid clotting, for blood collection (preferably sodium rather than lithium heparin and without preservative).
4. Incubation, unless otherwise specified, is at 37°C ± 1°C throughout, with gentle mixing of the cultures, preferably continuously (e.g. rocking platform). A humidified atmosphere of 5% CO_2 in air may be introduced at culture initiation. Lymphocytes are grown as suspension cultures in sterile, disposable centrifuge tubes (the caps on 15 mL BD Falcon tubes seal well; Fisher Scientific UK Ltd, Loughborough, Leics.).
5. Culture media: RPMI 1640 medium (Invitrogen Ltd, Paisley, Scotland), containing 25 mM HEPES-buffer and GlutaMAX™ (neither essential) or l-glutamine (if required), Penicillin–Streptomycin (100 IU/mL and 100 μg/mL, respectively), foetal calf serum (FCS, 10–20%), reagent grade PHA (2%). Supplements are added shortly before use, PHA just before the blood. Store media at 4°C ± 1°C. Since FCS is not of fixed composition, several batches, often sourced from different suppliers, are tested and sufficient of a suitable batch reserved.

2.1.2. Dosing Formulations

1. Cyclophosphamide: CPA, CAS no. 6055-19-2, formula weight 279.1 g/mol (Sigma-Aldrich Chemical Company, Poole, Dorset, UK).
2. Vehicle: anhydrous dimethyl sulphoxide (DMSO), puriss, absolute, CAS no. 67-68-5. Aliquot into air-tight glass bottles without an air-space, to prevent hydration. Alternative vehicle: water for injection (both from Sigma-Aldrich Chemical Company, Poole, Dorset, UK).
3. S-9 fraction: The post-mitochondrial fraction (Molecular Toxicology Incorporated, USA) is prepared from the liver of male Sprague Dawley rats, induced with Aroclor 1254. The MolTox™ S-9 is stored in aliquots at −80°C nominal. A quality control statement for sterility, protein content, ability to convert known promutagens to bacterial mutagens, and cytochrome P-450-catalysed enzyme activities (alkoxyresorufin *O*-dealkylase activities) accompanies each batch. Ampoules are stored at −80°C nominal in the dark and thawed just before use.

4. S-9 mix: Glucose-6-phosphate (180 mg/mL), NADP (25 mg/mL), and potassium chloride (KCl) (150 mM) aqueous solutions are filter sterilized. The solutions are mixed with rat liver S-9 fraction just before use in the ratio 1:1:1:2 and the mixture kept ice-cold.

2.1.3. Sampling, Slide Preparation, and Staining

1. KCl 0.075 M (0.56%) in water.
2. Methanol:glacial acetic acid (3:1, v/v). The methanol is pre-cooled in the deep freeze. The fixative should be prepared just before use and used whilst still cold, because the resulting acetates are short-lived.
3. Microscope slides: clean, ice-cold, grease-free labelled, at least two per culture, and coverslips (50 × 22 mm).
4. Gurr's Giemsa Improved R66, and Gurr's buffer tablets, pH 6.8 dissolved in water (BDH, Lennox Laboratory Supplies Ltd, Dublin, Ireland).

2.2. Estimation of AGT

1. 5-Bromo-2′-deoxyuridine (BrdU) molecular weight 307.4. Aliquot filtered (0.2 μm pore size) aqueous stock solution, 0.5 mM (0.1537 mg/mL) into bottles covered with foil to exclude light, and store at -80°C nominal. Use at 25 μM (0.5 mL per 9.5 mL).
2. MacIlvaine's buffer: comprises proportions of 0.1 M citric acid (21.01 g/L water, solution A), 0.2 M disodium hydrogen phosphate (35.6 g/L water, solution B) depending on pH. For pH 8.0, mix 2.25 mL solution A with 97.5 mL solution B.
3. Hoechst 33258 (bis Benzimide). Aliquots of a stock solution, prepared in MacIlvaine's buffer, pH 8.0, at 2.67 mg/mL may be stored at -80°C nominal in the dark. Dilute in buffer to 26.7 μg/mL for staining.

3. Methods

3.1. Conduct of Test

Good cell growth during the assay is critical to a successful outcome. High quality metaphase preparations, use of a trained and experienced observer who is familiar with the karyotype and robust acceptance criteria and evaluation of the data all help generate an accurate test result.

3.2. Safety

1. Health and safety guidance for handling human blood is provided for universities (13) and by the Advisory Committee on Dangerous Pathogens (14) and should be followed.
2. Tissue culture facilities, including cabinets where the air flow protects both the culture and operator, sterile equipment and

solutions must be available, and aseptic techniques are required to avoid microbial contamination. Appropriate safety clothing should be worn throughout.

3. Great care must be taken to avoid exposure to potential mutagens. The Material Data Safety Sheet should be consulted and any safety hazards noted. Known mutagens should be handled as little as possible, disposable materials used, and cross-contamination avoided.
4. Procedures involving the fixative should be carried out in a designated area where inhalation of the fumes is avoided (e.g. fume hood).

3.3. Study Design

A flow chart of the process (Fig. 1) and a study design for assessing the effects of an unknown test substance (see Note 4) may help when planning the experiment to test the effects of CPA.

The stage of the cycle when cells are exposed affects the sensitivity to mutagens, and so exposure to the test substance commences approximately 48 h after culture initiation, when cells are dividing and at all stages of the cell cycle, and the initial synchrony of the cells is less. After exposure to most mutagens, cells need to

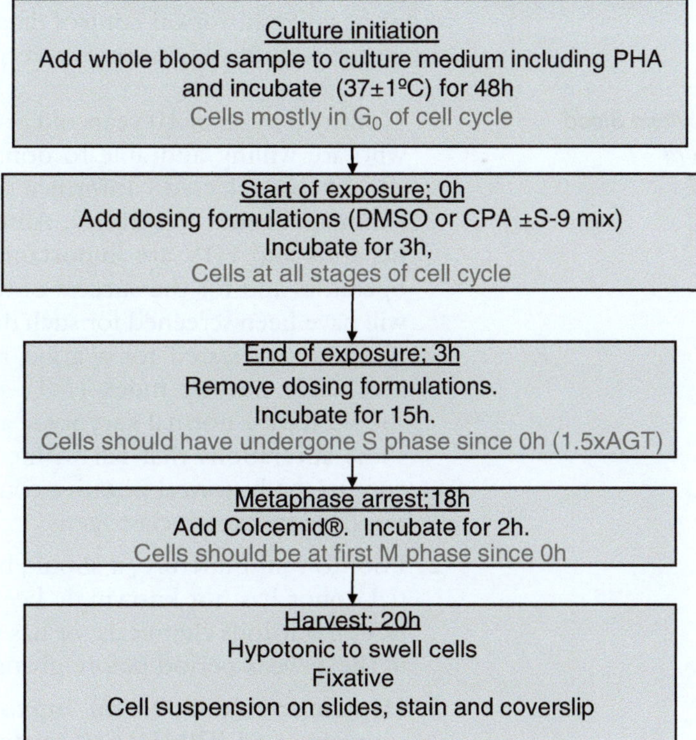

Fig. 1. Timings of in vitro human lymphocyte chromosome aberration test.

undergo S phase for chromosome damage to be manifest. Since many of the asymmetrical structural chromosome aberrations prevent unlimited division, it is important to harvest the cells when as many as possible are in their first division after the start of exposure. Cells with chromosome damage are likely to experience cell cycle delay and so more highly damaged cells will reach metaphase more slowly than their less highly damaged counterparts and it has been shown that the aberration yield increases with time (15). This is a reason why the effects of three concentrations of test substance that have shown high (50% mitotic inhibition), medium or low cytotoxicity are assessed where appropriate. Cells are harvested approximately 1½ the AGT after the start of exposure. The AGT should be measured before the effects of CPA are assessed, to ensure that the culture conditions and study timings are adequate.

Each concentration of test substance and the positive controls should be tested at least in duplicate cultures. Information from the solvent control cultures is used to monitor the background frequency of structural chromosome aberrations and to assess the clastogenic potential of a test substance. Very few structural chromosome aberrations are normally present in vehicle control cultures. Using quadruplicate negative control cultures will increase the ability of the test to detect significant changes. In laboratories where the test is performed infrequently, or there is no sizeable recent negative historical control data base, the use of quadruplicate negative control cultures is recommended.

3.4. Whole Blood Culture

1. Healthy donors of 40 years old or less, who do not smoke, and who are willing and able to donate blood on a regular basis should be selected. Informed consent must be obtained (Human Tissues Act 2004). Absence of viral diseases such as hepatitis and HIV are important, both for protection of the operators and for the success of the assay. NHS blood donors will have been screened for such diseases. Prior to use, the lymphocytes are tested for a good response to PHA, giving an acceptable mitotic index (MI) and AGT. The lymphocytes should have a normal karyotype and exhibit levels of chromosome aberrations that fall within a normal range, with reference to the historical negative control data base, and a suitable response to known clastogens.
2. Prior to veni-puncture, it should be ascertained that the potential donor has not knowingly been exposed to ionizing radiation, hazardous chemicals, or has suffered from viral infections in the 2-week period before giving blood.
3. Heparinised blood, 0.4 mL (up to 0.8 mL) is added to warmed supplemented RPMI 1640 medium so that the final volume following addition of S-9 mix/KCl and dosing formulation is

10 mL (5 mL cultures may also be used). These suspension cultures are incubated for approximately 48 h. Even small, but prolonged, reductions in incubator temperature can affect the AGT and this may not be obvious, as it is likely to be measured only periodically. Warm up media, etc., for the assays in an incubator or water bath separate from the main experiments.

3.5. Preparation, Addition, and Removal of Dosing Formulations

1. To reduce handling positive controls use the following procedure. Weigh a sterile, preferably glass, container. Tare the balance to zero and add approximately 10 mg CPA. Add sufficient DMSO to prepare a stock solution that is 100× as strong as required. Take into account the mass of a test substance when the concentration exceeds 5 mg/mL. An alternative is to prepare a stock solution that is 10× as strong as required and dissolve in water for injection, then filter sterilize (0.2 μm pore size) to avoid microbial contamination (see Notes 5–7 on selection on highest concentration, appropriate concentration range, and effects of osmolality and pH). Samples of excess stock solution may be stored at −80°C nominal in the dark, and thawed shortly before use.
2. The stock solution is diluted to provide a range of approximately six concentrations of CPA. In the presence of S-9 mix, the effective final concentration in media is likely to be in the range of 5–15 μg/mL. Higher concentrations will be tolerated in the absence of S-9 mix.
3. The S-9 fraction is thawed just before required and the S-9 mix is prepared.
4. The CPA dosing formulations or DMSO, 0.1 mL, are added to each culture to achieve the required final concentration of CPA in a total of 10 mL. Do not exceed a 1% v/v final concentration of DMSO in the media. If CPA has been dissolved in water, 1.0 mL will be added to each culture. The S-9 mix is added (0.5 mL) so that the final concentration of the liver homogenate in the test system is 2% (see Note 8). Cultures exposed to CPA in the absence of S-9 receive 0.5 mL of KCl (150 mM).
5. Incubation is continued for 3 h (see Note 8).
6. Culture tubes are centrifuged at approximately 300 ×g for 10 min, the supernatant carefully removed and cells re-suspended in 10 mL RPMI medium without supplements, pre-warmed to 37°C. This process is repeated, so that there have been two changes of medium. After the third centrifugation, cells are re-suspended in 10 mL complete RPMI medium, pre-warmed to 37°C and incubation is continued.

3.6. Sampling, Slide Preparation, and Staining

1. Metaphase, when chromosomes are condensed and can be seen as discrete entities, lasts a short time and so there are few such cells. Two to three hours before sampling, dividing cells at metaphase are arrested by adding Colcemid®, final concentration 0.1 μg/mL (or colchicine at 1 μg/mL). Incubation is continued.
2. At the sampling time, cultures are centrifuged at approximately 300 × *g* for 10 min, the supernatant carefully removed and cells re-suspended in 4 mL warmed KCl at approximately 37°C for 15 min to allow cells to swell.
3. Cells are fixed by gently mixing the suspension of cells with 6 mL fixative. The red blood cells form the vast majority of the cells mixture and, as these lyse, the suspension will turn brown. It is important to avoid clumping of the lymphocytes, which this slow mixing procedure should achieve. The tubes containing the cell suspension in fixative and hypotonic solution are centrifuged at approximately 300 × *g* for 10 min.
4. The supernatant is removed and discarded into a labelled container for disposal.
5. The cells are re-suspended by slow addition of fresh, cold fixative, mixing using a vortex mixer, to avoid cell clumping.
6. The tubes containing the suspension are centrifuged, when the speed can be increased to 1,250 × *g* and the time reduced to 2–3 min. This procedure is repeated so there have been at least three changes of fixative, until the cell pellets no longer contain traces of red blood cells.
7. Lymphocytes are usually kept at 1–10°C at least overnight to ensure adequate fixation.
8. Cell suspensions are centrifuged and re-suspended in a minimal amount of freshly prepared fixative, if required, to give a milky suspension. Several drops of cell suspension are placed on cold microscope slides (see Note 9 for methods of preparing high quality metaphase spreads).
9. After the slides have dried the cells are stained for 5 min in filtered 4% (v/v) Giemsa in pH 6.8 buffer.
10. The slides are rinsed, dried, and mounted with coverslips.

3.7. Estimation of AGT

The technique is based on work by Goto et al. (16), described by Marshall et al. (17). BrdU is added to cultures at least 48 h after culture initiation. The cultures are handled in subdued lighting conditions or a safelight once the BrdU has been added. This may be achieved by covering the cultures in metal foil. Exposure to BrdU should be for 1½–2 cell cycles (24 h). Colcemid is added for 3 h, and the cells are harvested (see steps 1–9 of Subheading 3.6 and note 9 of Subheading 4) and put onto microscope slides, which are air dried.

3.7.1. Staining

1. Stain slides in a solution of Hoechst 33258 for 25 min, protected from light.
2. Rinse slides thoroughly in MacIlvaine's buffer pH 8.0 twice, the second time being in buffer pre-warmed to 40°C. Place the slides flat in a tray and add sufficient pre-warmed buffer to cover the slides. The warm, high pH buffer relaxes the chromatin, causing the chromatid arms to swell.
3. Expose the slides to UV light at approximately 366 nm for 25–40 min, maintaining a temperature of 40°C. Other sources of white light can be used. Calibration may be necessary to ensure good differential staining. Natural daylight may be used, but prolonged periods (>12 h) are needed, and the results are dependent on amount of sunlight. The UV light preferentially photo-degrades the DNA that contains BrdU.
4. Remove the slides, and rinse thoroughly in PBS at pH 6.8, avoiding sudden differences in temperature.
5. Stain for 10 min in filtered 4% Giemsa stain in PBS at pH 6.8. Check that the staining is adequate and stain for longer if necessary. Rinse the slides thoroughly, first in PBS at pH 6.8 and then in water, shake off excess moisture, air dry, and mount with coverslips.

3.7.2. Analysis of Cells in First, Second, and Third Division

Figure 2 shows how BrdU is incorporated and how the chromosomes stain in the first to third metaphase. Cells where both chromatids of the chromosomes are stained purple are in their first division (M1) and these are practically indistinguishable from normal staining. Cells where one chromatid of the chromosome is lightly stained and the other is darkly stained are in their second division (Fig. 3). Where approximately one-fourth of the chromatids are darkly stained, and three-fourths are lightly stained, the cells are in their third division since BrdU was added (Fig. 3). The AGT is calculated as follows:

$$\text{Proliferative Index (PI)} = \frac{\%\text{cells M1} + 2(\%\text{cells M2}) + 3(\%\text{cells M3})}{100}$$

$$\text{Average Generation Time (AGT)} = \frac{\text{Hours in BrdU}}{\text{PI}}$$

The AGT should be in the region of 12–14 h for human lymphocytes. Longer than 16 h indicates that culture conditions are sub-optimal and the source of the problem should be identified and rectified.

3.8. Assessment of Mitotic Inhibition

See Note 10 for assessment of use of MI. Slides are examined for cell density and the presence of mitotic cells using a low power objective. If, after at least two scans across the slide after exposure to CPA, no mitotic cells are seen, this concentration is toxic.

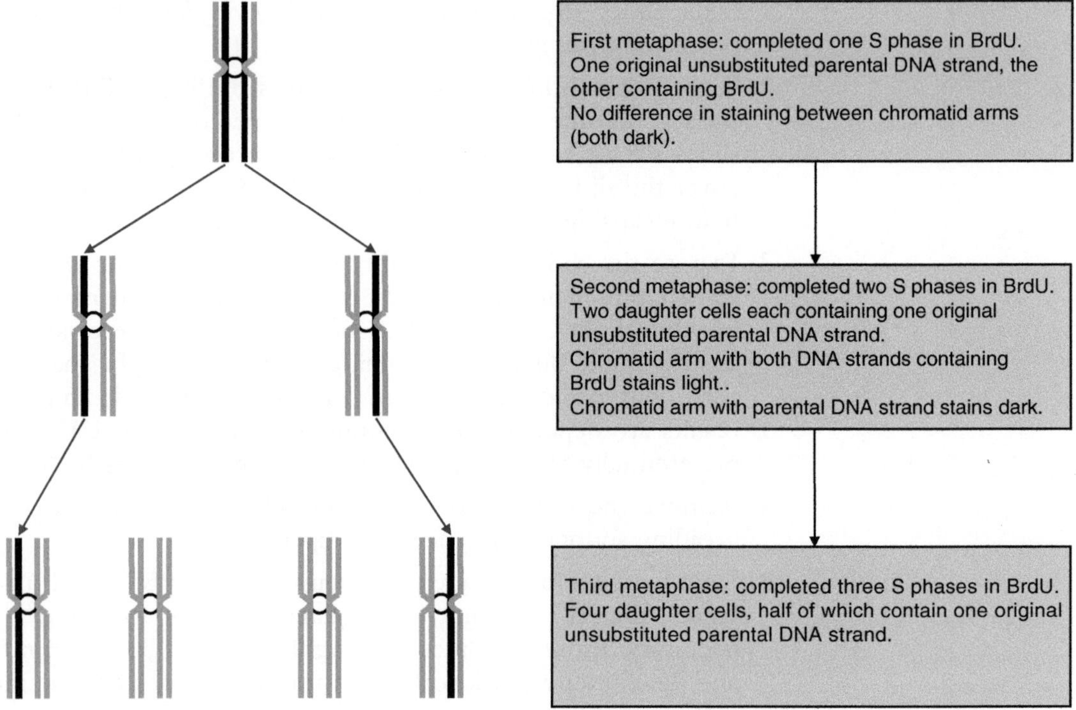

Fig. 2. Average generation time: staining of a chromosome through three metaphases.

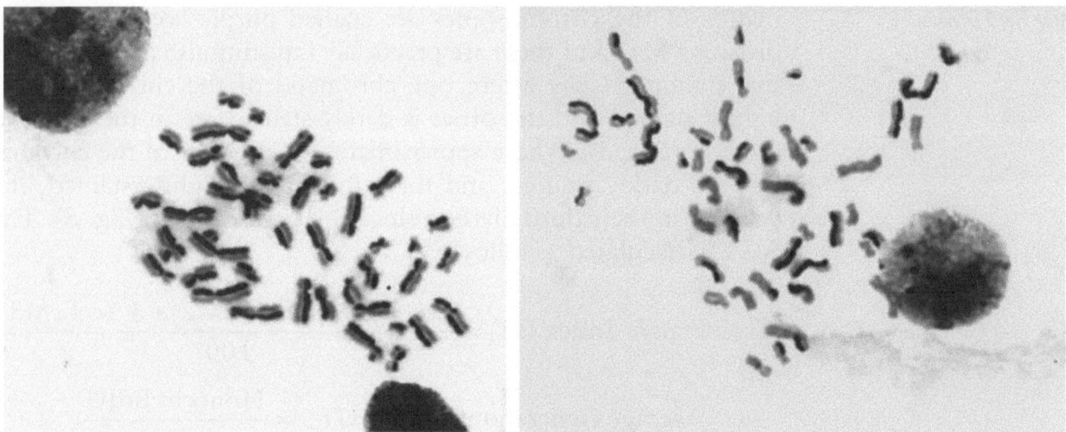

Fig. 3. Photographs of second and third division metaphases of human lymphocyte chromosomes stained with Hoechst 33258 and Giemsa (×1,000 magnification).

Often there will be very few cells and necrosis evident. Use a 40× objective, and move across the slide, avoiding clumped cells. Count cells with a smooth nucleus, irrespective of presence or absence of cytoplasm, and mitotic cells, regardless of whether the cell is suitable for chromosome analysis. The total number of cells counted per culture should be approximately 1,000. Calculate the MI per culture and per concentration as follows:

$$\text{Mitoticindex(MI)} = \frac{\text{Number of cells in mitosis}}{\text{Total number of cells observed}} \times 100\%$$

$$\text{Mitotic inhibition}(\%) = [1 - (\text{Mean}\,\text{MI}_{\text{T}} \,/\, \text{Mean}\,\text{MI}_{\text{C}})] \times 100\%$$

MI_T = MI in cells exposed to test substance, MI_C = MI in solvent control cells.

The MI in the vehicle controls should be in the region of 10% ideally, after metaphase arrest. A MI of less than 5% consistently in the vehicle controls indicates that growth could be improved and a repeat test may be warranted. Poor growth should be considered when evaluating the validity of results from a completed test. See Note 11 for alternative methods of measuring cytotoxicity and Note 3 for the use of cell lines.

3.9. Analysis of Chromosome Aberrations

The cytogenetic damage that is most easily seen using light microscopy and Giemsa staining is asymmetrical. This tends to prevent unlimited cell division and so is of limited long-term biological consequence. Nevertheless such damage represents the potential of test substances to induce persistent damage. For instance, chromosome exchanges can result in dicentric formation, where the re-arrangement contains two centromeres and is readily visible, or a balanced translocation where the chromosomes may appear normal using conventional staining techniques and allow normal cell division. Accumulation of translocations after exposure to ionizing radiation in vivo has been shown (18).

The slides of cells exposed to at least three concentrations of CPA, that have shown high (50% mitotic inhibition), medium, or low cytotoxicity, and the vehicle control are selected where appropriate levels of toxicity are observed. These slides may be examined by a person who will not score them, to ensure that there are sufficient cells at metaphase of a suitable quality. The metaphases should have chromosomes that are evenly stained, with very little cytoplasm visible and minimal overlapping of chromosomes, whilst ensuring that the majority of metaphases appear to be intact and not spread over too great an area. The centromeres and individual chromatids of each chromosome should be visible. Test substances may adversely affect chromosome morphology. If it is judged that it would be difficult to identify aberrations correctly, then it is worth getting a second opinion, since this is a subjective judgement. Substitution of slides of cells exposed to lower concentrations of test substance may be considered, and the reasons documented. The identity of the slides is concealed from the analyst, usually by applying a label with the experiment number and a random code to cover the existing label and a blank label on the back of the slide.

Slides are placed on the microscope stage the same way round each time. If electronic vernier positions are used, these are set to zero on a predetermined place on the slide. The slide is scanned in discontinuous steps to avoid analysing the same cell twice, using a low power objective (e.g. 10×). If a metaphase appears to be suitable for analysis, then an oil immersion objective is used, with a magnification of either 60× or 100×. Chromatic corrections of the highest quality are found in the plan-apochromatic range. The resolving power is expressed as numerical aperture (NA), and the highest possible is 1.4. Plan-apochromatic lenses allow separation of two objects approximately 0.25 μm apart.

The types of structural aberrations can be classified into six main, and two subsidiary, categories based on the ISCN classification (19). The definitions and some sub-divisions are given in the following table, with the ISCN abbreviations in italics. Examples of the chromatid and chromosome damage that may be seen are shown diagrammatically in Fig. 4a and 4b, a photograph of a cell exposed to CPA is shown in Fig. 5.

Damage	Chromatid	Chromosome
Gap	Non-staining region of a single chromatid, less than the width of the chromatid, in which there is minimal mis-alignment of the chromatid (*ctg*)	Non-staining region at the same locus of both the chromatids of a single chromosome, less than the width of the chromatid, in which there is minimal mis-alignment of the chromatids (*csg*)
Break	Discontinuity of a single chromatid in which there is a clear mis-alignment of one of the chromatids (*ctb*) Non-staining region of a single chromatid longer than the width of the chromatid (*ctdel*) Small single fragment appearing alone in the cytoplasm, considered to be a terminal deletion (*ctmin*)	Discontinuity, apparently at the same locus in both chromatids of a single chromosome, in which there is mis-alignment of the chromatids. This may be categorised according to whether the chromosome fragment is longer (*ace*) or shorter (*dmin*) than the width of one of the chromatids Non-staining region at the same locus of both the chromatids of a single chromosome, longer than the width of one of the chromatids (*csb*)
Exchange	Result of two or more chromatid lesions and the subsequent re-arrangement of chromatid material (*cte*). These may be between chromatids of different chromosomes (interchanges) or between or within chromatids of one chromosome (intrachanges). Even complex exchanges are classed as a single event (*cx*)	Result of two or more chromosome lesions and the subsequent relocation of both chromatids to a new position on the same or another chromosome (*cse*)
Pulverization	Cell contains both chromatid and/or chromosome gaps and breaks that are present in such numbers that they cannot be enumerated (*pvz*)	
Multiple aberrations	More than seven aberrations per cell, or too many aberrations to permit accurate analysis (mabs)	

a

Gap

Breaks

Assumed terminal deletion. Fragment may be displaced

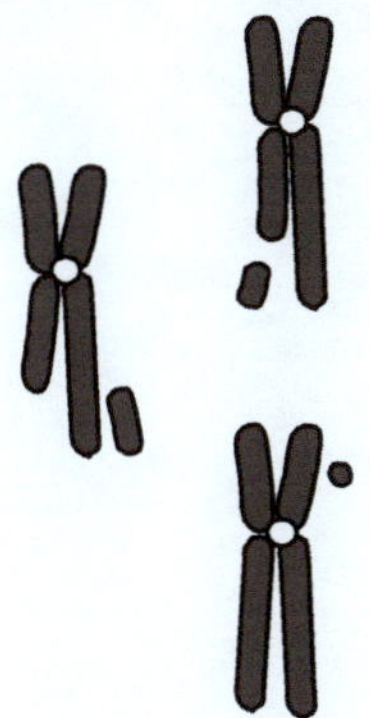

Exchanges; intra-chromosome

First example, intra-arm. Second two examples, inter-arms. Note single fragment closely associated with chromosomes as part of exchange

Exchanges; inter-chromosome

First example is an asymmetrical quadriradial. Second example is a monocentric triradial, resulting from an iso-chromatid/chromatid interchange and is more complex, but both examples are classed as single events

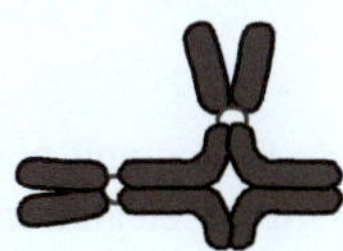

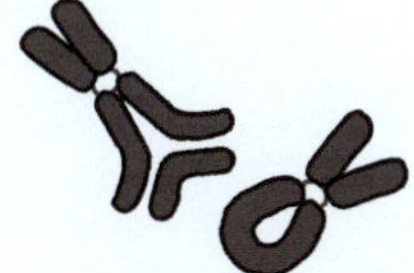

b

Gap

Breaks

Assumed terminal deletion. Fragment may be displaced, and smaller (double minute) or longer (acentric fragment) than width of chromatid arm.

Intra-chromosome exchanges

First example, an acentric ring. The second example is a centric ring which has an associated chromosomal fragment as part of the exchange.

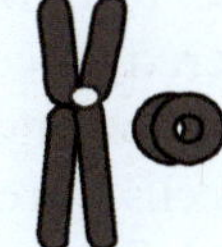

Inter-chromosome exchanges

A dicentric with an associated chromosomal fragment as part of exchange.

Fig. 4. (**a**) Examples of some types of asymmetrical chromatid aberrations. (**b**) Examples of some types of asymmetrical chromosome aberrations.

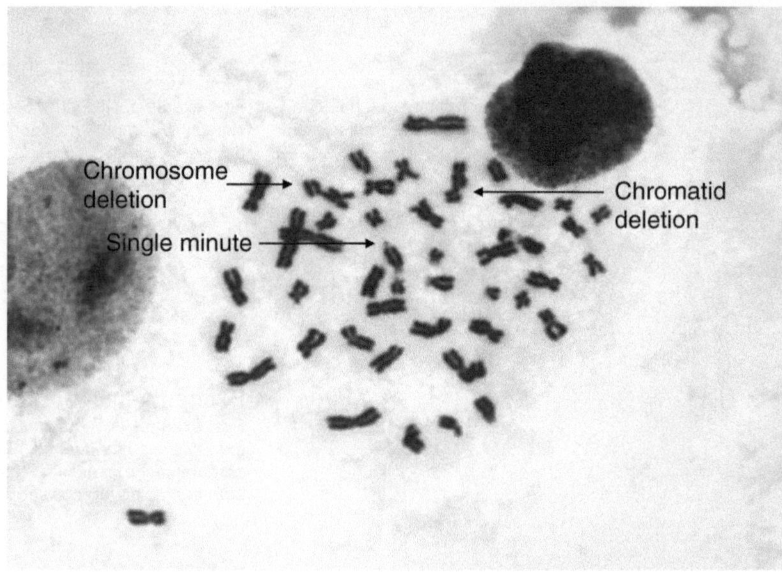

Fig. 5. Photograph of human lymphocyte chromosomes from cell exposed to CPA at 12.5 μg/mL, stained with Giemsa (×1,000 magnification).

100 Metaphases from each culture are analysed for chromosome aberrations, 50 per slide, making 200 per concentration of test substance. The exception is where 10 cells with structural aberrations (excluding gaps) have been noted on a slide, when analysis may be terminated. Only cells with 44–46 chromosomes are considered acceptable for analysis. Any cells with more than 46 chromosomes (that is, polyploid, hyperdiploid, or endoreduplicated cells) are recorded separately and may be quantified if required. The positions of all abnormal cells must be recorded. In the absence of a system to automatically record the position of all cells analysed, the area of the slide examined should be specified.

Various features of human chromosomes may appear like aberrations. Natural variation leading to increases in the length of secondary constrictions, for example, on the long arm of human chromosome 16, can occur, and these should be recognized as such. Satellites occur on human chromosome groups D and G. These satellites may be confused with breaks, and they may associate together at metaphase sometimes, giving the appearance of a chromatid exchange.

3.10. Analysis and Interpretation of Results

The culture is the experimental unit. After completion of analysis and decoding, the proportions of aberrant cells per culture (often expressed as a percentage, since 100 cells are routinely analysed) are tabulated as:

1. Cells with structural aberrations including gaps

2. Cells with structural aberrations excluding gaps (this information is used to draw a conclusion as to the clastogenic potential of a test substance as gaps may occur by non-genotoxic modes of action)
3. Polyploid, endoreduplicated, or hyperdiploid cells (An increase in polyploidy may indicate that a chemical has the potential to induce numerical aberrations, when use of the in vitro micronucleus test should be considered.)

The acceptance and evaluation criteria to determine the outcome of the test should be specified in advance. An acceptable test, assuming good cell growth, would be likely to have:

1. Homogeneity between replicate cultures (see below)
2. The proportion of cells with structural aberrations (excluding gaps) in vehicle control cultures should fall within the historical negative control range. In the absence of an historical negative control range, an approximate guide would be that the frequency of cells with aberrations should be <5%. This low frequency means that there may not be any aberrations seen in 50 or even 100 cells.
3. At least 160 cells out of an intended 200 should be suitable for analysis at each concentration of CPA, unless 10 or more cells per slide showing structural aberrations other than gaps only have been observed during analysis.

Structural chromosome aberrations are normally rare events and are distributed according to a Poisson distribution, which means that the variance is equal to the mean and the numbers of aberrant cells should vary between replicates to an extent that is compatible with binomial variation. This should be confirmed using a binomial dispersion test. Assuming homogeneity between cultures, the results from cultures can be combined to give a single proportion of aberrant cells per exposure group. A one-tailed Fisher's Exact test is used to compare each exposure group with the negative control (20).

The following evaluation criteria may be used for ascribing the potential of a test article to induce chromosome aberrations in a valid assay. They avoid ascribing small, random, statistical increases as a positive response, particularly when compared with zero values in the negative control cultures:

1. A proportion of cells with structural aberrations at one or more concentrations exceeds the concurrent vehicle control values and the historical negative control (normal) range (if available) in both replicate cultures.
2. A statistically significant increase in the proportion of cells with structural aberrations (excluding gaps) is observed ($p \leq 0.05$).

3. Evidence of a concentration-related trend in the proportion of cells with structural aberrations (excluding gaps) will be judged to support the conclusion. However this is not considered essential in the evaluation of a positive result.

Results that only partially satisfy the above criteria need to be dealt with on a case-by-case basis. Biological relevance must be taken into account, for example consistency of response within and between concentrations and between experiments, or effects occurring only at high or very toxic concentrations, and the types and distribution of aberrations. Generally, gaps and breaks are more common than exchanges, which are very rare events, and chromatid damage is more commonly seen than chromosome damage, after actively dividing cells have been exposed to a clastogen.

A test article that satisfies none of the above criteria may be considered not to have the potential to induce chromosome aberrations. The criteria to establish a negative response are more stringent, and two independent experiments are required currently.

CPA should elicit a positive response in the presence, and a negative response in the absence, of S-9 mix.

Chemicals may cause chromosome damage indirectly in mammalian cells, for example, by affecting cell division, rather than directly affecting DNA (21). If the primary effect of a chemical is on a sub-cellular target other than DNA, and there are multiple copies (e.g. aneuploidy) then the genotoxic effects may have a threshold below which damage will not occur (1). Detailed, comprehensive studies are necessary to demonstrate the occurrence of a threshold and the mode of action (e.g. ref. 22). The distinction between an indirect-acting and a direct-acting genotoxin is important in predicting in vivo risk arising from a positive in vitro assay.

For extrapolation to risk assessment for humans, the toxicokinetic profile, intended use and estimated exposure are important considerations, and the results need to be considered in context with the results of other genotoxicity tests.

4. Notes

1. Several transformed cell lines originated from the Chinese hamster, which has a relatively small number of distinctive chromosomes. Chinese hamster ovary (CHO) cells and Chinese hamster lung fibroblasts (CHL/IU) have been extensively used to assess potential chromosome damaging ability of a test substance (e.g. ref. 23). Transformed cell lines provide a readily available characterised population of cells that are easy to grow. Exposure to the test substance and removing it is easier from monolayers than suspension cultures, such as lymphocytes.

More comprehensive measures of cytotoxicity than measuring MI inhibition are available (see Note 11).

2. Those cells that lack functional p53, which is the case for many immortal cell lines, are not arrested at a cell cycle checkpoint. Cell cycle arrest provides an opportunity for DNA repair or for progression to apoptosis. Failure of cell cycle control is likely to result in a higher frequency of chromosomal aberrations induced by test substances than would be the case under normal circumstances. Cell lines appear easier to analyse than human chromosomes because they have small numbers of large chromosomes, for example, cells derived from the Chinese hamster, such as CHO (modal number in the region of 21, depending on the sub-line), CHL (modal number 25, depending on the sub-line). However, the karyotype is variable and experience is necessary before damage can be distinguished from natural variation accurately. Also, satellites and secondary constrictions are present, as in human cells.

3. Good tissue culture techniques and characterisation of transformed cell lines are critical for reliable results.
   (a) Clear records must be maintained of the culture procedures and characterisation, since the cells may be used many years after the master stock was received, and it is important to be able to trace to the source of the cells.
   (b) Buy from a reputable supplier, unless the provenance of the cells is known. Evidence that proves authenticity and adequate characterisation should be readily available. A problem with a culture split, for example, may mean that very few of the cells may have survived from one passage to the next, meaning that the subsequent population of cells is less representative of the cells from which it was derived.
   (c) The cells should be used at as early a passage number as possible after receipt. This should avoid the possibility of genetic drift which may manifest itself as a changed response to known genotoxins. Also it is more difficult to analyse cells for chromosome aberrations if there is a change in the chromosome constitution.
   (d) On receipt, there will need to be a limited number of passages of the cells to increase their number. Vials containing the cells will be frozen down into liquid nitrogen and designated as the master stock. To produce the working cultures, one master stock vial is grown up with the aim of producing as many cells using as few passages as possible. These will be frozen down as a working stock.
   (e) The success of storage of the cells depends on reliable maintenance of the liquid nitrogen facility, for example,

the Dewars are checked and topped up regularly as necessary. Viability from liquid nitrogen should be in the region of 90%.

(f) If more than a single cell type is used in a laboratory, cross-contamination can happen surprisingly easily. The most vigorously growing cell type will soon predominate. Use separate equipment, etc., for each cell line where possible and do not handle different cell lines at the same time. Apparent shifts in karyotype in a cell line should be investigated immediately, bearing this possibility in mind.

(g) Information needed before testing, besides the type and source of the cells and their suitability for the intended purpose, is given below.

- Number of passages and methods for maintenance of cultures.
- The chromosome number distribution, including the modal number and the frequency of polyploidy. Count chromosomes from approximately 200 cells at metaphase.
- The karyotype, including the sizes of chromosomes, the positions of centromeres, and the presence of marker chromosomes.
- MI, PD (population doubling), and AGT.
- Evidence that the master and working stock is mycoplasma free, by testing a representative vials.
- The frequency of structural chromosome aberrations in the negative and positive controls.

(h) Regular microscopic examination of cells means that warning signs, such as contamination, or cells rounding up and detaching from the surface of the culture vessel, can be detected and a strategy developed for dealing with any potential problem.

(i) Cells undergo a lag period after establishment, especially monolayer cultures, and need a period of recovery of at least 12 h before exposure to test substances.

4. The table below illustrates a study design for a regulatory assay. The timings are based on a cell cycle time of 12–14 h. A preliminary experiment determines the toxicity profile of an unknown test substance. The potential to the test substance to induce structural chromosome aberrations is assessed in two main independent experiments. If the first main experiment is valid and the test substance is clearly positive, a second experiment is unnecessary. Otherwise the second experiment comprises a continuous exposure for 1½ the AGT in the absence of S-9, since a number of chemicals have been reported as only exerting positive effects following prolonged treatment (24, 25).

Some chemicals (e.g. nucleoside/tide analogs or nitrosamides) may be more readily detected by treatment/sampling times longer than 1½ the AGT (26) when provision needs to be made for sampling at later times.

Positive controls are included in the main experiments to check that clastogenic effects can be detected in the presence and absence of S-9 mix. These include Mitomycin C (CAS no. 50-07-7) and 4-nitroquinoline N-oxide (CAS no. 56-57-5) in the absence of S-9 mix.

Experiment	S-9	Duration of exposure (h)	Hours between end of exposure and harvest
Range finder	–	3	17
	+	3	17
	–	20	0
Experiment 1	–	3	17
	+	3	17
Experiment 2	–	20	0
	+	3	0

5. The maximum concentration of an unknown test substance is 10 mM or 5 mg/mL, whichever is lower (3). For freely soluble test articles, where the molecular weight is unknown, this means that the highest concentration tested will be 5,000 μg/mL. Within the pharmaceutical industry there is a debate as to whether these concentrations are too high, as they may render the tests over-sensitive at the expense of specificity, interfere with S-9 enzyme activity and generally exceed human exposure by a large margin. The high rate of unique positive findings in the in vitro mammalian cell tests is likely to lead to a revision of the guidelines (27).
6. Changes in osmolality of more than 50 mOsm/kg and fluctuations in pH of more than one unit can give rise to chromosome aberrations (28, 29). Where the molecular weight of the test substance is unknown, osmolality should be assessed. The effect of the test substance on the pH of the culture medium should always be assessed. This may be limited to noting changes in the colour of the culture medium. If there are changes in the colour, or it is impossible to ascertain accurately, for example, owing to the colour of the test substance or haemolysis, then pH should be measured.
7. Concentrations should be separated by no more than a factor of √10. More closely spaced concentrations are advisable if the concentration-effect relationship is steep. The range of concentrations should include those associated with high, medium, and low toxicity.

8. Induced rat liver S-9 mix is the regulatory standard and has been used extensively in vitro. S-9 mix has elevated levels of P450 (CYP) enzymes. These catalyse oxidative metabolism to electrophilic metabolites, that may react with nucleic acid. S-9 mix is toxic to mammalian cells. S-9 fraction is usually used at concentrations within the range 1–10% v/v in the final test medium, with exposure to test substances in the presence of S-9 mix being limited to between 3 and 6 h. Despite wide use, S-9 mix only approximates to simulating metabolism in vivo. Structural knowledge of potential metabolites should be considered, such as using CYP2E1 proficient systems for small molecules and SULT proficient systems for amino, amido, and nitroarenes (30). Optimized or alternative metabolic activation sources, such as human microsomes, or genetically engineered test systems may be appropriate (30). Cells with inherent metabolic capacity, where potential mutagens are metabolized near the nuclear material, such as the human hepatoma cell line Hep G2 (31) or MCL5 (32) may be advantageous, particularly if the toxicokinetic profile of the test substance is known, but their response to non-genotoxins has not been fully assessed. This is a developing area.
9. When the drop of cell suspension comes into contact with the slide, the nuclear membranes of the cells at metaphase disintegrate and the chromosomes spread out, analogous to throwing dice out of a pot. The objective of successful slide preparation is to ensure that the metaphase spreads are kept intact, but the chromosomes are separate and there is as little cytoplasm as possible. This may be achieved by a number of cunning ruses. Firstly the slides must be free of grease. Several drops of 45% (v/v) aqueous acetic acid may be added to each suspension to enhance chromosome spreading. The drops of cell suspension may be dropped from a height onto to the microscope slides. Spreading may be improved also by blowing the drops, as they spread across the slide. Slides may be flamed to improve quality. Practice will be rewarded by high quality metaphase spreads. This will make it easy to analyse the chromosome damage.
10. The only measure of toxicity that is currently widely used for whole blood cultures is the MI, because T-lymphocytes respond to PHA stimulation and these form only a tiny fraction of the cells, the majority being red blood cells. The MI has the advantage of being linked with analysing cells at metaphase for chromosome damage, and most mutagens will delay cell division and cause MI reduction. However, account is taken only of cells that have survived to harvest and only the proportion of these surviving cells that are dividing during the period of metaphase arrest. Therefore cytotoxicity may be underestimated and

is subject to considerable variability. Also, measurement of PHA-stimulated lymphocyte sub-populations in culture reveals their behaviour to be complex (33) and the use of extended cultures of human T-lymphocytes has been proposed (34). Development of alternative ways of measuring cytotoxicity in lymphocytes, for example, using flow cytometry, would be a step in the right direction.

11. For cell lines, where the majority of cells divide and form the target population, more sensitive and reliable methods of measuring cytotoxicity are available, based on cell counts. The relative cell count (RCC) is a ratio of the final cell count in the exposed compared with the control cultures. It is not presented as it is not as sensitive as the two formulae below using standard protocols, and so offers no advantage. The relative increase in cell counts (RICC) and relative population doubling (RPD) take into account what has happened from just before the start of exposure to harvesting. Although neither can be used to distinguish between cell death and cell cycle delay, there is a measure of the resulting potential reduction in cell numbers. The RICC is the most sensitive measure of cytotoxicity, since the outcome is not expressed as a log ratio. The difference is greatest at the middle of the relative survival range, that is, where choosing the highest concentration from which to analyse chromosome damage. This is shown in Fig. 6, using a theoretical example, when the highest concentration selected would be 2.0 mM or 4.75 mM, depending whether RICC or RPD was used. The difference between RICC and RPD, which is independent of the number of cells plated, may not be as apparent in practice, owing to variability.

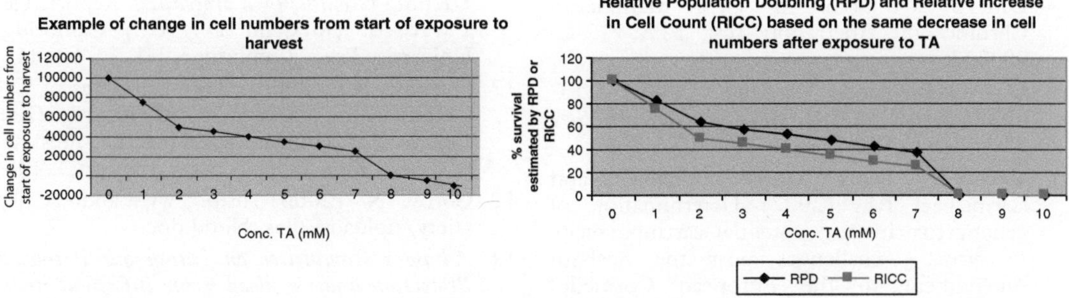

Fig. 6. Comparison of different methods of measuring cytotoxicity based on the same decrease in cell numbers after exposure to test article (TA). Example based on all cultures containing 5.0×10^{-5} cells at the start of exposure, an increase to 15.0×10^{-5} cells in the vehicle control at harvest, and a progressive decrease in the number of cells with increasing concentrations of TA, including cell loss at the highest concentrations (*left hand graph*). The *right hand graph* shows the effect of applying the formulae to estimate RPD and RICC to these numbers. The difference is greatest at approximately 50% survival, apparently achieved by 2.0 mM for RICC, 4.75 mM for RPD.

$$\text{Population Doubling (PD)} = [\log(\text{Post-treatment cell number} / \text{Initial cell number})] / \log 2$$

$$\text{RPD} = \frac{\text{No. of PDs in exposed cultures}}{\text{No. of PDs in negative control cultures}}$$

$$\text{RICC} = \frac{\text{Increase in number of cells in exposed cultures (final} - \text{starting)}}{\text{Increase in number of cells in negative control cultures (final} - \text{starting)}} \times 100$$

Acknowledgements

The author would like to thank J. Whitwell and M. Lloyd for their helpful comments, and Covance Laboratories Ltd for providing the photographs.

References

1. Committee on Mutagenicity of Chemicals in Food, Consumer Products and the Environment (2000) *Guidance on a Strategy for Testing of Chemicals for Mutagenicity. Introduction* Department of Health, UK pp 5–7.
2. European Agency for the Evaluation of Medicinal Products (1995) ICH Topic S 2 A. Genotoxicity: *Guidance on Specific Aspects of Genotoxicity Tests for Pharmaceuticals.* ICH Harmonised Tripartite Guideline.
3. OECD (1997) 'Genetic Toxicology: *In vitro* mammalian Cytogenetic Test'. In OECD *Guidelines for the testing of chemicals.* OECD Paris, Test Guideline 473.
4. US EPA (1998*) Health Effects Test Guideline* OPPTS 870.5375 *In Vitro* Mammalian Chromosome Aberration Test. EPA 712-C-98–223.
5. EC Commission Directive 2000/32/EC (2000) Annex 4A B.10. *Mutagenicity – In Vitro Mammalian Chromosome Aberration Test.*
6. Tweats D.J., Scott A.D., Westmorland C. and Carmichael P.L. (2007) Determination of genetic toxicity and potential carcinogenicity *in vitro* – challenges post the Seventh Amendment to the European Cosmetics Directive. *Mutagenesis* **22,** 5–13.
7. Kirkland D., Pfuhler S., Tweats D. *et al* (2007) How to reduce false positive results when undertaking *in vitro* genotoxicity testing and thus avoid unnecessary follow-up animal tests: Report of an ECVAM workshop. *Mutat. Res.* **628**, 31–55.
8. Nowell, P.C. (1960) Phytohaemagglutinin: an initiator of mitosis in cultures of normal human leukocytes. *Cancer Research*, **20**, 462.
9. Evans H.J. and O'Riordan M.L. (1975) Human peripheral blood lymphocytes for the analysis of chromosome aberrations in mutagen tests. *Mutation Research*, **31**, 135.
10. Evans H.J. (1984) Human peripheral blood lymphocytes for the analysis of chromosome aberrations in mutagen tests. In *Handbook of Mutagenicity Test Procedures* (Kilbey B.J. *et al* eds) Elsevier Science Publishers, pp 407–427.
11. Scott D., Dean B.J, Danford N.D. and Kirkland D.J. (1990) Metaphase chromosome aberration assays *in vitro.* In *Basic Mutagenicity Tests: UKEMS recommended procedures.* Report. Part 1 revised. (Kirkland D.J. ed.) Cambridge University Press, Cambridge, UK. pp 62–84.
12. Freshney R.I. (2006). Basic Principles of Cell Culture. In *Culture of Cells for Tissue Engineering.* (Vunjak-Novakovic G., Freshney R.I. eds.) John Wiley & Sons, Inc., pp 3–22.
13. Corby N (2000) http://www.abdn.ac.uk/safety/uploads/files/blood.doc.
14. *Advisory Committee on Dangerous Pathogens Protection against blood borne infections in the workplace: HIV and hepatitis.* The Stationery Office, 1996, ISBN 0113219539.
15. Hone P., Edwards A., Lloyd D., Moquet J. (2005) The yield of radiation-induced chromosomal aberrations in first division human lymphocytes depends on the culture time. *Int J Radiat. Biol.* **81**, 523–529.

16. Goto K., Maeda S., Kano Y. and Sugiyama T. (1978) Factors involved in differential Giemsa-staining of sister chromatids. *Chromosoma* (Berl.) **66**, 351–359.

17. Marshall R., Tricta F., Galanello R. *et al.* (2003) Chromosomal aberration frequencies in patients with thalassaemia major undergoing therapy with deferiprone and deferoxamine in a comparative crossover study. *Mutagenesis* **18**, 457–463.

18. Edwards A.A. *et al.* (2005) Review of translocations detected by FISH for retrospective biological dosimetry applications. *Radiation Protection Dosimetry* **113**, 396–402.

19. An International System for Cytogenetic Nomenclature (1995). *Report of the Standing Committee on Human Cytogenetic Nomenclature* S. (Hardnen D.G. and Klinger H.P. eds.) Karger A.G., Basel (Switzerland).

20. Richardson C., Williams D.A., Allen J.A., Amphlett G., Chanter, D.O. and Phillips B. (1989) Analysis of data from *in vitro* cytogenetic assays. In *Statistical Evaluation of Mutagenicity Test Data: UKEMS sub-committee on guidelines for mutagenicity testing. Report.* Part 111. (Kirkland D.J. ed.) Cambridge University Press, Cambridge, UK. pp 141–154.

21. Hilliard C.A., Armstrong M.J., Bradt C.I., Hill R.B. Greenwood S.K. and Galloway S.M. (1998) Chromosome aberrations *in vitro* related to cytotoxicity of nonmutagenic chemicals and metabolic poisons. *Environ and Mol. Mutagenesis* **31**, 316–326

22. Doak S.H., Jenkins G.J.S., Johnson G.E. *et al* (2007) Mechanistic influences for mutation induction curves after exposure to DNA-reactive carcinogens. *Cancer Res.* **67**, 3904–3911.

23. Ishidate M.Jr., Harnois M.C. and Sofuni T. (1988) A comparative analysis of data on the clastogenicity of 951 chemical substances tested in mammalian cell culture. *Mutat. Res.* **195**, 151–213.

24. Ishidate M Jr (1987) *Chromosomal aberration test in vitro* (Ed) L.I.C. Inc. Tokyo.

25. Galloway S M, Armstrong M J, Reuben C, Colman S, Brown B, Cannon C, Bloom A D, Nakamura F, Ahmed N, Duk S, Rimpo J, Margolin B H, Resnick MA, Anderson B and Zeiger E (1987) Chromosome aberrations and sister chromatid exchanges in Chinese hamster ovary cells: Evaluations of 108 chemicals. Environ Mol *Mutagenesis* **10**, Suppl 10, 1–175.

26. Galloway S M, Aardema M J, Ishidate M, Ivett J L, Kirkland D J, Morita T, Mosesso P, Sofuni T (1994) Report from working group on *in vitro* tests for chromosomal aberrations. In: Sheila M. Galloway (Ed), Report of the International Workshop on Standardisation of Genotoxicity Test Procedures. *Mutation Research* **312**, 241–261.

27. European Agency for the Evaluation of Medicinal Products (2006) ICH Topic S 2 R1: Final concept paper: *Guidance on genotoxicity testing and data interpretation for pharmaceuticals intended for human use.*

28. Scott D, Galloway S M, Marshall R R, Ishidate M, Brusick D, Ashby J and Myhr B C (1991) Genotoxicity under extreme culture conditions. A report from ICPEMC Task Group 9. *Mutation Research* **257**, 147–204.

29. Brusick D (1986) Genotoxic effects in cultured mammalian cells produced by low pH treatment conditions and increased ion concentrations. *Environ Mutagenesis* **8**, 879–886.

30. Ku W.W., Bigger A., Brambilla G. *et al* (2007) Strategy for genotoxicity testing – metabolic considerations. *Mutat. Res.* **627**, 59–77.

31. Knasmuller S., Parzefall W., Sanyal R. *et al* (1998) Use of metabolically competent human hepatoma cells for the detection of mutagens and antimutagens. *Mutat. Res.* **402**, 185–202.

32. Parry E.M., Parry J.M., Corso C., *et al* (2002) Detection and characterisation of mechanisms of action of aneugenic chemicals. *Mutagenesis* **17**, 509–521.

33. O'Donovan M.R.O., Johns S. and Wilcox P. (1995a) The effect of PHA stimulation on lymphocyte sub-populations in whole-blood cultures. *Mutagenesis* **10**, 371–374.

34. O'Donovan M.R.O., Freemantle M.R., Hull G. *et al* (1995b) Extended cultures of human T-lymphocytes: a practical alternative to primary human lymphocytes for use in genotoxicity testing. *Mutagenesis* **10**, 189–201.

Chapter 6

The Interpretation and Analysis of Cytogenetic Data

Natalie Danford

Abstract

Chromosome aberration analysis has been the basis of one of core tests in genetic toxicology since guidelines were first established (DHSS (1981) Guidelines for the Testing of Chemicals for Mutagenicity. Prepared by the Committee on Mutagenicity of Chemcials in Food, Consumer Products, and the Environment, Department of Health and Social Security. Report on Health and Social Subjects, No. 24. Her Majesty's Stationery Office, London; IPCS (1985) Guide to short-term tests for detecting mutagenic and carcinogenic chemicals prepared for the IPCS by the International Commission for Protection against Environmental Mutagens and Carcinogens. Geneva, WHO). The technique consists of microscope examination of preparations of chromosomes, usually mammalian, for clastogenicity (chromosome breakage events), and agents which induce such changes are considered genotoxic.

There are a number of different types of aberrations, and within types, considerable variation in their appearance. This chapter addresses aberration classification, their appearance, frequency and fate, and the range within aberration types, potential mis-classifications, and data recording and interpretation.

Key words: Chromosome, Chromatid, Chromosome aberration, Clastogenicity, Microscope analysis, Karyotype, ISCN

1. Introduction

Chromosome aberrations refer to changes in the visible structure of chromosomes following double-strand breakage. The use of this test in genetic toxicology (1, 2) and the preparation of material for microscope examination is described elsewhere in this volume, while this section is concerned with the observation, interpretation, and recording of aberrations.

It is well accepted that there is a subjective component in aberration analysis, and experienced analysts observing the same damaged chromosome may give it a different classification. However, this is unlikely to be the major cause of differences between analysts, and both selection of cells for analysis and recognition that

James M. Parry and Elizabeth M. Parry (eds.), *Genetic Toxicology: Principles and Methods*, Methods in Molecular Biology, vol. 817, DOI 10.1007/978-1-61779-421-6_6, © Springer Science+Business Media, LLC 2012

there is damage within a cell are likely to play a greater part in this variation.

The structure of this chapter does not conform to the standard outline of materials, methods, and so on, as these are described in the chapter on in vitro chromosome aberrations assays. Instead, it covers the following topics: aberration formation, classification and characteristics (including gaps), fate of aberrations, data recording, routine scoring and mis-classifications, toxicity, numerical aberrations, and data analysis and interpretation.

2. Aberration Formation

There is a substantial range of visible changes that may be observed, and a basic appreciation of how they arise is of considerable help in interpretation. Chromosome morphology and the appropriate terms are shown in Fig. 1. Aberrations which affect both sister chromatids are referred to as chromosome-type, while those affecting only one are called chromatid-type. However, these terms can be confusing as the term chromosome aberration can be applied to all types in general, and also, as many aberrations are caused by two breaks, they might appear to be chromosome-type aberrations when they are in fact two separate chromatid-type breaks at or near the same locus.

Aberrations fall into two main categories, breaks and exchanges, following double-stranded breaks, producing so-called "sticky ends" at both ends of the breaks, which are attracted to each other. If there is one break in a cell, there are only two sticky ends which

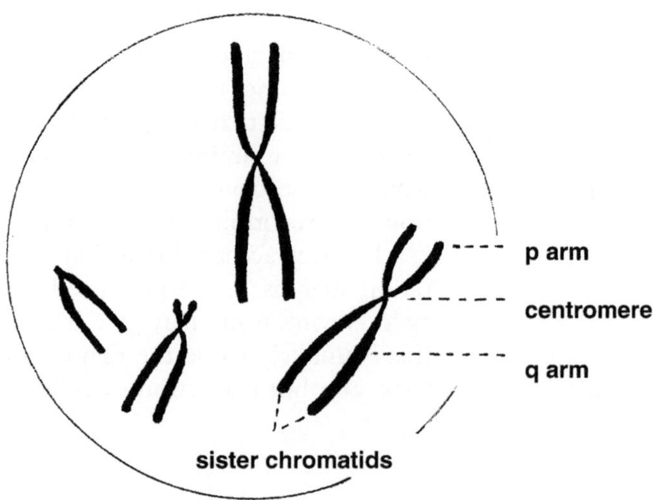

Fig. 1. Chromosome structure and classification: clockwise from top, metacentric, submetacentric, acrocentric, telocentric.

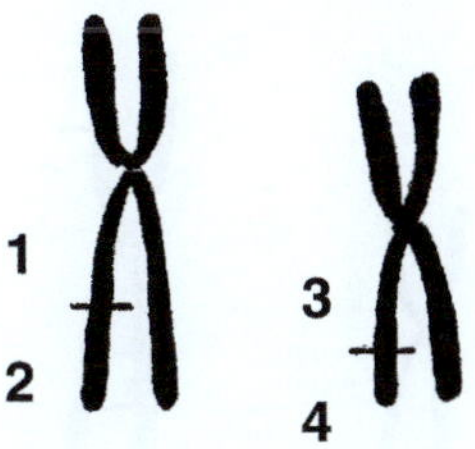

Fig. 2. Formation of exchange figures following two breaks, producing four sticky ends. Reunion is between 1 + 3 and 2 + 4 or 1 + 4 and 2 + 3.

will either not have rejoined when the cells are observed, giving a visible aberration, referred to as a break or deletion, or have rejoined, in which case no damage is observed. If there is more than one break in an individual cell, there will be twice as many sticky ends as breaks, and these may join in other than the original arrangement, giving a configuration which can then be identified down the microscope. These are referred to as exchanges because the sticky ends have rejoined differently (Fig. 2).

3. Classification and Characteristics

There are a number of different methods of classifying chromosome aberrations, of which the most practical is based on the appearance of the aberrations down the microscope, following interactions between the sticky ends (3, 4). This does not reflect the frequency of the aberration types but does greatly aid identification and depends on the relative positions of the breaks. If they are on two different chromosomes, the resulting aberration is described as an interchange (Fig. 3a). If they are on separate arms of the same chromosome, they are referred to as inter-arm intra-changes (Fig. 3b), and on the same arm of the same chromosome as intra-arm intra-changes (Fig. 3c, d). Each of these exchange events has a distinctive appearance, as described below. Simple breaks form a further category (Fig. 3e).

Before giving further details of the appearance of the aberrations, two systems by which the initial breaks are produced need to be considered (3). First, some agents, notably ionising radiation, produce double-strand breaks immediately on contact with the chromosomes. These will give rise to aberrations at the first metaphase after treatment, but if the breaks occurred before DNA synthesis took place, the damage is replicated and the resulting aberrations are chromosome-type (Fig. 4a). If the breakage occurred after DNA synthesis, then the aberrations are chromatid-type (Fig. 4b).

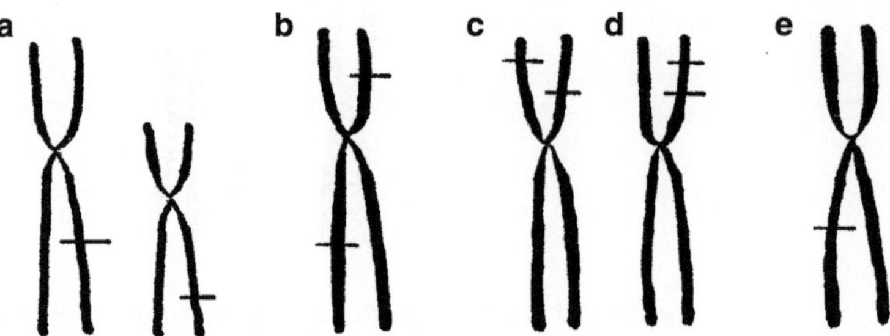

Fig. 3. Classification of chromatid aberrations based on position of break points. (**a**) Interchange (**b**) Inter-arm intra-change (**c**) Inter-chromatid intra-arm intra-change (**d**) Intra-chromatid intra-arm intra-change (**e**) Deletion (single break).

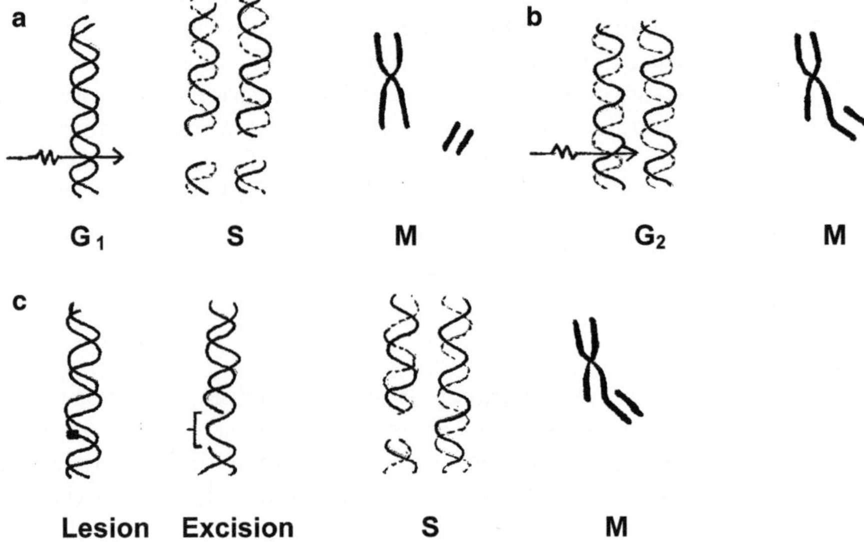

Fig. 4. Formation of aberrations. *Upper line*, ionising radiation and radiomimetic chemicals (**a**) Treatment during G1. (**b**) Treatment during G2. *Lower line* (**c**) S-phase-dependent chemicals.

The second mechanism, produced by the majority of chemicals, initially produces a lesion in the DNA, which is converted to a single-strand break during repair, when a section of DNA is removed. If DNA synthesis occurs before the repair is complete, this single-strand break can be converted to a double-strand break, leading to the production of a visible aberration. For this reason, the production of chromosome aberrations requires the cells to pass through an S-phase for aberration formation. In addition, these aberrations will be only chromatid-type at the first metaphase after development (Fig. 4c).

The fate of aberrations, following either type of induction, will be discussed following the descriptions of the full range of aberrations. But while only chromatid-type aberrations are produced initially from S-phase dependent chemical damage, analysts should be aware that after a further cell cycle, some of these aberrations can be converted into derived chromosome-type aberrations. The derived aberrations are identical in appearance to directly produced chromosome aberrations, although there may be differences such as the absence of an accompanying fragment. With the standard sampling time of one and a half cell cycles, the overwhelming majority of aberrations will be chromatid-type, but a small number of chromosome aberrations may be observed. In human lymphocytes, there is also the possibility of the occurrence of chromosome-type aberrations which were induced in vivo.

All four categories of aberrations occur in both chromosome- and chromatid-type forms and will be described below, without reference to their mode of induction, and with the chromosome- and chromatid-types detailed separately. Figs. 5–9 show simplified diagrams of their formation; more detailed illustrations can be found in refs. 3 and 4. It is important to be aware that following the induction of a chromatid aberration, the sister chromatids remain closely aligned until anaphase, a situation which is actually very helpful for microscope analysis, as it maintains the chromosome

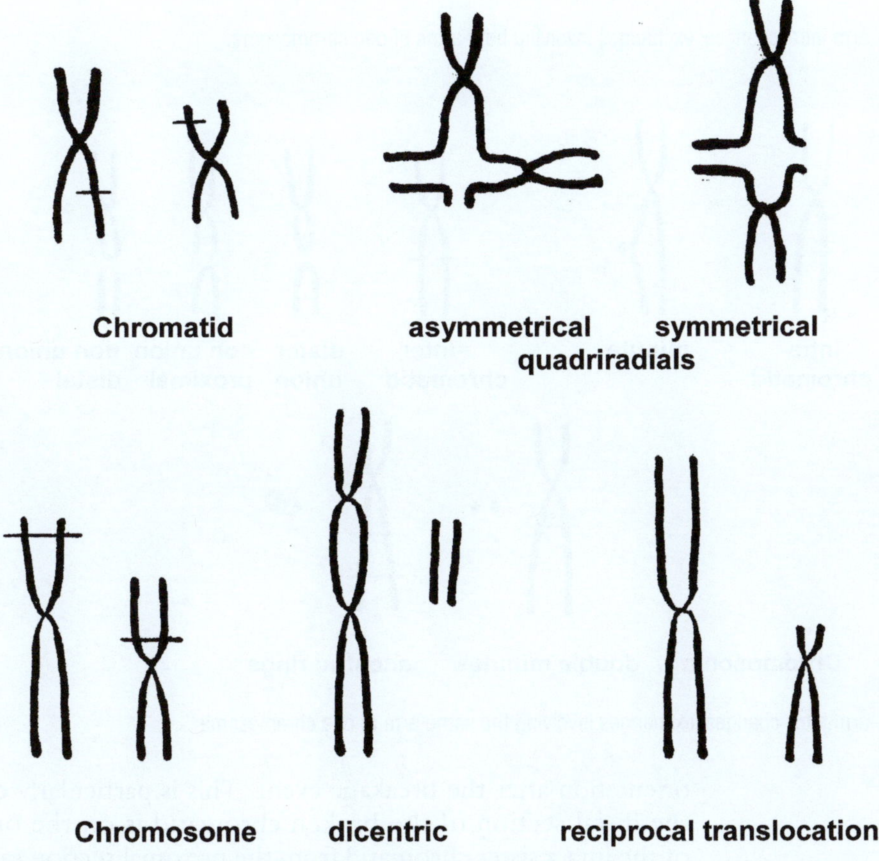

Fig. 5. Interchanges: exchanges involving two chromosomes.

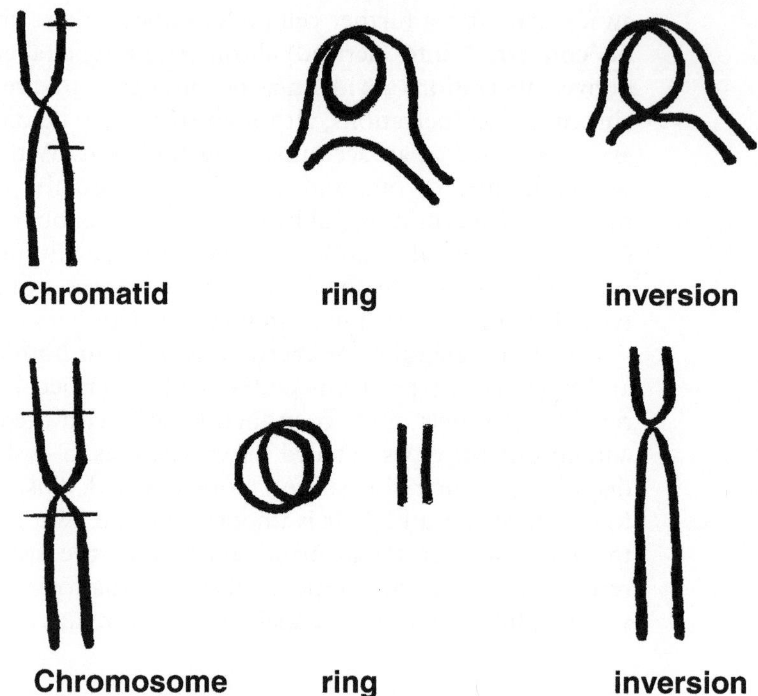

Fig. 6. Inter-arm intra-changes: exchanges involving both arms of one chromosome.

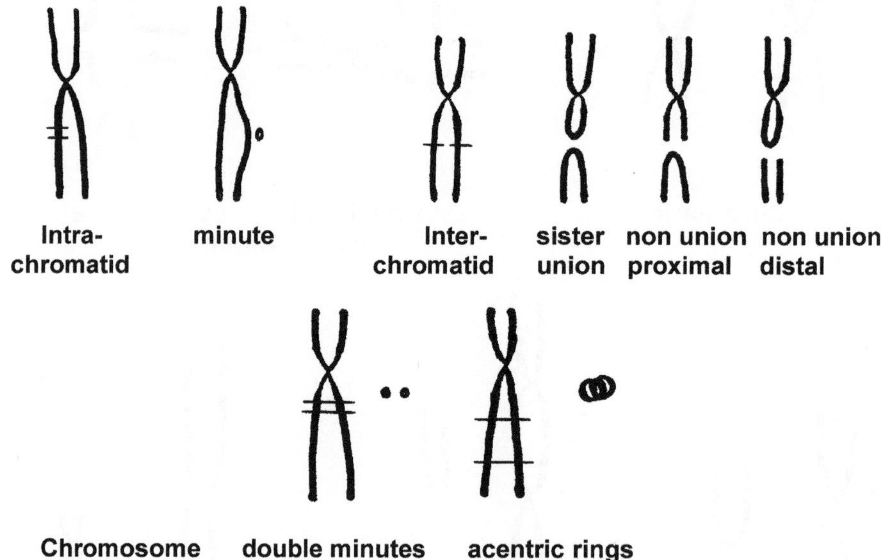

Fig. 7. Intra-arm intra-changes: exchanges involving the same arm of one chromosome.

orientation after the breakage event. This is particularly evident if the distal section of the broken chromatid is on the other side of the intact sister chromatid from the proximal section (as in "M" in Fig. 4b, c). Many of the chromatid events in Figs. 5–8 could be easily missed if the acentric parts of incomplete exchanges

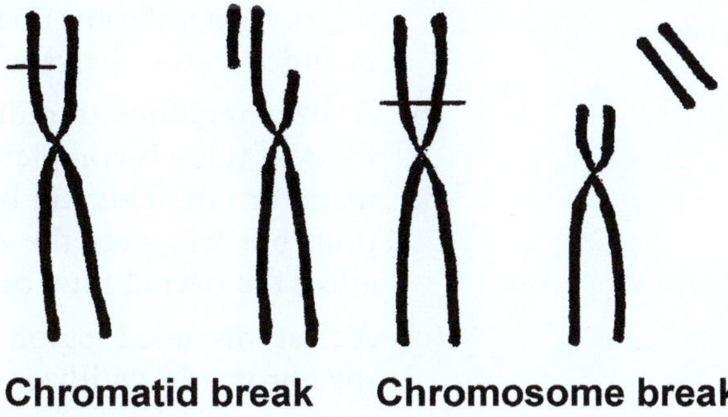

Fig. 8. Single break deletions.

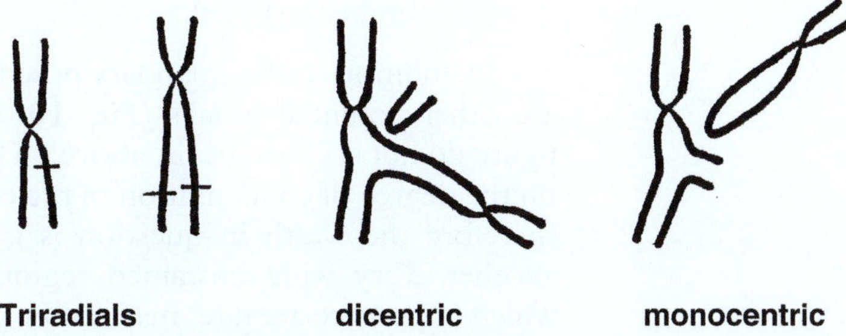

Fig. 9. Triradial exchange figures produced from two breaks on one chromosome and one break on a second chromosome.

or breaks were in another part of the cell, particularly if the breaks were close to the ends of the chromosomes.

Analysis for chromosome breakage is almost always carried out on unbanded chromosomes. Additional damage such as translocations and other rearrangements can be found by using banded chromosomes, but these are relatively rare events and considerably more time is required for this more detailed analysis. Some aberrations can actually be obscured by the banding.

4. Gaps

As well as true aberrations, there is a class of special aberrations which are not usually included in the total of aberrant cells determining whether a particular test agent is positive or not, and perversely can also be rather difficult to detect and classify (3, 4). These are referred to as gaps and constitute unstained regions on the chromosomes with no or minimal misalignment with the rest of the chromosome. Some laboratories use the chromatid width as a measurement to decide on the classification. Thus the following situations can arise:

(a) A clear unstained region, smaller than a chromatid width, with minimal misalignment is classified as a gap (chromatid or chromosome). These are enumerated, but not included in the aberration totals.

(b) A fuzzy region with reduced staining, where it is difficult to see if there is complete lack of staining across the chromatid or not. This will be an equivocal gap/normal situation, but whatever the decision, the cell is not going to affect the overall total of aberrant cells.

(c) A clear unstained region either the same or just wider than the chromatid width or with slight misalignment. This is either a gap or a break, and the decision could make the difference between a negative/equivocal/positive result, particularly if there are several cells like this in the total number analysed.

In addition to the quandary posed by (b) and (c) above, there are other potential pitfalls (Fig. 10; note that the letters in the figure do not refer to the list above). The chromatid width depends on the degree of condensation of the chromosomes in the cell, and therefore the width in question is going vary from one cell to another. Very wide unstained regions without any dislocation, which in practice are rare, may not be gaps at all, but intra-changes (Fig. 10d) (4). The chromatid width criterion is best applied as shown in Fig. 10c. Here although there is only slight misalignment, the acentric part of the chromosome has moved away from the centric part, widening the unstained region, so that a break seems the more likely event.

What is misalignment? With clear, sharp chromosomes, it may be possible to see whether the points where the staining ceases are

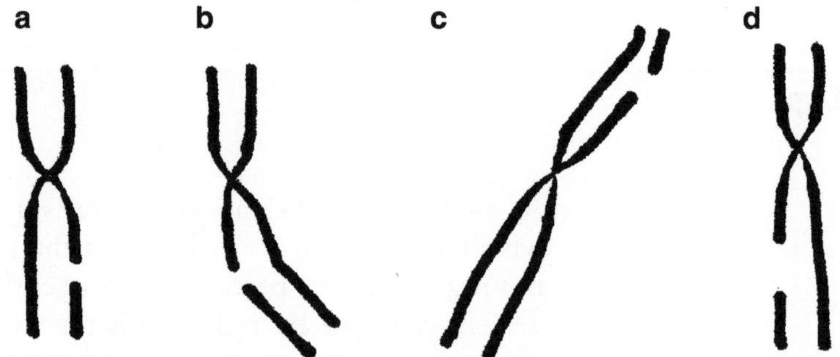

Fig. 10. Interpretation of chromatid gaps and breaks. (**a**) Unequivocal gap. (**b**) Dislocation but no misalignment, classify as gap. (**c**) Misalignment, gap wider than the chromatid width: classify as break. (**d**) Very rarely seen, no dislocation or misalignment, considerably wider than chromatid width. See text for discussion.

aligned or not, but it is often very difficult to determine. It has also been shown that some apparent gaps are in fact breaks, and vice versa (5, 6).

Having thus described gaps, it should be pointed out that normally at least nine out of ten gaps or breaks fall clearly and unequivocally into one or other classification. There are undoubtedly some situations where the only recourse is to not count the cell at all rather than mis-classify it. Otherwise, it is generally better to err on the side of caution and score the event as a gap if it is equivocal.

5. Fate of Aberrations

When considering the fate of aberrations, it is important to bear in mind that the preparations seen down the microscope represent a particular moment in time, when the cells were harvested and fixed. Otherwise they would have continued through the cell cycle, and the aberrations would have developed further. This is clearly illustrated by the incomplete nature of many chromatid exchanges, that is, where not all the sticky ends have rejoined. Incomplete exchanges are normally recorded as if they were complete events, apart from sister union, for which many laboratories record non-union proximally or distally as appropriate. Simple breaks are always incomplete events at the point of observation. It should be noted that chromosome-type events, whether induced directly or derived from chromatid aberrations, are not observed in an incomplete form.

There are several possible outcomes of the induction of aberrations. The aberrant cell may die as a result of the loss or damage of vital sections of the genome or problems during cell division, particularly if a bridge is formed during anaphase. The aberrations may no longer be visible in subsequent mitoses. Some chromatid aberrations can be converted into derived chromosome-type aberrations. The various consequences are shown in Table 1, which make it clear that the majority of aberrations are visible for one metaphase only.

While this emphasises the importance of the timing from treatment to harvest, so that first division metaphases are observed, this has to be balanced against the probability of cell cycle delay caused by the test agent, particularly at toxic doses which could result in examination of metaphases from cells which have not passed through S-phase. The standard harvest time, representing one and a half cell cycles if there is no delay, is considered to be the most appropriate time to take these opposing factors into account (7).

Table 1
Fate of aberrations

Event	Original aberration (and outcome)	
(1) Aberration visible for one division only	Chromatid break Chromatid minute (fragment) Non-union proximal Chromatid intra-change (?) Paired fragment Double minute Translocation	Acentric part of aberration lost at anaphase, remaining chromosome shows no detectable change. (Cell may not survive depending on lost/rearranged genetic material.)
(2) Conversion to chromosome- type aberration (50% of cells in each case)	Quadriradial: asymmetrical ⇒dicentric (no fragment) symmetrical ⇒reciprocal translocation Chromatid ring ⇒chromosome ring (no fragment) Chromatid inversion ⇒pericentric inversion	
(3) Maintained chromosome-type aberrations	Dicentric (50%); fragment lost after first division Chromosome ring; fragment lost after first division Reciprocal translocation and inversion, though only those visible at first division will be observed	
(4) Cell death	Sister union and non-union distal Triradial Chromatid intra-change (?) Complex exchange Aberrations in categories 1–3 may also not survive due to formation of bridges, duplication/deletion, or rearrangements damaging vital genes	

6. Data Recording

There are nearly 20 different aberration types, but many of these can be grouped together for recording purposes. A suggested system of recording the data is given in Table 2. Thus, a table with seven columns for aberration types is sufficient. The data can be recorded in the appropriate columns by their 1985 ISCN abbreviation (8); it is strongly recommended to adhere to cs/ct for chromosome/chromatid rather than the ambiguous and longer 2009 ISCN recommendations of chr/cht (9). Alternatively, aberrations can just be recorded numerically, which simplifies totalling the data; an additional column for details of the aberration types (and other comments) can be included if more information is deemed desirable. As the reunion of sticky ends is believed to be random, there is no significance in the formation of asymmetrical or symmetrical exchanges, even though the outcome would be different

Table 2
Scheme of aberration data recording

Type	Aberration	ISCN abbreviation
Gaps	Chromatid gap Chromosome gap	ctg csg
Chromatid breaks/ deletion	Break/deletion Single break "minute" (involving end of chromatid; not a true minute)	ctb ctb
Chromatid exchange	Quadriradial Ring Inversion Two-break minute (not involving end of chromatid) Sister union, non-union proximal, non-union distal Triradial (with sister union) Intra-change event not included above Complex chromatid exchange	qr ctr ctinv ctmin[a] su, nup, nud[b] tr(+su) cte cx
Chromosome breaks/ deletion	Paired fragment "Double minutes" if not definitely rings (not true double minutes)	f (or csf) f (or csf)
Chromosome exchange	Dicentric (with paired fragment) Translocation Ring, centric, or acentric including double minutes Pericentric inversion	dic(+f) t csr/acr/dmin inv
Multiple aberration	More than specified number of aberrations (e.g. five) in one cell excluding gaps	mabs[b]
Others	Pulverisation	pvz

Some specific ISCN abbreviations are formed from combinations of the basic abbreviations, e.g. ct + b = ctb = chromatid break
[a] This could also be abbreviated to ctf
[b] No ISCN abbreviation

(e.g. quadriradials: asymmetrical can develop into dicentrics; symmetrical into translocations). It is therefore not necessary to record this information.

The classification of the aberrations generally adheres to (7), but it is suggested here that unequivocal intra-chromatid events should be recorded as exchanges. It is unfortunate that chromatid minutes have a name which implies a very small piece of material, as this could result in an analyst including chromatid breaks near the end of the chromatid in this category. If breaks which do not involve the end of the chromosome are designated as chromatid minutes, these can correctly be classified as exchanges as they always involve two breaks, while bearing in mind that they may not actually be physically "minute". Similarly, sister union events are exchanges as delineated in Fig. 2.

The term minute was probably applied to a chromatid event as a result of a specific type of chromosome-type aberration which was introduced into the terminology through the ISCN, which deals predominantly with clinical aspects of human cytogenetics. True double minutes are a manifestation of gene amplification found in some types of tumour cells, and not only vary in the numbers found within a single cell, which can be considerable, but also in their size (5). Their origin is quite different from small paired rings that may occasionally be observed during chromosome aberration analysis after clastogenic treatment. However, the name is widely used in this connection, and if applied only to paired densely stained dots which make it clear they are rings, they can also be classified as exchanges.

All categories in the table have already been described, except for multiple aberrations and pulverisation. A suggested requirement for multiple aberrations is given in Table 2. It is not necessary to enumerate more than five aberrations in one cell, particularly as the cell is taken as the unit of damage anyway. Occasionally, a cell will be observed which contains very large numbers of gaps and breaks, but normally not exchanges. The chromosomes appear to be in pieces, and thus the cell is classified as pulverised. This type of event is also observed when cell fusion takes place between a mitotic and an S-phase cell, the mitotic cell causing the S-phase cell to enter mitosis prematurely (premature chromosome condensation or PCC) (5). This rare observation may not, therefore, be due to extreme breakage or fragmenting of the chromosomes. A single "pulverised" chromosome is also sometimes observed, but again this may not be due to breakage. Either the chromosome has unravelled, but more often, a count of the chromosomes in the rest of the cell shows that they are all there, and the apparently damaged chromosome is probably from another prometaphase or early metaphase cell.

7. Practical Aspects of Routine Analysis

The various types of aberrations can be described clearly and unequivocally, and the system given here covers all possibilities as it comprises all interactions, i.e. there are no "new" aberrations to be discovered. However, problems of identification and classification arise, partly because there is an accepted subjective component in the assessment. But it is worth enumerating some of the other factors involved, as well as considering possible mis-classifications (see below). The main difficulty both in identification and interpretation arises out of the considerable range which even an aberration as simple as a chromatid break can take. A break close to the end of a chromosome looks quite different from one close to the centromere, or halfway between. Any of these is easier to identify if the acentric section is on the other side of the intact sister chromatid from the centric section rather than dislocated but on the same side (Fig. 11). In some instances, one chromatid may appear shorter than the other, but as already described, the sister chromatids remain aligned, and close examination will normally establish that the end of the longer chromatid consist of the deleted section from the shorter one (see chromosome at the bottom of Fig. 11). Examples of the varied appearance of some aberrations with a range of break points are shown in Fig. 12.

The criteria for selection of metaphases for analysis are usually given as requiring well-stained, well-spread cells, with the chromosomes having distinct sister chromatids, but not overcondensed and with a minimum of overlapping. In practice, cell selection probably makes up a high proportion of the variation in scoring between analysts (P. Fowler, personal communication). Unfortunately, while the negative control samples may have an abundance of such cells, cultures with the required 50% toxicity may have far fewer ideal cells. It is also not uncommon for cells containing unequivocal aberrations to be of reduced quality, and discounting any but high quality cells could seriously reduce the aberration frequency. Although so-called “fuzzy chromosomes” should never be scored, cells of at least slightly reduced quality need to be accepted if aberrations are clearly present. As a rule of thumb, observation of the individual chromosomes should

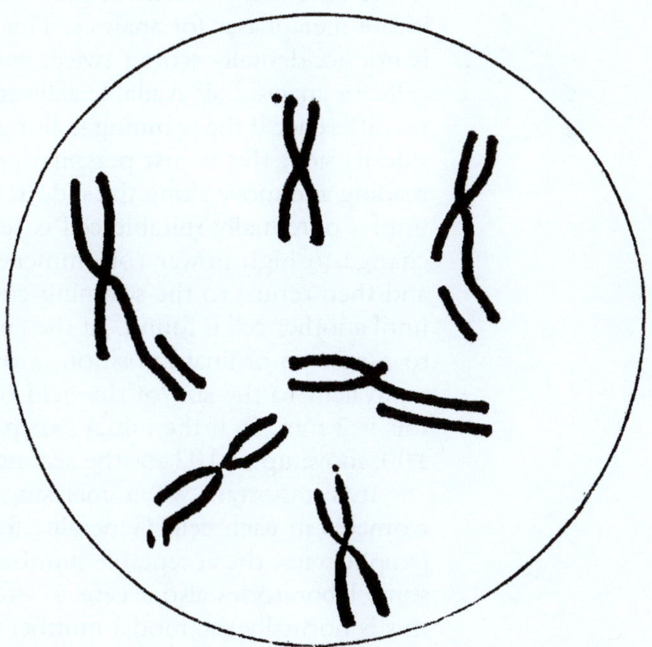

Fig. 11. Examples of the range of positions of chromatid breaks and their orientations.

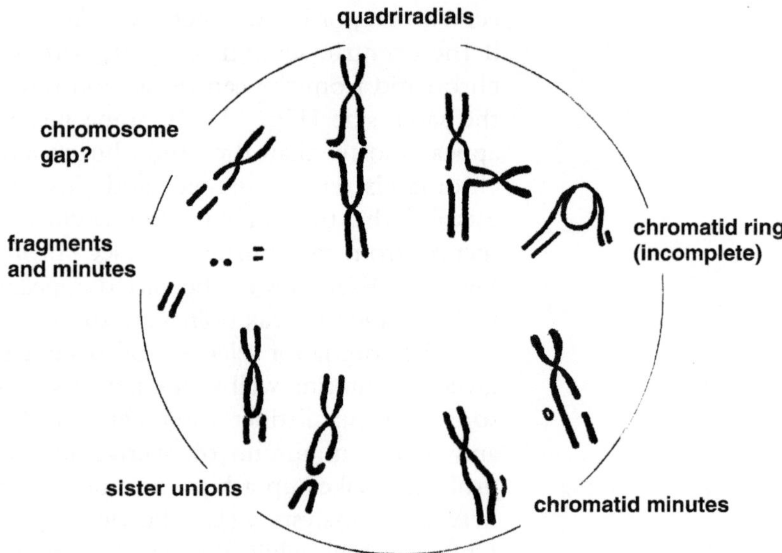

Fig. 12. Examples of variation in appearance of aberrations. See text for discussion.

consider whether aberrations would be visible if present, and if more than four of the chromosomes are not sufficiently clear to answer this, the cell should be excluded. This assumes there is the full complement of chromosomes; if any are missing, this total should be reduced accordingly.

A schematic system of scanning slides needs to be used to locate metaphases for analysis. This is to ensure that the same cell is not accidentally scored twice, and that if there is a shortage of cells for analysis, all available cells are located and scored. It makes no difference if the scanning is along the length of the slide or from side to side; this is just personal preference. Select a co-ordinate reading and move along the slide at low power (e.g. 100× or 200×) until a potentially suitable cell is seen. Move over to the cell and change to high power (oil immersion, typically 1,000×), analyse and then return to the scanning co-ordinate position, moving on until another cell is found. At the end of the sweep, move the stage to a new co-ordinate position, a distance from the previous one equivalent to the size of the field of view. On many microscopes, this is 2 mm, so if the initial sweep is on a co-ordinate reading of 100, move up to 102 for the second sweep.

It is important when analysing to count the number of centromeres in each cell. Generally, for human preparations such as lymphocytes, the acceptable number is 44–46 or 45–46, although some laboratories also accept 47–48. For cell lines such as CHO, this is normally the modal number ± 2. As well as confirming that sufficient chromosomes are present for analysis, each chromosome is observed individually, even if only briefly, which often alerts the analyst to possible damage which they can then examine in more detail. In most aberrant cells there is only one aberration, and so this has to be noticed amid 44 or 45 normal chromosomes.

As mentioned previously, unbanded chromosomes are generally used for aberration analysis, in which case detailed karyotyping is not possible. However, a certain amount of basic knowledge of the karyotype is extremely helpful in interpreting some situations. This is indicated below in Subheading 8, but at this stage a brief description of the human karyotype is given (8, 9).

The 46 chromosomes which make up the human genome consist of a mixture of metacentric, submetacentric and acrocentric chromosomes. There are no telocentric chromosomes. There is considerable size range, the largest chromosome making up over 9% of the total genome. Using the International System for Human Cytogenetic Nomenclature (8, 9), the chromosomes are arranged in seven groups, from A to G. See Fig. 13. The only chromosomes which can be individually recognised are the following: 1, 2, 3, 16, and Y, the remainder being allocated to their group only. As an aid to analysis, the chromosomes from groups D and G, the two acrocentric groups, are the most useful and counting the numbers of these chromosomes, six in group D and four plus the Y (in males) in group G, will in many instances greatly assist in resolving a query. The main source of normal variation within the karyotype is found in specific regions which are stained less intensely or even appear as gaps; these are referred to as secondary constrictions, the primary constriction being the centromere (5). The human chromosomes which may manifest secondary constrictions are discussed in the following section.

After human lymphocytes, the most widely used cell types are obtained from Chinese hamsters, which have a diploid number of 22 and have large, easily distinguished chromosomes. The cell lines used are usually CHL (lung in origin, modal number 25), CHO

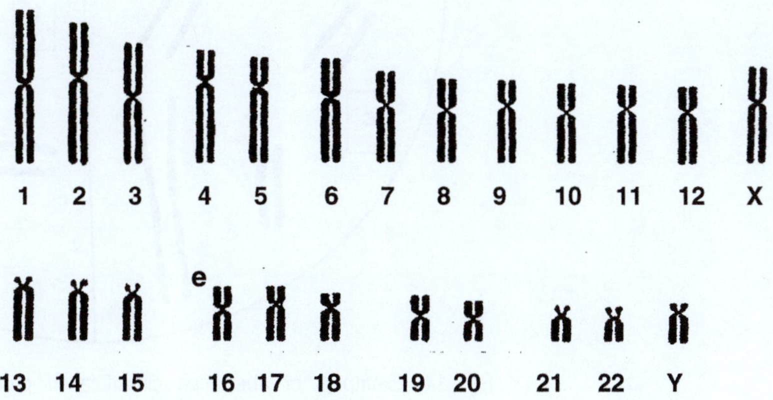

Fig. 13. Idiogram of human chromosomes.

(ovarian, modal number 20 or 21), and V79 (lung, modal number 21). Although it is possible to become familiar with the chromosomes in these lines, there is an intrinsic level of variation which both prevents the inclusion of chromosome rearrangements such as translocations in the aberration totals, and also makes the use of the karyotype to elucidate aberration interpretation impractical.

8. Misclassifications

While many, often the majority, of aberrations seen in the course of analysis are clear, unequivocal and easily observed, there are also many possibilities of misinterpretation. Figure 14 shows some examples of aberration classification problems. In Fig. 14a acrocentric chromosomes are orientated in such a way that they appear as either a dicentric chromosome or a triradial. In both cases, counting the other acrocentrics should clarify the situation. Figure 14b illustrates how careful observation is required to distinguish between quadriradials and crossed chromosomes. Figure 14c characterises the common human chromosome variations (8, 9); chromosome 1 with a secondary constriction close to the centromere,

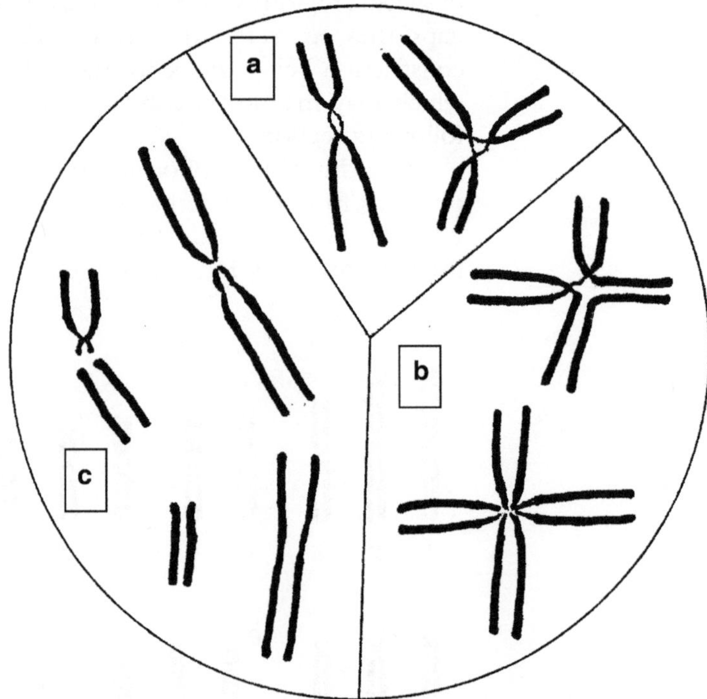

Fig. 14. Examples of aberration classification problems. (**a**) Satellite association. (**b**) Quadriradial/crossed chromatids. (**c**) Variation in human chromosomes. See text for discussion.

both the X and Y chromosomes sometimes lose their centromeric activity, and chromosome 9 has an unstained region, also a secondary constriction, on its q arm.

It is also useful to be able to consider the criteria for classification of specific aberrations. A scoring checklist is given in Tables 3 and 4, which recommends questions to apply to confirm or reject a particular aberration.

Table 3
Chromatid exchanges: checklist and potential mis-classifications

Aberration	ISCN abbreviation	Comments
Gap	ctg	True gaps are right across the chromatid, at right angles (though they may be wider at the outside of the chromosome than at the middle) and completely unstained, not just paler or fuzzy. There is no dislocation or misalignment, and the gap is not wider than the chromatid width. Scratches on the slide can be mistaken for gaps, but are usually not at right angles, and often extend across both chromatids and beyond.
Break/deletion	ctb	There is misalignment of the broken ends, or if the unstained part is wider than the chromatid width without misalignment, there is some dislocation (as in Fig. 11). The acentric part includes the end of the chromosome. If near the centromere, there is some visible chromosome material at the centromere separate from the broken end. Breaks very near the ends of chromosomes, or close to the centromere, are easily missed.
Quadriradial	qr	Check it is not two chromosomes crossed at or near the centromere (Fig. 14).
Ring/inversion	ctr/inv	Confirm the chromosome is not just twisted or curved.
Minute (fragment)	ct min (ctf)	The acentric part does not include the end of the chromosome. The fragment is next to a chromosome (with some evidence of bulging towards it unless very small). The colour and staining intensity is the same as the chromosome (i.e. not a stain particle).
Sister union	su, nud, nup	None of these should be scored without both parts (proximal and distal) being observed. In Chinese hamster CHO cells, be aware of the size of a small telocentric marker chromosome, which is like the distal part of a non-union proximal event.

(continued)

Table 3 (continued)

Aberration	ISCN abbreviation	Comments
Triradial	tr	Ensure it is not three acrocentrics in satellite association. Quadriradials with one break point close to the end of the chromosome (T-shaped) can also be mistaken for triradials (Y-shaped) (Fig. 12).
Unspecified intra-change	cte	Unusual event even with high levels of damage, so do not score if there are few aberrant cells (>10%; potential cells of this type can be checked and reclassified once analysis of the slide is complete). If bulging, twisting or other awkward shapes are observed, compare with other chromosomes in the cell for similar appearance.
Complex exchange	cx	Exchange involving three or more breaks excluding triradials. Very varied in appearance, only in cultures with high levels of damage.

Table 4
Chromosome-type aberrations: checklist and potential mis-classifications

Aberration	ISCN abbreviation	Comments
Gap (may be referred to as isochromatid gap)	csg	Relatively uncommon event; only normally scored when a fragment is too close to its point of origin to score as a break (see Fig. 12) or if there are two chromatid gaps at or close to the same locus. Be aware of normal variants such as human chromosome 9. Rarer variants can occur in individuals and may sometimes be seen within a study if one donor, rather than pooled blood from several, is used.
Acentric fragment (also referred to as isochromatid break)	f (ace)	The most frequently seen chromosome-type event following chemical treatment, and may also be due to two separate chromatid breaks. Human Y and occasionally X chromosomes can lose their centromeric activity and appear as fragments. The distal end of chromosome 9 can also look like a fragment if the secondary constriction (unstained region) is large.
Dicentric (with or without fragment)	dic (+f)	Any of the following can appear as dicentrics: crossed chromatids, secondary constrictions, satellite association, sister union and acentric fragments (all incorrect classification of aberrations) plus an acrocentric chromosome situated at the end of another chromosome (a non-aberrant situation classified as aberrant and therefore more serious). Conversely, a true dicentric plus fragment could be missed if one of the above was thought to have occurred and the fragment was taken to be a Y chromosome.

(continued)

Table 4 (continued)

Aberration	ISCN abbreviation	Comments
Translocation or inversion	t	Not normally detectable, and even when a rearranged chromosome is seen, it is not usually possible without banding to determine if it is a reciprocal translocation or an inversion. For this reason, it should just be recorded as an unspecified translocation or "t".
Centric/acentric ring	csr/acr	Check the chromosome is not just twisted or curved.
Double minute	dmin	This should only be scored if the density of staining makes it clear they are small rings. These are then classified as exchanges. If unsure, they should be scored as acentric fragments. See text.

It is important to be aware of the likely frequency of different aberration types within a study. Some of the salient points when analysing human lymphocyte cells following chemical treatment are:

1. Chromatid breaks (and gaps) are the most frequently observed aberrations and occur in many negative control samples, at less than 5% in total. However, cultures with no aberrations in 100 cells are also not uncommon, but not in all cultures within a set (excluding positive controls). An analyst who gets no aberrations of any sort in 16/20 cultures (i.e. a standard assay minus positive controls) may well be missing aberrations.
2. As spontaneous breaks arise at random, it is possible for two to occur by chance in the same cell. Although this is a relatively rare event, it is not unknown, and in laboratories where large numbers of studies are carried out, at about 4,000 cells per study, exchanges will be seen in negative control and test compounds from time to time. Sets of slides where several exchanges are recorded but with fewer breaks should be reviewed carefully to confirm or reject the exchanges, although occasionally the quality of the preparations means that, while exchanges are clear and unequivocal, some breaks may be missed or rejected because of reduced clarity.
3. Paired fragments are the most commonly observed chromosome-type aberrations. As they should not be able to result from replicated chromatid deletions, which are acentric and lost at anaphase, their frequency is rather higher than expected.

However, some will be due to isochromatid breaks, produced in the same way as sister union events, but with no sister union. They may also have originally been induced in vivo.

4. The most frequently missed aberration type is sister union, but as it requires two breaks on the same arm of the same chromosome, it will normally only be found in positive cultures.
5. There are a number of events which can be mistaken for dicentrics, but they do nevertheless occur even in negative control, if they are induced in vivo, for example by donor exposure to radiation. Dicentrics of this type should be accompanied by an acentric fragment, although this may not always be visible within the metaphase spread. Dicentrics derived from asymmetrical quadriradials will not have an accompanying fragment.
6. Normal chromosome variants, such as the secondary constriction on the long arm of human chromosome 9, are often mistaken for chromosome gaps, and analysts should be aware of their locations.
7. Unexpected aberrations such as a chromatid intra-change in a sample with a low aberration frequency should be examined very thoroughly, and the affected chromosome should be compared with the appearance of the chromosomes in the rest of the cell. Usually, other chromosomes will also be twisted or bulging and the suspected aberration should be rejected.

It is better to discount a cell altogether than score it incorrectly, so if an apparent aberration is unclear, do not score. However, if there is an unequivocal aberration in a cell, other, inconclusive events are not important and the cell can be scored.

Most of the above also apply to other cell types, such as Chinese hamster CHO and V79 cells. The background aberration levels in cultured cell lines tend to be higher, but any laboratory should build up a data base on the spontaneous aberration frequency. The karyotype is also less stable, so chromosome rearrangements would not normally be recorded, as mentioned previously.

Whatever cell type is used, it is expedient to look at the data as a whole when analysis is complete to identify unexpected or improbable results. Peer reviewing is particularly important in such cases.

9. Toxicity Assessment

There are a number of methods of measuring toxicity as part of a chromosome aberration assay, but most are applied to other tests as well. Mitotic index evaluation is discussed here, as this is widely used in cytogenetic tests and is particularly relevant.

Mitotic index is measured on microscope slides and involves counting a number of cells, usually 1,000 or 2,000, and recording how many mitotic cells are present in that number. Mitotic index (MI) is expressed as a percentage, for example, 53 mitotic cells in 1,000 gives an MI of 5.3%. The mitotic cells can be at any stage of mitosis, although if the same cells are used for its estimation and for chromosome aberration analysis, it will not include anaphase or telophase cells as the cells are arrested at metaphase. Because of the accumulation of the metaphase cells in this case, the MI will be higher than the spontaneous value found if cells are harvested without mitotic arrest.

The percentage value obtained indicates the proportion of the total cell cycle taken up by mitosis. A reduction in mitotic index results from another phase taking up more of the cell cycle, which is then longer than normal while mitosis still takes the same amount of time. Cell cycle delay is usually produced either when cells go into G_0, or when G_1 is extended. Normally, an MI reduction of about 50% is recommended to indicate sufficient toxicity. However, agents such as spindle poisons which arrest cells at metaphase can cause an increase in MI, so in some instances a reduction in MI cannot be expected even at toxic doses.

The measurement of MIs is relatively quick and easy both to learn and to perform. It can be measured on the same slides used for metaphase analysis, unless any cell selection has been carried out, such as mitotic shake-off preparations. One of its particularly valuable features is that any cell type suitable for metaphase analysis can be used for MI estimation.

There are also disadvantages to MI estimation to evaluate toxicity. It does not measure toxicity directly, but changes in the length of different stages of the cell cycle, generally G_1. Also, the interphase cells counted may be in the normal cell cycle (G_1, S, and G_2), in G_0, or dying, if not yet apoptotic, and no distinction is made of their longer-term survival. There is lack of conformity between laboratories as to which cells are counted in lymphocyte cultures, although this should not be of importance as long as the intra-laboratory scoring is consistent. It should also be borne in mind that there are differences between human donors in the response to the mitogen used. There is also a possibility of some synchrony within the cultures, and if there is also cell cycle delay at toxic doses, use of the same sampling time may result in some cultures being at a peak of division and others at a trough.

MI is a valuable method of assessing toxicity, and the drawbacks are more than outweighed by the advantages. It can also be used in combination with the differential staining method, culturing the cells with bromodeoxyuridine to stain for first, second, and third division cells. This allows for calculation of the proliferative index, and thus the average generation time (10).

10. Numerical Aberrations

There are two main groups of numerical aberrations, polyploidy, divided into simple polyploidy and endoreduplication, and aneuploidy, either with extra chromosomes (hyperdiploidy) or missing ones (hypodiploidy). It is quite common for chromosomes to be lost during slide preparation, so hypodiploidy is not normally considered to be a result of induction by test agents. It should be appreciated that the standard time interval from treatment to harvest is such that as high a proportion of cells as possible are at their first metaphase after passing through an S phase. This is not sufficient time for the majority of cells to reach the required metaphase for numerical aberrations to develop, and therefore, it is not possible to detect the induction of numerical aberrations accurately. It is, however, common practice to record both hyperdiploid and polyploid cells during analysis.

As the number of chromosome is counted during metaphase analysis, the number of additional cells with hyperdiploidy found while analysing the diploid (and hypodiploid) cells in human lymphocytes gives a true estimate of the hyperdiploidy levels. In practice, this is usually very low, and it is not unusual to see no hyperdiploid cells in an entire study. In cells which appear to have 47 chromosomes, look for a chromosome with a different degree of condensation, which may have come from another cell. With cell lines such as CHO and V79, estimations of hyperdiploidy cannot be made because of the intrinsic levels of hyperdiploidy, which are much higher than in human lymphocytes, and also difficulty in obtaining repeatable results.

Correct identification of numerically aberrant cells is usually straightforward, particularly in the case of endoreduplication. In these cells, the chromosomes are all grouped in pairs of identical copies (not with their homologues as in meiosis), which is often surmised even on low power from their thick, dark appearance before unmistakable confirmation on 1,000× magnification. Other types of polyploidy result in a random distribution of twice the diploid number of chromosomes (5). If a cell is thought to be polyploid, check if the chromosomes all have a similar degree of condensation, and lie in one, not two circles. If unsure, identify the positions of the acrocentric chromosomes, groups D and G; in polyploid cells, they are likely to be randomly distributed. In some laboratories, co-inicident metaphases are excluded from aberration analysis unless the degree of condensation is sufficiently different to ensure correct cell allocation. This prevents the possibility of two aberrations in one cell being attributed to two cells or vice versa.

Analysts in many laboratories note the presence of polyploid and endoreduplicated cells while scoring diploid cells for aberrations. It is important to be aware that while this may flag up

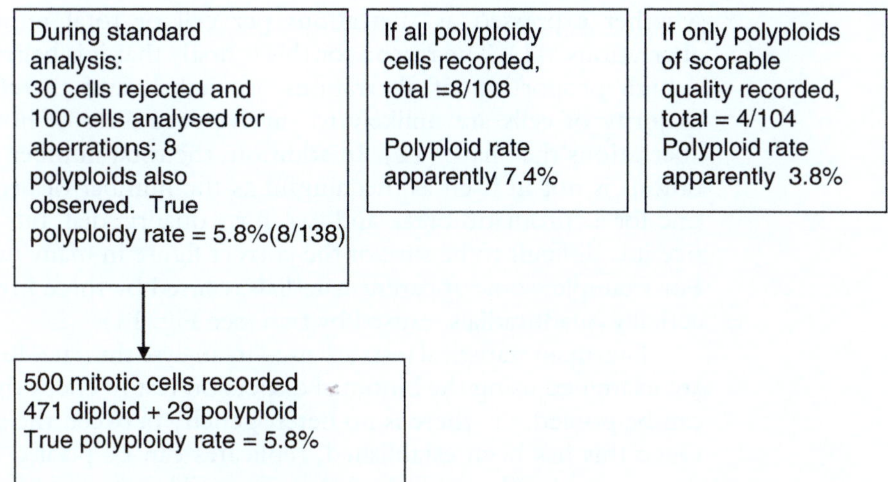

Fig. 15. Estimation of polyploidy frequency, showing the distortion of the apparent level if the inclusion criteria differ.

substantial increases in polyploidy (as a proportion of the cells will be at the second metaphase after treatment), it is not an accurate means of measuring the polyploid level, spontaneous or induced. This is because different criteria are being applied to the two types of mitotic cells. Many diploid cells are rejected because of overcondensed, undercondensed, or too many overlapping chromosomes amongst other reasons, so if all polyploids observed are included, the polyploid rate will be augmented. Conversely, if polyploid cells are similarly assessed, the level will be below the true value, as the chromosomes in polyploid cells are often not as clear and sharp, plus the greater chance of overlapping because of the numbers of chromosomes. If polyploidy is to be accurately measured, the same criteria must be applied to both diploid and polyploid cells. It is recommended to count the numbers in 500 mitotic cells in total for such an evaluation (Fig. 15).

11. Data Analysis and Interpretation

Experimental design of chromosome aberration assays is discussed elsewhere in this volume, but statistical analysis and interpretation of the results are addressed here. The most common body of data generated from chromosome analysis consists of two replicates at three concentration levels, negative (solvent) and positive controls, plus and minus metabolic activation, giving a total of 20 individual cultures with 100 cells being analysed per culture. The recommended procedure is to record the number of cells with aberrations excluding gaps, not the number of aberrations observed, whether expressed as aberrations per cell or total numbers of aberrations (11). The reason for this is firstly that it is believed that a high proportion of aberrations are lethal, and therefore the majority of cells are unlikely to survive regardless of how many aberrations they have (12). In addition, the total number of aberrations is not in itself as meaningful as the number of breaks, i.e. one for a chromatid break and two for a quadriradial, but in practice it is difficult to be sure of the correct figure in many instances. For example, some apparent triradials, caused by three breaks, are actually quadriradials, caused by two (see Fig. 14).

Two main statistical tests are used to analyse the data. Replicates are examined using the binomial dispersion test to check that these can be pooled, i.e. there is no heterogeneity between the samples. Once this has been established, replicates can be pooled and the data are analysed used Fisher's Exact test. This is a modification of the 2×2 X^2 test to take into account numbers less than ten. Because the numbers in the negative controls are usually very low, it is important not to use a test in which relatively small increases could give a statistically significant result, such as might be found with the standard X^2 test. Both tests can be rapidly and easily performed using a statistical package available, for example from SPSS (Statistical Package for the Social Sciences). A detailed consideration of the statistical analysis of chromosome aberration data can be found in ref. 11.

In the majority of cases, there will be a clear and unequivocal result, either negative or positive. Before discussing some problem situations, it is important to be aware that even conclusively positive compounds may not elicit a dose response. Because of cell cycle delay, cultures at high doses may not have reached their maximum response when harvested, resulting in a reverse dose response; alternatively, lower doses may not show any increase at all if the cells are at their second metaphase after treatment and induced aberrations have therefore been lost (7).

The main situations giving rise to problems in data interpretation are discussed below. Cytogenetic assays are routinely performed twice, so where a repeat experiment is referred to, it means exactly that, not experiment two in the standard protocol. Note also that, while it should be self-evident, it is vital to ensure that any further analysis is performed on different cells from those originally analysed. Where appropriate, it can also be useful to peer-review cells; see particularly Subheading 11.5 below.

11.1. Heterogeneity Between Cultures

If this is observed, there is little option but to repeat the experiment. However, if there is only marginal significance (probability close to 0.05), analysis of a further 100 cells for the relevant dose levels, with the slides still coded and the original data not known to the analyst, might reduce the differences sufficiently and would be an acceptable alternative. Statistically, a significance level of 0.05

indicates that although in 19 out of 20 times the result is expected to be within a particular range, once in 20 it is not, so marginally significant results will arise from time to time for this reason.

Provided all positive control samples give a positive response, there would be no need to repeat a study if they show heterogeneity, although if there were substantial differences between analysts within codes, this would require further investigation. See below Subheading 11.5 for discussion of analyst variation.

11.2. Negative Controls Outside Historical Database or Lack of Response from Positive Controls

A laboratory carrying out routine testing using this assay will over time build up a negative control database, which can then be used to ensure that negative control results do not fall outside this. Some data should be generated from trial runs of the test before testing of new compounds is carried out, but there are no specific guidelines as to what constitutes a database of a suitable size. It would be advisable to set the maximum limit of aberrant cells in the negative controls to less than 5% until at least 20 sets of negative data have been included in the database; in practice, this would be nine aberrant cells per 200 negative control cells as the highest acceptable. In practice, negative control levels may well be somewhat lower, particularly with human lymphocytes. Once the database is established, any test in which the historical levels were exceeded would have to be rejected.

The situation is different with positive controls. There have been proposals such as in ref. 10 to introduce requirements for relatively low responses in positive controls, based on good scientific principals, but in practice this is rather difficult. If a marginal increase were required in the positive control, rather than a strong response, the variation often seen between experiments would result in an unacceptably high number of experiments being rejected. The purpose of the positive controls is to ensure that the system works, particularly with metabolic activation, and also that analysts are recognising aberrations, and it not necessary to test either the system or analysts beyond a practicable level. In practice, there is likely to be quite a range of responses to the positive controls between experiments, but if there is not a significant increase in either positive controls (with and without metabolic activation) the experiment would require repeating. Usually this would only be with or without metabolic activation, i.e. half the experiment, as appropriate.

11.3. Insufficient Toxicity

In some instances, concentrations of a test agent which caused sufficient toxicity in preliminary tests do not cause the same in the main experiment. This is another situation in which a repeat will have to be done, unless higher doses were included in the experiment but not initially analysed. Some laboratories do include additional dose levels routinely, both above and below those intended for the main analysis, to cover both the above situation, and the converse, where there is greater toxicity than expected and insufficient cells might therefore be available for analysis. It should also be noted that at very high toxicity, chromosome aberrations may be observed which are a response to the toxicity and not a specifically genotoxic response. The appearance of such cells is quite characteristic although difficult to describe. The chromosomes tend to be very darkly stained and have irregular outlines ("blocky/smudgy"), and while being of marginal quality the aberrations, which are generally gaps and breaks only, are unequivocal. An experienced analyst will become familiar with this toxic effect, but it is a reminder that it is not only unnecessary but also inappropriate and inadvisable to exceed the 50% toxicity level to any great extent.

Fifty percent toxicity cannot always be achieved, even at the highest recommended concentrations (13) (10 mM or 5 mg/ml; ±1 pH unit; 100 mOsm/kg H_2O); experimental conditions specifying such cases are described in the chapter on in vitro chromosome aberration assays.

11.4. Equivocal Data: A Marginally Significant Increase or an Increase in a Single Replicate

The importance of a negative control database is evident here, as results which fall inside the historical control level while being marginally significantly higher than the negative control within the experiment can be discounted. This would only arise where an unusually low control level was observed, and even the use of Fisher's exact test will not always entire compensate for this.

For values outside the historical control level, analysis of additional cells at the appropriate doses may clarify the situation. This can also be applied to an increase in a single replicate, with both replicates included. If there is only one single replicate with a significant increase in aberration levels in both experiment one and two, a repeat with no increase in any replicate would allow the single result to be discounted, accepting that errors do occur during experimental procedure which can give incorrect results.

While it would be critical for any repeat experiment to include the concentration(s) showing the significant increase, the use of doses closer to these, or additional dose levels, rather than an exact repeat, needs to be considered.

Ultimately, a significant, repeated increase in aberration levels has to be accepted as positive for genotoxicity. The overall evaluation of such compounds, and indeed all positives, has to be taken on a case-by-case basis in terms of its characterisation and also what further different tests may be performed to assess its risk.

11.5. Analyst Variation

There is considerable variation between laboratories as to how slide analysis is apportioned, ranging from one analyst for a complete set of slides, through a 50-50 split between two analysts, to a panel of analysts simply taking slides consecutively from a set. Although there may be a "best practice" way, practicalities such as the number of available analysts, often determine the system used. It is essential that the slides are coded and then scored "blind", and it should also be ensured that there is no possibility of one analyst scoring only a specific subset such as the top concentration tested.

Occasionally the heterogeneity analysis will indicate that there are significant differences between analysts. Because of the relatively small number of cells analysed per code (50) if they are split between analysts, one or two discordant pairs of results is not a cause for concern unless highly significant ($p<0.01$). But if this does occur, or there is consistent variation in the same direction between analysts, the material will need to be either re-analysed or reassessed by appropriate means. Two scenarios are described here.

If one analyst has significantly higher aberration levels than the other, first of all the recorded aberrations should be relocated and checked. Confirmation of the aberrations implies that the analyst with lower values may be missing them, and their part of the analysis should be rescored, preferably by a third analyst. Alternatively, if a notable number of recorded aberrations are rejected, that analyst's scoring should be reassessed.

If an analyst misses a positive control picked up by the other analyst, their slides would need to be rescored. This does highlight the problem that, while aberrant cells can be relocated and checked by an independent analyst, it is very difficult to identify whether an analyst is missing aberrations. Even checking a sample of apparently normal cells, if the co-ordinates of all metaphases are recorded, is not a practical way of picking this up. If the true aberration level is 20%, a sample of 10 out of 50 cells would only be expected to find approximately two aberrant cells, and even checking half of them would only give rise to about five.

11.6. Gaps

Gaps are routinely recorded when scoring chromosome aberrations but are not normally included in statistical analysis and in classifying a compound as positive or negative, even if there are clearly increased numbers of them. Conversely, there is often no increase in their numbers in positive cultures, the level frequently being the same as in negative slides from the same experiment. The only situation in which they make a contribution is in an equivocal case, where the presence of increased numbers of gaps can be added to evidence of a positive response, as it has been shown that some gaps are in fact due to true discontinuities in the DNA (6).

12. Conclusions

There are a number of aspects of chromosome aberration analysis which need to be addressed to ensure that the data produced is as reliable an assessment as possible. A clear understanding of the underlying mechanisms, classification based on break and rearrangement and fate of aberrations all contribute to accurate scoring, but uniform selection criteria and knowledge of the potential mis-classifications, as well as training and, most importantly experience, will all help to produce accurate and consistent results.

Acknowledgements

The author is very grateful for discussions on practicalities of scoring chromosome aberrations from cytogeneticists in a number of laboratories, in particularly the following: CiToxLAB France, Covance Laboratories Ltd., Harrogate, UK, Sanofi-Aventis Alfortville, France and Frankfurt, Deutchland, and Sequani Ltd., Ledbury, UK.

Thanks also to Paul Fowler (Covance) for information on variation in cell selection between analysts.

For information on training courses in chromosome aberration analysis, go to http://www.microptic.com.

References

1. DHSS (1981) Guidelines for the Testing of Chemicals for Mutagenicity. Prepared by the Committee on Mutagenicity of Chemcials in Food, Consumer Products and the Environment, Department of Health and Social Security. *Report on Health and Social Subjects, No. 24.* Her Majesty's Stationery Office, London.
2. IPCS (1985) *Guide to short-term tests for detecting mutagenic and carcinogenic chemicals prepared for the IPCS by the International Commission for Protection against Environmental Mutagens and Carcinogens.* Geneva, WHO.
3. Savage, J. R. K. (1975) Annotation: Classification and relationships of induced chromosomal structural changes *J Med Genet* **12** 103–22.
4. Savage, J. R. K. (1983) Some practical notes on chromosomal aberrations *Clin Cytogenet Bull* **1** 64–76.
5. Therman, E. (1986) *Human Chromosomes, Structure, Behaviour, Effects* Second Ed. Springer-Verlag, New York.
6. Satya-Prakash, K. L., Hsu, T. C. and Pathak, S. (1981) Chromosome lesions and chromosome core morphology *Cytogenet Cell Genet* 30 *248–52.*
7. Scott, D., Danford, N. D., Dean, B. J. and Kirkland, D. J. (1990) Metaphase chromosome aberration assays *in vitro* In *UKEMS Sub-committee on Guidelines for Mutagenicity Testing. Report. Part 1 revised. Basic mutagenicity tests: UKEMS recommended procedures,* ed. D. J. Kirkland *et al,* Cambridge University Press, Cambridge, 62–86.
8. ISCN (1985) *An International System for Human Cytogenetic Nomenclature,* ed. D. G. Harnden *et al* S. Karger, Switzerland.
9. ISCN (2009) *An International System for Human Cytogenetic Nomenclature (2009)* ed. L. G. Shaffer *et al* S. Karger, Switzerland.
10. Preston, R. J., San Sebastian, J. R. and McFee, A. F. (1987) The in vitro human lymphocyte assay for assessing the clastogenicity of chemical agents *Mutat Res* **189** 175–83.
11. Richardson, C., Williams, D. A., Allen, J. A., Amphlett, G., Chanter, D. O. and Phillips, B. (1990) Analysis of data from *in vitro* cytogenetic assays In *UKEMS Sub-committee on Guidelines for Mutagenicity Testing. Report. Part 111. Statistical evaluation of mutagenicity test data,* ed. D. J. Kirkland *et al,* Cambridge University Press, Cambridge, 141–54.
12. Brewen, G. (1982) Chromosome aberrations in mammals as genetic parameters in determining mutagenic potential and assessory genetic risk. *Prog Mutat Res* **3** Eds. K. C. Bora *et al.* Elsevier Biomedical Press, Amsterdam.
13. Scott, D., Galloway, S. M., Marshall, R. R., Ishidate, M. Jr., Brusick, D., Ashby, J. and Myhr, B. (1991) ICPEMC, Genotoxicity under extreme culture conditions. *Mutat Res* **257** 147–204.

Chapter 7

The In Vitro Micronucleus Assay

Ann T. Doherty

Abstract

The in vitro micronucleus test detects genotoxic damage in interphase cells. The in vitro micronucleus test provides an alterative to the chromosome aberration test, and because the in vitro micronucleus test examines cells at interphase, the assessment of micronuclei can be scored faster, as the analysis of damage is thought to be less subjective and is more amenable to automation.

Micronuclei may be the result of aneugenic (whole chromosome) or clastogenic (chromosome breakage) damage. This chapter provides methods for mononucleate and binucleate micronucleus tests and the addition of centromeric labelling and a non-disjunction assay to investigate any potential aneugenic mode of action.

Key words: Micronuclei, Centromeric labelling, Non-disjunction, Clastogenicity

1. Introduction

Micronuclei in erythrocytes have been identified for over 100 years as the Howell-Jolly bodies seen in haematology. The first induced micronuclei were first seen in *Vicia faba* root tips exposed to X-rays (1). Bone marrow micronucleus assays in which micronuclei are examined in immature erythrocytes have been in use since 1975. (2). Cell division is an essential requirement for the expression of micronuclei; therefore, to show chromosome damage following exposure to a genotoxic agent in vitro, the cell must undergo a nuclear division. Since 1985, cytochalasin B (CB) an actin inhibitor has been added to cells in vitro to block cytokinesis, thus resulting in the formation of binucleate cells. The presence of a binucleate cell demonstrates a cell that has divided in or following the presence of the test agent, and therefore, micronuclei are scored for these cells (3).

James M. Parry and Elizabeth M. Parry (eds.), *Genetic Toxicology: Principles and Methods*, Methods in Molecular Biology, vol. 817, DOI 10.1007/978-1-61779-421-6_7, © Springer Science+Business Media, LLC 2012

Micronuclei are DNA fragments that are separate from the main nucleus and have originated from centric or acentric chromosome fragments. The micronuclei may be the result of clastogenic or aneugenic damage. The in vitro micronucleus test is an umbrella term for many differing micronucleus tests, such as those with and without CB and a variety of treatment and recovery schedules (Plates 1 and 2). Rapidly dividing cells can be used for the mononucleate micronucleus protocol and require a robust measurement of cytotoxicity, such as population doubling.

Micronuclei represent damage that has been transmitted to daughter cells. Micronuclei in interphase cells can be assessed relatively objectively, as a result, the preparations can be scored rapidly and analysis can be automated. This makes it practical to score thousands instead of hundreds of cells per treatment, increasing the statistical power of the assay. Finally, as micronuclei may arise from lagging chromosomes, there is the potential to detect aneuploidy-inducing agents that are difficult to study in conventional chromosomal aberration tests, e.g. OECD Test Guideline 473 ((4, 5), Fig. 1).

The presence of a centromere signal in micronuclei is assumed to indicate the presence of a whole chromosome rather than a fragment. Chromosome paints are commercially available for the centromeres of human and mouse chromosomes and have fluorescent labels incorporated (6–8).

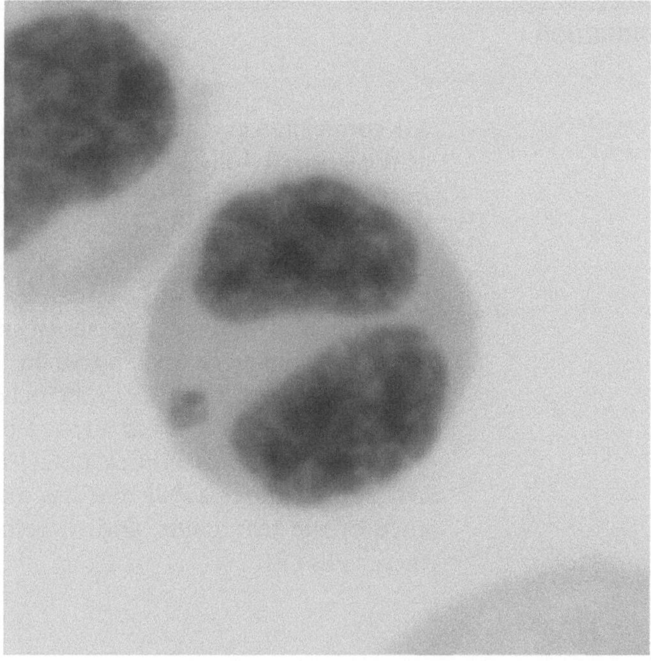

Plate 1. Photograph of a micronucleus in a binucleate human lymphocyte cell (the image was captured from an acridine orange preparation in fluorescent colours and then negative image was used to convert it into grey scale).

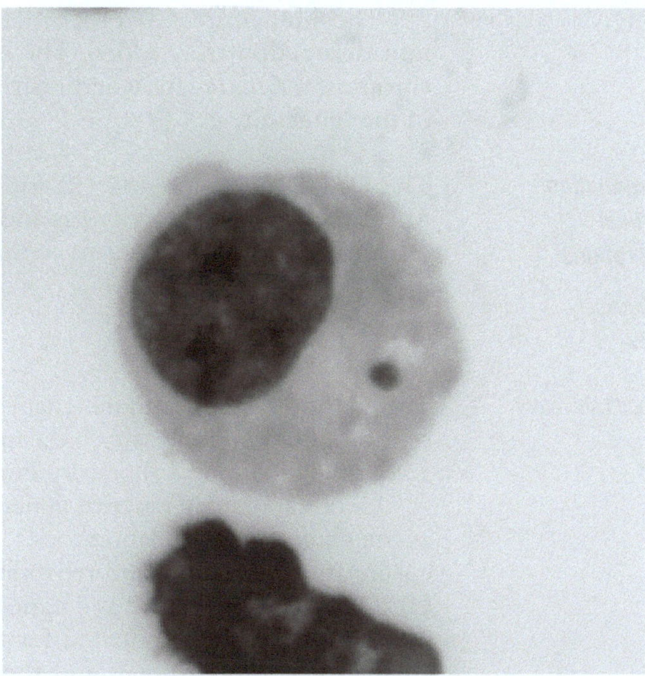

Plate 2. Photograph of a micronucleus in a mononucleate L5178Y cell (the image was captured from an acridine orange preparation in fluorescent colours and then negative image was used to convert it into grey scale).

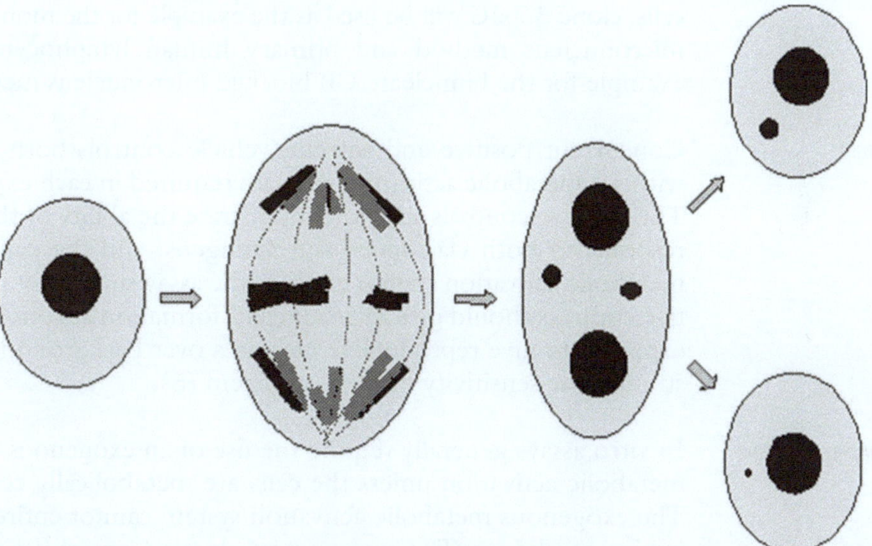

Fig. 1. Overview of micronucleus formation with micronuclei originating from either a whole chromosome or a chromosomal fragment, in binucleate and mononucleate cells.

The fluorescent in situ hybridisation (FISH) procedures can be used when there is an increase in micronucleus formation and the investigator wishes to determine if the increase was the result of clastogenic and/or aneugenic events.

The binucleate CB blocked micronucleus test can have a non-disjunction assay added. The non-disjunction assay examines chromosome segregation and distribution in the individual nuclei of the binucleate cell (9).

1.1. Introduction to Practical Considerations

1.1.1. Regulatory Guidelines

The OECD guideline 487 (5) for the In Vitro Mammalian Cell Micronucleus Test (MNvit) was adopted in July 2010. Some information used in this chapter has come from the draft version.

1.1.2. Good Laboratory Practice

The methods described in this chapter have all come from a pharmaceutical laboratory, and as such the methods have all been conducted accordingly to Good Laboratory Practice (GLP) requirements.

Good laboratory practice generally refers to a system of management controls for laboratories and research organisations to ensure the consistency and reliability of results as outlined in the OECD Principles of GLP and national regulations. GLP applies to non-clinical studies conducted for the assessment of the safety of chemicals to man, animals and the environment. A definition of GLP is found in the in vivo micronucleus chapter.

1.1.3. Cell Lines

Many primary or transformed cell lines are appropriate to use for in vitro micronucleus testing. The mouse lymphoma cell line L5178Y cells, clone 3.7.2C will be used as the example for the mononucleate micronucleus method and primary human lymphocytes as the example for the binucleate CB blocked micronucleus method.

1.1.4. Controls

Concurrent positive and solvent/vehicle controls both with and without metabolic activation (S9) are required in each experiment. The positive controls should demonstrate the ability of the cells to respond to both clastogens and aneugens, and the capability of metabolic activation system used for the assay such as S9. The positive controls should give micronucleus formation at concentrations expected to give reproducible increases over background, demonstrating the sensitivity of the test system (5).

1.1.5. Metabolic Activation

In vitro assays generally require the use of an exogenous source of metabolic activation unless the cells are metabolically competent. The exogenous metabolic activation system cannot entirely mimic in vivo conditions. The most commonly used system is a co-factor-supplemented post-mitochondrial fraction (S9) prepared from the livers of rodents treated with enzyme-inducing agents such as Aroclor 1254 (10) or a combination of phenobarbital and β-naphthoflavone, which is as effective as Aroclor 1254 for inducing mixed-function oxidases (11).

1.1.6. S9 Rat Liver Homogenate

The post-mitochondrial fraction (S9) of a rat liver homogenate is purchased from Mol Tox Inc. (Boone, North Carolina, USA). Its metabolic capacity is demonstrated by key enzyme assays and the ability to activate reference agents to bacterial mutagens. S9 is stored in liquid nitrogen or at −80°C until used. S9 is added to a co-factor solution and then added to cultures.

1.1.7. Cytotoxicity Measure

Appropriate cytotoxicity measurements are required for all in vitro genotoxicity assays. A comparison of cytotoxicity measures for the in vitro micronucleus tests has been undertaken by Fellows et al. (12) and Lorge et al. (13). They compared relative cell counts (RCC), relative increase in cell counts (RICC) and relative population doubling (RPD) for treatments without cytokinesis block with replication index (RI) for treatments with cytokinesis block, when evaluating the corresponding induction of micronucleated cells. No two cytotoxicity end points give the same result and RCC markedly underestimated the extent of cytotoxicity when compared with several other measures, such as RICC, RPD, and RI (12).

Furthermore, using these estimations of cytotoxicity and the limit of 50% survival, all the mutagens and aneugens tested were appropriately identified as positive in the in vitro micronucleus assay. Accordingly, it was clear that testing beyond 50% survival was not necessary to identify the potential of these agents to induce micronuclei (12).

Methods Used to Determine Cytotoxicity in the Absence of Cytochalasin B (12), Reproduced with Permission of the Author)

(a) *Relative cell count (RCC)*
RCC was determined as:

$$\frac{\text{Final count treated cultures}}{\text{Final count control cultures}} \times 100$$

(b) *Relative increase in cell count (RICC)*
RICC was determined as:

$$\frac{\text{Increase in number of cells in treated cultures (final} - \text{starting)}}{\text{Increases in number of cells in control cultures (final} - \text{starting)}} \times 100$$

(c) *Relative population doubling (RPD)*
RPD was determined as:

$$\frac{\text{Number of population doublings in treated cultures}}{\text{Number of population doublings in control cultures}} \times 100$$

where

$$\text{Population doubling} = \left[\begin{array}{r}\text{log (Post - treatment cell number} \\ \text{/ Initial cell number)}\end{array}\right] / \log 2$$

(d) *Relative cloning efficiency (RCE)*

To estimate cloning efficiency (CE), cells were plated out at 1.6 cells per well in 96-well microtitre plates and CE was calculated from the zero term of the Poisson distribution, P(0).

P(0) is calculated from the proportion of wells in which a colony has not grown.

$$CE = \frac{-\ln P(0)}{\text{Number of cells per well}}$$

RCE for an individual culture was the CE expressed as a percentage of the mean control CE. It should be noted that the CE values are not corrected for the numbers of cells lost during the treatment period.

Method Used to Determine Cytotoxicity in the Presence of Cytochalasin B

Replicative index (RI)
RI was determined as:

$$\frac{\text{(No. binucleated cells} + 2\times \text{ No. multinucleate cells)} / \text{Total number of cells treated cultures}}{\text{(No. binucleated cells} + 2\times\text{No. multinucleate cells)} / \text{Total number of cells control cultures}} \times 100$$

1.2. Mononucleate Assay

The mononucleate micronucleus assay is described in Mouse lymphoma L5178Y cells, clone 3.7.2C, obtained from Dr J. Cole, (MRC Cell Mutation Unit, University of Sussex, Brighton, UK). The cells used were confirmed to have the expected karyotype (14), including two copies of chromosome 11 detected by FISH. The average doubling time of the cells was approximately 9–10 h.

1.3. Binucleate Assay

The binucleate micronucleus assay is described in separated human lymphocytes and the average cell cycle of 18–20 h was found in pooled human lymphocytes from at least two donors of the same sex.

1.4. Centromeric Labelling

Centromeric labelling provides a method of identifying if a micronucleus contains a centromere. The presence of a centromere signal in the micronucleus is assumed to indicate the presence of a whole chromosome. The FISH technique is used with commercially available probes for the centromeres of mouse and human chromosomes that have fluorescent labels incorporated; the CY3 label is red and the FITC label is green when viewed through a fluorescent microscope with the appropriate filters (Fig. 2).

1.5. Non-disjunction Assay

Non-disjunction can be examined in binucleate human lymphocytes by using centromere-specific probes to examine distribution of pairs of chromosomes between the two nuclei (9).

CB blocks cells at cytokinesis, leading to an accumulation of binucleate cells. The incorporation of centromere-specific probes allows visualisation of chromosome segregation and distribution in the individual nuclei of the binucleate cell. When using two centromere-specific probes the normal distribution of chromosomes would be two copies in each nucleus written as 2:2 distribution.

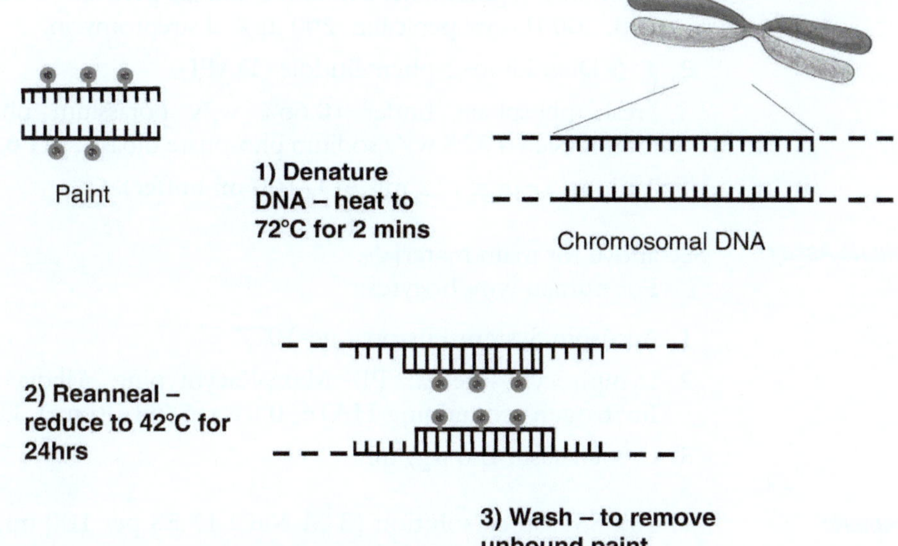

Fig. 2. Fluorescent in situ hybridisation (FISH) cartoon of the FISH technique showing denaturation and reannealing steps.

Non-disjunction is the malsegregation of chromosomes due to the failure of chromosomes on the metaphase plate to divide to each daughter nuclei and may be determined by a 3:1 or 4:0 distribution of centromere-specific signals.

2. Materials

2.1. S9 Rat Liver Homogenate

1. *Buffer solution A*: 14.2 g Na_2HPO_4 in 250 ml ultra pur water.
2. *Buffer solution B*: 3.12 g $NaH_2PO_42H_2O$ in 50 ml ultra pur water.
3. *Magnesium chloride solution*: 8.14 g $MgCl_2{\cdot}6H_2O$ and 12.3 g KCl in 100 ml distilled water.
4. 79 mg nicotinamide adenine dinucleotide phosphate (NADP).
5. 38 mg sodium salt and glucose-6-phosphate (G6P).

The sterilised co-factor solution may be formulated as a stock batch and aliquoted into sterile 20-ml vials and stored at −80°C for 12 months.

2.2. Mononucleate Assay

1. Media for L5178Y cells: RPMI 1640 medium (Invitrogen, Paisley, UK) supplemented with 10% heat-inactivated donor horse serum,

2 mmol/L L-glutamine, 2 mmol/L sodium pyruvate, 1% Pluronic F68, 200 IU/ml penicillin, 200 µg/ml streptomycin.

2. 4′-6-Diamidino-2-phenylindole (DAPI).
3. Fresh phosphate buffer (0.66% w/v potassium phosphate monobasic + 0.32% w/v sodium phosphate dibasic, pH 6.4–6.5).
4. Acridine orange (12 mg AO/100 ml buffer).

2.3. Binucleate Assay

See above for main materials.

For human lymphocytes:

1. Accuspin System Histopaque-1077.
2. Lymphocyte media: PB Max Karyotyping Media (Gibco Invitrogen) containing HA16, 0.01 mg/ml (Remel, UK).
3. Cytochalasin B 6 µg/ml.

2.4. Centromeric Labelling

1. 20× SSC stock solution (3 M NaCl 17.53 per 100 ml, 0.3 M trisodium citrate 8.82 g per 100 ml).
2. 2× SSC (1 in 10 dilution of stock).
3. 2× SSC with 0.1% Tween-20 (100 µl of tween-20 in 100 ml of 2× SSC).
4. 0.4× SSC with 0.3% Tween-20 (20 ml 2× SSC + 80 ml water + 300 µl Tween-20).
5. The pan-centromeric paint either human for binucleate assay or mouse for mononuclear assay is available from Cambio UK Star FISH paints (light sensitive).

All solutions are prepared by dilution from stock of 20× SSC, which is stable for 1 year and stored at room temperature.

2.5. Non-disjunction Assay

As the non-disjunction assay described is for primary human lymphocytes and uses the culture media for lymphocytes this is described in the binucleate assay (Subheading 2.3). Non-disjunction is determined by examining the segregation of centromere-specific probes in the binucleate lymphocyte; therefore, the centromere probe materials are same as those in the centromeric labelling (Subheading 2.4).

1. Human centromere-specific probes are available from four suppliers; Cambio, Abbot, MP Biomedicals (formerly Qbiogene) and Poseidon DNA Probes (Kreatech Biotechnology NL). Concentrated probes are needed to allow mixing of individual chromosomes (light sensitive).
2. The program used for centromere-specific probes has denaturation step at 72°C for 2 min followed by 16–40 h at 42°C.

Table 1
Co-factor solution

Co-factor constituent		Amount	Final concentration in co-factor solution
Phosphate buffer	NaH_2PO_4	12 ml	0.11 mmol/L
	$NaH_2PO_4 \cdot 2H_2O$		
Magnesium chloride	$MgCl_2 \cdot 6H_2O$	0.5 ml	19 mmol/L
	KCl		37.5 mmol/L
NADP		79 mg	4.7 mmol/L
G6P		38 mg	6.12 mmol/L
Distilled water		9.5 ml	
Rat S9 (30 mg/ml)		5 ml	20%

3. Methods

3.1. S9 Rat Liver Homogenate

To prepare a 0.2 M Phosphate buffer (pH 7.4)

Mix 60 ml of buffer solution A and 440 ml of buffer solution B. If necessary, adjust to pH 7.4 and sterilise using 0.22 μm filter unit.

To prepare co-factor

Weigh 79 mg nicotinamide adenine dinucleotide phosphate (NADP) and 38 mg sodium salt and glucose-6-phosphate (G6P) and dissolved in:

12 ml of Phosphate buffer.

0.5 ml Magnesium chloride.

9.5 ml Distilled water.

The co-factor solution is filter-sterilised using a 0.22-μm filter.

Immediately prior to use, snap thaw the S9 fraction and add 5 ml to 20 ml co-factor solution and mix well. Then add 0.5 ml of the S9/co-factor solution to 10 ml cultures.

See Table 1.

3.2. Mononucleate Assay

3.2.1. Treatment Schedules

In order detect an aneugen or clastogen acting at a specific stage in the cell cycle it is important that cells are treated with the test substance during all stages of their cell cycle (Table 2).

3.2.2. Cell Culture and Treatment

Cells are cultured in supplemented RPMI 1640 and maintained at 37°C in a humidified atmosphere of 5% CO_2 in air.

Remove cells from liquid nitrogen wash in media and resuspended in fresh media and allow growth for 3–4 days to get sufficient cells for a micronucleus test.

Table 2
Treatment schedules for the mononucleate micronucleus assay

Cell lines treated without cytoB	+ S9	Treat for 3–6 h in the presence of S9; remove the S9 and treatment medium; add fresh medium and harvest 1.5–2.0 normal cell cycles later.
	– S9 Short exposure	Treat for 3–6 h; remove the treatment medium; add fresh medium and harvest 1.5–2.0 normal cell cycles later.
	– S9 Extended exposure	*Option A:* Treat for 1.5–2.0 normal cell cycles; harvest at the end of the exposure period. *Option B:* Treat for 1.5–2.0 normal cell cycles; remove the treatment medium; add fresh medium and harvest 1.5–2.0 normal cell cycles later.

Reproduced from OECD Guideline 487 (5)

To set up test cultures cells are disaggregated and counted, and the volume is then adjusted with fresh media to give an appropriate concentration, (usually 1×10^4 to 2×10^5 cells per ml), and 10 ml aliquots dispensed into tissue culture flasks and incubated until required for cytotoxicity measurements and micronucleus frequency.

Cells are then exposed to the test compound for 3 or 24 h see Table 2. The cells for the micronucleus test are removed from the culture for cytospinning and remaining cells grown on in culture for cell counts to determine population doubling.

1. Individual cultures are vortexed and where possible 850 μl dispensed into a Megafunnel™ and centrifuged at 1,000 rpm (113×g) for 8 min using a Shandon Cytospin 4.
2. Slides are removed from the cytocentrifuge and left to air dry completely.
3. Slides are fixed with 90% methanol for 10 min (see Note 1).
4. For automated scoring using the MicroNuc ™ module of the Metafer system, slides may be stained immediately with DAPI or stored until ready for analysis and then stained. Slides are stained by adding antifade containing DAPI counterstain and mounted with large (22 × 40 mm) coverslips and placed in card trays and stored flat and protected from light prior to scoring on Metafer (15).
5. Following addition of antifade and prior to scoring, slides are examined under the microscope for the presence of nucleated cells (blue from DAPI counterstain).

6. For manual scoring, slides are dipped in fresh phosphate buffer and stained in a solution of acridine orange (AO) for 1 min. Slides are then placed in buffer for 10 min followed by a further 15 min in fresh buffer.
7. After staining, slides are air-dried and stored protected from light.
8. Slides are analysed using an automated scoring system or scored "by eye". Whatever method is used results are recorded appropriately.

3.2.3. Coding of Slides

To prevent bias in the micronucleus scoring, slides may be coded prior to scoring.

1. A code sheet is generated. All cultures are allocated a slide code.
2. The slide codes are written or printed on adhesive labels together with the study number.
3. The code labels are applied to the appropriate slides according to the code sheet. A blank label is also placed on the reverse of the frosted end of the slide to cover all identification marks.
4. Ideally, someone not involved in the micronucleus analysis should perform the coding but when this is not possible it will not invalidate the study.

The code sheet must then sealed and only opened after all analysis is complete for de-coding.

3.2.4. Analysis of Slides (Microscope)

1. Slides are "wet" mounted (carefully avoiding air bubbles) prior to scoring with phosphate buffer and a glass coverslip. Microscopic analysis is performed using a fluorescence microscope with BG-12 excitation filter and 0-530 barrier filter.
2. The cells are identified by the following staining properties of acridine orange: nuclei and micronuclei (DNA) are stained yellow/green and the cytoplasm is stained red. Micronuclei are identified according to the criteria of Countryman and Heddle (16).
3. Counting is performed on electronic digital counters and data recorded on paper.
4. Relative proportions of micronuclei are determined in a total of 1,000 mononuclear cells per culture.
5. 2,000 mononuclear cells are scored for each dose.
6. In the event of an equivocal result, analysis may be extended up to 4,000 mononuclear.
7. Peer review of slide analysis may be undertaken.

3.2.5. Analysis of Slides (Semi-automated Scoring)

The automated system used in our laboratory is the MicroNuc program by MetaSystems (15, 17). This program has been written to detect micronuclei in binucleate cells and contains classifiers that are easily modified to determine the size and shape of nuclei and micronuclei and adaptable to mononuclear screening. The automated system is only used on non-GLP studies.

The slides are scanned on Metafer at 20× magnification on the eight slide automatic stage. The classifier was developed by MetaSystems to score binucleate cells has been modified to score mononucleated L5178Y cells and is deliberately oversensitive to detect all aberrant divisions.

Slides are scanned using the Metafer 4 master station, comprising of a Zeiss Axioplan Imager Z1, equipped with a Maerzhaeuser stepping motor stage that can scan eight slides unattended. The MicroNuc module is run on the Metafer MSearch platform v3.4.102 (MetaSystems GmbH, Altlussheim Germany). Images were acquired on a peltier cooled greyscale digital CCD camera Axiocam MRm (Carl Zeiss). The plane of focus is determined at a number of grid positions that are distributed evenly across the scan area. A predetermined scan area for the Shandon Megafunnel is used for all slides. The scan area is to deliberately avoid the outside margins of the cell preparation area, as this area contains cellular clumps that cannot be scanned accurately.

The nuclei classfier for scanning mononuclear preparations is set to the following criteria: object threshold, 20%; minimum area, 10 μm^2; maximum area, 400 μm^2; maximum relative concavity of depth, 0.9; aspect ratio, 2.5; maximum distance between nuclei, 0 (as this feature is designed for binucleate scoring capabilities); maximum area asymmetry, 90%; region of interest radius, 40 μm; maximum object area in region of interest, 90 μm^2.

The criteria for the micronuclei are set to: object threshold, 10%; minimum area, 1 μm^2; maximum area, 55 μm^2; maximum relative concavity of depth, 1; aspect ratio, 3.5; maximum distance, 35 μm.

Once the slides are scanned, the images collected in the "gallery" were arranged in order of the number of micronuclei they contain. Images were then visually assessed on screen and the number of cells containing true micronuclei counted, thus allowing rejection of artefacts or cell debris from cytotoxicity. The automated scoring was set to capture 2,500 cells (Fig. 3).

Criteria for Evaluation

Micronuclei evaluated should be less than a third of the diameter of the main nucleus, separate from the main nucleus with intact cytoplasmic membrane, and located within the cytoplasmic area. Cells containing three or less micronuclei were assessed.

3.2.6. Flow Cytometry

Micronuclei have been identified by flow cytometry methods by preparing a suspension of nuclei and micronuclei using the method of Nusse and Kramer (18). In brief, the method involves removing the

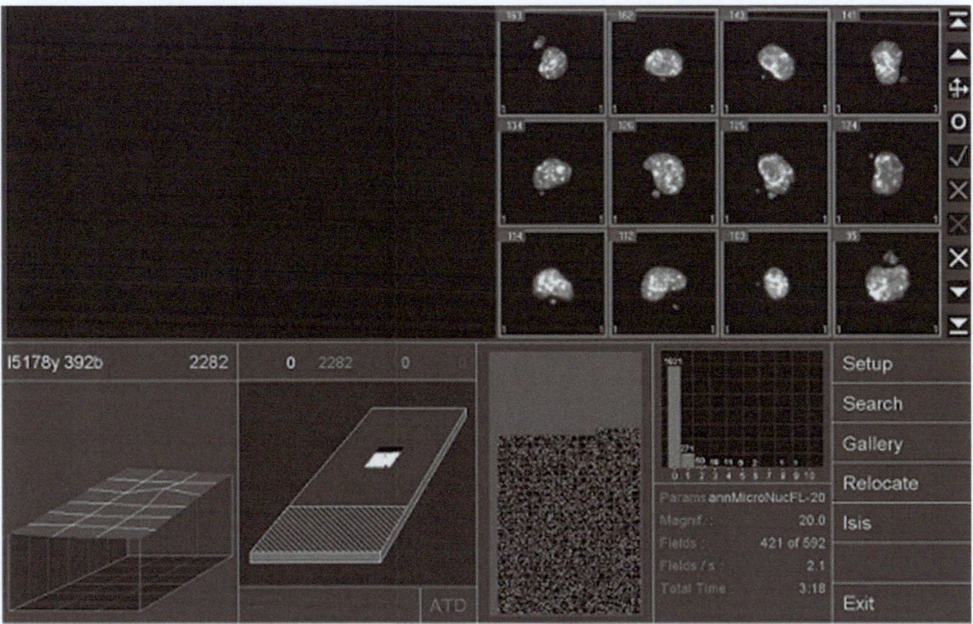

Fig. 3. Screen shot of the automated micronucleus system, Metafer™. Showing the gallery of mononuclear cells with micronuclei.

cell membrane and cytoplasm and staining the cells nuclei and micronuclei with ethidium bromide and analysing by FACS up to 10,000 events. A good correlation has been seen between conventional microscopy and the flow method for five compounds tested (19).

3.3. Binucleate Assay

3.3.1. Treatment Schedules

One of the most important considerations in the performance of the in vitro binucleate micronucleus test is ensuring that the cells being scored have completed mitosis during the treatment or the post-treatment incubation period, if one is used Table 3).

3.3.2. Human Peripheral Blood Lymphocytes

Human lymphocytes have been extensively used in the in vitro binucleate micronucleus assay (20–22). CB is required when human lymphocytes are used because cell cycle times will be variable within cultures and among donors and because not all lymphocytes will respond to phytohaemagglutinin (PHA) (5). See Note 2.

3.3.3. Donors

Blood donors should be less than 45 years, healthy, non-smoking individuals with no known recent exposures to genotoxic chemicals or radiation. Micronucleus frequency increases with age (23).

3.3.4. Lymphocyte Culture

Lymphocytes may be separated prior to culture initiation or at the end of treatment; however, it is faster and simpler to separate the lymphocytes at culture initiation.

Table 3
Treatment schedules for the binucleate micronucleus assay

Lymphocytes, primary cells and cell lines treated with cytoB	+ S9	Treat for 3–6 h in the presence of S9; remove the S9 and treatment medium; add fresh medium and cytoB and harvest 1.5–2.0 normal cell cycles later.
	– S9 Short exposure	Treat for 3–6 h; remove the treatment medium; add fresh medium and cytoB and harvest 1.5–2.0 normal cell cycles later.
	– S9 Extended exposure	*Option A:* Treat for 1.5–2.0 normal cell cycles in the presence of cytoB; harvest at the end of the exposure period. *Option B:* Treat for 1.5–2.0 normal cell cycles; remove the test substance; add fresh medium and cytoB and harvest 1.5–2.0 normal cell cycles later.

Reproduced from OECD Guideline 487 (5)

Lymphocytes are separated by layering fresh whole blood mixed 1:1 with media onto Accuspin System Histopaque-1077 and centrifuged at 800 × g for 15 min or 1,000 × g for 10 min. The mononuclear cell layer is then removed and washed in PBS or media twice and the cell number determined. At this stage cells should be counted on a haemocytometer (not a coulter counter), as the condition of the cells can be seen while counting. Separated lymphocytes are seeded at a density of 2×10^5 cells in 9.9 ml of media.

Human lymphocytes (from both separated and whole blood) are initiated and cultured in lymphocyte media.

Whole blood cultures are initiated by addition of 0.5 ml whole blood to 9.5 ml of media in 25-cm^2 flasks and placed upright in an incubator set at 37°C.

Forty-four hours after initiation whole blood and separated lymphocyte cultures are treated with test compound or vehicle control and simultaneously with addition of CB 6 μg/ml (24).

Sixty-eight hours following initiation whole blood cultures are harvested by separation of lymphocytes (as above). Seventy-two hours following initiation separated lymphocyte cultures are harvested. Slides from both culture methods are prepared by cytocentrifuge (150 × g 5 min) and air-dried. Slides are then fixed in 100% methanol for 8 min and stored at room temperature.

A more detailed human lymphocyte protocol for problem solving can be found in Nature Methods (25).

3.3.5. Staining and Analysis

Slides are dipped in fresh phosphate buffer [0.66%w/v potassium phosphate monobasic + 0.32%w/v sodium phosphate dibasic, pH 6.4–6.5] and stained in a solution of acridine orange (AO), (12 mg AO/100 ml buffer), for 1 min. Slides are then placed in buffer for 10 min followed by a further 15 min in a fresh batch of buffer. After staining, slides are air-dried and stored protected from light.

3.3.6. Coding of Slides

All slides are coded prior to being scored for micronuclei. As described in Subheading 3.2.3.

3.3.7. Analysis of Slides

1. Slides are "wet" mounted (carefully avoiding air bubbles) prior to scoring with phosphate buffer and a glass coverslip. Microscopic analysis is performed using a fluorescence microscope appropriate triple band pass filter.
2. The cells are identified by the following staining properties of acridine orange: nuclei and micronuclei (DNA) are stained yellow/green and the cytoplasm is stained red. Micronuclei are identified according to the criteria of Countryman and Heddle (16) and Fenech (26).
3. Scoring is performed on electronic digital counters and data recorded on paper.
4. Relative proportions of micronuclei will be determined in a total of 1,000 Binucleate cells per culture to give a total of 2,000 binucleates per dose.
5. The number of mononucleate and multinucleated cells are scored alongside the binucleate count to calculate the replicative index. In addition, the number of necrotic and apoptotic cells are noted.
6. In the event of an equivocal result, analysis may be extended up to 4,000 binucleate cells, see Note 3.

3.3.8. Evaluation of Results (Acceptance Criteria and Statistics)

There are several criteria for determining a positive result, such as a concentration-related increase or a reproducible increase in the number of cells containing micronuclei. The biological relevance of the results should be considered first. Consideration of whether the observed values are within or outside of the historical control range can provide guidance when evaluating the biological significance of the response. Appropriate statistical methods should be used to evaluate the test results (27, 28), but should not be the only determinant of a positive response (5). The experimental unit is the cell and reproducibility and biological relevance are paramount in evaluation of results (29).

3.3.9. Criteria for a Valid Assay

For a test to be considered valid, the following criteria should be fulfilled:

1. The mean concurrent vehicle control values fall within the acceptable limits as defined in the Laboratory historical control

data. The positive control group data clearly demonstrates a statistically and biologically significant increase when compared with the concurrent vehicle control group.

2. The test compound should be tested at a dose level equivalent a reduction of cytotoxicity index of 50%.

3.3.10. Evaluation of Data

The result of the test will be assessed using the following criteria:

1. The test will be regarded as clearly negative if there are no increases of either statistical or biological significance in the number of micronuclei at any dose compared with concurrent vehicle control lymphocytes.
2. The test will be regarded as clearly positive if there is an increase in micronuclei that is of statistical and biological significance and that clearly demonstrates a positive trend.
3. If an increase is seen that is statistically different from the concurrent control but does not fulfil the criteria for a positive result as defined in step 2), further statistical and/or microscopic analyses may be performed. However, biological relevance will remain the primary consideration.

3.4. Centromeric Labelling

Care should be taken at all stages of using the chromosome paints to minimise light exposure, as they are photo-degraded.

3.4.1. Fluorescent In Situ Hybridisation Using a Programmable Hotplate Such as HYBrite™ or Thermobrite™

Slides should be aged for at least 24 h at room temperature prior to centromeric labelling to dehydrate. The procedure is the same for both human and mouse probes.

HYBrite™ (A programmable hotplate) is switched on prior to use to allow hotplate to reach 42°C (approx. 10 min). The slides to be painted are warmed by placing them on the hotplate surface prior to use. See Note 4.

The pan-centromeric paint (Cambio, UK) is taken from freezer to thaw prior to use (at least 15 min).

15–20 μl of mouse or human centromeric chromosome paint is added to each slide with a small coverslip. The coverslip is then sealed by adding rubber cement glue around the edges of the coverslip. When the glue has dried the slides are placed on the HYBrite™ hotplate. The program has a denaturation step of 69°C for 5 min followed by hybridisation at 42°C for between 16 and 40 h.

Prepare 0.4× SSC with 0.3% Tween-20 in glass Coplin jar and place in the water bath with water covering approximately ¾ of the height of the Coplin jar.

Heat to 73 ± 1°C and allow approximately 1 h to get to temperature. Place 2× SSC 0.1% Tween 20 in a Coplin jar at room temperature.

Check the temperature of the water bath, see Note 5.

Remove slides from HYBrite$_{TM}$ remove glue and gently slide off the coverslips. Place 2 slides only once (see Note 6) in the

Coplin jar in the water bath, agitate for 3–4 s, and leave in Coplin jar for 1 min.

Transfer slides to Coplin jar containing 2× SSC 0.1% Tween-20 for 2 min. Drain the slides (N.B. do not allow to dry) and add antifade containing DAPI counterstain, add large (22 × 40 mm, or 22 × 50 mm) coverslip, and place in card tray store flat and in the dark prior to scoring on a suitable fluorescent microscope.

Allow temperature in Coplin jar to return to previous level (approx. 5–10 min) before washing the next two slides.

3.4.2. Alternative Protocol for FISH

If a HYBrite™ machine is not available, an alternative method must be used for probe hybridisation (as described in "StarFISH™ Catalogue and Protocols", available from Cambio).

Initially slides are dehydrated via serial ethanol washing in 70%, 80%, 90%, (v/v) ethanol for 2 min each, followed by 5 min in 100% ethanol.

Then slides are denatured in pre-warmed 70% formamide (70 ml formamide + 30 ml 2× SSC solution) at 65°C for 1.5–2 min.

Slides are then quenched in ice-cold 70% (v/v) ethanol for 4 min, followed by subsequent dehydration, as described above.

The whole chromosome probe, in hybridisation buffer, is then warmed to 37°C and denatured at 65°C for 10 min.

The pan-centromeric probe in hybridisation buffer is denatured at 85°C for 10 min and then immediately put on ice.

Finally, the probes would be combined, applied to the slide and then allowed to hybridise at 37°C for approximately 16 h in an airtight, humidified box.

Following hybridisation, slides would be washed twice for 5 min in 50% formamide/2× SSC at 37°C and then twice in 2× SSC for 5 min.

3.4.3. Slide Checking

Check slides after counterstain and antifade are added to determine the presence of centromeric signals (either red or green) under the microscope in any nucleated cells (blue from DAPI counterstain), to determine whether the hybridisation has taken place.

3.4.4. Slide Scoring

Slides are scored using an appropriate fluorescence microscope with triple band pass filter and individual single filters for CY3 and FITC. Scoring is performed on electronic digital counters. Data are recorded on paper.

100 micronuclei should be assessed. In control cultures or low levels of micronuclei induction, this may not be possible, so stop scoring at 20,000 cells.

3.5. Non-disjunction Assay

3.5.1. FISH Method

Slides are aged for at least 24 h at room temperature prior to centromeric labelling, see Note 7.

Methods for FISH are same as in centromeric labelling (Subheading 3.4). However, concentrated probes are used to allow

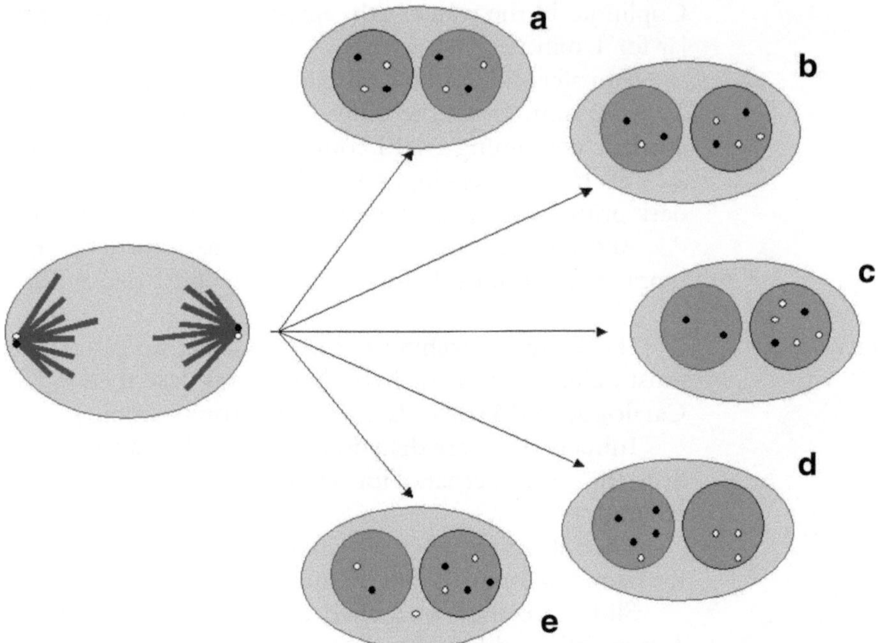

Fig. 4. Cartoon of cell division with *dark and light small circles* representing individual centromere-specific probes. (**a**) 2:2 normal distribution of chromosomes to daughter nuclei. (**b**) 3:1 non-disjunction of the chromosome represented by a *light circle*. (**c**) 4:0 non-disjunction of the chromosome represented by a *light circle*. (**d**) Both a 3:1 non-disjunction of the chromosome represented by a *light circle* and 4:0 non-disjunction of the chromosome represented by a *dark circle*. (**e**) both a 3:1 non-disjunction of the chromosome represented by a *dark circle* and loss of one copy of the chromosome represented by a *light circle* in a micronuclei.

mixing of two (or three) probes specific to individual chromosomes. For example, a probe for centromere of chromosome 2 labelled with FITC may be mixed with a probe for the centromere of chromosome 7 labelled with CY3; the quantities of concentrated probe are approximately 3 μl per individual centromeric paint added to 10–12 μl hybridisation buffer mix (supplied with paint and containing deionised formamide). See Note 8.

The program used for centromere-specific probes has denaturation step at 72°C for 2 min followed by 16–40 h at 42°C.

3.5.2. Slide Checking

Check slides after counterstain and antifade have been added to determine the presence of centromeric signals (both red CY3 and green FITC) under the microscope in any nucleated cells (blue from DAPI counterstain), to determine whether the hybridisation has taken place.

3.5.3. Slide Scoring

100 binucleate cells in which all four signals (2 for each paint) can be seen are selected for scoring and the distribution of centromeric signals in both nuclei recorded; any additional aberrant divisions will also be recorded on paper (Fig. 4).

4. Notes

1. Micronuclei may be investigated for the presence of aneugenic damage. Slides may be fixed in 90% methanol at -20°C and when air dried stored at -20°C if Kinetochore labelling is to be done, i.e. for cells lines that are from species for which centromeric probes are not commercially available such as the rat.
2. PHA obtained from Remel is HA15 or HA16; this is used to stimulate T cell division; however, HA16 purified form will give more consistent stimulation of lymphocytes and should be batch tested along with the serum used to obtain optimal growth.
3. To resolve equivocal data, the number of cells scored may be extended; however, the number to score should be determined with statistical assistance in the cell line used and with reference to the historical control for the individual lab. The ultimate resolution of equivocal data remains a repeat test.
4. An important element of the programmable hotplate method is the humidified atmosphere that is archived by placing water in the wells of the HYBrite machine. This water must never be allowed to seep onto the hotplate surface; or else, a good contact between the hotplate and slide will not be obtained and the hybridisation will fail.
5. The temperature of the water bath is the important one, not the temperature inside the Coplin jar.
6. Do not wash more than two slides at one time, as each slide reduces the temperature of Coplin jar by 1°C. The water bath should be allowed to get back to temperature for at least 20 min prior to washing additional slides.
7. Slides may be aged artificially by baking them in an oven at 60°C for an hour prior to labelling; this step is just to provide adequate dehydration.
8. The amounts of probe given are a starting place as batches of probes vary; it may be possible to reduce the amount of probe, and this should be tested on each new batch.

Acknowledgements

The author would like to thank Julie Hayes and Mick Fellows for the helpful discussions and Karolina Chocian, Vicky Howarth, and Katie Clare for the technical assistance.

References

1. Evans, H.J., Neary, C.J. and Williamson, F.S. (1959) The relative biological efficiency of single doses of fast neutrons and gamma rays on Vicia faba roots and the effect of oxygen, Part II. Chromosome damage: the production of micronuclei. *Int Jnl Rad Biol* **3**, 216–229.
2. Schmid, W. (1975) The micronucleus test. *Mutation Res* **31**, 9–15.
3. Fenech, M. and Morley, A.A. (1985) Measurement of micronuclei in lymphocytes. *Mutation Res* **147**, 29–36.
4. OECD Guideline for Testing of Chemicals. No. 473: *In Vitro* Mammalian Chromosome Aberration Test (1997) www.oecd.org/env/testguidelines
5. OECD Guideline for Testing of Chemicals. No 487 In Vitro Mammalian Cell Micronucleus Test (2010) www.oecd.org/env/testguidelines
6. Migliore, L., Bocciardi, R., Macri, C. and Lo Jacono, F. (1993) Cytogenetic damage induced in human lymphocytes by four vanadium compounds and micronucleus analysis by fluorescence in situ hybridization with a centromeric probe. *Mutation Res* **319**, 205–213.
7. Marshall, R.R., Murphy, M., Kirkland, D.J. and Bentley, K.S. (1996) Fluorescence *in situ* hybridisation (FISH) with chromosome-specific centromeric probes: a sensitive method to detect aneuploidy. *Mutation Res* **372**, 233–245.
8. Zijno, P., Leopardi, F., Marcon, R. and Crebelli, R. (1996) Analysis of chromosome segregation by means of fluorescence *in situ* hybridization: application to cytokinesis-blocked human lymphocytes. *Mutation Res* **372,** 211–219.
9. Doherty, A.T., Ellard, S., Parry, E.M. and Parry, J.M. (1996) A study of aneugenic activity of trichlorofon detected by centromere-specific probes in human lymphoblastoid cell lines. *Mutation Res* **372**, 221–231.
10. Ong, T., Mukhtar, M., Wolf, C.R. and Zeiger, E. (1980) Differential effects of cytochrome P450-inducers on promutagen activation capabilities and enzymatic activities of S-9 from rat liver. *J Environ Pathol Toxicol* **4**, 55–65
11. Elliott, B.M., Combes, R.D., Elcombe, C.R., Gatehouse, D.G., Gibson, G.G., Mackay, J.M. and Wolf, R.C. (1992) Alternatives to Aroclor 1254-induced S9 in *in-vitro* genotoxicity assays. *Mutagenesis* 7, 175–177.
12. Fellows, M.D., O'Donovan, M., Lorge, E. and Kirkland, D. (2008) Comparison of different methods for an accurate assessment of cytotoxicity in the in vitro micronucleus test. II. Practical Aspects with Toxic Agents. *Mutation Res* **655**, 4–21.
13. Lorge, E., Hayashi, M., Albertini, S. and Kirkland, D. (2008) Comparison of different methods for an accurate assessment of cytotoxicity in the in vitro micronucleus test. Theoretical Aspects. *Mutation Res* **655**, 1–3.
14. Hozier, J., Sawyer, J., Moore, M., Howard, B. and Clive, D. (1998) Cytogenetic analysis of the L5178Y/TK$^{+/-}$ mouse lymphoma mutagenesis assay system. *Mutation Res* **84**, 169–181
15. Varga, D., Johannes, T., Jainta, S., Suchuster, S., Schwarz-Boeger, U., Kiechle, M., Garcia, B.P. and Vogel, W. (2004) An automated scoring procedure for the micronucleus test. *Mutagenesis* **19**, 391–397.
16. Countryman, P.I. and Heddle, J.A. (1976) The production of micronuclei from chromosome aberrations in irradiated cultures of human lymphocytes. *Mutation Res* **41**, 321–332.
17. Schunck, C., Johannes, T., Varga, D., Lörch, T. and Plesch, A. (2004) New developments in automated cytogenetic imaging: unattended scoring of dicentric chromosomes, micronuclei, single cell gel electrophoresis, and fluorescence signals. *Cytogenet Genome Res* **104**, 383–389.
18. Nüsse, M. and Kramer, J. (1984) Flow cytometric analysis of micronuclei found in cells after irradiation. *Cytometry* **5**, 20–25.
19. Roman, D., Locher, F., Suter, W., Cordier, A., Bobadilla, M. (1998) Evaluation of a New Procedure for the Flow Cytometric Analysis of In Vitro, Chemically Induced Micronuclei in V79 Cells. *Environmental and Molecular Mutagenesis* **32**, 387–396.
20. Lorge, E., Thybaud, V., Aardema, M.J., Oliver, J., Wakata, A., Lorenzon, G., Marzin, D. (2006) SFTG International collaborative Study on in vitro micronucleus test. I. General conditions of the study. *Mutation Res* **607**, 13–36.
21. Clare, G., Lorenzon, G., Akhurst, L.C., Marzin, D., van Delft, J., Montero, R., Botta, A., Bertens, A., Cinelli, S., Thybaud, V. and Lorge, E. (2006) SFTG International collaborative study on the in vitro micronucleus test. II. Using human lymphocytes. *Mutation Res* **607**, 37–60.
22. Miller, B., Potter-Locher, F., Seelbach, A., Stopper, H., Utesch, D. and Madle, S. (1998) Evaluation of the in vitro micronucleus test as an alternative to the in vitro chromosomal aberration assay: position of the GUM Working Group on the in vitro micronucleus test. Gesellschaft fur Umwelt-Mutations-forschung. *Mutation Res* **410**, 81–116.

23. Bonassi, S., Fenech, M., Lando, C., Lin, Y.P., Ceppi, M., Chang, W.P., Holland, N., Kirsch-Volders, M., Zeiger, E., Ban, S., Barale, R., Bigatti, M.P., Bolognesi, C., Jia, C., Di Giorgio, M., Ferguson, L.R., Fucic, A., Lima, O.G., Hrelia, P., Krishnaja, A.P., Lee, T.K., Migliore, L., Mikhalevich, L., Mirkova, E., Mosesso, P., Muller, W.U., Odagiri, Y., Scarffi, M.R., Szabova, E., Vorobtsova, I., Vral, A. and Zijno, A. (2001) Human MicroNucleus Project: international database comparison for results with the cytokinesis-block micronucleus assay in human lymphocytes. I. Effect of laboratory protocol, scoring criteria and host factors on the frequency of micronuclei. *Environ Mol Mutagen* **37**, 31–45.
24. Ellard, S. and Parry, E.M. (1993) A modified protocol for the cytochalasin B in vitro micronucleus assay using whole human blood or separated lymphocyte cultures. *Mutagenesis* **4**, 317–320.
25. Fenech, M. (2007) Cytokinesis-block micronucleus cytome assay. *Nature Protocols* **2**, 1084–1104.
26. Fenech, M. (2000) The in vitro micronucleus technique. *Mutation Res* **455**, 81–95.
27. Hoffman, W.P., Garriott, M.L. and Lee, C. (2003) *In vitro* micronucleus test, in *Encyclopedia of Biopharmaceutical Statistics*, (Chow, S., ed.), Marcel Dekker, Inc. New York, NY, pp. 463–467.
28. Garriott, M.L., Phelps, J.B. and Hoffman, W.P. (2002) A protocol for the in vitro micronucleus test I. Contributions to the development of a protocol suitable for regulatory submissions from an examination of 16 chemicals with different mechanisms of action and different levels of activity. *Mutation Res* **517**, 123–134.
29. Kirsch-Volders, M., Sofuni, T., Aardema M., Albertini, S., Eastmond, D., Fenech, M., Ishidate Jr., M., Kirchner, S., Lorge, E., Morita, T., Norppa, H., Surrallés, J., Vanhauwaert, A. and Wakata, A., (2003) Report of the in vitro micronucleus assay working group. *Mutat. Res.* **540**, 153–163

Chapter 8

The In Vitro and In Vivo Comet Assays

Brian Burlinson

Abstract

The strategy for testing for genotoxicity covers three main areas, namely gene mutation, chromosome aberration or breakage (clastogenicity), and chromosome loss or gain (aneuploidy). The current generalized strategy consists of assays capable of detecting all of these endpoints using in vitro assays such as the Ames test for detecting gene mutations in bacteria, the human peripheral lymphocyte chromosome aberration (CA) test for detecting clastogenicity, and the in vitro micronucleus test for clastogenicity and aneuploidy. The primary in vivo assay, and generally the only in vivo assay required, is the in vivo rodent bone marrow micronucleus assay. However, there are instances when these assays alone are inadequate and further testing is required, especially in vivo. Historically, the preferred second assay has been the rodent liver unscheduled DNA synthesis assay but recently this has been superseded by the rodent single cell gel electrophoresis or Comet assay. This assay has numerous advantages especially in vivo, where virtually any tissue can be examined. The status of the in vitro comet assay in regulatory testing is much less clear although a preliminary review of data from the assay has shown it to be more specific than other in vitro genotoxicity tests and less prone to false positives.

Detailed here are general protocols for both the in vitro and in vivo comet assays which will form the basis of the pending OECD guideline for the assay.

Key words: In vivo comet, In vitro comet, Genotoxicity, DNA strand breaks, Gel electrophoresis, Alkali labile sites, DNA adduct, GLP

1. Introduction

The single cell gel electrophoresis assay, i.e., the Comet assay, is a fairly simple procedure based upon the original idea of Östling and Johanson (1) where they used a microgel and electrophoresis to detect DNA damage. Their technique was further developed by Singh et al. (2) who introduced the use of high alkaline conditions (>pH 13). This step increased the ability of the assay to detect not only the double-strand breaks (DSB) of the Östling and Johanson method but also alkali labile sites (ALS) and single-strand breaks (SSB). Since its development there have been numerous publications

James M. Parry and Elizabeth M. Parry (eds.), *Genetic Toxicology: Principles and Methods*, Methods in Molecular Biology, vol. 817, DOI 10.1007/978-1-61779-421-6_8,

both on defining a suitable internationally acceptable protocol for the assay and the application of the assay to various genotoxicity testing strategies (3–14).

The in vivo comet assay is probably the most useful because it requires only a few hundred cells and so can be used to investigate virtually any tissue. This makes it a useful adjunct to the in vivo chromosome assay in cases such as

1. A positive in vitro assay.
2. When the bone marrow may not be exposed to the test substance.
3. When the site of first contact with the material is of interest, e.g., the stomach after oral dosing.
4. When the standard battery of genetic toxicology tests are negative but neoplastic changes are seen in particular tissues after long-term testing.

The assay involves treating animals or cells with a test substance and leaving them for a set of exposure period. For single cells, this generally includes treating the cells with the test material in the presence and absence of an exogenous metabolizing system, e.g., the postmitochondrial fraction (S9) of rat livers induced with Arochlor 451 (15) or a mixture of phenobarbitone sodium and ß napthaflavone (16, 17). Isolated cell suspensions are then mixed with agar and used to form the central layer of an agar "sandwich" on a microscope slide. Once set, the slides are immersed in chilled complete lysis solution, in a light proof box and refrigerated for a minimum of 1 h. This lysis solution removes the cellular and nuclear membranes leaving the tightly coiled DNA in what is referred to as the nucleoid.

For electrophoresis, the slides are placed onto a level platform of an horizontal electrophoresis unit containing chilled electrophoresis buffer. The nucleoids are left to unwind at 2–8°C for approximately 20 min, depending on the cell type under investigation. This unwinding, as its name suggests, allows any breaks in the DNA to release the tension in the tightly coiled DNA and also release fragments of DNA produced by double-strand breaks.

After alkali unwinding, the slides are electrophoresed at 18–25 V and approx. 300 mA (between 0.7 and 1.0 V/cm) for between 15 and 40 min. The electrophoresis pulls the negatively charged DNA toward the anode, so forming the distinctive "comet" shape. When electrophoresis is complete, the slides are washed with neutralization buffer and stored, refrigerated. Although methods for assessing comet slides by eye are available (18), analysis of slides for GLP or regulatory submission is usually via a CCD camera and associated image analysis software. Proprietary software, such as the Comet IV system by Perceptive Instruments, can measure a number of parameters, e.g., tail length, tail moment, and tail intensity

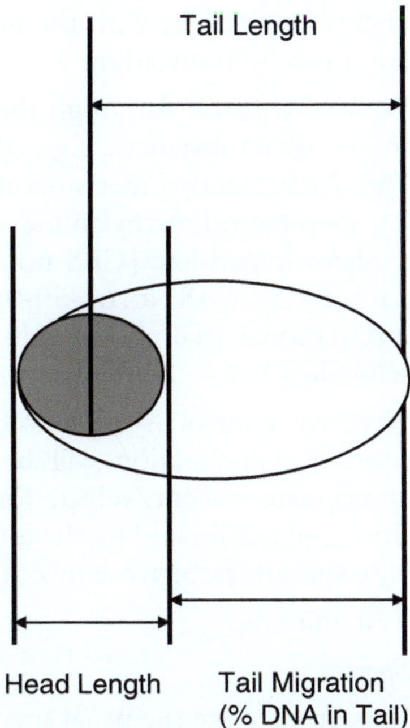

Tail Moment = product of Tail length and % DNA in Tail

Fig. 1. Potential comet measurements by image analysis.

(% of DNA in the "tail" of the comet) see Fig. 1, with the tail intensity being the measure most favoured for comparison between studies and laboratories (19).

Currently, an international workgroup led by the Japanese Committee on the Validation of Alternative Methods (JaCVAM), which includes representatives from the Interagency Coordinating Committee on the Validation of Alternative Methods (ICCVAM) and the European Centre for the Validation of Alternative methods (ECVAM) are working to produce an OECD guideline (20). The protocols given here are those which have been developed or adopted for use by participating laboratories in the JaCVAM initiative.

2. Materials

2.1. *In Vivo* Comet Assay

1. Test substances and positive/negative controls: Solid test substances should be dissolved or suspended in appropriate solvents or vehicles and diluted, if appropriate, prior to dosing of the animals. The solvent/vehicle should not produce toxic effects at the dose levels used and should not be suspected of

chemical reactivity with the test substance. Suggested vehicles are given in Subheading 2.1 item 3.

2. Positive control: Although there are numerous potential positive control substances, e.g., ethyl nitrosourea (ENU) [CAS no. 759-73-9); methyl methanesulfonate (MMS) [CAS no. 66-27-3); *N*-nitrosodimethylamine (N-DMA) [CAS no. 62-75-9); 1-nitrosopiperidine [CAS no. 100-75-4), ethylmethyl sulphonate (EMS), CAS No. 62-50-0, is recommended in the JaCVAM international trial and should be formulated in physiological saline just before administration and used within 2 h.
3. Negative control (solvent/vehicle): Solvents/vehicles for test substance preparation will be used as negative controls. An appropriate solvent/vehicle for a test substance can be chosen from, but not limited to, the following: physiological saline, 0.5% w/v sodium carboxymethylcellulose aqua solution, corn oil.
4. Test animals:

   Species

   Rats or mice are the usual species, although under special conditions other laboratory animals may be considered. Justification for the choice of species must be given. The 3Rs-initiative for experimental animal use must be considered when designing the experiment.

   Sex

   Males and females may be used in the comet assay.

   Strain

   Any of the standard laboratory strains can be used, e.g.,

   Mouse: Crl:CD1™(ICR) (CD-1)

   Rat: Crl:CD (SD), or Han Wistar

   Source

   Animals can be obtained from any of the reputable laboratory animal breeders, e.g., Charles River Laboratories, Inc.

   Age

   At the time of purchase: Rats: 6–8 weeks of age (body weight 150–320 g)

   Mice: 6–8 weeks of age (body weight 22–32 g).

   At the time of dosing: Rats and Mice: 7–9 weeks of age.

   Body weight

   The weight variation of animals should be ±20% of the mean weight at the time of dosing.

   Animal quarantine and acclimatization

   Animals should be quarantined and acclimatized for at least 5 days prior to the start of the study, according to standard operating procedures (SOP) in each testing facility. Only healthy animals approved by the Study Director and/or the Animal Facility Veterinarian should be used.

Animal maintenance

Animals will be reared under appropriate housing and feeding conditions according to the SOP in each testing facility. Appropriate national and/or international regulations on animal welfare must be followed.

Animals will be fed ad libitum with a commercially available pellet diet and be given free access to tap water ad libitum.

Animal identification and group assignment

Animals will be identified uniquely and assigned to groups by randomization and identified according to normal practice for that laboratory.

2.2. In Vitro Comet Assay

Cells:

(a) L5178Y mouse lymphoma cells, subline 3.7.2.c, are heterozygous for the thymidine kinase locus ($TK^{+/-}$) American Type Culture Collection (ATCC), Virginia.

(b) TK6 human lymphoblast cell, ATCC, Virginia.

(c) Human peripheral lymphocytes.

Culture Media:

(a) L5178Y: RPMI 1640, Sigma Cell Culture Ltd., Life Technologies,

Heat-inactivated donor horse serum (HiDHS). Biosera East Sussex.

(b) TK6: RPMI 1640,

Heat-inactivated foetal bovine serum (FBS). Biosera East Sussex
L-Glutamine – SAFC biosciences Ltd., Andover, England
Gentamicin and Sodium Pyruvate – Gibco Invitrogen, Paisley, Scotland
Pen/Strep, HiDHS, and FBS – Sigma-Aldrich, Poole, England
Synperonic F68 – Serva Electrophoresis, Heidelberg, Germany.

(c) Human peripheral Lymphocytes

RPMI 1640
Foetal calf serum, Sigma-Aldrich, Poole, England
heparin,
L-glutamine.

S9 Mix:

S9 Fraction. KCl-buffered, Phenobarbital/5,6-benzoflavone induced S9, Molecular Toxicology Incorporated, USA.

Co-factors.
Glucose-6-phosphate.
NADP.
R0.
NaOH.

Positive Controls:

Methylmethane sulphonate (MMS) – Sigma/Aldrich Poole, England.

Ethylmethane sulphonate (EMS) – Sigma/Aldrich.

Cyclophosphamide (CPA) – Sigma/Aldrich.

2.3. Lysis and Electrophoresis

Lysis and Electrophoresis Solutions

1. Mincing buffer (see Note 1)
   20 mM EDTA (disodium) (Sigma/Aldrich, Gillingham, Dorset).
   10% DMSO (Sigma/Aldrich, Gillingham, Dorset).
   Hank's Balanced Salt Solution (HBSS) (Ca^{2+}, Mg^{2+} free, and phenol red free if available), (Gibco and Invitrogen Ltd., Paisley).
2. Lysing solution (see Note 2)
   100 mM EDTA (disodium) (Sigma/Aldrich, Gillingham, Dorset).
   2.5 M sodium chloride (Fisher Scientific, Loughborough).
   10 mM tris hydroxymethyl (Sigma/Aldrich, Gillingham, Dorset).
   1 M sodium hydroxide (Fisher Scientific, Loughborough).
   1 M hydrochloric acid (Sigma/Aldrich, Gillingham, Dorset).
   1% (v/v) of Triton-X100 (VWR, Lutterworth, Leicestershire).
   10% (v/v) DMSO (Sigma/Aldrich, Gillingham, Dorset).
3. Alkaline solution for unwinding and electrophoresis (see Note 3).
   300 mM sodium hydroxide (Fisher Scientific, Loughborough)
   1 mM EDTA (disodium). (Sigma/Aldrich, Gillingham, Dorset).
4. Neutralization solution (see Note 4).
   0.4 M tris hydroxymethyl aminomethane (Tris).
5. Staining solution
   SYBR Gold (Invitrogen-Molecular Probes). (Gibco and Invitrogen Ltd., Paisley).

Agarose Gel: (see Note 5)

1. 1.0–1.5% (w/v) standard agarose gel (Sigma/Aldrich, Gillingham, Dorset).
   Dulbecco's phosphate buffer (Ca^{2+}, Mg^{2+} free, and phenol free (Gibco and Invitrogen Ltd., Paisley)).

2. 0.5% (w/v) low-melting agarose (Lonza, NuSieve GTG Agarose) (Lonza Biologics, Slough, Berks).

   Dulbecco's phosphate buffer (Ca^{2+}, Mg^{2+} free, and phenol free (Gibco and Invitrogen Ltd., Paisley)).

Electrophoresis Equipment:

Any standard flat bed gel electrophoresis tank can be used. However, the more slides that can be run in one go the better. A tank capable of holding 40 slides is supplied by Geneflow, Fradely, Staffordshire and the power supply for the tanks is supplied by Fisher Scientific, Loughborough.

Image Analysis:

Although it is not absolutely necessary to measure the comets by image analysis it is accepted that this will give data which can be more easily compared to data from other laboratories and will be more acceptable to regulatory agencies. Details of how slides can be scored without image analysis are given in ref 17.

Although there are other sources of image analysis systems and software the Perceptive Instruments COMET IV™ image analysis system is recommended.

3. Methods

A flow diagram of the overall method for both in vivo and in vitro assays is given in Fig. 2.

3.1. In Vivo Comet Assay

3.1.1. Experimental Design

The protocol described here is derived from that used in the JaCVAM international trial which will form the basis of the OECD guideline. In this design, five animals per sex per group are used, see Table 1. However, if data are available from the same species using the same route of exposure which show there are no substantial differences in toxicity or metabolic profile between the two sexes then only one sex, generally males are required.

The high dose level of a test compound will be selected as the dose producing signs of toxicity such that a higher dose level, based on the same dosing regimen, would be expected to produce mortality or an unacceptable level of animal distress. In the absence of signs of toxicity, current guidelines recommend a maximum dose level of 2,000 mg/kg (21, 22) although this is under review and may lead to the maximum dose being limited to 1,000 mg/kg for pharmaceuticals.

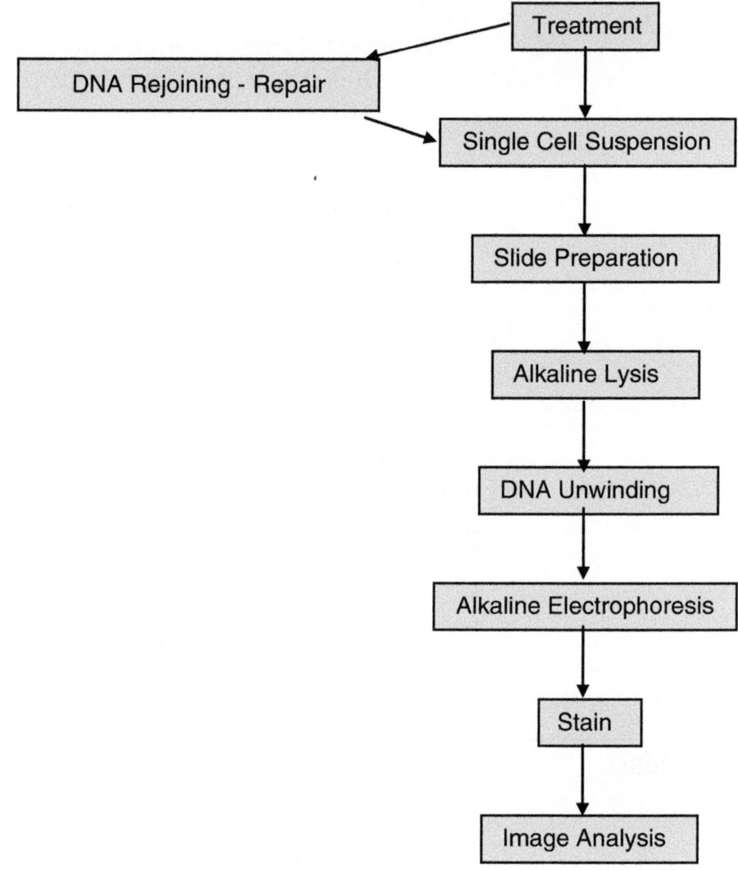

Fig. 2. Flow diagram of the comet assay.

Table 1
In vivo comet assay protocol

Compound	Dose (mg/kg/day)	Number of animals per sex
Vehicle (negative control)	0	5
EMS (positive control)	200	5
Test compound	Low (1/4 of high)	5
Test compound	Medium (1/2 of high)	5
Test compound	High (see text)	5

The test substance will be administered as a single or repeated treatment. Various dosing regimes have been suggested and these are outlined below:

1. Single dose with animals being killed and tissue/organs sampled at 2–6 and 16–26 h after dosing. This regimen is more costly in animals since control animals are needed for the two kill times.
2. Repeated dose, either two, three times at 24 h intervals with a kill 2–6 h after the last dose. With this option there is the ability to also sample bone marrow for micronucleus formation.
3. A recent development has been the inclusion of genotoxicity endpoint analysis at the end of a longer term, e.g., 28 day general toxicity study (23). These "bolt on" studies can be comet or blood and bone marrow micronucleus assays.

3.1.2. Dose Administration

The test substance is usually administered orally by gavage.

The maximum volume of liquid that can be administered by gavage at one time should be based on the size of the test animal and should not exceed 20 mL/kg body wt. The use of higher volumes must be justified. Routes of exposure other than oral (e.g., dermal or inhalation) are acceptable, where justified. However, the i.p. route is not recommended when examining tissues such as the liver that could be exposed directly to the test substance rather than via the circulatory system.

Inhalation exposures may be appropriate for testing gases, dusts, vapors, or aerosols and depending on the nature of the test substance and the tissue(s) to be sampled, either whole body or nose only exposure may be appropriate.

3.1.3. Clinical Observations and Bodyweight

Individual body weights should be measured in accordance with local SOPs and just prior to administration of the test and control material and at the time of termination. The animals should be observed for clinical signs from just after dosing to just before tissue removal with an appropriate interval according to the SOP in each testing facility.

3.1.4. Tissue Sampling

Animals should be humanely killed 3 h after the third administration of a test substance and at 3 h after the second treatment of EMS. Although virtually any tissue can be investigated using this assay, detailed procedures for the liver, stomach, and bone marrow are given here (see Note 6). The stomach and the liver will be removed. Tissues are then placed into ice-cold mincing buffer, rinsed sufficiently with the cold mincing buffer to remove residual blood (more rinses would likely be needed if exsanguination is not used), and stored on ice until processed.

In the event of a positive response in the comet assay, a sample of tissue should be taken, fixed in formalin, and sent for histopathological assessment for the presence and incidence of apoptotic and/or necrotic cells.

3.1.5. Cell Preparation

Single cell preparation should be done within 1 h after animal sacrifice. The liver and the stomach and bone marrow are processed as follows:

1. *Bone Marrow*
   Foetal calf serum (FCS) is completely defrosted, mixed, and passed through a 0.2-µm millipore filter to remove any particles. Filtered FCS is dispensed into 2 mL aliquots (for mice) or 3 mL aliquots (for rats) into prelabelled tubes.

   One femur is dissected from each animal and the proximal heads are removed. The bone marrow from the femur is pooled in suspension by drawing the FCS from the appropriate tube through the bone using a 2-mL disposable syringe fitted with a 21-g needle. The liquid is passed back through the needle and the bone into the sample tube until all the bone marrow has been removed. A single cell suspension is ensured by aspiration of the liquid into the syringe and back into the sample tube. The syringe and needle are replaced for each group of animals.

   The cell suspension is centrifuged at ca. 1,000 rpm ($150 \times g$) for 5 min, the supernatant serum removed and the cells are resuspended in 2 mL of prefiltered FCS.

   The sample is stored on ice until slide preparation.

2. *Liver*
   Approx. 0.5 cm^3 of liver is cut into several small pieces and washed in fresh Merchants solution until as much blood as possible has been removed.

   The pieces are then transferred to 150 µm bolting cloth held over a 50-mL falcon tube.

   2 mL of Merchants solution is added to the tissue and the liver pushed through the cloth using a plunger from a disposable syringe.

   Another 2 mL of Merchants is added and any remaining liver pushed through the bolting cloth.

   The sample is stored on ice until slide preparation.

3. *Stomach*
   The forestomach (nonglandular) is removed and discarded unless it is a target tissue for the assay. The glandular section is cut open and washed free from food using ice-cold saline solution.

The stomach is then placed into a suitable container, covered with approx. 2 mL fresh Merchants, and incubated on ice for 30 min ± 10 min.

After incubation, the stomach is removed and the surface epithelia is gently scraped a couple of times using a cell scraper. This layer is discarded and the gastric mucosa rinsed with fresh Merchants solution.

2 mL of Merchants is added to a clean petri dish and the stomach is carefully scraped four to five times using a cell scraper to release the cells.

The cells are collected into a clean test tube using a disposable pipette and the sample is stored on ice until slide preparation.

3.2. In Vitro Comet Assay

3.2.1. Cell Culture and Maintenance

The L5178Y mouse lymphoma cells, subline 3.7.2.c, are heterozygous for the thymidine kinase locus ($TK^{+/-}$). The method for removal of spontaneously occurring $TK^{-/-}$ mutants and subsequent preparation of frozen stocks of L5178Y cells are described by Amacher, Paillet, and Ray, 1979 (24).

The TK6 human lymphoblast cell lines are stored at −196°C, in heat-inactivated FBS containing 10% DMSO.

L5178Y and TK6 cells are stored in polypropylene ampoules in liquid nitrogen. To establish a cell culture from the frozen stock, one ampoule is rapidly thawed. The ampoule should be swabbed with 70% alcohol, before the contents are removed.

For L5178Y cells the cell suspension is added to RPMI 1640, buffered with 2 mg/mL sodium bicarbonate and supplemented with 2.0 mM L_glutamine, 50 g/mL gentamicin, 0.1% v/v Synperonic F68, 1.0 mM sodium pyruvate and heat_inactivated donor horse serum (HiDHS) at 10% v/v.

For TK6 cells the cell suspension is added to 30 mL RPMI 1640, supplemented with 200 ug/ml sodium pyruvate, 100 U/ml penicillin-100 ug/ml streptomycin and heat-inactivated fetal bovine serum (FBS) at 10 % v/v..

Cells may be centrifuged to remove the DMSO, if deemed necessary. The cells are transferred to a vented tissue culture flask with a filtered lid. Flasks are placed in a static CO_2 incubator (37°C, 5% CO_2).

The cells grow in suspension culture with a population doubling time of approximately 12 h. Cultures should be maintained at cell population densities of between 10^4 and 10^6 cells/mL, with subculturing at least every 2–3 days. It is important to maintain the cells in an actively dividing state as much as possible throughout each test. Cell cultures must be used within 10 days of recovery from frozen stock.

L5178Y and TK6 cells tend to grow in small clumps, which must be disaggregated before accurate cell counts can be made. Disaggregation is achieved by shaking the flask vigorously and/or vigorously aspirating the contents of the flask by pipette approximately five times. Immediately after disaggregation 1 mL of cell suspension is transferred to a dilution cup containing 9 mL Isoton II for counting. The cells may then be counted using a Coulter electronic particle counter or on a haemocytometer.

Human peripheral lymphocytes are taken by venupuncture from healthy nonsmoking donors and diluted with tissue culture medium (RPMI 1640) containing 10% FCS, heparin, L-glutamine, and antibiotics. The cultures are prepared as 5 mL aliquots (0.4 mL blood: 4.5 mL medium in sterile universal containers and incubated at 37°C). The cultures will be occasionally shaken to resuspend the cells. In some cases, the lymphocytes may be isolated from whole blood but this is generally not necessary.

3.2.2. In Vitro Comet Assay Experimental Design

At least four dose levels (half dilutions) of the test substance are tested, together with at least two solvent controls and a known mutagen (positive controls).

The test substance is dissolved and serial dilutions prepared, usually at 100× the desired final concentrations, shortly before addition to the cell suspensions. The solubility of each test substance is assessed on a case-by-case basis and some substances may be prepared at either more or less than 100× the desired final concentration. In such cases, the volume of test substance dilution added to the cell cultures will be decreased or increased, as appropriate.

If a test substance has been prepared at more or less than 100× the desired final concentration, then the volume of solvent or test substance solution, and the population density of the cell cultures and volumes may be decreased or increased accordingly. Solvents that are used must be compatible with the test system and include, water, DMSO, saline, or culture media.

S9 Mix

Cells should be exposed to the test chemicals both in the presence and absence of the metabolic activation system (S9-mix). The S9 is prepared from the livers of rats treated with Aroclor 1254 or a

Table 2
In vitro comet assay protocol

Components	Composition	Concentration in S9-mix	[a]Concentration in culture
S9	4 mL	40 vol%	2 vol%
G-6-P	2 mL of 180 mg/mL sol.	118 mM	5.90 mM
NADP	2 mL of 25 mg/mL sol.	6.4 mM	0.32 mM
KCl	2 mL of 150 mM sol.	30 mM	1.50 mM

[a]Treatment in culture: 5% S9-mix

combination of phenobarbitone and beta-naphthoflavon (generally preferred). The standard S9-mix is prepared by combining 4 mL S9 and 2 mL each 180 mg/mL glucose-6-phosphate, 25 mg/mL NADP, and 150 mM KCl. The concentration of S9-mix is 5% during treatment and the final concentration of S9 is 2%. The preparation of S9-mix is shown in Table 2.

Positive Controls: Any positive control can be used in this assay as long as the chosen material gives a clear unambiguous positive response. Given below are materials and dose levels which are known to work well in this assay.

L5178Y cells: In the absence of S9 mix, the positive control compound is methyl methanesulphonate (MMS), at a final concentration of 12.5 μg/mL (3 h exposure).

TK6 cells: In the absence of S9 mix, the positive control compound is ethyl methanesulphonate (EMS), at a final concentration of 250 μg/mL (3 h exposure).

L5178Y and TK6 cells: In the presence of S9 mix, the positive control compound is CPA, at a final concentration ranging between 25 and 100 μg/mL (3 h exposure).

Two 25 cm^2 vented tissue culture flasks are required for each concentration of the test compound and for the positive and solvent controls. Each flask should be labelled with the treatment code.

Cell population of 2×10^5 cells (9 mL).

1 mL KCl or 1 mL S9 mix.

1–10% v/v of the test substance formulation (at 100 times the desired final concentration) solvent or positive control.

After the addition of the test substance formulations, the cultures should be examined by eye for the presence of precipitate and colour change in media (due to pH changes) and any observations noted.

Flasks are placed in a static CO_2 incubator (37°C, 5% CO_2) for between 3 and 6 h. The JaCVAM international trial recommends 4 h exposure.

At the end of exposure, posttreatment observations are taken. The cultures should be examined by eye for the presence of precipitate and colour change in the media and any observations noted.

At the end of the 3-h exposure period, 1 mL of each cell culture will be transferred to the corresponding sterile universal. The cells are centrifuged at 1,000 rpm (150×*g*) and washed with 1 mL of phosphate-buffered saline, centrifuged again at 1,000 rpm (150×*g*) and resuspended in 0.5 mL of phosphate-buffered saline.

Cytotoxicity determination

A 200-μL sample of each culture will be used to measure cytotoxicity by trypan blue exclusion. Two hundred cells per culture are counted in order to determine cytotoxicity.

A further measure of toxicity is the relative cell growth. After samples have been taken for the slide preparation, the remaining culture should be centrifuged at low speed (approximately 1,000 rpm (150×*g*) for 5 min), and the supernatant discarded. Each culture is then washed with 5 mL of fresh medium once by resuspension and centrifugation. The cells are then resuspended in 10 mL of fresh medium and transferred to culture bottles (TS-25) or culture dishes. The cell density is measured by a haemocytometer or an automatic cell counter before starting culture. The cultures are incubated at 37°C in a humidified incubator gassed with 5% CO_2 and in air. Twenty-four hours later, the cell density is measured again. The relative cell growth compared to the solvent control is then calculated.

3.3. Slide Preparation, Lysis, and Electrophoresis

3.3.1. Slide Preparation

Frosted-end glass slides are dipped in 1% normal melting point agarose (NMPA) and left to air dry prior to the addition of the cell suspension layer. It is important to ensure that the frosted end of the slide is dipped into the agar to prevent agar loss from the slides during lysis. These predipped slides can be stored in an airtight container for approximately 1 month.

At least three slides per tissue or cell culture are prepared and labelled with the study number and a code number which refers to the tissue type, animal number, or the culture and the date.

For each tissue/culture, an appropriate dilution of the cell suspensions is made and mixed with the appropriate concentration of 0.5% low melting point agarose (LMPA). 75 μL of the cell/agar mix is dispensed onto the predipped slide and covered with a clean coverslip.

Once the agar has set, the coverslips are carefully removed and the slides are immersed in chilled complete lysis solution, in a light proof box, for a minimum of 1 h refrigerated. If necessary, the slides can be stored refrigerated in lysis solution for 14 days prior to electrophoresis.

3.3.2. Lysis

The lysing solution consists of 100 mM EDTA (disodium), 2.5 M sodium chloride, and 10 mM tris hydroxymethyl aminomethane in purified water, with the pH adjusted to 10.0 with 1 M sodium hydroxide and/or hydrochloric acid. This solution can be kept refrigerated at <10°C until use. On the day of use, 1% (v/v) of Triton-X100 and 10% (v/v) DMSO is added to make the complete lysing solution which should be kept refrigerated at <10°C for at least 30 min prior to use.

Once prepared, the slides are immersed in chilled lysing solution for at least 1 h or overnight in a refrigerator in a light proof container. After this incubation period, the slides are rinsed in purified water or neutralization solution to remove residual detergent and salts prior to the alkali unwinding step.

3.3.3. Unwinding and Electrophoresis

The slides are randomly placed onto a dry, level platform of a horizontal electrophoresis unit. The slides from each treatment group should be spread across the platform to avoid any positional effects. If more than one electrophoresis unit is required, then the slides from one tissue should be split between separate units.

The buffer reservoir of the unit is topped up with electrophoresis buffer until the surfaces of the slides are covered. The nucleoids are left to relax and unwind at 2–10°C for 20 min depending on the cell type under investigation.

After alkali unwinding, the slides are electrophoresed at 18 V and approx. 300 mA (between 0.7 and 1.0 V/cm) for 20 min although experience may show that shorter or longer times may be needed for different cell types (see Note 7). As DNA carries a net negative charge, the nonsuper coiled loops and single strand fragments migrate toward the anode.

The temperature of the electrophoresis solution at the start of unwinding, the start of electrophoresis, and the end of electrophoresis should be recorded.

3.3.4. Neutralization and Dehydration of Slides

After completion of electrophoresis, the slides are immersed in the neutralization buffer for at least 5 min. All slides are dehydrated by immersion into absolute ethanol (≥99.6%) for at least 5 min if slides will not be scored soon, allowed to air dry, and then stored until scored at room temperature, protected from humidity.

3.3.5. DNA Staining, Comet Visualization, and Analysis

The slides are labelled with a random code number and are "scored blind" to prevent operator bias.

The slides are first stained with 45 μL of SYBR GOLD and the comets visualized using a fluorescence microscope linked to a computer via a CCD camera, and measured using an image analysis system, e.g., Perceptive Instruments COMET IV™.

Although it is not absolutely necessary to measure the comets by image analysis, it is accepted that this will give data which can be more easily compared to data from other laboratories and will be more acceptable to regulatory agencies.

Heavily damaged cells exhibiting a microscopic image, commonly referred to as hedgehogs or ghost cells (see Fig. 3) consisting of small or nonexistent head and large, diffuse tails will be excluded from data collection if the image analysis system cannot properly score them (see Note 8). However, the frequency of such comets should be determined per sample, based on the visual scoring of 100 cells per sample. Care should also be taken to avoid any selection bias, counting of overlapping cells, and edge areas of slides. Examples of positive and negative comets are given in Figs. 4 and 5, respectively.

Where possible, 50 cells are scored per slide to give a total number of 150 cells per culture or per tissue per animal.

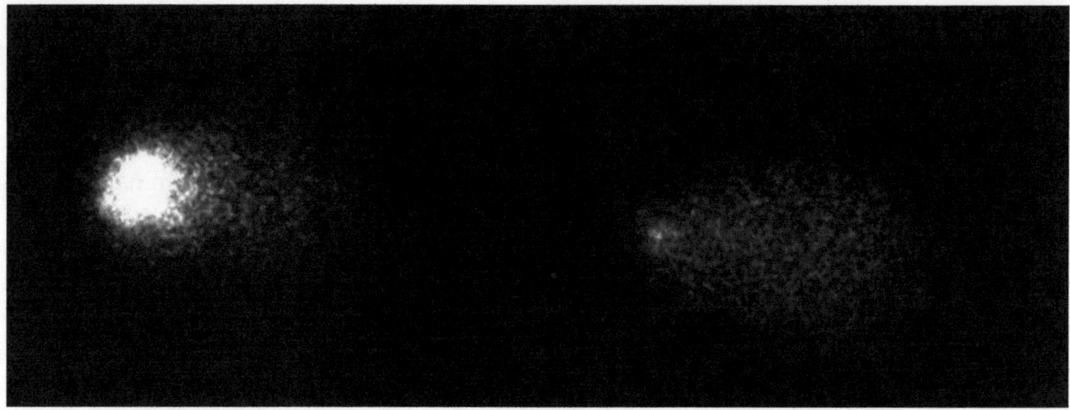

Fig. 3. Photograph showing a positive (P) comet *on the left* and a "hedgehog" (H) comet *on the right.*

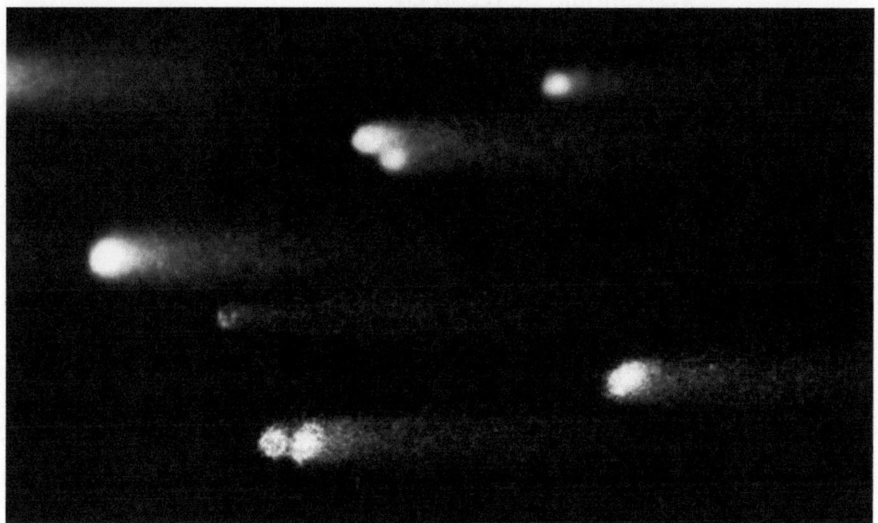

Fig. 4. Photograph of positive comets from rat hepatocytes.

Fig. 5. Photograph of negative comets from rat hepatocytes.

Using the image analysis system three different measurements of the comet can be made:

Tail length, defined as a measurement from the point of greatest intensity within the comet head, to the end of the fluorescing tail.

Tail moment, defined as the product of the tail length and the fraction of total DNA present within the tail.

Tail intensity, defined as the fluorescence detected by image analysis in the tail, which is proportional to the amount of DNA that has moved from the head region into the comet tail.

Of the three measurements, the tail intensity has been shown to be the measure most easily compared either between experiments in the same laboratory and also between laboratories (19).

3.4. Data Analysis

Vehicle and positive control animals or cultures should have an individual and group mean tail intensity, tail moment, and tail length value within or close to their respective laboratory historical ranges. Table 3 gives examples of the positive and negative control ranges expected for the stomach, liver, and bone marrow in rats.

The biological relevance of the results should be considered first, e.g., toxicity seen either in terms of debris on the slide, reduced numbers of scorable cells, or increased numbers of "hedgehogs." For in vivo studies, the samples taken for histopathology should be examined for evidence of apoptosis, necrosis, hyperplasia, or any disturbance in the overall tissue morphology. The presence of any of the above must be taken into account when interpreting the comet data.

In the in vitro assay, the viability of the test cultures should be no less than 50% of the negative control viability.

A negative result is normally indicated where individual and group mean incidences of tail intensity, tail moment, and tail length for the group treated with the test substance are not significantly greater than incidences for the vehicle control group and where these values fall within the historical control range.

An equivocal response is obtained when the results do not meet the criteria specified for a positive or negative response. Where results remain equivocal, further testing and/or modification of the study design may be required.

A positive response is normally indicated by a statistically significant increase in the tail intensity, tail moment, and tail length for the treatment group compared with the vehicle control group ($P<0.01$); individual and/or group mean values should exceed the current laboratory historical control range. Different statistical methods have been proposed for the comet assay although there has been no decision so far on the most acceptable (25–27).

Table 3
Example of positive and negative control data from the rat liver, stomach, and bone marrow

	Vehicle control data									Positive control data								
	Tail intensity (%)			Tail moment (arbitrary)			Tail length (μm)			Tail intensity (%)			Tail moment (arbitrary)			Tail length (μm)		
	Min	Max	Mean	Min	Max	Mean	Min	Max	Mean	Min	Max	Mean	Min	Max	Mean	Min	Max	Mean
Liver																		
Male	0.29	8.31	2.68	0.03	2.46	0.47	17.31	73.58	33.12	12.06	60.14	37.41	1.61	38.28	11.97	32.25	187.15	92.79
Female	1.20	7.79	3.92	0.22	1.94	0.67	21.66	59.07	34.63	20.28	73.76	46.64	4.36	46.07	16.23	55.43	183.41	106.79
Combined			3.30			0.57			33.88			42.03			14.10			99.79
Glandular stomach																		
Male	3.48	20.44	9.00	0.45	6.63	2.12	24.81	90.07	51.62	31.22	78.41	54.82	7.70	42.01	22.43	61.48	166.87	112.84
Female	2.58	12.36	7.70	0.39	3.32	1.70	24.95	72.88	47.42	32.73	66.42	46.81	8.46	28.24	15.69	63.47	187.21	102.11
Combined			8.35			1.91			49.52			50.82			19.06			107.48
Bone marrow																		
Male	0.83	8.59	3.08	0.08	1.48	0.48	20.01	43.02	30.73	10.70	57.83	30.38	1.79	18.27	7.76	40.45	119.50	75.87
Female	0.69	9.80	2.96	0.09	2.92	0.54	17.73	57.35	29.00	8.38	49.76	29.64	1.53	15.58	6.98	34.51	107.72	68.83
Combined			4.56			0.51			29.87			30.01			7.37			72.35

4. Notes

1. The mincing buffer consists of 20 mM EDTA (disodium) and 10% DMSO in HBSS (Ca^{2+}, Mg^{2+} free, and phenol red free if available), pH 7.5 (DMSO will be added immediately before use). This solution will be refrigerated at <10°C until use.
2. The lysing solution consists of 100 mM EDTA (disodium), 2.5 M sodium chloride, and 10 mM tris hydroxymethyl aminomethane in purified water, with the pH adjusted to 10.0 with 1 M sodium hydroxide and/or hydrochloric acid. This solution may be refrigerated at <10°C until use. On the same day of use, 1% (v/v) of triton-X100 and 10% (v/v) DMSO will be added to this solution and the complete lysing solution will be refrigerated at <10°C for at least 30 min prior to use.
3. The alkaline solution consists of 300 mM sodium hydroxide and 1 mM EDTA (disodium) in purified water, pH > 13. This solution will be refrigerated at <10°C until use. The pH of the solution will be measured just prior to use.
4. The neutralization solution consists of 0.4 M tris hydroxymethyl aminomethane in purified water, pH 7.5. This solution will be either refrigerated at <10°C or stored consistent with manufacturer's specifications until use.
5. 1.0–1.5% (w/v) standard agarose gel for the bottom layer. Regular melting agarose will be dissolved at 1.0–1.5% (w/v) in Dulbecco's phosphate buffer (Ca^{2+}, Mg^{2+} free, and phenol free) by heating in a microwave. 0.5% (w/v) low-melting agarose (Lonza, NuSieve GTG Agarose) gel for the cell-containing layer and, if used, a top layer. Low-melting agarose will be dissolved at 0.5% (w/v) in Dulbecco's phosphate buffer (Ca^{2+}, Mg^{2+} free, and phenol free) by heating in a microwave. During the study, this solution will be kept at 37–45°C and discarded afterward.
6. For any novel tissues, time should be taken to ensure that the cell isolation is as easy as possible especially if more than one tissue is being processed. Note that any new guidelines will stipulate that a target and reference tissue should be investigated.
7. The electrophoresis time required for different cells may change. However, after extensive investigations of more than 15 tissues an unwinding time of 20 min and electrophoresis of 30 min have given reliable and consistent results. More limited data from a range of in vitro assays show that 20 min unwinding followed by 20 min electrophoresis works well.
8. Full agreement on the nature of the "hedgehogs" has not been reached and there is still a level of uncertainty as to when a comet is a hedgehog. This is confounded by the fact that

some image analysis systems can "score" hedgehogs by predicting where the head would be. The recommendation is that if the image analysis system can score the comet and give a realistic "head" then it should be scored. If not it should be regarded as a hedgehog and not be scored. Checking for debris and reduced numbers of scorable comets is also a good measure of toxicity.

References

1. Ostling, O. and Johanson, K. J. (1984). Microelectrophoretic study of radiation induced DNA damages in individual mammalian cells. *Biochem. Biophy.s Res. Commun.* **123**, 291–298.
2. Singh, N. P., McCoy, M. T., Tice, R. R., and Schneider, E. L. (1988). A simple technique for quantitation of low levels of DNA damage in individual cells. *Exp. Cell Res.* **175**, 184 –191.
3. Tice, R. R., Andrews, P. W., Hirai, O., and Singh, N. P. (1991). The single cell gel (SCG) assay: an electrophoretic technique for the detection of DNA damage in individual cells. In: Witmer, C. R., Snyder, R. R., Jollow, D. J., Kalf, G. F., Kocsis, J. J., Sipes, I. G., editors. Biological reactive intermediates IV. Molecular and cellular effects and their impact on human health. New York: Plenum Press. p157–164.
4. Tice, R. R. (1995). The single cell gel/comet assay: A microgel electrophoretic technique for the detection of DNA damage and repair in individual cells. In: Phillips, D. H., Venitt, S, editors. Environmental mutagenesis. Oxford: Bios Scientific Publishers. p315–339.
5. Fairbairn, D. W., Olive, P. L., and O'Neill, K. L. (1995). The Comet assay: a comprehensive review. *Mutat. Res.* **339**, 37–59.
6. Anderson, D., Yu, T-W., and McGregor, D. B. (1998). Comet assay responses as indicators of carcinogenic exposure. *Mutagen* **13**, 539–555.
7. Rojas, E., Lopez, M. C., and Valverde, M. (1999). Single cell gel electrophoresis assay: methodology and applications. *J. Chromat.* **B 722**, 225–254.
8. Speit, G., and Hartmann, A. (1999). The comet assay (single-cell gel test): a sensitive genotoxicity test for the detection of DNA damage and repair. *DNA Repair Protocols* **113**, 203–212.
9. Tice, R. R., Agurell, E., Anderson, D., Burlinson, B., Hartmann, A., Kobayashi, H., Miyamae, Y., Rojas, E., Ryu, J.-C., Sasaki, Y. F. (2000). The single cell gel/comet assay: Guidelines for in vitro and in vivo genetic toxicology testing, *Environ. Mol. Mutagen.* **35**, 206–221.
10. Hartmann, A., Agurell, E., Beevers, C., Brendler-Schwaab, S., Burlinson, B., Clay, P., Collins, A. R., Smith, A., Speit, G., Thybaud, V., and Tice, R. R. (2003). Recommendations for conducting the in vivo alkaline comet assay, *Mutagenesis* **18**, 45–51.
11. Brendler-Schwaab, S. Y., Hartmann, A., Pfuhler, S., and Speit, G. (2005). The in vivo comet assay: use and status in genotoxicity testing, *Mutagenesis* **20,** 245–254.
12. Burlinson B, et al., (2007). Fourth International Workgroup on Genotoxicity Testing: result of the in vivo comet assay workgroup. *Mutation Res.* **627**, 31–35.
13. Hartmann, A., Schumacher, M., Plappert-Helbig, U., Lowe, P., Suter, W., and Mueller, L. (2004). Use of the alkaline in vivo comet assay for mechanistic genotoxicity investigations, *Mutagenesis* **19,** 51–59.
14. COM (2000) (Committee on Mutagenicity of Chemicals in Food, Consumer Products and the Environment), COM guidance on a strategy for testing of chemicals for mutagenicity, United Kingdom, December. (http://www.advisorybodies.doh.gov.uk/com/).
15. Ames, B. N., McCann, J., Yamasaki, E. (1975). Methods for detecting carcinogens and mutagens with the Salmonella/mammalian microsome mutagenicity test. *Mutation Res.* **31**, 347–364.
16. Elliot, B. M., Combes, R. D., Elcombe, C. R., Gatehouse, D. G., Gibson, G .G., Mackay, J. M., and Wolf, R. C. (1992). Report of UK Environmental Mutagen Society Working Party: Alternatives to Aroclor 1254-induced S9 in *in vitro* genotoxicity assays. *Mutagenesis* 7, 175–177.
17. Galloway, S. M., Aardema, M. J., Ishidate, M. Jr., Ivett, J. L., Kirkland, D. J., Morita, T., Mosesso, P., Sofuni, T. (1994). Report from working group on in vitro tests for chromosomal aberrations. *Mutation Res.* **312**, 241–261.

18. Collins, A. R., Ai-guo, A., and Duthie, S. J. (1995). The kinetics of repair of oxidative DNA damage (strand breaks and oxidised pyrimidines) in human cells, *Mutation Res.* **336**, 69–77.
19. Kumaravel, T. S., and Jha, A. N. (2006). Reliable Comet assay measurements for detecting DNA damage induced by ionising radiation and chemicals, *Mutation Res.* **605**, 7–16.
20. OECD (1986). (Organisation for Economic Co-operation and Development), OECD Guidelines for Testing of Chemicals. Introduction to the OECD Guidelines on genetic toxicology testing and guidance on the selection and application of assays, OECD, Paris, France.
21. ICH (1995). (International Cooperation on Harmonization), ICH Harmonized Tripartite Guideline. S2A. Guidance on Specific Aspects of Regulatory Genotoxicity Tests for Pharmaceuticals, July. (http://www.ich.org).
22. ICH (1997). (International Cooperation on Harmonization), ICH Harmonized Tripartite Guideline. S2B. Genotoxicity: A Standard Battery for Genotoxicity Testing of Pharmaceuticals, July. (http://www.ich.org).
23. FDA (2006). Guidance for Industry and Review Staff Recommended Approaches to Integration of Genetic Toxicology Study Results, FDA. http://www.fda.gov/cder/guidance/index.htm.
24. Amacher, D E, Paillet S and Ray V A (1979) Point mutations at the thymidine kinase locus in L5178Y Mouse Lymphoma cells. I Application to Genetic Toxicological testing. *Mutat Res* **64**: 391–406.
25. Lovell, D. P., Thomas, G., and Dubow., G. R. (1999). Issues related to the experimental design and subsequent statistical analysis of in vivo and in vitro comet studies. *Teratog Carcinog Mutagen.* **19(2)**, 109–119.
26. Wiklund, S. J., and Agurell, E., (2003). Aspects of design and statistical analysis in the Comet assay. *Mutagenesis* **18(2)**, 167–175.
27. Smith, C. C., Adkins, D. J., Martin E. A., and O'Donovan, M. R., (2008) Recommendations for design of the rat comet assay. *Mutagenesis* **23(3)**, 233–240.

Chapter 9

Assessment of DNA Interstrand Crosslinks Using the Modified Alkaline Comet Assay

Jian Hong Wu and Nigel J. Jones

Abstract

The single cell gel electrophoresis (SCGE) assay, more commonly known as the comet assay, due to the "comet-like" appearance of the cells, was originally developed as a technique to measure the presence of DNA single-strand breaks. The assay is performed on single cells embedded in agar and placed in an electrical field at alkaline pH, so that fragments of negatively charged single-stranded DNA move through the gel toward the positively charged anode. Undamaged DNA moves relatively slowly, forming the head of the comet, while DNA fragmented due to the presence of single-strand breaks, moves more quickly giving the appearance of the tail. The extent of DNA migration is a measure of the DNA damage present. Since it was first developed, the comet assay has been adapted for measuring other types of DNA damage. The neutral comet assay has been employed for DNA double-strand breaks, while techniques using DNA repair enzymes to cleave specific adducts, UvrABC for ultraviolet radiation induced adducts, for example, have also been described. Here, we describe a modified version of the comet assay for the measurement of interstrand crosslinks (ICLs). Interstrand crosslinking agents include the chemotherapeutic agents mitomycin C and cis-platin, psoralen plus UVA light (PUVA) used to treat hyperproliferative skin disorders and diepoxybutane, a metabolite of 1,3-butadiene used in industrial processes and an environmental pollutant. ICLs are a potent and cytotoxic form of DNA damage as they prevent DNA strand separation, thereby preventing DNA replication. Their removal requires several different DNA repair processes including translesion synthesis and homologous recombination. As ICLs prevent separation of the DNA strands, their presence results in less DNA migration in the comet assay. To successfully measure ICLs, it is necessary to incorporate a step that induces single-strand breaks (using a defined dose of ionizing radiation) that allows the crosslinked DNA to migrate.

Key words: Comet assay, Interstrand crosslink, Psoralen plus ultraviolet A light, 8-Methoxy-psoralen, Mitomycin C, γ-Irradiation-induced DNA strand breaks

1. Introduction

Interstrand crosslinks (ICLs) are a type of DNA damage, in which the two complementary DNA strands are covalently connected by a cross linking agent, such as mitomycin C (MMC) and psoralen

James M. Parry and Elizabeth M. Parry (eds.), *Genetic Toxicology: Principles and Methods*, Methods in Molecular Biology, vol. 817, DOI 10.1007/978-1-61779-421-6_9, © Springer Science+Business Media, LLC 2012

plus ultraviolet A (PUVA). Since interstrand crosslinking prevents DNA strand separation, and therefore blocks DNA replication and transcription, it is believed to be one of the most detrimental DNA lesions. It has been demonstrated that a single ICL in bacteria, or as few as 20 ICLs in mammalian genome, can be lethal to cells that are deficient in the repair of ICL damage (1, 2). Removal of ICLs is a complex process and involves a number of DNA repair mechanisms, including translesion synthesis, homologous recombination, and nucleotide excision repair (3).

Methods for the assessment of ICLs rely on the fact that the two complementary strands of DNA are covalently linked by the ICL and are thus prevented from complete denaturation by heat or alkali. Available techniques to detect ICLs include alkaline elution, high-performance liquid chromatography (HPLC), and the more recently developed modified alkaline comet assay (4–7).

The comet assay, also called single cell gel electrophoresis (SCGE) and microgel electrophoresis (MGE), is a gel electrophoresis method initially developed to measure DNA strand breaks in individual cells (8, 9). It is based on the principle that fragments of negatively charged DNA migrate toward the anode in a weak electric field. Undamaged DNA migrates more slowly than that DNA containing strand breaks (due to its larger size). The extension of the DNA migration reflects the frequency of DNA strand breaks; therefore, assessment of DNA strand breaks is achieved by evaluation of the DNA migration pattern. The general procedure (shown in Fig. 1) of the alkaline comet assay involves embedding cells in a thin layer of agarose gel on a microscope slide and lysing the immobilized cells with high salt and detergent to remove cell membranes, cytoplasm, and nucleoplasm. This is followed by electrophoresis of the liberated DNA under alkaline condition (pH > 13). The alkaline condition allows predictable movement of DNA in the agarose gel, as it disrupts DNA secondary/tertiary structure and further denatures the double-stranded DNA. It also degrades contaminating RNA and breaks DNA–protein crosslinks (10). After staining with a fluorescent dye, the slide is visualized under a fluorescent microscope and the resulting image shows the damaged DNA in a cell with a comet-like appearance (Fig. 2).

A modified alkaline comet assay incorporating induction of a fixed level of random DNA strand breaks by γ-ray irradiation has been developed to detect ICLs induced by crosslinking agents, such as MMC (6) and PUVA (7). The random DNA strand breaks create free DNA fragments, whereas the presence of the ICLs prevents the complete denaturation of two DNA strands in alkaline conditions, resulting in the decreased number of free DNA fragments. As the number of ICLs increases, the amount of DNA able to migrate in the electrophoresis decreases.

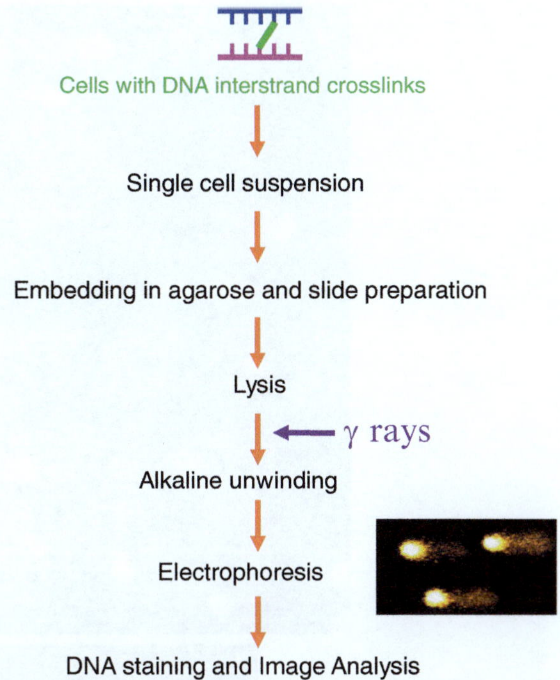

Fig. 1. Schematic representation of the critical steps in the comet assay. These involve embedding cells in agarose gel on a microscope slide and lysing the cells to liberate the DNA. A short period of electrophoresis enables DNA to move toward the anode. The fluorescent dye stained image is viewed under a microscope showing the damaged DNA in a cell having a comet-like appearance. An extra γ-irradiation step enables the comet assay to detect interstrand crosslinks.

Therefore, the relative reduction of γ-irradiation-induced DNA migration can be used as an indicator of ICL formation in the modified assay. The modified alkaline comet assay, owing to its simplicity, sensitivity, rapidity, small cell sample, and low cost, has been recommended as the most valuable method in assessment of ICLs (7).

Two crosslinking agents, MMC and PUVA, are employed in this chapter for the description of the modified alkaline comet assay protocols for the assessment of ICL formation. MMC is a bifunctional alkylating agent widely used for the treatment of solid tumors, such as breast cancer, gastric cancer, and nonsmall cell lung cancer. It functions as an inhibitor of DNA replication as a result of ICL formation with DNA between the two complementary strands. MMC forms ICLs between the N^2 position of two adjacent guanines (11). PUVA is commonly used for the treatment of hyperproliferative skin disorders such as psoriasis and vitiligo. Its ability of inhibiting cellular proliferation is believed to

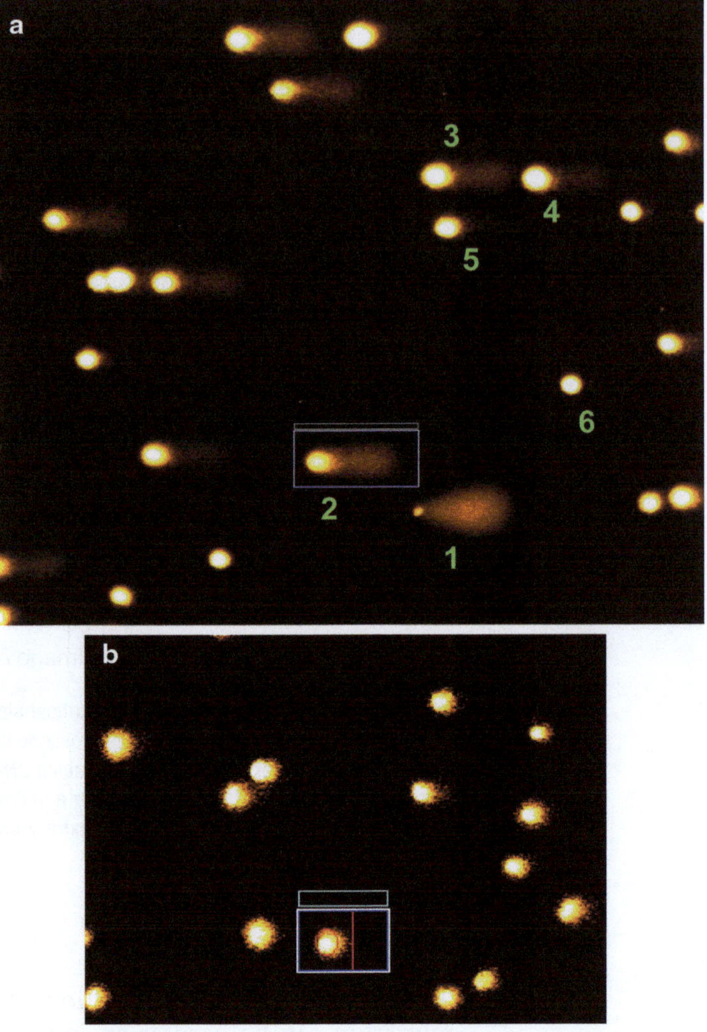

Fig. 2. A: A single field of a comet assay image showing various degrees of DNA damage of the human keratinocyte cell, HaCaT, treated with bleomycin (**a**) which induces DNA strand breaks. Comet 1 shows highly fragmented DNA with a large tail and little DNA remaining in the comet head (so-called "hedgehog" comet). Comet 6 shows the least amount of damage with almost all the DNA remaining in the comet head (numbers 2, 3, 4, and 5 show decreasing and intermediate levels of DNA damage). B: HaCaT cells treated with crosslinking agent PUVA (10 μM 8-MOP plus 0.05 J/cm^2 UVA and irradiated with 9 Gy γ-ray) tend to give rise to more homogenous damage levels (**b**).

be mediated by the ICL formation in DNA when psoralens are irradiated with UVA at the wavelengths of 320–400 nm. The ICL formed by psoralen are between two thymines (12), as shown in Fig. 3. One of the most widely used psoralens is 8-methoxy-psoralen (8-MOP).

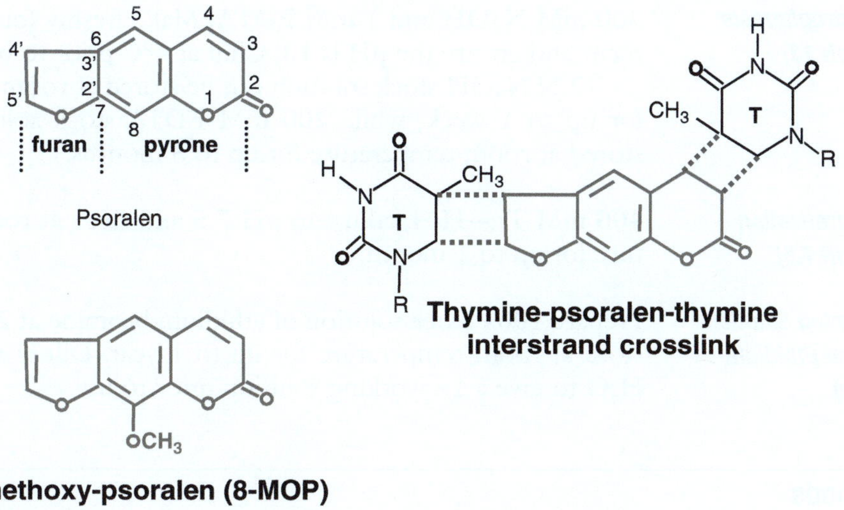

Fig. 3. Structures of psoralen, 8-methoxy-psoralen (8-MOP), and a psoralen interstrand crosslink.

2. Materials

2.1. MMC (1 mM) and 8-MOP (50 nM) Stock Solutions

1. Dissolve MMC at 1 mg/3 ml in Hank's balanced salt solution (HBSS). Mix well to dissolve MMC completely and sterilize by filtration (0.2 μM). Dispense in 500 μl aliquots and store at −20°C.
2. Dissolve 8-MOP at 10 mg/ml in dimethylsulfoxide (DMSO). Dispense in 100 μl aliquots and store at −20°C (see Note 1).

   Prior to experimentation, aliquots of the stock solutions are diluted in Dulbecco's modified Eagles medium (DMEM), prewarmed to 37°C, to the recommended working concentration of 10, 20, 50, 100 μM, with a final volume of 1 ml (see Note 2).

2.2. 0.75% W/V Normal Melting Point Agarose and 0.5% W/V LMP Agarose

Dissolve 0.75 g normal melting point (NMP) and 0.5 g low melting point (LMP) agarose, respectively, in 100 ml of phosphate-buffered saline (PBS). Warm in a microwave until fully dissolved, then dispense into 10 ml aliquots and store at 4°C for up to 1 month. On the day of use, gently microwave the required amount of agarose until fully liquefied and hold at 42°C in a water bath until needed.

2.3. Lysis Solution (pH 10)

2.5 M NaCl, 100 mM Na_2EDTA, 10 mM Trisma base, 0.2 N NaOH. Adjust to pH 10 and store at 4°C for up to 1 month. Add 10% DMSO and 1% Triton X-100 immediately prior to use.

2.4. Electrophoresis Buffer (pH 13)

300 mM NaOH and 1 mM EDTA. Make freshly for each experiment and ensure the pH is 13. Chill at 4°C prior to use.

(2 N NaOH stock solution can be stored at room temperature for up to 1 week, while 200 mM EDTA stock solution can be stored at room temperature for up to 3 months.)

2.5. Neutralization Buffer (pH 7.5)

400 mM Tris–HCl, adjust to pH 7.5 and store at room temperature for up to 1 month.

2.6. Staining Solution (20 μg/ml Ethidium Bromide)

Prepare 100× stock solution of ethidium bromide at 2 mg/ml and store at room temperature for up to 1 year. Dilute with distilled H_2O to give a 1× working solution prior to use.

3. Methods

The basic steps of the modified alkaline comet assay include:

- Treatment of cells with a test substance.
- Preparation of single cell suspensions.
- Preparation of microscope slides with cells imbedded in agarose gel.
- Lysis of the immobilized cells to liberate DNA.
- Exposure of cellular DNA to γ-irradiation to generate random strand breaks.
- Electrophoresis under alkaline conditions.
- Neutralization of alkali.
- DNA staining and comet visualization.
- Image analysis.

3.1. Treatment of Cells with MMC and PUVA

The comet assay is developed to assess DNA damage in individual cells and, in theory, any eukaryotic cell can be utilized as starting materials for the assay provided a single cell suspension can be obtained. The most widely used materials are cultured cell lines generated from human or animal tissues.

1. Cell preparation
   Inoculate individual wells of a 6-well tissue culture plate (35 mm) with the appropriate number of cells (depending on the cell type) in 2 ml of the complete growth medium, such as DMEM with supplements (fetal calf serum, antibiotics, etc.). Incubate at 37°C for 2–4 h until most of the cells are attached to the bottom of the well. Replace the medium containing any floating cells with 2 ml of fresh medium and culture the cells for further 20 h to allow cells to reach an exponential phase of

growth. Wash the cells once with 2 ml of DMEM prior to the chemical treatment (see Note 3).

2. Treatment of cells with MMC or PUVA
Treat the cells with 1 ml of MMC working solution freshly made with DMEM (without supplements) at 37°C for 1 h in dimmed light.

   Treat the cells with 1 ml 8-MOP working solution freshly made with DMEM (without supplements) at 37°C for 1 h in dimmed light. Carefully remove the 8-MOP solution and then irradiate the treated cells on ice at 365 nm with a dose of 0.05 J/cm^2 using an UV lamp. The output of the lamp should be determined using a UV radiometer with a 365 nm sensor.

3.2. Preparation of Single Cell Suspensions

After the treatment, wash the cells three times in 2 ml of chilled DMEM (without supplements), each for 5 min, and harvest the cells using 750 μl of trypsin (0.05%) – EDTA (0.02%) at room temperature. Add an equal volume of complete growth medium (DMEM with supplements) to quench the action of the trypsin and then transfer the cell suspension to a 1.5 ml Eppendorf tube. Centrifuge at $1000 \times g$ for 1 min in a desk-top centrifuge at 4°C and wash the pellets once with chilled DMEM. Resuspend the cells in chilled DMEM at a density of 1.5×10^5 to 1×10^7 cells/ml, depending on whether the comets are later scored at real-time (see Notes 4 and 5).

Remove two 10 μl aliquots from each Eppendorf tube and transfer to fresh Eppendorf tubes. Hold the tubes on ice to prevent DNA repair before being processed for the comet assay.

3.3. Trypan Blue Exclusion (see Note 6)

Mix 10 μl of cell suspension with equal volume of trypan blue and load the mixture onto a hemocytometer. Cell viability is measured by counting the number of bright cells (live cells) versus dark-blue cells (dead cells):

$$\text{Cell viability:} \frac{\text{Number of live cells} \times 2 \times 100\%}{\text{Number of live cells} + \text{Number of dead cells}}$$

3.4. Preparation of Microscope Slides with Cells Imbedded in Agarose Gel

1. Slide preparation (see Note 7)
A slide precoating step is employed to reinforce the microgel attachment to the slide during the subsequent assay procedure. The precoating is achieved by dipping the custom-made MGE slides (Erie Scientific, Menzel GmbH & Co KG) (Fig. 4) vertically in 0.75% NMP agarose kept at 42°C and wiping off the excess agarose from the back of the slide, then laying the slide on a piece of tissue paper to dry to a thin film (about 30 min at room temperature). The precoated slides can be kept at 4°C for up to 1 month.

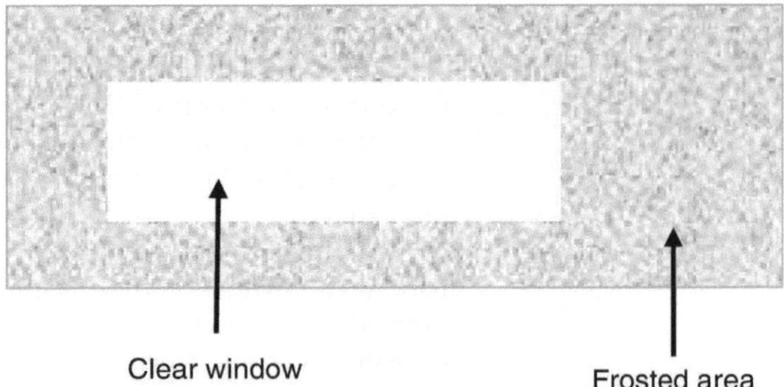

Fig. 4. The frosted slide with a clear window supplied by Erie Scientific Inc. The central clear window gives minimal background noise and the frosted area provides firm gel bonding to the slide.

2. Microgel preparation
   Rapidly mix 80 μl of 0.5% LMP agarose, held at 42°C, with 10 μl of cell suspension containing 1.5×10^3 to 1×10^5 cells (see Note 4). Pipette immediately the cell/agarose mixture onto the slide precoated with NMP agarose and place a 22 × 50 mm slide coverslip on top of the mixture to ensure an even layer. Lay the slide on ice for 5 min to allow the agarose to solidify and then return to room temperature for 5 min. (Place a piece of aluminum foil on ice and lay the slide on the foil to avoid water generated from thawing ice coming into contact with the slide.) Remove the coverslip by gently sliding it sideways from the slide.

3.5. Lysis of Cells to Liberate DNA

Immerse the slides in chilled lysis solution and leave overnight at 4°C (see Note 8). Cell lysis should be performed in light-proof boxes (e.g., aluminum foil-wrapped) to avoid fluorescent light, as under high pH conditions (the pH value for the lysis buffer is 10), fluorescent light can cause DNA strand breaks (13).

3.6. γ-Irradiation of Cellular DNA to Generate Random Strand Breaks

Remove the slides from the lysis solution and rinse gently in chilled electrophoresis buffer. Irradiate the slides with 5–10 Gy of γ-irradiation using an appropriate γ-ray source (see Notes 9 and 10).

3.7. Electrophoresis Under Alkaline Conditions and Neutralization of Alkali

Place the slides in a bath of chilled electrophoresis buffer for 15 min to further rinse the slides and to allow DNA denaturation. Then transfer the slides into a horizontal electrophoresis tank containing fresh chilled electrophoresis buffer and incubate for another 30 min to equilibrate and further denature DNA. Perform electrophoresis for 30 min at 0.85 V/cm and 300 mA (see Notes 11–13). The gel tank should be kept chilled during unwinding and electrophoresis, either by placing the tank in a cold room at 4°C or running ice-cold water through central chamber of the tank via an ice bath (if available).

Remove the slides from the electrophoresis tank and submerse the slides in a bath of neutralization buffer to neutralize the alkali. The neutralization is carried out for three times, each for 5 min.

3.8. DNA Staining and Comet Visualization

Drain excess neutralization buffer by leaving the slide upright at a slight angle on a piece of tissue paper for 15 min. Stain the slides with 60 μl 1× ethidium bromide and replace the coverslip. Store the slides for up to 3 days in a humidified light-proof box at 4°C prior to image analysis (see Note 14).

3.9. Image Analysis and Selection of Parameters

Analysis of duplicate slides for each treatment should be routinely performed. Fifty cells are randomly selected and analyzed from each slide.

1. Selection of comets
   It is crucial to select comets without bias during image analysis and the selected comets must represent the whole gel. The edges and areas around air bubbles should be avoided, as they tend to contain highly damaged comets. Overlapping comets should be avoided as their analysis is difficult. It is controversial as to whether to include the "hedgehog" comets (see Note 15).

   It should be noted that treatment of cells with high doses of MMC or PUVA induces a significant reduction of comet tail intensity (below the baseline level) as a result of DNA aggregation (tightening effect) (Fig. 5). This effect has only been observed in cells treated with high doses of crosslinking agents.
2. Selection of parameters
   Automated comet image analysis software is available commercially and the most commonly used parameters for comet image analysis are comet tail length, relative fluorescence intensity of the comet tail (% tail intensity), and tail moment (see Notes 16–18)

   There is still no general agreement regarding the most relevant parameters to be used in the comet assay. Although it has been shown that both tail moment and % tail intensity are significantly correlated with DNA damage (14), based on the fact that the tail moment is measured in arbitrary units and different image-analysis systems give different values, it is suggested that the % tail intensity is more meaningful and comparable.
3. The degree of ICL formation after treatment of cells with a crosslinking agent is described by comparing the % tail intensity (TI) of the treated samples (TI-test) with that of the untreated controls (TI-control) (6, 15).

$$\%\text{Relative tail intensity (RTI)} = (\text{TI - test}/\text{TI - control}) \times 100.$$

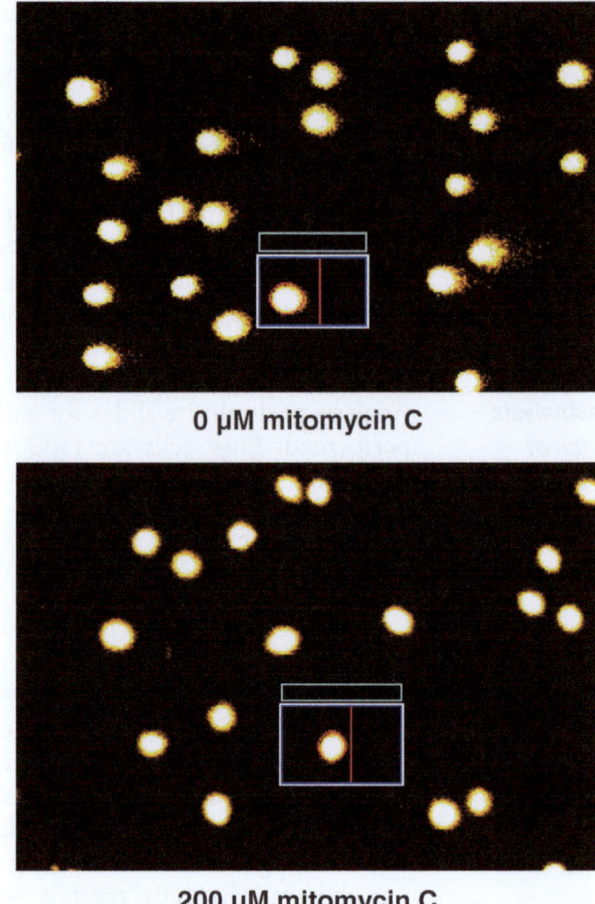

Fig. 5. Effect of DNA aggregation caused by MMC. HaCaT cells were treated with MMC at 0 μM (*top*) or 200 μM (*bottom*) at 37°C for 4 h.

4. Notes

1. The stock solution preparation of both MMC and 8-MOP, and the subsequent cell treatment should be carried out in dimmed light due to the light-sensitive nature of the chemicals.
2. It is recommended that the stock solutions be diluted in DMEM (without any supplementation), rather than in HBSS, due to the fact that HBSS appears to cause cytotoxicity for some cell types during the 1-h treatment period.
3. To obtain healthy cells, it is critical to replace the medium containing floating cells with fresh medium after incubating the cells for a few hours at 37°C.

4. An appropriate cell density in agarose is important to ensure a successful comet analysis, since a high cell density results in increased numbers of overlapping comets, especially at high levels of DNA migration. Analysis of comet images with excessive overlapping comets is impossible, whereas analysis of comet images with a low level of overlapping comets generates inaccurate results. This is due to the fact that comets with bigger tails, thus containing more breaks, are more likely to overlap with other comets, and the overlapping comets have to be excluded during image analysis. The exclusion of the more badly damaged comets results in the data showing less damage than the actual level.

   Cell densities ranging from 1×10^3 to 1×10^5 cells/slide have been used for the comet assay (13, 16). While a cell number ranging from 1.5×10^3 to 2×10^4 per slide is optimal for real-time image analysis, approximately five times this cell number is needed for stored image analysis, depending on the type of a microscope/camera system used to take and store the image. In the latter case, an optimization of the cell number is recommended to generate an optimal cell density that is not too dense to affect the comet scoring and at the same time not so scarce that too many images are needed for each slide. For example, with a cell density of 2×10^4cells per slide, approximately ten images have to be captured for scoring 50 comets on each slide, which is impractical. On the other hand, a higher cell density of 5×10^5 cells per slide gives rise to overlapping comets that are not analyzable (Fig. 6).

5. It is important to start the comet assay with evenly dispersed single cells, as cells that are not well separated give rise to overlapping comets, making comet scoring impossible (Fig. 7). The cell separation can be facilitated by passing the solution through a 1 ml Gilson tip several times during trypsinization process.

6. Trypan blue exclusion is a means of assessing cell viability (17). It is believed that trypan blue enters dead cells through a damaged cell membrane, thus the dead cells appear dark-blue in color, while live cells remain unstained when examined under a microscope. It is a common practice in the comet assay to carry out trypan blue exclusion and the cells are considered healthy for the assay when viability is more than 80%.

7. It is important that slides are prepared correctly in order to obtain agarose gel layers firmly attached to the microscope slides during the comet assay procedure, as well as to ensure comets with minimal background noise during microscopic visualization. Loss of gels from slides at different assay stages is a major problem when conventional microscope slides are used. Various approaches, such as scoring the edge of the slides

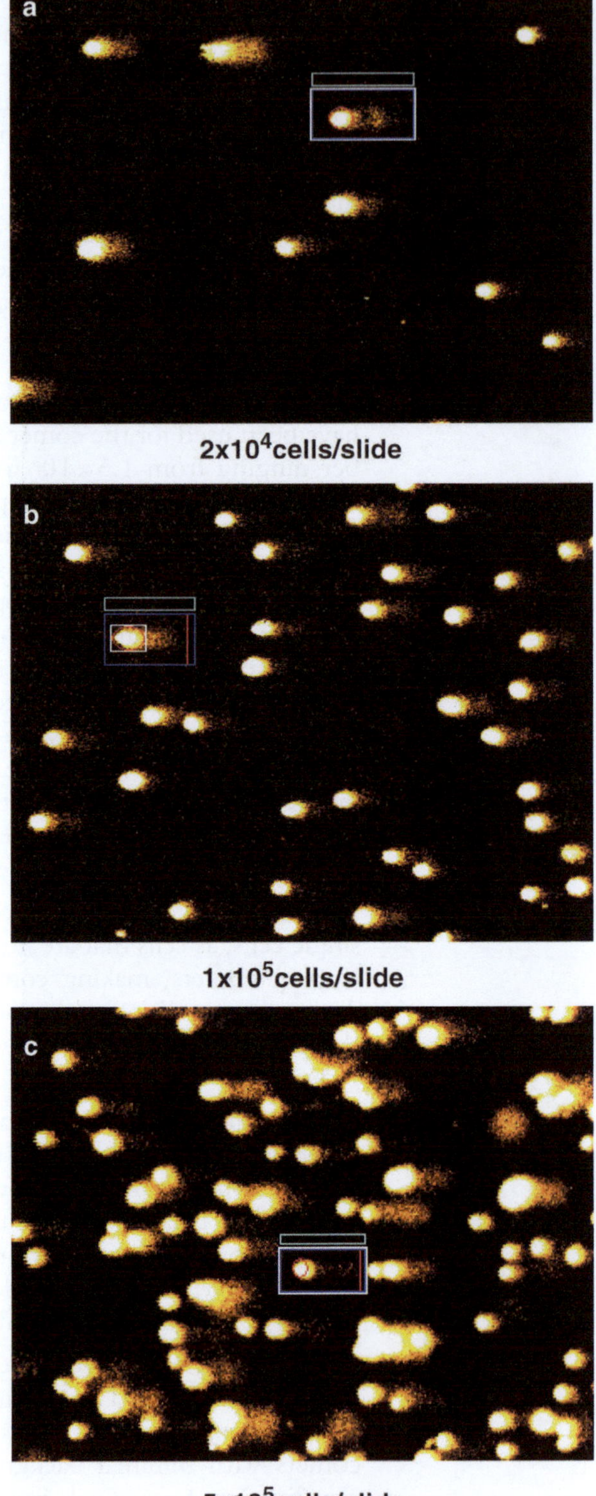

Fig. 6. Optimization of cell density. Images of V79 cells treated 50 μM bleomycin at 37°C for 1 h showing different cell densities on slides. (**a**) Low cell density – 2×10^4 cells/slide; (**b**) Optimal cell density – 1×10^5 cells/slide; and (**c**) High cell density – 5×10^5 cells/slide.

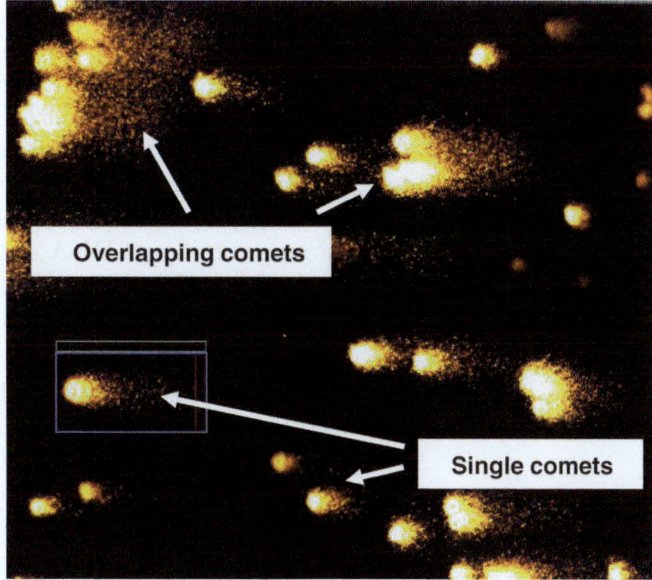

Fig. 7. Image of Chinese hamster V79 cells treated with 50 μM bleomycin at 37°C for 1 h. Cells that are not well separated give rise to overlapping comets and tails, making the scoring of comets difficult or impossible.

using a glass-cutting diamond pen, cleaning the slides with methanol or ethanol before use, increasing the NMP agarose concentration for the first layer, have been employed to increase the gel bonding ability. However, we have found that those methods are not efficient in solving the problem. Although fully frosted slides can be used to avoid gel detachment, higher background noise during comet analysis makes them problematic (13). The custom-made MGE slides, which are frosted slides with a clear window in the centre (1 cm × 3 cm) for each slide, are proven to overcome the problems described above in the modified alkaline comet assay. The clear window in the middle gives minimal background noise for later comet scoring and the surrounding frosted area provides firm gel bonding to the slide.

8. The lysis duration used by different research groups varies considerably, from 1 h to weeks. The minimal time needed to liberate the DNA is believed to be 1 h (18) and a longer lysis at 4°C allows more time for DNA unwinding, and therefore increases sensitivity for detecting breaks (by a factor of approximately 2 for lysis times greater than 6 h) (13, 19). Nevertheless, prolonged lysis (e.g., 10 days at 4°C) has been observed to cause an increase in DNA damage unrelated to the treatment that the cells receive (Wu and Jones, unpublished data). Therefore, overnight lysis at 4°C is recommended.

9. Construction of a calibration curve of dose–response for a particular cell type to γ-irradiation is recommended. This assesses the extent of DNA migration by random strand breaks in that cell line for the source of ionizing radiation utilized. This is followed by determination of the optimal γ-irradiation dose using different dose combinations of test substance and γ-irradiation. Low doses of γ-irradiation may be insufficient for generating significant DNA migration that is retarded by the presence of the ICLs, thus it is difficult to differentiate ICL formation at different doses of crosslinking agent.
10. γ-Irradiation has also been performed before the cell lysis step (20, 21) but the irradiation of the chemical-treated cells could affect other cellular processes and protein functions, which interferes with the effect of the crosslinking agent. The postlysis γ-irradiation overcomes the problem and is considered a better strategy for irradiation (6, 7).
11. Due to the fact that the sizes of the commercially available electrophoresis units vary considerably, it is more accurate to present the voltage in V/cm, rather than only in V. The optimal voltage and electrophoresis duration applied should be able to induce DNA migration of approximately 10% for the control cells (22). The reported voltages applied in the comet assay range from 0.6 to 1.0 V/cm and the electrophoresis duration from 5 to 40 min (23). A low voltage and short duration does not separate DNA fragments efficiently, while a high voltage may cause horizontally overlapping comets, due to longer comet tails. A combination of a voltage and duration of electrophoresis should be optimized to give an appropriate comet size for image analysis.
12. The position of slides in the electrophoresis tank can impact on assay variability, especially where a tank is too small. The reason is that the electrophoretic current varies at different positions of the tank. Nevertheless, it is generally considered that the current variation is not a significant factor when the tank is sufficiently large.
13. The source of distilled H_2O used for electrophoresis is critical in the comet assay. In theory, well-purified H_2O is more suitable for making the electrophoresis buffer, but it has been noted that the use of double distilled H_2O to make the electrophoresis buffer leads to little or no DNA migration for unknown reasons.
14. The stained comets retain strong optical intensity when stored like this for up to 3 days, whereas DNA diffusion is apparent after storage for longer times (Fig. 8). Ideally, comet images should be viewed and analyzed immediately using computerized image analysis system (real-time analysis). Optionally, the

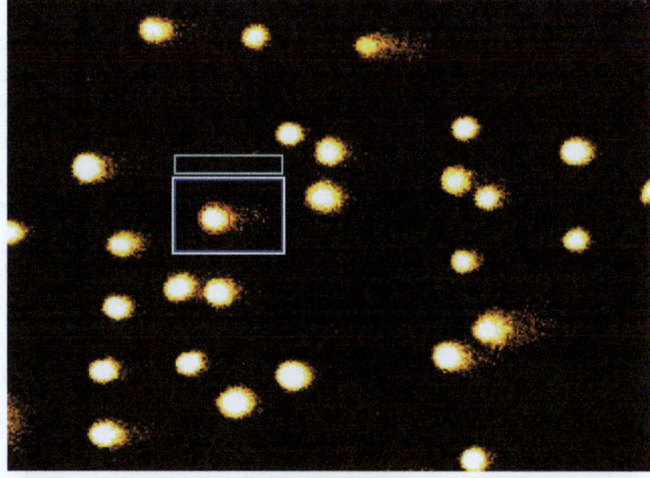

3 days at 4°C

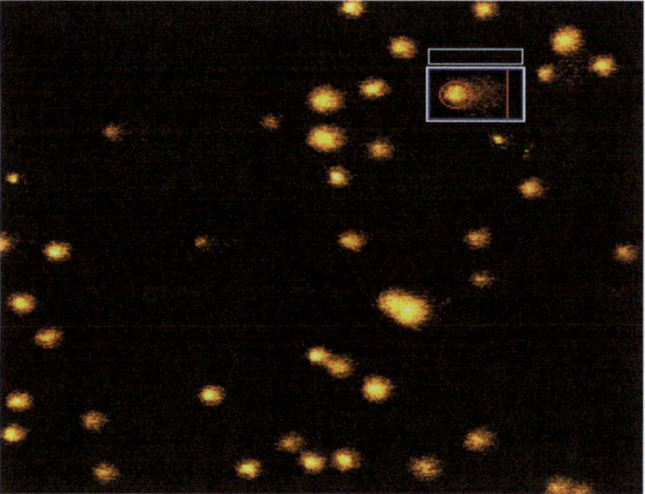

10 days at 4°C

Fig. 8. Effect of slide storage. The *upper panel* shows slides stored at 4°C for 3 days. The stained comets still retain strong optical intensity. In the *lower panel*, slide storage was for 10 days at 4°C. Significant diffusion of the DNA diffusion is apparent (*note "fuzzy" comet heads*).

images can be captured, stored, and analyzed later if a real-time analysis system is not available. In this case, the cell density for each slide should be approximately five times more than that for the real-time analysis (see Note 4).

15. When almost all the DNA is in the comet tail with little DNA remaining in the head, the image is referred to as a "hedgehog" comet or "ghost" cell (Fig. 2). It is believed that this specific type of comet represents apoptotic or necrotic cells and it is commonly excluded in the image analysis to avoid bias (23).

However, Collins (22) argued that, although it is possible that some of these severely damaged cells will subsequently go through programmed cell death, the cells cannot all be described as apoptotic cells for two reasons: (a) Apoptosis is an irreversible process, but the "hedgehog" cells appear capable of repair of their damage; (b) Apoptosis is characterized by fragmentation of DNA to the size of nucleosome oligomers and DNA of such small fragments would disappear either during lysis or electrophoresis, while the "hedgehog" comet may represent a residue of high-molecular-weight DNA. It is therefore recommended that the "hedgehog" cells be treated as severely damaged cells and be included in the comet analysis.

16. Comet tail length is largely dependent on voltage and duration of electrophoresis. It increases only while tails are first becoming established at relatively low damage levels, and subsequently the tail increases in intensity, rather than in length, as the dose of damage increases (13). Tail length is also sensitive to the background or threshold setting of the image analysis program, as the end of the tail is defined by a certain excess of fluorescence over background (22). Therefore, tail length is not considered a very good indication of damage levels.
17. Relative fluorescence intensity of the comet tail, or % tail intensity, measures the percentage of DNA migrated from the comet head into the tail. It is considered to be the most useful parameter, as it shows a linear relationship to DNA strand break frequency, and is much less dependent upon electrophoresis voltage and time. It is also relatively unaffected by threshold setting of the image analysis program (22). It is, however, suggested that the % tail intensity is less sensitive for detecting changes at low damage levels because it does not take into consideration the distribution of DNA damage in the tail.
18. Tail moment is defined as the percentage of DNA in the tail multiplied by the tail length (24). It is recommended to be used when the damage levels are low, although some research groups believe that it is not a better parameter than the relative tail intensity, as it is not linear with respect to dose and does not give impression of the comet's appearance, thus the damage level (22).

References

1. Murnane, J. P., and Byfield, J. E. (1981) Irreparable DNA cross-links and mammalian cell lethality with bifunctional alkylating agents, *Chem Biol Interact* 38, 75–86.
2. Lawley, P. D., and Phillips, D. H. (1996) DNA adducts from chemotherapeutic agents, *Mutat Res* 355, 13–40.
3. Dronkert, M. L. G., and Kanaar, R. (2001) Repair of DNA interstrand cross-links, *Mutation Research-DNA Repair* 486, 217–247.
4. Clingen, P. H., De Silva, I. U., McHugh, P. J., Ghadessy, F. J., Tilby, M. J., Thurston, D. E., and Hartley, J. A. (2005) The XPF-ERCC1

endonuclease and homologous recombination contribute to the repair of minor groove DNA interstrand crosslinks in mammalian cells produced by the pyrrolo[2,1-c][1,4]benzodiazepine dimer SJG-136, *Nucleic Acids Res* 33, 3283–3291.

5. Spanswick, V. J., Craddock, C., Sekhar, M., Mahendra, P., Shankaranarayana, P., Hughes, R. G., Hochhauser, D., and Hartley, J. A. (2002) Repair of DNA interstrand crosslinks as a mechanism of clinical resistance to melphalan in multiple myeloma, *Blood* 100, 224–229.
6. McKenna, D. J., Gallus, M., McKeown, S. R., Downes, C. S., and McKelvey-Martin, V. J. (2003) Modification of the alkaline Comet assay to allow simultaneous evaluation of mitomycin C-induced DNA cross-link damage and repair of specific DNA sequences in RT4 cells, *DNA Repair (Amst)* 2, 879–890.
7. Wu, J. H., Wilson, J. B., Wolfreys, A. M., Scott, A., and Jones, N. J. (2009) Optimization of the comet assay for the sensitive detection of PUVA-induced DNA interstrand cross-links, *Mutagenesis* 24, 173–181.
8. Ostling, O., and Johanson, K. J. (1984) Microelectrophoretic study of radiation-induced DNA damages in individual mammalian cells, *Biochem Biophys Res Commun* 123, 291–298.
9. Singh, N. P., McCoy, M. T., Tice, R. R., and Schneider, E. L. (1988) A simple technique for quantitation of low levels of DNA damage in individual cells, *Exp Cell Res* 175, 184–191.
10. Singh, N. P., Stephens, R. E., and Schneider, E. L. (1994) Modifications of alkaline microgel electrophoresis for sensitive detection of DNA damage, *Int J Radiat Biol* 66, 23–28.
11. Tomasz, M. (1995) Mitomycin C – Small, fast and deadly (but very selective), *Chemistry & Biology* 2, 575–579.
12. Chiou, C. C., and Yang, J. L. (1995) Mutagenicity and specific mutation spectrum induced by 8-methoxypsoralen plus a low dose of UVA in the hprt gene in diploid human fibroblasts, *Carcinogenesis* 16, 1357–1362.
13. Olive, P. L. (2002) The comet assay. An overview of techniques, *Methods Mol Biol* 203, 179–194.
14. Kumaravel, T. S., and Jha, A. N. (2006) Reliable Comet assay measurements for detecting DNA damage induced by ionising radiation and chemicals, *Mutat Res* 605, 7–16.
15. Rothfuss, A., and Grompe, M. (2004) Repair kinetics of genomic interstrand DNA crosslinks: evidence for DNA double-strand break-dependent activation of the Fanconi anemia/BRCA pathway, *Mol Cell Biol* 24, 123–134.
16. Rojas, E., Lopez, M. C., and Valverde, M. (1999) Single cell gel electrophoresis assay: methodology and applications, *J Chromatogr B Biomed Sci Appl* 722, 225–254.
17. Altman, S. A., Randers, L., and Rao, G. (1993) Comparison of trypan blue dye exclusion and fluorometric assays for mammalian cell viability determinations, *Biotechnol Prog* 9, 671–674.
18. Singh, N. P. (2000) Microgels for estimation of DNA strand breaks, DNA protein crosslinks and apoptosis, *Mutat Res* 455, 111–127.
19. Banath, J. P., Kim, A., and Olive, P. L. (2001) Overnight lysis improves the efficiency of detection of DNA damage in the alkaline comet assay, *Radiat Res* 155, 564–571.
20. Merk, O., and Speit, G. (1999) Detection of crosslinks with the comet assay in relationship to genotoxicity and cytotoxicity, *Environ Mol Mutagen* 33, 167–172.
21. Hartley, J. M., Spanswick, V. J., Gander, M., Giacomini, G., Whelan, J., Souhami, R. L., and Hartley, J. A. (1999) Measurement of DNA cross-linking in patients on ifosfamide therapy using the single cell gel electrophoresis (comet) assay, *Clin Cancer Res* 5, 507–512.
22. Collins, A. R. (2004) The comet assay for DNA damage and repair: principles, applications, and limitations, *Mol Biotechnol* 26, 249–261.
23. Tice, R. R., Agurell, E., Anderson, D., Burlinson, B., Hartmann, A., Kobayashi, H., Miyamae, Y., Rojas, E., Ryu, J. C., and Sasaki, Y. F. (2000) Single cell gel/comet assay: guidelines for in vitro and in vivo genetic toxicology testing, *Environ Mol Mutagen* 35, 206–221.
24. Olive, P. L., and Banath, J. P. (1993) Induction and rejoining of radiation-induced DNA single-strand breaks: "tail moment" as a function of position in the cell cycle, *Mutat Res* 294, 275–283.

Chapter 10

^{32}P-postlabelling for the Sensitive Detection of DNA Adducts

Nigel J. Jones

Abstract

^{32}P-postlabelling is a technique originally described by Kurt Randerath and colleagues for the sensitive detection of damage produced in DNA by reactive chemicals or genotoxins. The procedure essentially entails the enzymatic digestion of DNA to nucleoside 3′-monophosphates which are then radioactively labelled using T4 polynucleotide kinase and [γ^{32}P]-adenosine triphosphate. Adducted nucleoside-3′-5′-bisphosphates are then separated from their normal counterparts by thin layer chromatography. Prior to the development of the assay, quantification of DNA adducts was confined to studies that utilised compounds synthesised to be isotopically labelled with tritium or carbon-14. As such, these studies were limited to specific and recognised genotoxins that could be administered only in the laboratory to cultures or animals. With ^{32}P-postlabelling it was possible not only to determine DNA adduct induction by a relatively uncharacterised suspected carcinogen, but also following exposure to complex mixtures containing a multitude of known and unknown potential genotoxins. The small amount of DNA required to perform the ^{32}P-postlabelling assay also meant that human biomonitoring studies using readily obtainable tissues, such as lymphocytes, were possible. Using the standard ^{32}P-postlabelling method, it is possible to detect a single DNA adduct in 10^7 to 10^8 normal nucleotides. The subsequent development of several enhancement methods improved this detection rate to one adduct in 10^{10} nucleotides. For these reasons, the ^{32}-postlabelling assay represents an extremely versatile and extremely sensitive method to detect and monitor DNA damage.

Key words: DNA adducts, ^{32}P-postlabelling, Human biomonitoring, Genotoxicity testing, DNA damage and repair, Occupational exposures, Environmental pollution monitoring

1. Introduction

1.1. The ^{32}P-postlabelling Assay

The ^{32}P-postlabelling method for the detection of DNA adducts was originally conceived and developed in the laboratory of Randerath and co-workers (1, 2). The origins of the procedure were methods previously described by them for the analysis of RNA and DNA composition (3–5). The hallmarks of the assay are its extreme sensitivity (the basic protocol detects one adduct per 10^7 to 10^8 normal nucleotides) and its versatility in (1) detecting structurally diverse DNA adducts and (2) being applicable to a

James M. Parry and Elizabeth M. Parry (eds.), *Genetic Toxicology: Principles and Methods*, Methods in Molecular Biology, vol. 817, DOI 10.1007/978-1-61779-421-6_10,

wide range of samples. It is possible to detect DNA damage induced by a wide variety of genotoxins using the ^{32}P-postlabelling assay, including polycyclic aromatic hydrocarbons (PAHs), heterocyclic polyaromatics, aromatic amines, nitrated PAHs, alkenyl benzenes, quinines, mycotoxins, pyrolysis products, alkylating agents and complex mixtures such as tobacco smoke and vehicle emissions (6). Given that the method has a low requirement for DNA (the basic protocol is performed on ~0.2 μg DNA, while enhancement techniques generally use 5–20 μg), the assay may be successfully applied to virtually any biological sample from which DNA can be extracted. Since its inception in the early 1980s, several enhancement or adduct enrichment procedures have been described. These include (a) an ATP-deficient adduct intensification method; (b) enrichment with nuclease P1 digestion (or nuclease S1); (c) a dinucleotide/nucleoside 5′-monophosphate method; (d) butanol extraction of adducted nucleosides and (e) HPLC enrichment. The most widely used of these enhancement methods are butanol extraction and nuclease P1 digestion, which each increase the sensitivity of the assay to one adduct in 10^{10} normal nucleotides.

1.2. The Standard Protocol

The standard or basic protocol (shown in Fig. 1) entails the enzymatic cleavage of DNA with micrococcal nuclease and calf spleen phosphodiesterase to give a mixture of normal/unmodified nucleoside 3′-monophosphates (Nps) and adducted nucleoside 3′-monophosphates (Xps). The Xps and Nps are then radioactively labelled by attachment of a ^{32}P-label via T4 polynucleotide kinase (T4-PNK) catalysed transfer from (γ^{32}P]-adenosine triphosphate ((γ^{32}P]-ATP) to yield nucleoside-3′-5′-bisphosphates (pNps and pXps). To separate the adducted pXps from their normal counterparts (pNps) multidimensional anion exchange thin layer chromatography (TLC) on polyethylenimine (PEI)-cellulose plates is employed. Typical solvent conditions used for the chromatographic purification and resolution of adducted nucleoside-3′-5′-bisphosphates are shown in Fig. 2. The labelled digest (~0.2 μg DNA for the standard protocol) is applied to the origin of a 20 × 20 cm plastic-backed PEI-cellulose plate. The initial developments using aqueous solvents in the D1 and D2 directions are intended to remove normal nucleotides, residual ATP and inorganic phosphate from the plate onto the attached filter paper wicks. Modified pXps are retained at the origin due to their greater affinity for PEI-cellulose. To move and separate structurally distinct pXps from one another, and to create ^{32}P-adduct maps, solvents containing high molarity urea are employed in the D3 and D4 developments. Finally a fifth development is sometimes utilised in order to remove residual non-adduct material. ^{32}P-labelled DNA adducts are visualised by autoradiography and subsequently quantified by excision and scintillation counting of adduct spots or zones. Alternatively, imaging and quantification may be performed by directly scanning the TLC

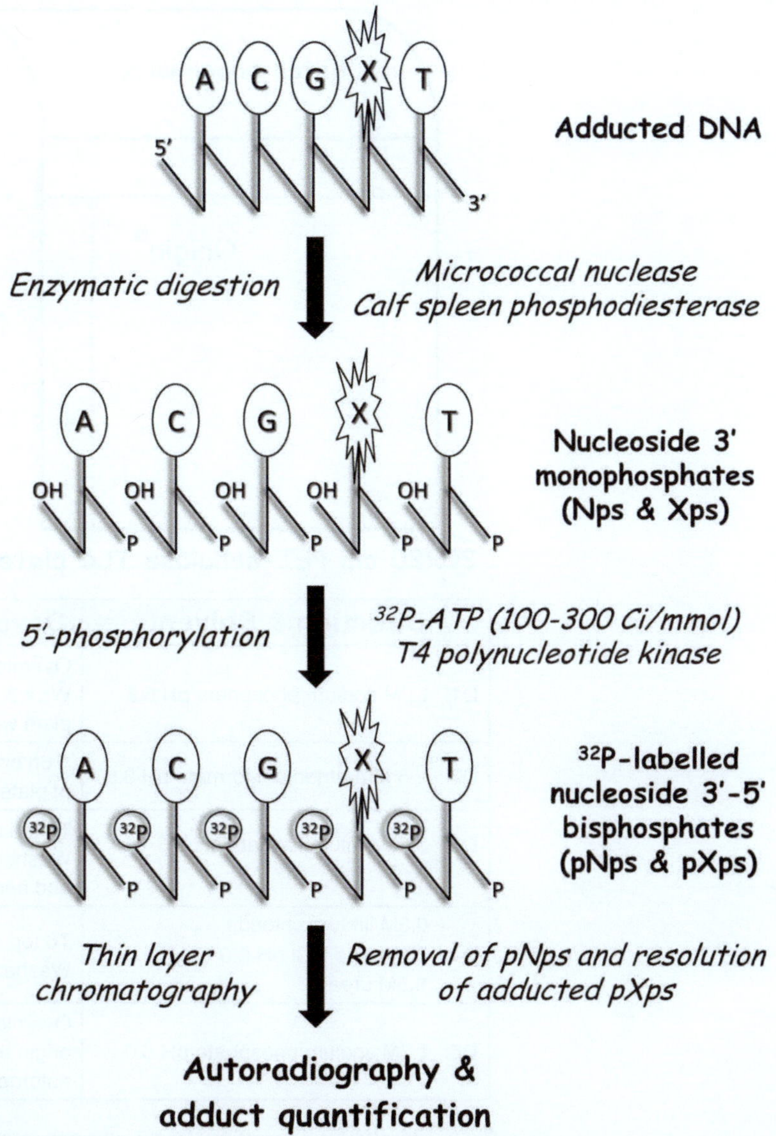

Fig. 1. Standard protocol for the ^{32}P-postlabelling of DNA adducts.

plate with a radioanalytical imaging system, or indirectly by exposing a phosphoimager plate to the chromatogram and subsequent analysis with a molecular imager. The resulting ^{32}P-adduct maps produced will depend on the resolution of adducts in the D3 and D4 developments and will be determined by urea concentration, pH, ionic strength and the anions. The chromatographic conditions shown in Fig. 2 are those typically employed for DNA adducts containing large aromatic moieties such as those induced by benzo[*a*]pyrene and these can be adjusted for smaller aromatic or non-aromatic adducts (7–9). It should be noted that in the standard

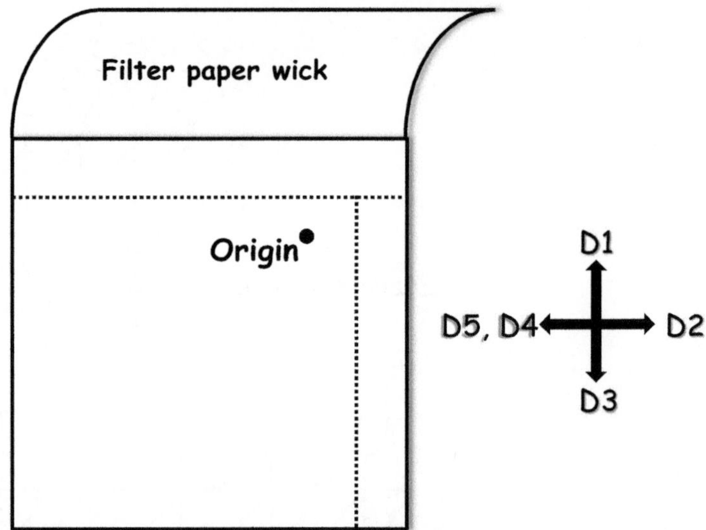

Direction & Solvent	Development & washes
D1: 1.0M sodium phosphate pH 6.8	Overnight onto wick Wick & top 3cm of plate removed; plate washed in H_2O
D2: 2.5 M ammonium formate pH 3.5	~ 6h onto wick; Wick & top 3cm of plate removed; washed in H_2O
D3: 3.5M lithium formate pH 3.5 8.5M urea	To top of plate (~ 16h) Washed once in 13mM Tris base and once in H_2O
D4: 0.8M lithium chloride 0.5M Tris-HCl pH 8.0 8.5M urea	To top of plate (~ 5.5h) Washed twice in H_2O
D5: 1.7M sodium phosphate pH 6.0	Overnight onto wick. Wick & origin removed; plate dried & autoradiography performed

Fig. 2. Typical solvent conditions for the chromatographic purification and resolution of DNA adducts containing aromatic moieties. The labelled adduct digest is spotted onto an origin 5 cm from the top of the plate in both the D1 and D2 directions. Solvents D1 and D2 remove normal nucleosides, residual [γ^{32}P]-ATP and inorganic phosphate onto a filter paper wick attached to the top of the plate. After chromatographic development, the wick and the top 3 cm of the plates are removed resulting in a 17 × 17 cm plate. Solvents D3 and D4 move the DNA adducts (pXps) in the directions indicated. D5 is used to remove residual non-adduct material thereby reducing the background.

protocol both modified (pXps) and unmodified (pNps) are ^{32}P-labelled (Fig. 1). As such, the specific activity of the [γ^{32}P]-ATP that can be used is between 100 and 300 Ci/mmol and the amount of DNA is limited to ≤0.2 μg.

1.3. Enhancement and Enrichment Techniques

To increase the sensitivity of the standard assay, several enhancement techniques have been developed. (a) In the adduct intensification method, the concentration of ATP is limited, so that certain adducts are labelled in preference to normal nucleosides. Under such ATP-limiting conditions some bulky adducts such as those induced by 7,12-dimethylbenz[*a*]anthracene (DMBA) are labelled at increased levels due to the substrate preference of T4-PNK (10). The use of this modification is rather restricted as only certain adducts are preferentially labelled, the labelling is not quantitative and the method does not represent an improvement in sensitivity compared with the standard technique. (b–e) Significant increases in the assay's sensitivity are only achieved when an additional step is incorporated into the procedure. In each case, the extra step essentially removes the vast bulk of normal nucleotides prior to the ^{32}P-labelling reaction and therefore allows the use of larger amounts of DNA (up to 20 μg) and a molar excess of carrier-free ATP with a very high specific activity (4,000–9,000 Ci/mmol). The normal nucleosides are either physically removed (by butanol extraction or HPLC) prior to labelling or they undergo an enzymatic digestion so that they are no longer substrates for T4 PNK (nuclease P1 digestion of nucleoside-3′-monophosphates or of DNA itself in the dinucleotide/nucleoside 5′-monophosphate assay). (b) In the nuclease P1 (nuclease S1 has been used with similar results) enhancement procedure (11), the nucleoside-3′-monophosphates produced by the initial micrococcal nuclease/calf spleen phosphodiesterase digestion are further digested by incubation with nuclease P1. Normal nucleoside-3′-monophosphates (Nps) are cleaved to nucleosides (Ns) which do not then serve as substrates for T4-PNK. On the other hand, most aromatic adducted nucleosides are partially or totally resistant to nuclease P1's dephosphorylating action and are therefore available for ^{32}P-labelling (Fig. 3). (c) Nuclease P1 is also utilised in the dinucleotide/nucleoside 5′-monophosphate assay (12). Here DNA is initially digested with nuclease P1 (as opposed to micrococcal nuclease and calf spleen phosphodiesterase) to generate normal 5′-nucleosides and 5′-phosphate dinucleotides (adduct/normal base; pXpN), that then undergo further hydrolysis with prostatic acid phosphatase to yield normal nucleosides and adducted dinucleotides (XpN). Only adduct-containing dinucleotides are substrates for subsequent ^{32}P-labelling with T4-PNK. (d) Adduct enrichment by the butanol extraction technique (Fig. 3) involves the selective organic extraction of adducted nucleoside-3′-monophosphates (dXps) following the initial micrococcal nuclease/calf spleen phosphodiesterase digestion (13). Hydrophobic adducts are partitioned into butanol at acid pH in the presence of tetrabutyl ammonium chloride, a phase transfer agent. Residual normal nucleoside-3′-monophosphates are removed from the butanol phase by back extracting with water. (e) High performance liquid chromatography (HPLC) can also be utilised to separate unmodified and modified nucleosides prior to the ^{32}P-labelling.

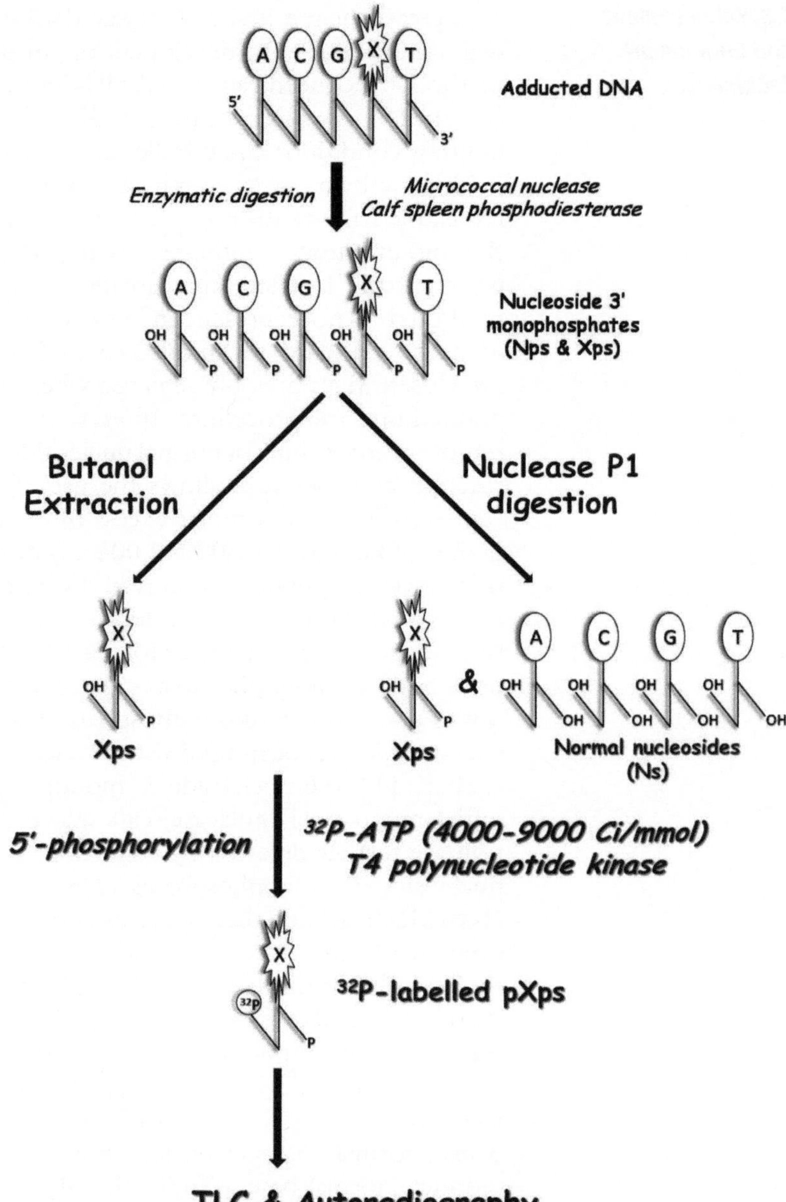

Fig. 3. The nuclease P1 digestion and butanol extraction enhancement procedures. In the nuclease P1 method, normal 3′-nucleosides (Nps) are digested to nucleosides (Ns) that are not substrates for T4-PNK, whilst most adducted 3′ monophosphates (Xps) are resistant or partially resistant to the digestion. With butanol extraction, adducted 3′ monophosphates (Xps) are physically partitioned from the normal 3′ monophosphates (Nps). The removal of the normal nucleosides from the labelling reaction allows the use of high specific activity [γ^{32}P]-ATP and greatly increases the sensitivity of the assay.

Reverse phase HPLC enrichment of bulky hydrophobic DNA adducts using methanol gradients has been described (14). Several HPLC-^{32}P-postlabelling methods have been developed for the detection of specific and smaller DNA lesions including alkylations (O^6-methyldeoxyguanosine and *N7*-methyldeoxyguanosine) and etheno adducts, as reviewed by Gorelick (15).

1.4. Nuclease P1 Digestion Versus Butanol Extraction Enhancement

Of the enhancement methods described above, the two that are most prevalently utilised and reported in the literature are nuclease P1 digestion and butanol extraction (Fig. 3). It is these techniques that are described in detail in Subheading 3 of this chapter. While the two techniques enhance the detection of broadly similar classes of DNA adducts, it has been clearly demonstrated that the range of adducts recovered does not exactly coincide (16, 17). While the two methods exhibit comparable enrichment of most PAH-adducts (e.g. deoxyguanosine-N_2-benzo(*a*)-pyrene diol-epoxide), butanol extraction shows a more efficient enhancement of adducts resulting from aromatic amines and nitrated PAHs. Deoxyguanosine-C8-aromatic amine adducts and *N*-substituted aromatic adducts (e.g. *N*-(deoxyguanosin-8-yl)-1-aminopyrene), for example, are refractive to enhancement following nuclease P1 digestion (16). On the other hand, aceanthrylene adducts show better enhancement with nuclease P1 (6). Even for carcinogens that show very similar enhancement with the two methods, there can be subtle differences in the adducts detected. Figure 4 shows the adduct

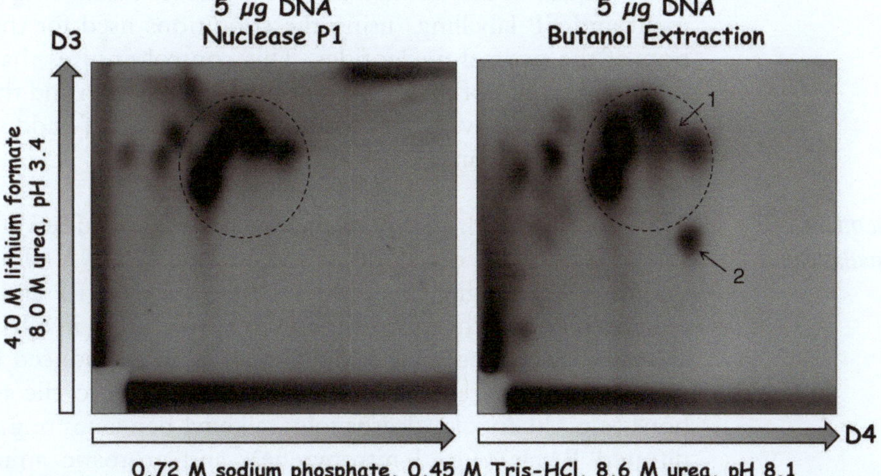

Fig. 4. ^{32}P-labelled adduct maps obtained following the postlabelling of DNA extracted from the skin of a 9–10 week old mouse exposed 24 h earlier to a single dose (40 µg) of DMBA (7,12-dimethylbenz[*a*]anthracene). Autoradiography was for 48 h with intensifying screens at –70°C. Such autoradiographs with discrete spots are typical following exposure to a single genotoxin. The different enhancement methods result in broadly similar adduct maps (a group of the five more prevalent adducts detected with nuclease P1 enrichment are indicated by the *circle*). However, the adducts detected by the two methods do not exactly coincide. *Arrow 1* indicates an adduct that is recovered at reduced levels with butanol extraction, while *arrow 2* indicates an adduct that was not detected following nuclease P1 digestion.

profiles generated by the two enhancement techniques following the ^{32}P-postlabelling of DNA from the skin of a mouse exposed to DMBA (18). While the adduct maps and the major adducts detected are largely comparable, there are differences in the more minor adducts and one adduct in particular was only observed after butanol extraction (N.J. Jones, unpublished). As a consequence of this differential enhancement, it is strongly recommended that new studies utilise a number of enhancement protocols until it is determined which is the most appropriate. This is particularly true when a new uncharacterised compound is assessed for genotoxicity or for monitoring exposure to complex mixtures of unknown composition. In addition, use of the distinct enhancement procedures may also provide insights into the nature of the DNA damage detected. A number of studies on the oral tissue of smokers have demonstrated that butanol extraction reveals a much wider range and substantially higher levels of DNA adducts than obtained following nuclease P1 enhancement, suggesting that aromatic amines and nitroaromatics may be the source of these adducts (19, 20). A major adduct found in human bladder biopsies of tobacco smokers using butanol extraction was identified as the deoxyguanosine-C8 adduct of 4-aminobiphenyl was undetectable following nuclease P1 digestion (21).

When these enhancement techniques are employed, it is also important to monitor the efficiency of 3′-dephosphorylation by nuclease P1 or removal of normal nucleosides by butanol extraction. This may be assessed by performing chromatography on a small aliquot of the labelled adduct digest (i.e. following enhancement and ^{32}P-labelling) using the conditions used for the separation of the normal nucleotides. This control ensures that normal nucleotides are not present in the labelling reaction and that excess [γ^{32}P]-ATP is available for the labelling of adducts (see Subheadings 3.9 and 3.12).

1.5. Applications of the ^{32}P-postlabelling Assay

The versatility of the ^{32}P-postlabelling assay has resulted in its widespread use for the detection of a vast range of DNA adducts. The nature of the original method and the nuclease P1 and butanol enhancement methods lend themselves to studies involving bulky aromatic or hydrophobic adducts such as those induced by PAHs (e.g. benzo[*a*]pyrene, DMBA), aromatic heterocyclic hydrocarbons (e.g. dibenzo[*c,g*]carbazole), alkenyl benzenes (e.g. safrole), nitrated PAHs (e.g. 1-nitropyrene), and aromatic amines (e.g. 4-aminobiphenyl) (6, 11, 13, 16). However, more specialised methods for the detection of other adduct types have been described. These include alkyl adducts (e.g. O^6-methyldeoxyguanosine and *N7*-methyldeoxyguanosine), pyrimidine dimers (induced by ultraviolet radiation), apurinic sites, oxidative DNA damage (e.g. thymine glycols, 8-hydroxydeoxyguanosine, 8,5′-Cyclopurine-2′-deoxynucleotides), intrastrand crosslinks and even DNA strand breaks (7, 22–28). The extreme sensitivity of the assay has also enabled the analysis of endogenous DNA damage. Randerath and co-workers described indigenous DNA modifications (or so-called I-compounds) in both unexposed laboratory animals and in humans, that accumulate in an age-dependent manner (29). A particular advantage of the ^{32}P-postlabelling assay is that its promiscuous detection of adducts, enables the analysis of the DNA damage induced by complex mixtures that consist of an assortment of known and unknown genotoxins (e.g. tobacco smoke).

Given the ability to detect such a plethora of DNA lesions, it is not surprising that an important application of the assay is as a means of determining whether a compound is a suspect genotoxin by testing its ability to induce DNA adducts both *in vitro* and *in vivo* (30). A major utilisation of ^{32}P-postlabelling is the evaluation in laboratory animals or cultured cells of the mechanism of action of known or potential carcinogenic compounds. Such studies include investigation into the metabolic activation, tissue specificity and the structural basis for a compound's genotoxicity (18, 31–35). ^{32}P-postlabelling has been used to analyse the induction of DNA damage by numerous pharmaceutical drugs. For example, tamoxifen, an oestrogen receptor antagonist used in the treatment of breast cancer, was itself found to be a carcinogen. It induces liver cancer in rats and results in an increased incidence of endometrial cancer in treated patients (36). While it was originally believed that the carcinogenicity of tamoxifen was an outcome of its oestrogenic activity (and thus acting as a non-genotoxic carcinogen), ^{32}P-postlabelling analysis established that metabolic derivatives of tamoxifen induced DNA adducts in the livers of rats suggesting that the drug might instead be acting as a genotoxic carcinogen. However, adduct levels in patients receiving the drug are low (36). Whereas, many ^{32}P-postlabelling studies have primarily focused on the induction of DNA damage, the assay may also be used to investigate the removal of adducts by DNA repair (37–39).

The extreme sensitivity and wide variety of DNA adducts detected by ^{32}P-postlabelling make it the ideal analytical tool for human biomonitoring and the detection of DNA damage resulting from environmental or occupational exposures. Studies on humans are very extensive and tissues examined include lung, heart, bladder, breast, bone marrow and more readily obtainable and less invasive tissues such as lymphocytes, buccal mucosa, and placenta (6). Investigation of occupational exposures include those made in individuals working in iron foundries and aluminium plants, operating coke ovens, or exposed to urban air pollution (40, 41). Tobacco-smoking associated DNA adducts have been found in several tissues including lung, bladder, and placenta (42). In cases of exposure to complex mixtures, such as tobacco smoke or other products of combustion such as diesel exhaust, DNA damage is

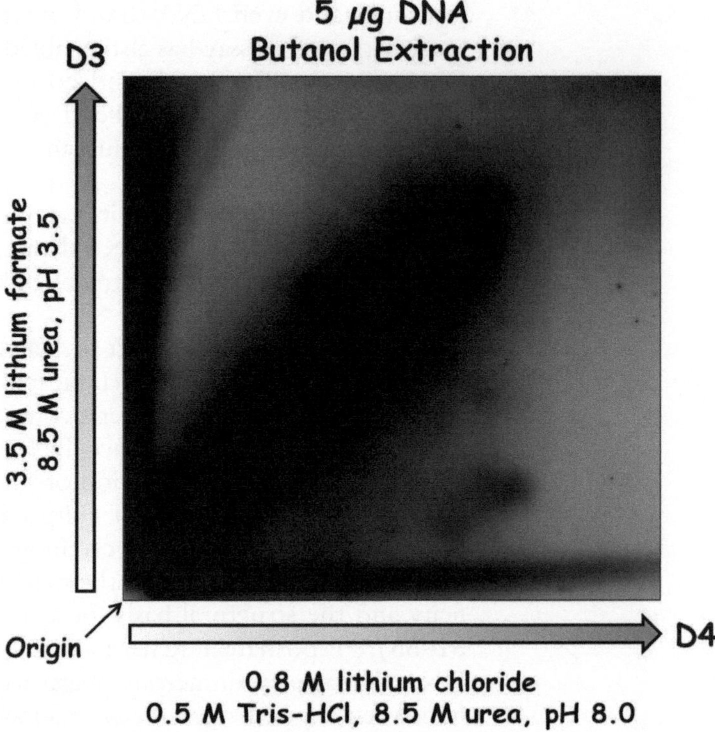

Fig. 5. Autoradiograph of DNA adducts detected in the oral tissue of a heavy cigarette smoker. DNA was extracted from clinically normal oral tissue obtained from a patient who underwent surgery for intraoral squamous cell carcinoma and analysed using butanol extraction enhancement. Autoradiography was for 96 h with intensifying screens at −70°C. The diagonal radioactive zone (DRZ) is typical following exposure to complex mixtures of genotoxic substances.

generally visualised as diagonal radioactive zones (DRZ) on autoradiographs. Such a DRZ is shown in Fig. 5, obtained following the ^{32}P-postlabelling of DNA extracted from the oral tissue of a heavy cigarette smoker. Studies on environmental exposures are not limited to humans and the assessment of environmental contamination has been made in a number of indicator organisms, including both aquatic and terrestrial species (43–46).

1.6. Quantification of Adduct Levels

Following the production of adduct maps (by autoradiography, scanning or phospho-imaging), quantification of radioactivity in individual adduct spots or adduct zones may be determined. Radioactivity is also determined in adjacent blank areas (of equivalent size) and subtracted when calculating adduct levels. The amount of radioactivity in normal nucleotides (dpNp) is determined in a similar way by performing TLC on a diluted aliquot of the labelled digest (~0.01 ng DNA) using solvents which separate the normal nucleotides from each other and inorganic phosphate (Fig. 6). In the standard assay, using the assumption that normal and adducted nucleosides are labelled to an equal extent, the relative adduct labelling may be calculated (13):

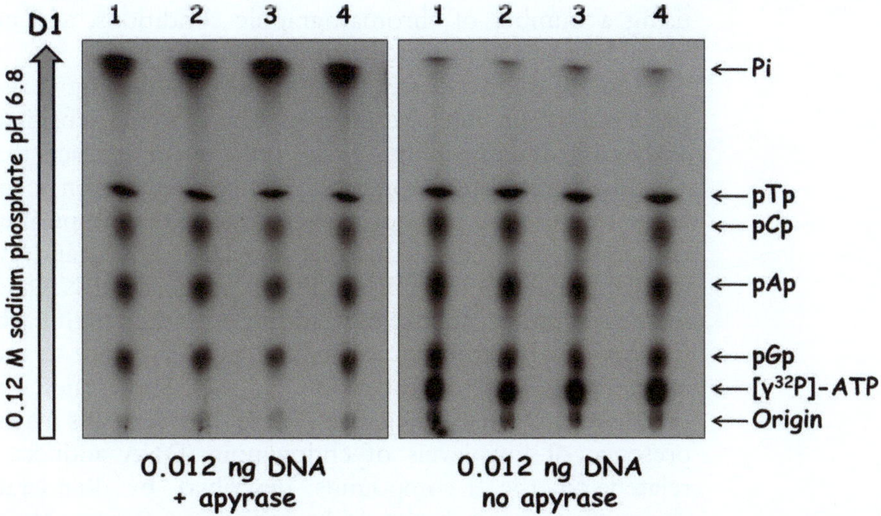

Fig. 6. ^{32}P-labelling and one-dimensional separation of normal nucleoside 3′-5′-bisphosphates from four individual samples. An aliquot of the DNA digest is removed prior to enhancement, diluted, and labelled in parallel with the enriched adducts. Autoradiography was for 60 min at room temperature. The level of radioactivity is determined in the four-labelled normal bisphosphates in the aliquot treated with apyrase and used when calculating adduct levels. The presence of [γ^{32}P]-ATP in the aliquot without apyrase indicates that sufficient ATP was available in the labelling reaction.

$$\text{RAL} = \frac{\text{decays per minute (dpm) in adduct (s)}}{\text{dpm in total (normal) nucleosides} \times \text{dilution factor}}$$

Alternatively RAL may be calculated based on the specific activity of the [γ^{32}P]-ATP (expressed as dpm/pmol) and the amount of DNA used (1 μg DNA = 3,240 pmol dNp)

$$\text{RAL} = \frac{\text{dpm in adduct (s)}}{\text{specific activity of ATP} \times \text{pmol dNp used for analysis}}$$

RAL values can then be translated into adduct frequencies (1/RAL) or into attomol adduct per μg DNA by assuming that 1 μg DNA is equivalent to 3.24×10^9 attomol nucleotides. These calculations are appropriate when the standard protocol without enhancement is used; however, if nuclease P1 or butanol extraction is utilised, it is necessary to remove and ^{32}P-label normal nucleotides in parallel to the labelling of the samples processed for enhancement (see Subheading 3.5 below). To do this, an aliquot of the micrococcal nuclease/calf spleen phosphodiesterase digest is removed prior to the nuclease P1 reaction or butanol extraction.

1.7. Weaknesses and Limitations

While the ^{32}P-postlabelling assay is very useful in quantifying levels of adducts by known or unknown carcinogens, the identification of particular adducts is more challenging. In most cases, this depends on the co-migration of an adduct spot with characterised standards using a number of chromatographic conditions, and even then identification cannot be certain.

One of the major strengths of ^{32}P-postlabelling is also potentially one of its chief weaknesses. The extreme sensitivity of the assay raises the possibility of false positives or artefacts that appear as adducts on chromatograms. Artefacts have been seen to arise from the use of particular microfuge tubes, impurities in the reagents used to extract, digest and label DNA, and even from individual batches of [γ^{32}P]-ATP. Contaminating RNA and protein in DNA samples can give false adduct spots and high backgrounds respectively. Butanol extraction is particularly prone to the appearance of so-called "test-tube spots" (6, 47). Some studies have used calf thymus DNA as a negative control but even this may show the presence of low levels of endogenous DNA adducts, possibly related to the I-compounds described by Randerath (29). Appropriate controls should be utilised to monitor the presence of artefact spots. These may include running, in parallel to test samples, blank samples (no DNA), and/or labelling DNA samples previously analysed and quantified by different enhancement techniques that have known low levels of adducts.

The appearance of artefact spots or false positives is likely to present a bigger problem in human or environmental biomonitoring. Conversely, false negatives may be of greater concern when testing the genotoxicity of a suspected carcinogen or the mechanism of action of a particular compound. This might result from incomplete digestion of the DNA by micrococcal nuclease/calf spleen phosphodiesterease, over digestion by nuclease P1, adduct instability or use of inappropriate solvent conditions. The use of positive controls may help limit such problems. Use of DNA treated with a known carcinogen or the incorporation of an adduct standard (e.g. (+)-N^2-(7R.8S.9R-dihydroxy-7,8,9,10-tetrahydro benzo[*a*]pyrene-10s-yl)2′-deoxyguanosine) may be used to monitor and quantify adduct labelling (39, 48).

1.8. Safety

The ^{32}P-postlabelling assay is, by its very nature, dependent on the use of considerable amounts of radioactivity due to the use of high specific activity [γ^{32}P]-ATP. As such, an extremely important consideration when utilising the assay is safety. The use of a room dedicated to the technique is highly recommended. The handling, storage, and disposal of all ^{32}P-containing material (disposable plastics, liquids, TLC plates etc.) should conform to HSE and local rules for the use of radioactivity and all radiation workers must receive appropriate and mandatory training. All procedures should be approved by local radiation protection supervisors and safety representatives prior to work commencing. All workers must wear safety glasses and appropriate devices for monitoring their radiation exposure (film badges, finger badges etc.) All manipulations must be performed behind 1-cm acrylic screens or other appropriate shielding; acrylic racks should be used for holding tubes and acrylic boxes with lids used for discard. Tubes containing ^{32}P must be handled with forceps. Disposable plastic aprons and several layers of disposable gloves should be worn at all times. Lead aprons and lead gloves should be available and used when deemed appropriate. A suitable radiation monitor (Geiger counter) is mandatory and frequent monitoring of gloved hands, equipment and work areas is imperative to detect contamination and to avoid possible cross-contamination.

2. Materials

2.1. DNA Digestion

1. SSCC buffer for DNA digestion: 100 mM sodium succinate, 50 mM $CaCl_2$, pH 6.0. Store at 4°C for short periods or –20°C for long periods.
2. Micrococcal nuclease (Sigma-Aldrich): Dialysed extensively in distilled water overnight at 4°C, diluted with water to give final concentration of 0.2 units/μl. Store at –20°C in small aliquots and avoid repeated freeze-thawing.
3. Calf spleen phosphodiesterase (Boehringer Mannheim or Calbiochem): Dialysed extensively in distilled water overnight at 4 °C, diluted with water to give final concentration of 2.0 mU/μl. Store at –20°C in small aliquots and avoid repeated freeze-thawing.

2.2. Enhancement Procedures

1. Nuclease P1 (Sigma-Aldrich): 2 μg/μl dissolved in 0.28 M sodium acetate, 0.5 mM zinc chloride, pH 5.0. Store at –20°C in small aliquots and avoid repeated freeze-thawing.
2. 1.0 M Tris base. Store at 4°C for short periods or –20°C for long periods.
3. 1-Butanol (water saturated) for butanol extraction. Store at 4°C.
4. 10 mM tetrabutylammonium chloride. Store at 4°C.
5. 100 mM ammonium formate, pH 3.5. Store at 4°C.
6. Water saturated with butanol. Store at 4°C.
7. 200 mM Tris–HCl pH 9.5 or 200 mM Tris–HCl pH 7.6 (see Note 1). Store at 4°C for short periods or –20°C for long periods.

2.3. ³²P-labelling of Adducts and Normal Nucleosides

1. Labelling buffer: 100 mM bicine, pH 9.5, 100 mM dithiothereitol, 10 mM spermidine. Store at –20°C. If using recombinant T4-PNK without 3′-phosphatase activity the pH of the labelling buffer may be altered to pH 7.6 to allow labelling at optimal pH (see T4-PNK below and Note 1).

2. T4-PNK (various suppliers): T4 polynucleotide kinase (at 10 units/μl) or 3′-phosphatase T4 polynucleotide kinase (see Note 1). Store at −20°C.
3. [γ^{32}P]-ATP (various suppliers): [γ^{32}P]-ATP (at 5,000–9,000 Ci/mmol, 10 μCi/μl). Store at −20°C. *[γ32P]-ATP is highly radioactive (beta-particle emitter), all manipulations involving this material must conform to the proper safety standards as outlined in Subheading 1.8 above.*
4. Potato apyrase (Sigma-Aldrich): Dissolved in water at 40 mU/μl. Store at −20°C.
5. 10 mM Tris–HCl, 5 mM EDTA pH 9.5 for dilution of normal nucleosides prior to chromatography.

2.4. Chromatography

1. 20 × 20 cm plastic-backed PEI-cellulose TLC plates (Macherey Nagal): Plates should be pre-developed in distilled water and air-dried before use. Store at 4°C.
2. Glass chromatography tanks with lids.
3. D1 solvent: 1.0 M sodium phosphate pH 6.0.
4. D2 solvent: 2.5 M ammonium formate pH 3.5.
5. D3 solvent: 3.5 M lithium formate, 8.5 M urea, pH 3.5.
6. 13 mM Tris base.
7. D4 solvent: 0.8 M lithium chloride, 0.5 M Tris–HCl, 8.5 M urea, pH 8.0.
8. D5 solvent: 1.7 M sodium phosphate pH 6.0.
9. Solvent for resolution of normal nucleosides: 0.12 M sodium phosphate, pH 6.8.

3. Methods

The methods described in this chapter are most suitable for the ^{32}P-postlabelling of aromatic DNA adducts and employ the nuclease P1 and butanol extraction modifications of the original standard assay. They were primarily developed for biomonitoring in humans and other species, although they have been successfully applied to analyses in laboratory rodents and cultured cells (18, 20, 47–51). When embarking on a new study, it is important to be aware that it may be necessary to optimise conditions for different DNA damaging agents or tissues as the DNA adducts generated may vary. New investigators should also refer to a collaborative study by a consortium of investigators that reported on attempts to devise a standardised protocol (52, 53). However, it should be borne in mind that while such standardised protocols (including the one described below) are useful, they may need to be adapted out of necessity for certain applications. Where appropriate, notes have been given to provide guidance or to emphasise the importance of particular controls (see Subheading 4).

3.1. Digestion of DNA to Nucleoside-3′-Monophosphates (Nps and Xps)

1. To 10 μg of DNA add 2 μl micrococcal nuclease (0.2 units/μl), 2 μl calf spleen phosphodiesterase (2 mU/μl), 2 μl SSCC buffer (100 mM sodium succinate, 50 mM $CaCl_2$, pH 6.0), and water to a final volume of 20 μl (see Note 2).
2. After incubation at 37°C for 2 h, add a further 1 μl SSCC, 2 μl micrococcal nuclease, 2 μl calf spleen phosphodiesterase, and 5 μl water to the digest and incubate for a further 4 h (see Note 3).
3. Of the final volume of 30 μl, 25 μl is used for enhancement (Subheadings 3.2 and 3.3 below) and subsequent ^{32}P-labelling of adducts (Subheading 3.4 below), whilet 5 μl is retained for the labelling of normal nucleosides (see Subheading 3.5 below).

3.2. Nuclease P1 Digestion

1. Normal nucleoside-3′-monophosphates (dNps) are converted to nucleosides using nuclease P1. To 25 μl of the DNA digest (from Subheading 3.1 above) add 7.6 μl of nuclease P1 (2 μg/μl in 0.28 M sodium acetate, 0.5 mM zinc chloride pH 5.0) and 4 μl water and incubate for at 37°C for 1 h (see Note 4).
2. After incubation, adjust the pH of the nuclease P1 digest by adding 4.0 μl of 1.0 M Tris base and evaporate to dryness in a vacuum centrifuge (see Note 5).

3.3. Butanol Extraction (see Note 6)

1. Dilute 25 μl of the DNA digest (from Subheading 3.1 above) to the equivalent of 0.032 μg/μl in water (final volume 260 μl when starting with 10 μg of DNA) containing tetrabutylammonium chloride and ammonium formate pH 3.5 (final concentrations 1 mM and 10 mM respectively).
2. Add an equal volume of butanol, vortex and separate the phases by centrifugation (1 min).
3. Remove the butanol phase to a fresh tube. Add a further equal volume of butanol, repeat the extraction, and combine the butanol phases.
4. Remove any residual dNps by back extracting the butanol phase three times with an equal volume of water (saturated with butanol).
5. Adjust the pH of the butanol phase by adding 3 μl of 200 mM Tris–HCl pH 9.5 (or pH 7.6, see Note 1) and evaporate to dryness.

3.4. ^{32}P-labelling of Adducts with T4 Polynucleotide Kinase

1. Reconstitute adduct residues (from either nuclease P1 digestion Subheading 3.2 or butanol extraction Subheading 3.3 above) in 9.5 μl of water.
2. Add 2 μl of labelling buffer (100 mM bicine NaOH, pH 9.5 or 7.6 (see Note 1); 100 mM $MgCl_2$, 100 mM dithiothreitol, 10 mM spermidine), 0.5 μl T4 polynucleotide kinase (10 units/μl) (see Note 7), and 8 μl [γ^{32}P]-ATP (>5,000 Ci/mmol, 10 μCi/μl).
3. Incubate at 37°C for 2 h for the 5′-phosphorylation of the adducted Xps with ^{32}P.
4. Of the resulting 20 μl of labelled adduct digest, 12–18 μl is used for chromatography of the ^{32}P-labelled adducts (see Subheading 3.7 below), while 2 μl is used to monitor the efficiency of the nuclease P1 and butanol extraction enrichments (see Subheading 3.8 below).

3.5. ^{32}P-labelling of Total/Normal Nucleosides

1. Dilute the 5-μl aliquot retained after the digestion of DNA to dNps (from Subheading 3.1), to the equivalent of 0.1 ng/μl DNA with water (add 1,662 μl of water and then perform a ten-fold dilution).
2. To 12 μl of this solution, add 2 μl labelling buffer, 0.5 μl T4 PNK, 2 μl [γ^{32}P]-ATP, and 3.5 μl of water.
3. Incubate at 37°C for 2 h in parallel with the labelling of adducts.

3.6. Use of PEI-Cellulose Plates

1. Thin layer chromatography is performed on 20 × 20 cm plastic-backed PEI-cellulose TLC plates (see Note 8). All plates must be pre-developed in water and air-dried.
2. When required, filter paper wicks should be attached to these plates with staples and lines of excision and the position of the origin should be lightly marked in pencil. For example, for the chromatography of ^{32}P-labelled adducts, excision lines drawn 3 cm from the top of the plate in both the D1 and D2 directions and the origin can be marked with a soft pencil prior to the application of the labelled digest (see Fig. 3).

3.7. Chromatography of ^{32}P-labelled Adducts

3.7.1. Removal of Non-adduct Material

1. Chromatography with the D1 and D2 solvents is intended to remove all non-adduct material, including any remaining undamaged nucleotides, residual [γ^{32}P]-ATP and inorganic phosphate, whilst the ^{32}P-labelled adducts remain at, or very close to, the origin.
2. Apply 12–18 μl of the labelled adduct digest (from Subheading 3.4 and equivalent to 5–7.5 μg of DNA) to an origin marked 5 cm from the top of the TLC plate (and 2 cm from the lines of excision) in both the D1 and D2 directions (*see* Fig. 10.3). Apply the labelled adduct digest in 2 μl aliquots

and allow to dry between applications in order to minimise the size of the origin spot.

3. Place the plate into a glass chromatography tank containing ~1 cm of 1.0 M sodium phosphate pH 6.8 (the D1 solvent) at the bottom of the tank, and with the filter paper wick extending out from the lid at the top of the tank. Wrap the lid and the top of the tank with cling film to avoid contamination.
4. Develop the plate overnight so that the solvent runs onto the wick. Remove the wick and the top 3 cm of the plate by cutting along the pre-marked pencil line with scissors and discard.
5. Wash the plate in water and allow to air dry. Attach a further filter paper wick at the top of the plate in the D2 direction and develop the plate with 2.5 M ammonium formate pH 3.5 for approximately 6 h.
6. Remove the wick and the top 3 cm of the plate and discard, wash the plate with water and air dry.

3.7.2. Adduct Resolution (see Note 9)

1. Adducts are separated on the resulting 17 × 17 cm plates.
2. To move and separate the adducts, develop first with 3.5 M lithium formate, 8.5 M urea, pH 3.5 to the top of the plate in the D3 direction (~16 h), wash once in 13 mM Tris base and once in water, and air dry.
3. Then develop to the top of the plate in the D4 direction with 0.8 M lithium chloride, 0.5 M Tris–HCl, 8.5 M urea, pH 8.0 (~6 h), wash the plate twice in water and air dry.
4. Finally, attach a filter paper wick to the top of the plate in the D4/D5 direction and develop overnight onto the wick using 1.7 M sodium phosphate pH 6.0 to remove any residual non-adduct material.
5. Prior to autoradiography, cut off the very top of the plate with the wick, excise the origin and thoroughly air dry the plate.

3.8. Chromatography of ^{32}P-labelled Total/ Normal Nucleosides

1. Split the 20 μl sample (from Subheading 3.5 above) into two 10 μl aliquots and add 2 μl of potato apyrase (40 mU/μl) to one aliquot and incubate at 37°C for 30 min to destroy the excess [γ^{32}P]-ATP.
2. Make up the volume of both aliquots (with and without potato apyrase) to 100 μl with 10 mM Tris–HCl, 5 mM EDTA (pH 9.5).
3. Apply 2 μl from each aliquot (equivalent to 0.012 ng of DNA) to an origin 2 cm from the bottom edge of a 20-cm long PEI-cellulose plate and develop in a single direction with 0.12 M sodium phosphate (pH 6.8). Dry the plate prior to autoradiography.

3.9. Chromatography of ³²P-labelled Adduct Digest to Monitor the Removal of Normal Nucleosides by Nuclease P1 Digestion or Butanol Extraction

1. To 5 μl of the labelled adduct digest (from Subheading 3.4 above) add 81.3 μl of 10 mM Tris–HCl, 5 mM EDTA (pH 9.5).
2. Apply 2 μl (equivalent to 20 ng of DNA) to an origin 2 cm from the bottom edge of a 20-cm long PEI-cellulose plate and develop in a single direction with 0.12 M sodium phosphate (pH 6.8) as described for the chromatography of the normal nucleosides (see Subheading 3.8 above).

3.10. ³²P-labelled Adduct Maps and Quantification of Radioactivity in Adduct Spots/Zones (see Note 10)

1. Visualise the ^{32}P-labelled adducts on the chromatograms by performing autoradiography with intensifying screens at −70–80°C for 24–150 h (depending on adduct levels). The appearance of adduct maps will depend of the nature of the genotoxin inducing the DNA damage. Exposure to a single carcinogen will often reveal discrete adducts spots (Fig. 4), while induction of DNA damage by complex mixtures will generate diagonal radioactive zones (DRZ) indicating the presence of a wide variety of DNA adducts (Fig. 5).
2. For quantification (see Note 10), cut out adduct spots or adduct-containing areas/zones of the TLC plates (with scissors), after precisely marking their location (using a pencil) by aligning the plates with the autoradiographs using a light box.
3. Measure the radioactivity by liquid scintillation counting.
4. To allow for background radioactivity on the plate, excise an equal sized area from an adjacent blank portion of the plate, and subtract when determining adducts levels.

3.11. Autoradiography and Quantification of Radioactivity in Total/Normal Nucleosides

1. Perform autoradiography of the chromatograms run to separate ^{32}P-labelled normal nucleosides (from Subheading 3.8) for 45–90 min. The chromatographic conditions used separate the four normal ^{32}P-labelled nucleoside-3′-5′-bisphosphates (pNps) from each other, and from any remaining [γ^{32}P]-ATP and inorganic phosphate (see Fig. 6) (see Note 11)
2. To enable the quantification of adducts, excise the normal nucleosides for the aliquot that was incubated with potato apyrase (as [γ^{32}P]-ATP runs quite close to the spot for dpGp) and determine the level of radioactivity by scintillation counting.
3. Chromatography and autoradiography of the aliquot without potato apyrase ensures that there was an excess of [γ^{32}P]-ATP in the labelling reaction.

3.12. Monitoring the Efficiency of Nuclease P1 Digestion or Butanol Extraction

1. Autoradiography of the chromatogram run to monitor the efficacy of the enhancement techniques (from Subheading 3.9 above) is also for 45–90 min. This control tests whether the vast bulk of undamaged nucleosides have been removed prior to the adduct labelling-reaction (either biochemically by their

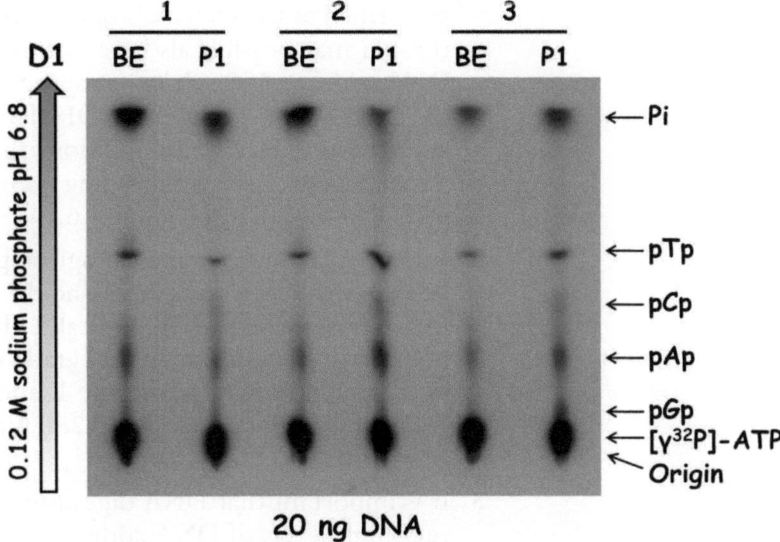

Fig. 7. Control to monitor the efficiency of the nuclease P1 and butanol extraction enhancements for three individual samples. An aliquot of the labelled adduct digest is removed, diluted, and chromatography performed using the conditions used to separate the labelled normal nucleosides (Fig. 6). Autoradiography was for 60 min at room temperature. This important control ensures the vast majority of normal nucleosides were removed prior to the labelling and that there was sufficient [γ^{32}P]-ATP present to label the adducted bisphosphates. Note that the amount of DNA loaded was 20 ng and therefore 1,667× more than that loaded onto the chromatogram shown in Fig. 6.

3′ dephosphorylation by nuclease P1 or physically by butanol extraction), and that excess [γ^{32}P]-ATP was available for adduct labelling.

2. The autoradiograph should show a spot of [γ^{32}P]-ATP for every sample and either no or very faint spots representing residual normal nucleoside-3′-5′-bisphosphates (Fig. 7).
3. If the spot of [γ^{32}P]-ATP is absent and/or there is excessive labelling of the normal nucleosides then the sample should be rejected.

4. Notes

1. The labelling reaction is generally performed at pH 9.5 as T4-PNK has a 3′-phosphatase activity in addition to its primary 5′-hydroxyl kinase activity. The alkaline pH of the reaction inhibits the 3′-phosphatase activity. Recombinant T4-PNK without 3′-phosphatase activity is now commercially available and can be substituted to allow the labelling reaction to be

performed at the enzyme's optimal pH (pH 7.6). Labelling at this pH may possibly also prevent the loss of some alkali-labile DNA adducts. If 3′-phosphatase-free T4-PNK is used, the pH of the 100 mM bicine NaOH labelling buffer should be adjusted to pH 7.6. In addition, the pH adjustment of the butanol phase (see Subheading 3.3 above) should be made with 200 mM Tris–HCl pH 7.6.

2. Micrococcal nuclease and calf spleen phosphodiesterase should be dialysed before use. DNA should be of high purity and be free of protein and RNA. Typically, phenol–chloroform extraction methods (using analytical grade reagents) have been used in conjunction with proteinase K and RNases. Absorbance of DNA preparations at 260 and 280 nm should be used to measure purity.
3. It is important that DNA digestion is completed – DNA with very high levels of DNA adduction may be resistant to digestion. In some protocols, DNA is digested overnight (53).
4. Some DNA adducts are only partially resistant to nuclease P1, so it is important not to over digest with the enzyme as this may reduce the recovery of certain adducts. Some studies have titrated nuclease P1 to optimise adduct recovery (41).
5. Many protocols do not evaporate the P1 digest and proceed directly to ^{32}P-postlabelling, but the evaporation does have the advantage of reducing the final volume of the labelling reaction. Evaporation of nuclease P1 digests was found not to adversely affect the recovery of benzo[*a*]pyrene adducts or of the adduct standard (+)-N^2-(7R.8S.9R-dihydroxy-7,8,9,10-tetrahydrobenzo[*a*]pyrene-10s-yl)2′-deoxyguanosine (48).
6. Butanol extraction can occasionally be found to generate artefacts related to some batches of plastic tubes or solvent impurities. For example, Harvey and Parry (1998) found coloured Eppendorf tubes generated artefact adduct spots while clear tubes did not (47).
7. If T4-PNK is supplied at a concentration other than 10 units/μl, adjust volumes accordingly.
8. PEI-cellulose TLC plates from commercial sources may be quite variable and this can influence the times of development for some solvents. This can be tested in advance using blank plates for different sources/batches. In addition, it has been found that the PEI-cellulose layer can detach from the plastic backing using the solvents described here for some batches/manufactures. Such batches should be rejected.
9. The D3 and D4 solvents may have to be changed to achieve the required purpose. For example, if a genotoxin of interest yields major adducts that have similar chromatographic properties using the solvents described here, and that need to be

quantified individually, it may be possible to adjust D3/D4 to allow their resolution. Also, as the conditions described here are most suitable for the resolution of DNA adducts containing large aromatic moieties, it will be necessary to change them for smaller adducts. For instance, Phillips et al. (9) used 2.45 M lithium formate, 5.95 M urea, pH 3.5 and 0.4 M LiCl, 0.25 M Tris–HCl, 4.25 M urea, pH 8.0 as the D3 and D4 solvents for the separation of alkenyl benzene adducts.

10. Visualization and quantification of radioactivity in adducts (and normal nucleosides, see Subheading 3.11 above) has more recently been analysed either using scanning radioanalytical imaging systems or using phospho(r)-imaging. These techniques provide a much more straightforward (and safer) method of quantification and should be used in preference to excision and scintillation counting where available.
11. In addition to allowing the calculation of adduct levels, the labelling of the normal nucleotides acts as a monitor for RNA contamination (there should be no bisphosphate for uracil).

References

1. Randerath, K., Reddy, M. V., and Gupta, R. C. (1981) ^{32}P-Labeling test for DNA damage, *Proceedings of the National Academy of Sciences of the United States of America-Biological Sciences* **78**, 6126–6129.
2. Reddy, M. V., Gupta, R. C., Randerath, E., and Randerath, K. (1984) ^{32}P-Postlabeling test for covalent DNA binding of chemicals *in vivo* – application to a variety of aromatic carcinogens and methylating agents, *Carcinogenesis* **5**, 231–243.
3. Randerath, K., and Randerath, E. (1973) Chemical postlabeling methods for base composition and sequence analysis of RNA, *Journal of Chromatography* **82**, 59–74.
4. Gupta, R. C., and Randerath, K. (1979) Rapid print-readout technique for sequencing of RNAs containing modified nucleotides, *Nucleic Acids Research 6*, 3443–3458.
5. Reddy, M. V., Gupta, R. C., and Randerath, K. (1981) ^{32}P-Base analysis of DNA, *Analytical Biochemistry* **117**, 271–279.
6. Beach, A. C., and Gupta, R. C. (1992) Human biomonitoring and the ^{32}P-postlabeling assay, *Carcinogenesis* **13**, 1053–1074.
7. Wilson, V. L., Basu, A. K., Essigmann, J. M., Smith, R. A., and Harris, C. C. (1988) O^{6} Alkyldeoxyguanosine detection by ^{32}P-postlabeling and nucleotide chromatographic analysis, *Cancer Research* **48**, 2156–2161.
8. Leuratti, C., Jones, N. J., Marafante, E., Kostiainen, R., Peltonen, K., and Waters, R. (1994) DNA damage induced by the environmental carcinogen butadiene: Identification of a diepoxybutane-adenine adduct and its detection by ^{32}P postlabeling, *Carcinogenesis* **15**, 1903–1910.
9. Phillips, D. H., Reddy, M. V., and Randerath, K. (1984) ^{32}P Post-labeling analysis of DNA adducts formed in the livers of animals treated with safrole, estragole and other naturally-occurring alkenylbenzenes: 2. newborn male B6C3F1 mice, *Carcinogenesis* **5**, 1623–1628.
10. Randerath, E., Agrawal, H. P., Weaver, J. A., Bordelon, C. B., and Randerath, K. (1985) ^{32}P-postlabeling analysis of DNA adducts persisting for up to 42 weeks in the skin, epidermis and dermis of mice treated topically with 7,12-dimethylbenz[a]anthracene, *Carcinogenesis* **6**, 1117–1126.
11. Reddy, M. V., and Randerath, K. (1986) Nuclease-P1-mediated enhancement of sensitivity of ^{32}P postlabeling test for structurally diverse DNA adducts, *Carcinogenesis* 7, 1543–1551.
12. Randerath, K., Randerath, E., Danna, T. F., Vangolen, K. L., and Putman, K. L. (1989) A new sensitive ^{32}P-postlabeling assay based on the specific enzymatic conversion of bulky DNA lesions to radiolabeled dinucleotides and nucleoside 5′-monophosphates, *Carcinogenesis* **10**, 1231–1239.

13. Gupta, R. C. (1985) Enhanced sensitivity of ^{32}P-postlabeling analysis of aromatic carcinogen-DNA adducts, *Cancer Research* **45**, 5656–5662.
14. Dunn, B. P., and San, R. H. C. (1988) HPLC enrichment of hydrophobic DNA-carcinogen adducts for enhanced sensitivity of ^{32}P-postlabeling analysis, *Carcinogenesis* **9**, 1055–1060.
15. Gorelick, N. J. (1993) Application of HPLC in the ^{32}P-postlabeling assay, *Mutation Research* **288**, 5–18.
16. Gupta, R. C., and Earley, K. (1988) ^{32}P-adduct assay – comparative recoveries of structurally diverse DNA adducts in the various enhancement procedures, *Carcinogenesis* **9**, 1687–1693.
17. Gallagher, J. E., Jackson, M. A., George, M. H., Lewtas, J., and Robertson, I. G. C. (1989) Differences in detection of DNA adducts in the ^{32}P-postlabeling assay after either 1-butanol extraction or nuclease P1-treatment, *Cancer Letters* **45**, 7–12.
18. Ashby, J., Brusick, D., Myhr, B. C., Jones, N. J., Parry, J. M., Nesnow, S., Paton, D., Tinwell, H., Rosenkranz, H. S., Curti, S., Gilman, D., and Callander, R. D. (1993) Correlation of carcinogenic potency with mouse-skin ^{32}P-postlabeling and muta-mouse lac Z-mutation data for DMBA and its K-region sulfur isostere: comparison with activities observed in standard genotoxicity assays, *Mutation Research* **292**, 25–40.
19. Chacko, M., and Gupta, R. C. (1988) Evaluation of DNA damage in the oral-mucosa of tobacco users and non-users by ^{32}P-adduct assay, *Carcinogenesis* **9**, 2309–2313.
20. Jones, N. J., McGregor, A. D., and Waters, R. (1993) Detection of DNA adducts in human oral tissue: Correlation of adduct levels with tobacco smoking and differential enhancement of adducts using the butanol extraction and nuclease P1 versions of ^{32}P-postlabeling, *Cancer Research* **53**, 1522–1528.
21. Talaska, G., Aljuburi, A., and Kadlubar, F. F. (1991) Smoking related carcinogen DNA adducts in biopsy samples of human urinary-bladder: Identification of N-(deoxyguanosin-8-yl)-4-aminobiphenyl as a major adduct, *Proceedings of the National Academy of Sciences of the United States of America* **88**, 5350–5354.
22. Haque, K., Cooper, D. P., and Povey, A. C. (1994) Optimization of ^{32}P postlabeling assays for the quantitation of O^{6}-methyl and N7-methyldeoxyguanosine-3′-monophosphate in human DNA, *Carcinogenesis* **15**, 2485–2490.
23. Weinfeld, M., Liuzzi, M., and Paterson, M. C. (1989) Enzymatic analysis of isomeric trithymidylates containing ultraviolet light-induced cyclobutane pyrimidine dimers: 2. Phosphorylation by phage T4 polynucleotide kinase, *Journal of Biological Chemistry* **264**, 6364–6370.
24. Weinfeld, M., Liuzzi, M., and Paterson, M. C. (1990) Response of phage-T4 polynucleotide kinase toward dinucleotides containing apurinic sites – design of a ^{32}P postlabeling assay for apurinic sites in DNA, *Biochemistry* **29**, 1737–1743.
25. Weinfeld, M., and Soderlind, K. J. M. (1991) ^{32}P Postlabeling detection of radiation-induced DNA damage – identification and estimation of thymine glycols and phosphoglycolate termini, *Biochemistry* **30**, 1091–1097.
26. Gupta, R. C., and Arif, J. M. (2001) An improved ^{32}P-postlabeling assay for the sensitive detection of 8-oxodeoxyguanosine in tissue DNA, *Chemical Research in Toxicology* **14**, 951–957.
27. Randerath, K., Zhou, G. D., Somers, R. L., Robbins, J. H., and Brooks, P. J. (2001) A ^{32}P-Postlabeling assay for the oxidative DNA lesion 8,5′-cyclo-2′-deoxyadenosine in mammalian tissues – Evidence that four type II I-compounds are dinucleotides containing the lesion in the 3′ nucleotide, *Journal of Biological Chemistry* **276**, 36051–36057.
28. Pluim, D., Maliepaard, M., van Waardenburg, R., Beijnen, J. H., and Schellens, J. H. M. (1999) ^{32}P-postlabeling assay for the quantification of the major platinum-DNA adducts, *Analytical Biochemistry* **275**, 30–38.
29. Randerath, K., Li, D. H., Nath, R., and Randerath, E. (1992) Exogenous and endogenous DNA modifications as monitored by ^{32}P postlabeling – relationships to cancer and aging, *Experimental Gerontology* **27**, 533–549.
30. Phillips, D. H., Farmer, P. B., Beland, F. A., Nath, R. G., Poirier, M. C., Reddy, M. V., and Turteltaub, K. W. (1999) Methods of DNA adduct determination and their application to testing compounds for genotoxicity, *Environmental and Molecular Mutagenesis* **39**, 222–233.
31. Moorthy, B., Sriram, P., Pathak, D. N., Bodell, W. J., and Randerath, K. (1996) Tamoxifen metabolic activation: Comparison of DNA adducts formed by microsomal and chemical activation of tamoxifen and 4-hydroxytamoxifen with DNA adducts formed in vivo, *Cancer Research* **56**, 53–57.
32. Kaderlik, K. R., Minchin, R. F., Mulder, G. J., Ilett, K. F., Daugaardjenson, M., Teitel, C. H., and Kadlubar, F. F. (1994) Metabolic activa-

tion pathway for the formation of DNA adducts of the carcinogen 2-amino-1-methyl-6-phenylimidazo[4,5-b]pyridine (PHIP) in rat extrahepatic tissues, *Carcinogenesis* **15**, 1703–1709.

33. Ramakrishna, N. V. S., Devanesan, P. D., Rogan, E. G., Cavalieri, E. L., Jeong, H., Jankowiak, R., and Small, G. J. (1992) Mechanism of metabolic activation of the potent carcinogen 7,12-dimethylbenz[a] anthracene, *Chemical Research in Toxicology* **5**, 220–226.
34. Stiborova, M., Simanek, V., Frei, E., Hobza, P., and Ulrichova, J. (2002) DNA adduct formation from quaternary benzo[c]phenanthridine alkaloids sanguinarine and chelerythrine as revealed by the ^{32}P-postlabeling technique, *Chemico-Biological Interactions* **140**, 231–242.
35. Arlt, V. M., Glatt, H., Muckel, E., Pabel, U., Sorg, B. L., Seidel, A., Frank, H., Schmeiser, H. H., and Phillips, D. H. (2003) Activation of 3-nitrobenzanthrone and its metabolites by human acetyltransferases, sulfotransferases and cytochrome P450 expressed in Chinese hamster V79 cells, *International Journal of Cancer* **105**, 583–592.
36. Phillips, D. H. (2001) Understanding the genotoxicity of tamoxifen? *Carcinogenesis* **22**, 839–849.
37. Weinfeld, M., Lee, J., Gu, R. Q., KarimiBusheri, F., Chen, D., and AllalunisTurner, J. (1997) Use of a postlabelling assay to examine the removal of radiation-induced DNA lesions by purified enzymes and human cell extracts, *Mutation Research-Fundamental and Molecular Mechanisms of Mutagenesis* **378**, 127–137.
38. Lloyd, D. R., and Hanawalt, P. C. (2000) p53-dependent global genomic repair of benzo[a] pyrene-7,8-diol-9,10-epoxide adducts in human cells, *Cancer Research* **60**, 517–521.
39. Lagerqvist, A., Hakansson, D., Prochazka, G., Lundin, C., Dreij, K., Segerback, D., Jernstrom, B., Tornqvist, M., Seidel, A., Erixon, K., and Jenssen, D. (2008) Both replication bypass fidelity and repair efficiency influence the yield of mutations per target dose in intact mammalian cells induced by benzo[a,]pyrene-diol-epoxide and dibenzo[a,l]pyrene-diol-epoxide, *DNA Repair* 7, 1202–1212.
40. Schoket, B., Poirier, M. C., Mayer, G., Torok, G., Kolozsi-Ringelhann, A., Bognar, G., Bigbee, W. L., and Vincze, I. (1997) Biomonitoring of human genotoxicity induced by complex occupational exposures, *Mutation Research – Genetic Toxicology and Environmental Mutagenesis* **445**, 193–203.
41. Savela, K., Hemminki, K., Hewer, A., Phillips, D. H., Putman, K. L., and Randerath, K. (1989) Interlaboratory comparison of the ^{32}P-postlabeling assay for aromatic DNA adducts in white blood cells of iron foundry workers, *Mutation Research* **224**, 485–492.
42. Phillips, D. H. (2002) Smoking-related DNA and protein adducts in human tissues, *Carcinogenesis* **23**, 1979–2004.
43. Jones, N. J., and Parry, J. M. (1992) The detection of DNA adducts, DNA-base changes and chromosome damage for the assessment of exposure to genotoxic pollutants, *Aquatic Toxicology* **22**, 323–344.
44. Dunn, B. P., Black, J. J., and MacCubbin, A. (1987) ^{32}P-postlabeling analysis of aromatic DNA adducts in fish from polluted areas, *Cancer Research 47*, 6543–6548.
45. Harvey, J. S., Lyons, B. P., Waldock, M., and Parry, J. M. (1997) The application of the ^{32}P-postlabelling assay to aquatic biomonitoring, *Mutation Research-Fundamental and Molecular Mechanisms of Mutagenesis* **378**, 77–88.
46. VanSchooten, F. J., Maas, L. M., Moonen, E. J. C., Kleinjans, J. C. S., and Vanderoost, R. (1995) DNA dosimetry in biological indicator species living on PAH contaminated soils and sediments, *Ecotoxicology and Environmental Safety* **30**, 171–179.
47. Harvey, J. S., and Parry, J. M. (1998) Application of the ^{32}P-postlabelling assay for the detection of DNA adducts: False positives and artefacts and their implications for environmental biomonitoring, *Aquatic Toxicology* **40**, 293–308.
48. Jones, N. J., Kadhim, M. A., Hoskins, P. L., Parry, J. M., and Waters, R. (1991) ^{32}P-Postlabeling analysis and micronuclei induction in primary Chinese hamster lung cells exposed to tobacco particulate matter, *Carcinogenesis* **12**, 1507–1514.
49. Marafie, E. M., Marafie, I., Emery, S. J., Waters, R., and Jones, N. J. (2000) Biomonitoring the human population exposed to pollution from the oil fires in Kuwait: Analysis of placental tissue using ^{32}P-postlabeling, *Environmental and Molecular Mutagenesis* **36**, 274–282.
50. Morse, H. R., Jones, N. J., Peltonen, K., Harvey, R. G., and Waters, R. (1996) The identification and repair of DNA adducts induced by waterborne benzo[a]pyrene in developing Xenopus laevis larvae, *Mutagenesis* **11**, 101–109.
51. Stone, J. G., Jones, N. J., McGregor, A. D., and Waters, R. (1995) Development of a human biomonitoring assay using buccal mucosa: Comparison of smoking related DNA

adducts in mucosa versus biopsies, *Cancer Research* **55**, 1267–1270.

52. Phillips, D. H., and Castegnaro, M. (1999) Standardization and validation of DNA adduct postlabelling methods: report of interlaboratory trials and production of recommended protocols, *Mutagenesis* **14**, 301–315.
53. Phillips, D. H., and Arlt, V. M. (2007) The ^{32}P-postlabeling assay for DNA adducts, *Nature Protocols* **2**, 2772–2781.

Chapter 11

Methods for the Detection of DNA Adducts

Karen Brown

Abstract

The detection and characterisation of DNA adducts can provide mechanistic information on mode of action for genotoxic chemicals and in this context is vital for human risk assessments. Adducts are measured extensively in biomonitoring studies to examine exposure to environmental, dietary, and occupational chemicals and as biomarkers of efficacy for cancer chemotherapeutic drugs and chemopreventive agents. Methods used for adduct analysis must possess a certain degree of specificity and be sufficiently sensitive to detect lesions in the model system under investigation. A variety of techniques have been established for this purpose, which are capable of detecting and quantifying adducts in DNA isolated from animal or human tissues, cells, and biofluids as well as naked DNA from in vitro studies. These can be grouped as those involving ^{32}P-post-labelling, mass spectrometry, physical detection methods, immunological assays and radiolabelled compounds. Each approach presents different advantages and limitations and the most appropriate method depends on the type of sample, level of damage, and nature of the investigation as well as practical considerations. In this chapter, the basic principles of the most commonly used quantitative methods are described and their strengths and weaknesses discussed.

Key words: P-post-labelling, Mass spectrometry, Accelerator mass spectrometry, Immunoassay

1. Introduction

The critical first step in the initiation of cancer by genotoxic carcinogens involves reaction with DNA and the formation of covalent DNA adducts. If not repaired prior to replication by DNA polymerases, the presence of an adduct may cause misincorporation of an incorrect base and/or frameshift mutations. The ability of a chemical to bind to DNA, either directly or after metabolic activation, is, therefore, taken as evidence of mutagenic and carcinogenic potential. The analysis and characterisation of DNA adducts can provide mechanistic information on toxicological mode of action and is vital for conducting human risk assessments for genotoxic chemicals (1, 2). Measures of specific DNA adducts

James M. Parry and Elizabeth M. Parry (eds.), *Genetic Toxicology: Principles and Methods*, Methods in Molecular Biology, vol. 817, DOI 10.1007/978-1-61779-421-6_11, © Springer Science+Business Media, LLC 2012

are also used extensively in biomonitoring studies examining exposure to environmental/dietary/occupational chemicals and as biomarkers of efficacy for cancer chemotherapeutic drugs with DNA binding as a mechanism of action (e.g. cisplatin) as well as for chemopreventive agents that may interfere with carcinogenesis by reducing the burden of pro-mutagenic DNA damage (3, 4). For these reasons methods have been established for the detection and quantification of adducts in DNA isolated from tissues, cells, blood and saliva as well as naked DNA from in vitro studies. Additionally, numbers of free base lesions formed during DNA repair processes or spontaneous depurination can also be determined; this approach is most commonly adopted when analysing urine samples; for example, the concentration of aflatoxin-B_1-*N7*-guanine excreted in urine is used as an indicator of exposure to the mycotoxin (5).

Some genotoxins may be intrinsically active but for most chemicals that bind to DNA metabolism to an electrophilic intermediate is a prerequisite. The ultimate reactive species can then covalently bind to nucleophilic atoms in DNA, with the primary sites of reaction dependent on the nature of the electrophile involved and whether it acts through an S_N1 or S_N2 mechanism. Most chemicals have multiple targets, producing a mixture of structurally distinct adducts. Moreover, as well as numerous endogenous reactive species, humans are exposed to a plethora of environmental and dietary genotoxins which together account for the complex array of modifications existing in genomic DNA. It therefore, follows that methods used for adduct detection must possess a certain degree of specificity, particularly when analysing human samples. Additionally, in the majority of cases, any individual adduct derived from an exogenous exposure will be present at very low levels in human cells; for example, using ^{32}P-post-labelling, heterocyclic amine DNA adducts have been reported at frequencies in the region of ~0.02–3 adducts/10^8 nucleotides (6), whilst aromatic lesions at levels of ~11/10^8 have been described in white blood cells of current smokers (7). From these figures, it is clear that for any method to be capable of detecting DNA adducts, particularly in human samples, it must extremely sensitive.

A variety of techniques exist for DNA adduct analysis; these can be broadly grouped as those involving ^{32}P-post-labelling, mass spectrometry, physical detection methods, immunological assays, and radiolabelled compounds. Whilst each approach offers a different degree of specificity and sensitivity, as summarised in Table 1, the choice depends on the type of sample, level of damage, and nature of the investigation as well as practical considerations such as instrument/reagent availability and cost (8). In this chapter, the basic principles of each quantitative method are described and their advantages and limitations discussed.

Table 1
Comparison of the main quantitative methods used for DNA adduct detection

Method	Maximum sensitivity (adducts/nucleotide)	DNA required (μg)	Advantages	Limitations
HPLC-ECD/ fluorescence	~1/10^8	1–100	Simplicity	Only fluorescent/electrochemically active adducts, standards required
Immunoassay	~1/10^8	1–200	Specificity, localisation of adducts	Antibodies needed, cross-reactivity
^{32}P-post-labelling	~1/10^{10}	1–10	Sensitivity and versatility, screening possible	High levels of radioactivity, no structural information
Mass spectrometry	~1/10^8	10–100	Highest specificity, structural information	Adduct standards required
AMS	~1/10^{12}	1–2,000	Highest sensitivity, specificity of radiolabel	^{14}C/^{3}H-labelled compound needed, expensive, instrument availability

2. ^{32}P-Post-labelling

The term ^{32}P-post-labelling refers to the fact that adducts are isotope labelled *after* formation in DNA as opposed to studies using ^{14}C or ^{3}H-labelled forms of a compound to directly assess their DNA binding capability. Post-labelling is achieved by transfer of a ^{32}P containing phosphate group from [γ-^{32}P]ATP to adducted deoxyribonucleotides, which means samples can be analysed retrospectively using an assay optimised for the specific adduct or class of lesions induced by the exposure of interest. The ^{32}P-post-labelling assay was first developed in the early 1980s (9), and since then several modifications have been introduced to broaden the type of adducts that can be efficiently detected. In general, the standard technique is probably best suited to the analysis of bulky, aromatic, or hydrophobic lesions such as those formed by polycyclic aromatic hydrocarbons, acetylaminofluorene and tamoxifen, although smaller alkylated adducts and damage formed by reactive oxygen species can also be detected (10). Alternative versions of the assay designed to permit detection of base or phosphotriester lesions in a different form, such as ^{32}P-labelled dinucleotides, are beyond the scope of this chapter but are outlined in refs. 11, 12.

As illustrated in Fig. 1, the basic method consists of four main steps; DNA digestion, adduct enrichment, ^{32}P-labelling, and finally, separation and detection of the ^{32}P-labelled lesions. Detailed protocols for conducting the assay can be found in ref. 13. Typically, isolated DNA is digested to nucleoside 3′-monophosphates using the enzymes micrococcal nuclease and spleen phosphodiesterase. The former is an endonuclease that hydrolyses 5′-phosphodiester bonds, yielding 3′-phosphate mononucleotides or oligonucleotides, whilst spleen phosphodiesterase also attacks the 5′ terminal ends, liberating 3′-mononucleotides. This incubation generates a mixture of modified and normal nucleotides, with the latter vastly exceeding the numbers of adducts present. In order to enrich the sample, enabling preferential adduct labelling, several enhancement options have been devised, namely, nuclease P1 digestion, butanol extraction, and HPLC or immunoaffinity isolation. The enzyme nuclease P1 removes the 3′-phosphate group from 3′-mononucleotides, generating 2′-deoxyribonucleosides, which do not serve as substrates in the labelling reaction. By contrast, the presence of a bulky group on a DNA base can cause steric hindrance, thereby protecting the adduct from dephosphorylation (14). This approach is commonly employed for PAH derived lesions but is less successful for aromatic amines, which are better isolated by extraction into butanol. In this procedure, butanol is added to the DNA digest (in aqueous buffer) and after mixing the more polar normal nucleotides remain in the aqueous phase whilst the hydrophobic adducts are selectively extracted into the organic layer,

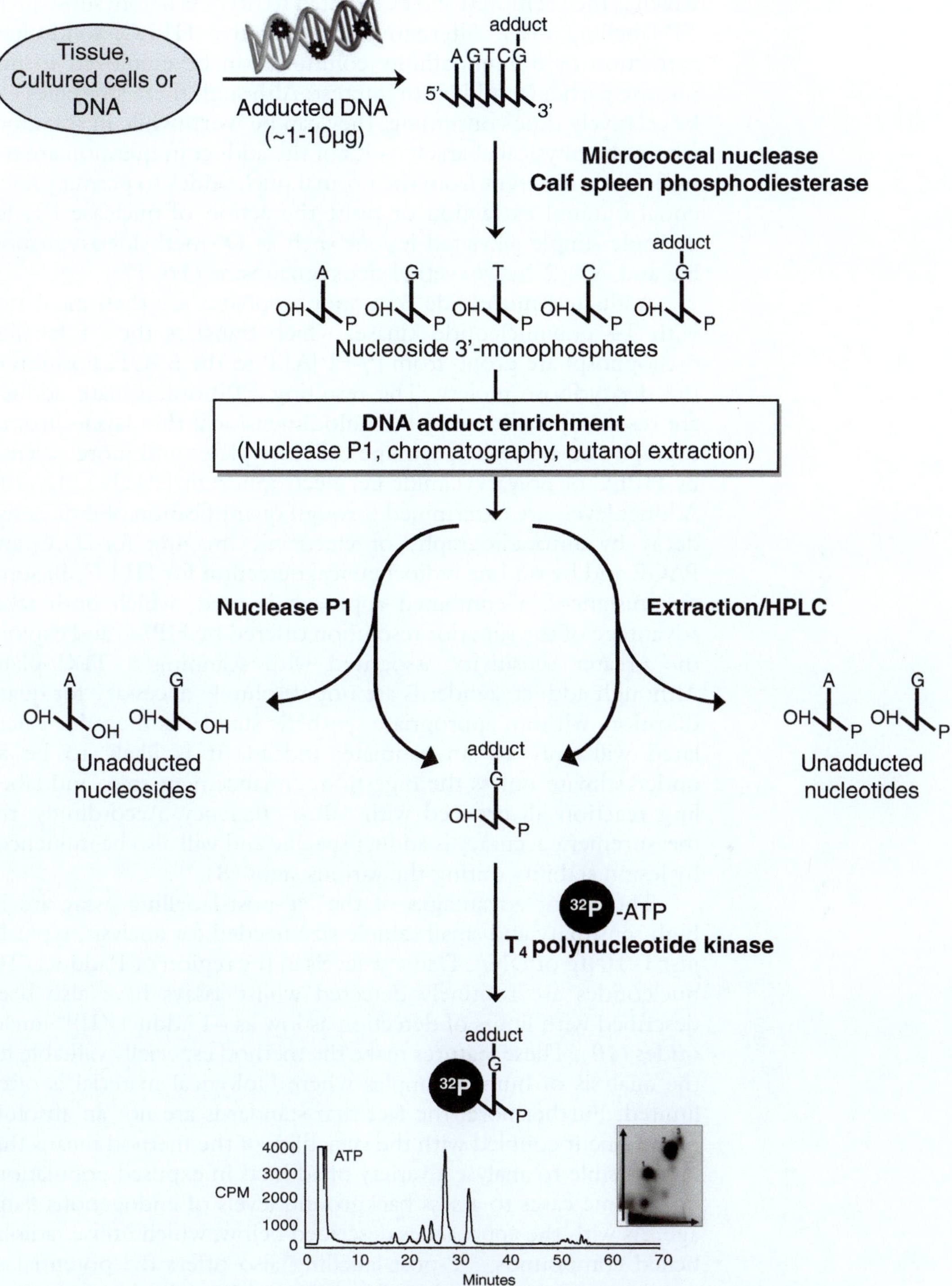

Fig. 1. Flow diagram illustrating the main steps of the standard ^{32}P-postlabelling assay.

which is then removed and evaporated to dryness before subsequent ^{32}P-labelling (15). Alternatively, preparative HPLC, solid-phase extraction or immunoaffinity columns, can be employed to pull out the particular adduct of interest. Although these strategies can be relatively time consuming, they can be worthwhile in situations where the physical characteristics of the adduct in question are not sufficiently different from the normal nucleotides to permit preferential butanol extraction or resist the action of nuclease P1, for example simple alkylated lesions such as *O*^{6}-methyldeoxyguanosine and *N7*-(2-hydroxyethyl)deoxyguanosine (16, 17).

Adducted nucleoside 3′-monophosphates are then incubated with T4 polynucleotide kinase, which transfers the ^{32}P-labelled orthophosphate group from [γ-^{32}P]ATP to the 5′-OH position of the deoxyribose moiety. The resulting ^{32}P-bisphosphate adducts are resolved, traditionally by multidimensional thin layer chromatography on polyethyleneimine cellulose plates and more recently by HPLC or polyacrylamide gel electrophoresis (PAGE) (18–20). Adduct levels are determined through quantification of radioactive decay by autoradiography or electronic imaging for TLC and PAGE and by on-line radiochemical detection for HPLC. In some circumstances, a combined approach is used, which both takes advantage of the superior resolution offered by HPLC and exploits the greater sensitivity associated with scanning a TLC plate. Although adduct standards are not absolutely necessary for quantification, without appropriate synthetic standards the value calculated will only be an estimate; indeed, it is likely to be an underestimate unless the digestion, enhancement step, and labelling reaction all proceed with 100% efficiency. Accordingly, the measurement accuracy is adduct specific and will also be influenced by lesion stability during the various steps (8).

The major advantages of the ^{32}P-post-labelling assay are its high sensitivity and small sample size needed for analysis, typically just 1–10 μg of DNA. Damage levels in the region of 1 adduct/10^8 nucleotides are routinely detected whilst assays have also been described with limits of detection as low as ~1 adduct/10^{10} nucleotides (10). These features make the method especially valuable for the analysis of human samples where biological material is often limited. Furthermore, the fact that standards are not an absolute requirement coupled with the versatility of the method means that it is possible to analyse a variety of adducts in exposed populations or in some cases to assess background levels of endogenous damage. As with the approaches described below, which utilise radiolabelled compounds, ^{32}P-post-labelling also offers the potential to screen a sample for adduct formation without having first elucidated the mechanisms and characterised the adduct structures.

One of the obvious downsides to using ^{32}P-post-labelling is the need to handle a high specific activity radioisotope; the protocols are also rather labour intensive and low-throughput. Even with an adduct standard available structural assignment can never be unequivocal, although co-elution in several different chromatographic systems can provide a reasonably high degree of confidence. Another potential issue that should be taken into account when interpreting results is the possible presence of I-compounds (8). These endogenous DNA lesions can be present at relatively high levels and may interfere with the detection of adducts formed by the test compound due to co-migration or if the signal greatly exceeds that produced by the test adducts.

3. Fluorescence and Electrochemical Based Detection

Certain intrinsic properties of some adducts, specifically fluorescence or redox status can be exploited as a means for their analysis using a fluorescence spectrometer or electrochemical (EC) detector, respectively. Typically, such methods are used in combination with HPLC or capillary electrophoresis (CE) separation. Numerous studies have utilised HPLC coupled with on-line fluorescence for the detection and quantification of a variety of bulky lesions formed by polycyclic aromatic hydrocarbons, aflatoxin and tamoxifen for example; for the latter compound, derivatisation to a phenanthrene species is required prior to detection as the parent compound is not sufficiently fluorescent (21–23). Enhanced specificity can be obtained through the use of synchronous fluorescence spectroscopy, in which the excitation and emission wavelengths are scanned synchronously (24). A more recently developed approach, which reportedly can detect attomole quantities of adducts, is to chemically derivatise the lesion with fluorescent dyes such as BODIPY FL. The labelled adducts are then separated by LC or EC and detected by laser-induced fluorescence. This strategy has been employed successfully in the analysis of a diverse range of modified nucleosides including those formed by heterocyclic amines, 4-aminobiphenyl and aflatoxin (25).

The type of adducts with electrochemical properties amenable to detection are exemplified by damage induced by reactive oxygen and nitrogen species. In particular, in vivo levels of 8-oxo-7,8-dihydro-2′-deoxyguanosine have been measured in various different cell types and in urine by this approach (26, 27), whilst levels of urinary 8-nitroguanine quantified using an HPLC-ECD assay with immunoaffinity purification have been shown to correlate with smoking status (28). Limitations of both fluorescence and EC techniques include the requirement for prior knowledge of the physical characteristics of the adduct under investigation and the fact that they are only applicable to a few classes of lesions with the desired properties. Furthermore, although assays are relatively simple and cheap to perform, to reach the levels of sensitivity attainable with other methods can require large amounts of DNA (29, 30).

4. Immunological Methods

Antisera raised against carcinogen modified DNA can be used to develop immunoassays for quantifying damage in biological samples as well as performing immunohistochemical staining for semi-quantitation and adduct localisation in nuclei of cells or tissues. Additionally, antibodies bound to a solid support can be used to enrich DNA digests or a biological matrix such as urine for a specific type of adduct prior to detection by another method (31, 32). For instance, O^4-ethylthymidine has been detected in cells arising in the lower respiratory tract of smokers using an assay comprising immunoenrichment of O^4-ethylthymidine 3′-monophosphate from digested DNA, followed by ^{32}P-post-labelling with HPLC separation (33).

Two types of antigens have been used for generating antibodies to adducted DNA: either intact DNA modified to a high level (~1 adduct/100 nucleotides) by incubation with a suitably reactive intermediate of the carcinogen concerned then electrostatically complexed to methylated carrier protein or the specific monoadduct of interest, synthesised as a base or nucleoside and coupled to a carrier protein (34). The particular choice of antigen is governed by the eventual application and ease of production; for example, if synthetic routes are not yet established for preparing the individual base/nucleoside adduct in sufficient yields, then reaction of an ultimate carcinogen with DNA may prove the best option, providing the required level of adduction can be achieved. Furthermore, if the intended use is immunoaffinity purification from a DNA digest then the monoadduct approach would be preferable, if however, the goal is immunohistochemical detection in cells or tissues, modified DNA would usually be more appropriate. Both polyclonal and monoclonal antibodies have been developed against chemically modified DNA and in general, similar specificities and sensitivity can be achieved with each type. The former are quicker to generate and more economical to produce, whilst monoclonals require culturing and screening of hybrid clones to identify the highest affinity clone, which can add several months to the development time.

Quantitative immunoassays for DNA adducts are most commonly performed in a competitive binding mode on microtitre plates, although alternatives such as slot blot methods have also been described for a small number of lesions, in which the DNA is immobilised on a nitrocellulose filter (35). There are many variations on the basic competitive enzyme-linked immunosorbent assay (ELISA) methodology, which utilise peroxidase or alkaline phosphatase-conjugated secondary antisera to detect the primary antibody, and the numerous fluorescent, coloured or radioactive substrates available for detection (36). The general procedure is depicted in Fig.2 and involves first coating each well with the antigen, either as the protein coupled monoadduct or adducted DNA, then blocking non-specific binding by incubation with a weak protein solution. Calibration samples for constructing a standard curve are prepared using serial dilutions of the antigen (monoadduct/adducted DNA), mixed with the primary antibody. Each test sample, containing unknown levels of damage is also mixed with the antibody and added to the plate. The assay works on the principle that antigen in solution competes with the bound antigen. Consequently, the higher the adduct load in a sample, the lower the concentration of free antibody remaining to bind to the plate; this results in an inverse relationship between damage level and strength of signal detected in the final step. After incubation, all non-bound material is removed by washing, then the amount of primary antibody is quantified by addition of an enzyme-linked secondary antibody and an appropriate substrate. Alternatively, the more sensitive chemiluminescent assays such as those developed for the analysis of tamoxifen and benzo[*a*]pyrene adducts contain an additional amplification step; after incubation with the secondary

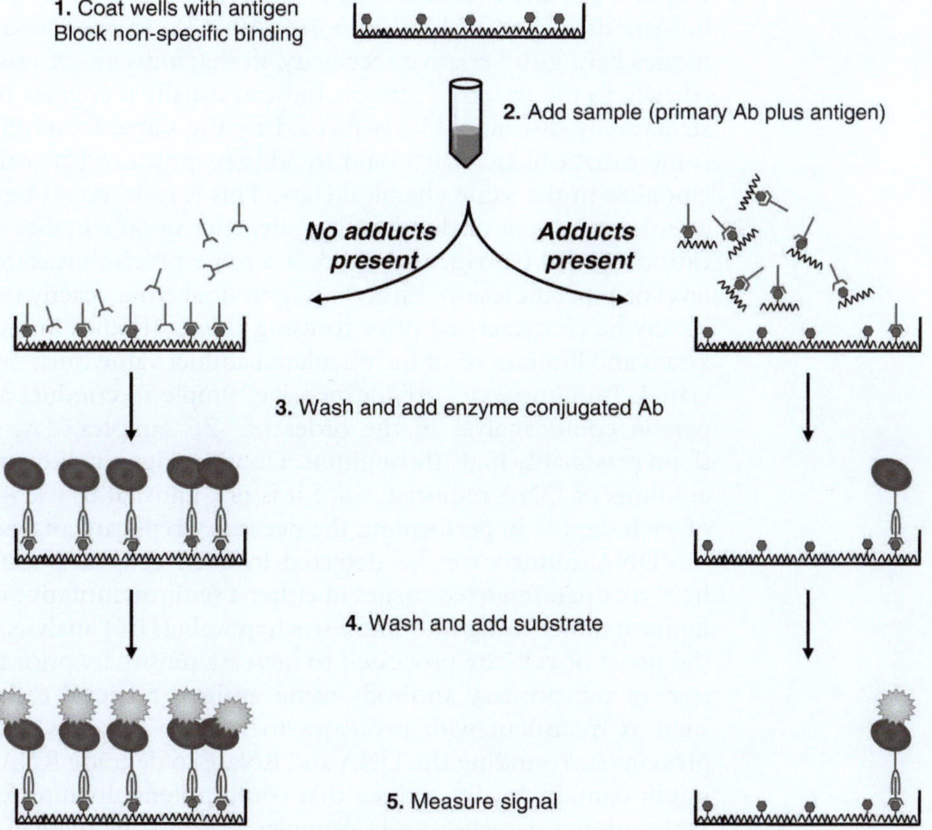

Fig. 2. Outline of the steps involved in a basic competitive immunoassay.

antibody (biotinylated anti-rabbit IgG), streptavidin-alkaline phosphatase is added. Streptavidin binds to the biotin and the enzyme dephosphorylates a chemiluminescent dioxetane substrate, resulting in light emission which is read using a microplate luminometer (37, 38). For accurate quantitation, it is important to first validate the assay prior to analysing biological specimens, ideally using an independent method to verify the number of adducts in a series of test samples. In addition, the standard curve produced for an experiment should cover the range of damage anticipated in the biological samples, since it is possible that for some antibodies the affinity for adduct binding may be dependent on the extent of modification (39).

A particular strength of immunoassays is their sensitivity, which is largely influenced by the last detection step. Typically, 1 adduct/10^8 nucleotides can be detected but with improvements such as incorporation of the streptavidin-biotin amplification and chemiluminescent substrate described above, as little as ~2–3 adducts/10^9 nucleotides have been detected for tamoxifen and benzo[*a*]pyrene. For the latter carcinogen, this degree of sensitivity has allowed quantitation of adducts in peripheral blood lymphocytes of humans. Indeed, immunoassays have been successfully employed in the measurement of a variety of chemical adducts in human tissues and blood samples (40–43). Immunoassay techniques have good relative specificity, in that antisera are raised specifically to the lesion of interest but can usually recognise multiple structurally distinct adducts formed by the same chemical and in some situations may also bind to adducts produced by other carcinogens in the same chemical class. This feature could be viewed as either a benefit or disadvantage, depending on whether an indication of total damage is desired or a more precise measure of the level of a specific lesion. Either way, potential cross-reactivity should ideally be characterised prior to using the antibodies in biological assays and limitations of the calculated adduct value must be appreciated. Immunoassays are inexpensive, simple to conduct and one person could analyse in the order of ~25 samples/day, making them reasonably high-throughput. Disadvantages include the large amounts of DNA required, since it is not unusual to use ~200 μg of each sample in performing the necessary replicate analyses (8).

DNA adducts can be detected in fixed cells and sections of frozen or paraffin fixed tissues in either a semi-quantitative or qualitative manner using immunohistochemical (IHC) analysis. Often, the tissue or cells are processed to increase sensitivity prior to addition of the primary antibody using antigen retrieval techniques, such as treatment with proteases to remove histones and other proteins surrounding the DNA and RNase to degrade RNA, which might contain modified bases that could potentially interfere with DNA adduct detection (34). Samples may also be incubated with acid or base solutions to denature the DNA, increasing adduct exposure to the antisera. The most commonly used detection systems for the adduct-bound primary antibody are immunofluorescence and immunoperoxidase staining. Qualitative analysis then involves counting the number of positively stained cells and/or subjective assessment of the staining intensity, whilst quantitative data are obtained by measuring fluorescence or absorption in the cell nuclei using automated cellular imaging systems equipped with appropriate software. This approach has been used extensively to detect a variety of adducts in animal and human samples; for example, polycyclic aromatic hydrocarbon-DNA adducts have been reported in oral mucosa cells, prostate, oesophageal, and breast tissue (44–47), and correlations have been demonstrated between levels of carboplatin DNA adducts in buccal cells and disease response in patients undergoing chemotherapy (48).

The major advantage of IHC is the ability to detect damage in specific cell types within human and animal tissues, including archived samples, using small amounts of material; generally much less tissue is needed for IHC compared to an ELISA assay for the same adduct. Drawbacks of the technique are essentially the same as those described above for ELISA, but additionally IHC typically offers lower sensitivity (29). It is important to confirm that the adduct of interest is stable under the processing conditions employed and that appropriate controls are performed alongside samples.

5. Mass Spectrometry

Mass spectrometry (MS) based assays are increasingly being used for the detection and quantification of a diverse range of DNA adducts and MS is also an invaluable tool for structurally identifying new adducts formed by reactive carcinogens. Of all the methods currently in use, MS offers the ultimate in chemical specificity, since adducts are detected directly on the basis of their molecular mass and in the majority of cases, their characteristic chromatographic properties. In the past this specificity has been offset by lower sensitivity, compared to ^{32}P-post-labelling for example, but with ongoing advances in interface technology, ionisation sources and detectors, the sensitivity of MS is continually improving and limits of detection in the order of ~1 adduct/10^8 nucleotides are now regularly attained (49). Adducts are routinely measured in the base or 2′-deoxynucleoside form, with 2′-deoxynucleotides studied to a lesser extent; adducted oligonucleotides can also be detected, but currently this type of analysis is normally performed to gain information on preferential binding sites for a carcinogen within a given DNA sequence, rather than for quantitation purposes (50, 51). MS is usually coupled to a liquid chromatography

(HPLC and UPLC) or gas chromatography (GC) system to help separate out the adduct of interest from residual unmodified bases/nucleosides and other types of interfering lesions. Before the development of LC-MS, GC-MS with electron impact or chemical ionisation was the main MS based method for DNA adduct detection; it is still used today but to a lesser extent than LC-coupled assays (52). For a molecule to be detectable by GC-MS it must be non-polar and volatile. Since most adducts do not satisfy these criteria they must undergo derivatisation at high temperatures prior to analysis, which constitutes the main limitation of the technique. Introduction of the electrospray ionisation (ESI) source enabled polar compounds to be analysed directly, allowing LC systems to be linked to the mass spectrometer and this is now the most widely adopted MS approach for assaying DNA adducts (for details of the ESI process see ref. 52).

Preparation of samples for LC-ESI-MS analysis involves isolation of DNA followed by digestion, to generate a mixture of normal and unmodified 2′-deoxynucleosides or nucleotides. Adducts are then normally enriched by solid-phase extraction, preparative HPLC, or use of an immunoaffinity column. Alternatively, sample clean up may be performed on-line using column-switching, which involves selectively directing the LC flow containing interfering unmodified 2′-deoxynucleosides/nucleotides to waste rather than into the MS instrument (53). In situations where the adduct is to be measured in the base form, they can be liberated by DNA hydrolysis and collected by filtration. This strategy is particularly useful for adducts that tend to readily depurinate such as alkylation products of guanine and adenine at the *N7* and *N3* position respectively (54). Not only does the method of isolation minimise loss of labile adducts during processing but it also acts as an enrichment step because normal bases are generally stable under the hydrolysis conditions employed. The major adduct formed by reaction of ethylene oxide with DNA, *N7*-(2-hydroxethyl)guanine (*N7*-HEG) is detected using such a protocol. Humans are exposed to this chemical through a variety of occupational and environmental sources and it is also generated endogenously by metabolism of ethylene, which arises in vivo through normal physiological processes. Using an LC-MS/MS assay with selected reaction monitoring and a limit of detection of ~6 adducts/10^9 nucleotides, background levels of the 2-hydroxyethylated base have been detected in DNA isolated from tissues of control rats in the region of ~1–4 adduct/10^8 nucleotides, as illustrated for the liver, in Fig. 3 (55).

The type of instrument most commonly employed in DNA adduct analysis is a triple quadrupole, which is a tandem mass spectrometer consisting of two quadrupole mass analysers in series separated by a second quadrupole that acts as a collision cell, filled with an inert gas such as argon. The first and third quadrupoles

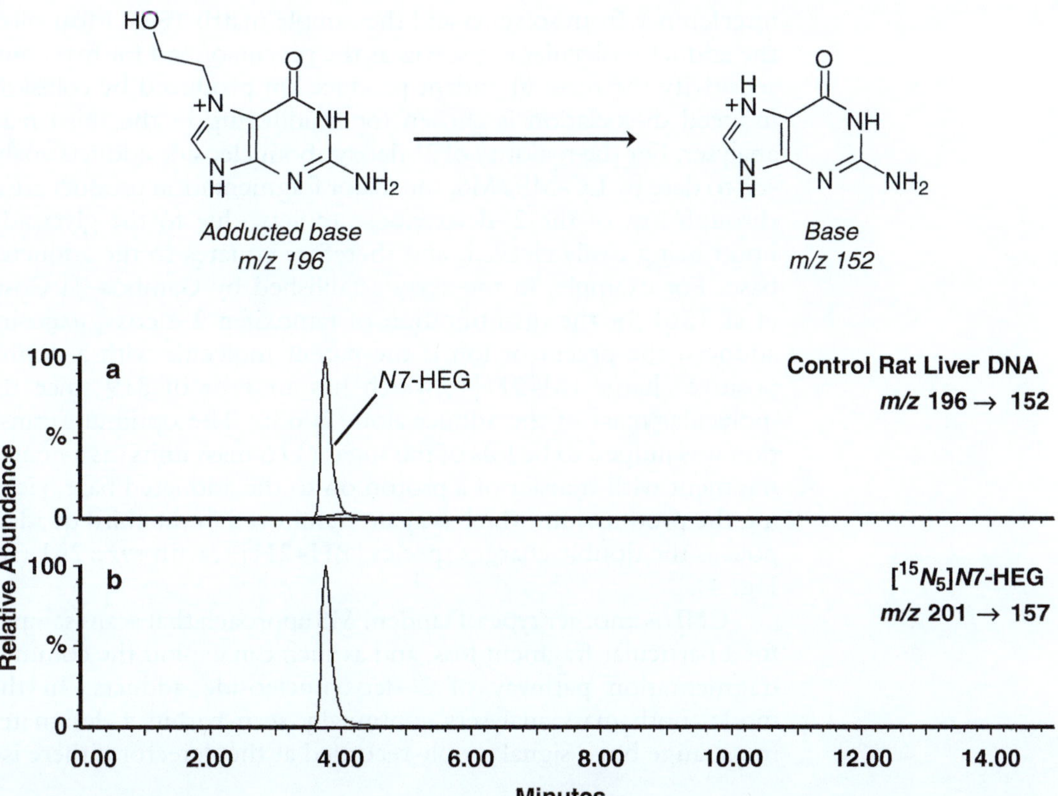

Fig. 3. Typical LC-MS/MS selected reaction monitoring ion chromatogram (positive ionisation mode) showing the analysis of (**a**) endogenous *N*7-(2-hydroxethyl)guanine adducts in the liver of a control (untreated) rat and (**b**) the [$^{15}N_5$]-labelled internal standard, which has a mass five units higher than the adduct of interest. In each case the transition monitored corresponds to loss of the 2-hydroxethyl group from the adducted base. Data are adapted from ref. 55.

contain four circular rods aligned in parallel, to which voltages are applied to generate oscillating electric fields; these essentially filter out the ions of interest on the basis of their mass-to-charge ratio (m/z). In the collision cell, analyte ions selected by the first mass analyser can undergo collision-induced dissociation producing characteristic fragments, which are subsequently analysed in the third quadrupole. There are several different analysis modes employed in the detection of adducts by LC-ESI-MS: single ion monitoring (SIM), selected reaction monitoring (SRM) and constant neutral loss (CNL). In SIM, the first quadrupole is set to allow passage only of the ions of interest and these continue all the way through to the detector. In SRM, which is sometimes referred to as multiple reaction monitoring (MRM), the first analyser selects a precursor ion, this is then fragmented in the collision cell to yield a product ion characteristic of the adduct molecule, which is then filtered out by the third quadrupole and transmitted through to the detector. SRM is more specific and offers significantly greater sensitivity (≥100-fold higher) compared to SIM due to reduced interference from solvents and the sample matrix (52). Most often the adduct molecular ion serves as the precursor and for maximum sensitivity the most abundant product ion produced by collision-induced dissociation is chosen for monitoring in the third mass analyser. For the majority of 2′-deoxyribonucleoside adducts analysed to date by LC-MS/MS, the major fragmentation product arises through loss of the 2′-deoxyribose moiety, due to the glycosidic bond being easily cleaved, and therefore equates to the adducted base. For example, in the assay established by Gamboa da Costa et al. (56) for the quantification of tamoxifen 2′-deoxyguanosine adducts, the precursor ion is the parent molecule with a double positive charge $[M+2H]^{2+}$, which has an *m/z* of 319 since the molecular mass of the adduct alone is 635. The optimum transition was judged to be loss of the sugar (116 mass units) as a neutral fragment with transfer of a proton on to the adducted base, yielding the product ion, which is again monitored in the third quadrupole as the double charges species $[BH+2H]^{2+}$, with *m/z* 261 (see Fig. 4).

CNL is another type of tandem MS approach that scans samples for a particular fragment loss, and as such can exploit the common fragmentation pathway of 2′-deoxynucleoside adducts. In this mode, both mass analysers continually scan within a designated mass range but a signal is only recorded at the detector if there is a

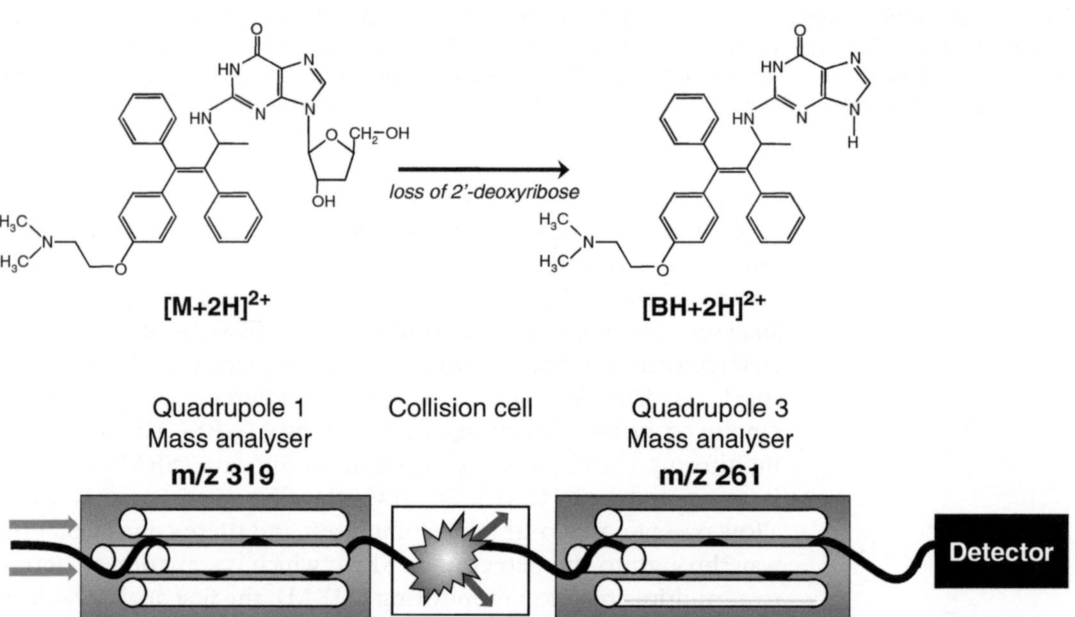

Fig. 4. Analysis of tamoxifen 2′-deoxyguanosine adducts by LC-ESI-MS/MS with SRM according to the method developed by Gamboa da Costa et al. [56]. The transition monitored corresponds to loss of the 2′-deoxyribose moiety from the double protonated molecular ion $[M+2H]^{2+}$. The precursor ion is selected by the first quadrupole, which is set to transmit ions with a *m/z* of 319, the third quadrupole allows the passage of ions with *m/z* 261, equivalent to the product ion $[BH+2H]^{2+}$.

specified difference in mass between the precursor and resulting product ions. For example, Bessette et al. (57) have used this approach (although with a different type of instrument to the standard triple quadrupole MS described above) to simultaneously screen DNA samples for multiple adducts from different classes of carcinogens, based on loss of the 2′-deoxyribose group. In a proof of principle study they demonstrated detection of five structurally distinct 4-aminobiphenyl adducts in human hepatocytes incubated with the chemical, whilst three different adducts formed by the food mutagen 2-amino-3,8-dimethylimidazo(4,5-*f*]quinoxaline (MeIQx) were identified in liver tissue of treated rats. Although current CNL assays are much less sensitive than SRM, ultimately this type of analysis may offer a new route to map entire profiles of damage present in human tissues and surrogate cells, or what has recently been referred to as the "adductome", with a view to correlating damage with endpoints such as disease risk or efficacy of a therapeutic or preventive intervention (58).

The most accurate means of quantifying adduct levels is through the use of a stable isotope-labelled internal standard. These adduct standards, which are most often labelled in the purine or pyrimidine moiety with ^{13}C or ^{15}N, are identical to the adduct of interest in terms of chromatographic and ionisation properties, but have a slightly higher mass, enabling differentiation by MS. Typically, a known amount of a labelled adduct standard is added to the sample prior to processing so that any losses can be accounted for. If stable isotope standards are not available, the next best option is to use an external calibration line constructed using dilutions of an unlabelled adduct standard spiked in to the relevant matrix and processed in exactly the same manner as the test samples.

Traditionally, one of the main limitations of using MS based techniques for adduct analysis is the inability to screen unknown mixtures, for example, to test the ability of a compound to bind to DNA, or to examine what types of lesions are present in the DNA of a human exposed to multiple unidentified carcinogens. With recent advances in the scanning (CNL) approaches described above this may be less of an issue in the future (59), although adduct standards will still be needed for quantification, which obviously requires that the structure is known and that the standards can be readily synthesised or purchased. The amount of DNA needed can also be a problem, especially if tissue or cells are in short supply, since relatively large quantities, typically 10–100 μg can be required for a single analysis. However, the unique feature of MS, compared to the other adduct analysis methods available, is that the sensitivity is essentially still improving with each new generation of instruments developed. Consequently, it is likely that it will be possible to achieve accurate adduct quantification with much smaller sample sizes in the future, which will further increase the applicability of the technique.

6. Radiolabelled Compounds and Accelerator Mass Spectrometry

The use of a radiolabelled form of a test compound is perhaps the simplest way to establish whether a chemical binds to DNA and can be employed in situations where nothing is known about the potential mechanisms of activation or nature of the adducts formed, for example, as a follow-up to unusual responses in statutory genetic toxicology assays or rodent bioassays. Covalent DNA binding can be assessed using both in vitro systems (naked DNA or cultured cells) and in vivo animal models after exposure to single doses of a ^{14}C or ^{3}H-labelled compound. Purified DNA is then isolated and the increase in radioactivity due to the presence of the bound $^{14}C/^{3}H$-compound, relative to appropriate control DNA samples, is quantified by liquid scintillation counting (LSC) (2, 8, 60). The disadvantages of this straightforward approach include poor sensitivity compared to the other methods available, which are now more commonly used, and the need for high amounts of relatively hazardous radioactivity for binding to be detectable. Radiochemicals can also be expensive to purchase and waste disposal is costly, which can make it difficult to perform multiple dose in vivo experiments. In addition, adduct data should be interpreted with appropriate caution due to the reasons described below, which are applicable to any study performed with a radiolabelled compound.

An alternative, far more sensitive technique for detecting DNA adducts formed by a radiolabelled compound is accelerator mass spectrometry (AMS), the most sensitive method available for detecting and quantifying rare, long-lived isotopes with high precision. Traditionally AMS has been employed in the geological and environmental sciences and is perhaps most noted for its application to radiocarbon dating (61). Since the 1980s, however, AMS has been increasingly exploited in biological and medical research to investigate the absorption, distribution, metabolism, and excretion of radiolabelled chemicals, nutrients, and drugs, and in the detection of covalently adducted DNA and proteins in animal models and humans (62). In terms of DNA adducts, the methodology is capable of quantifying levels of 1–10 lesions/10^{12} nucleotides, following administration of a [^{14}C]-labelled carcinogen, which corresponds to less than one modification per cell (63). The exquisite sensitivity of AMS translates to the use of low chemical and radioisotope doses and relatively small sample sizes (1–2,000 μg DNA), enabling studies to be performed safely in humans, using relevant doses, whilst generating little radioactive waste. This is the major advantage of the technique, and in this respect, AMS has proven especially valuable in demonstrating that carcinogenic compounds bind to DNA in potential target tissues of humans following exposure to a low dietary, environmental, or therapeutic dose.

One of the first biomedical applications of AMS was to determine if the heterocyclic amine MeIQx, which is formed in cooked meat, binds to DNA in rodent tissues at low doses and whether the relationship between adduct formation and dose is linear (64). Subsequent investigations went on to show that MeIQx has a greater propensity for binding to DNA in human colon tissue compared to the rat, after administration of a single [^{14}C]-labelled dose equivalent to a typical dietary intake (65). The radioisotope dose in this instance was less than the estimated daily exposure to background ionising radiation (3 μSv). Similar in vitro and in vivo studies have been performed for a wide variety of chemicals including benzene, trichloroethylene, ethylene oxide, the antioestrogen tamoxifen, which is extensively used in the treatment of breast cancer, and the chemotherapeutic drug adriamycin (62, 66–68). Just as importantly, AMS has also provided convincing data illustrating the absence of detectable DNA binding for the [^{14}C]-labelled carcinogens 2,3,7,8-tetrachlorodibenzo-*p*-dioxin and ochratoxin, adding to the weight of evidence that DNA adduction is unlikely to be involved in the mechanism of tumour formation by these chemicals (64, 69).

The general strategy underlying AMS based experiments is outlined in Fig. 5. Essentially the fate of any ^{14}C or ^{3}H-labelled

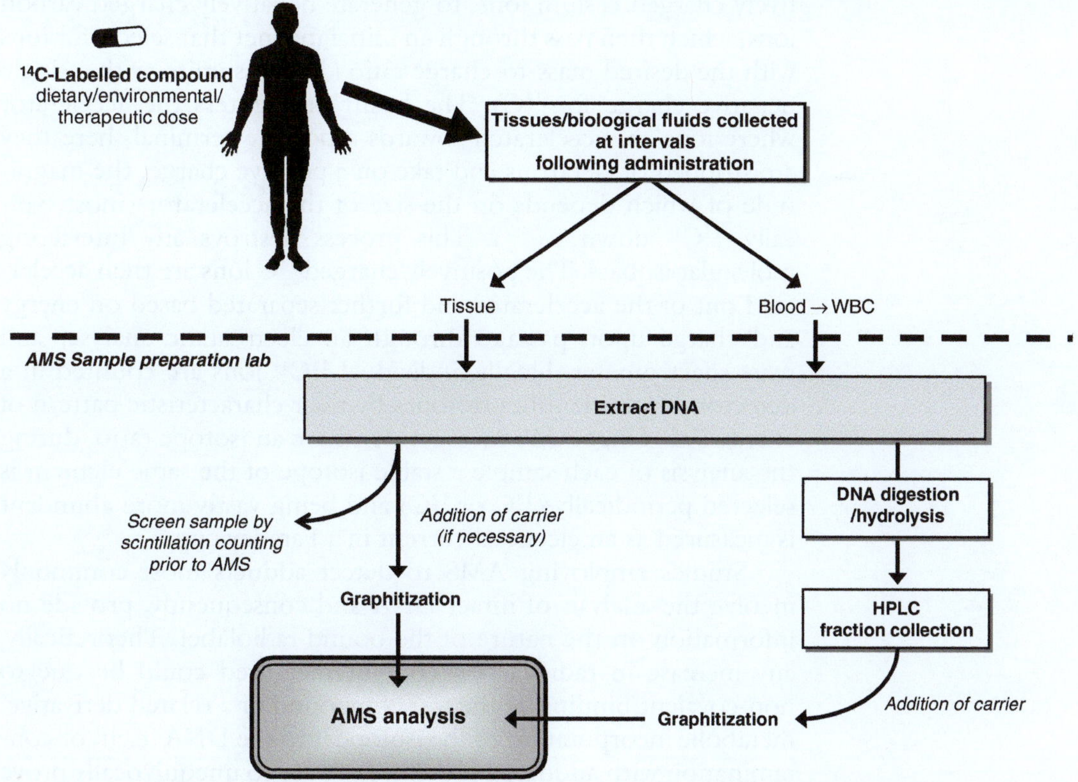

Fig. 5. General protocol for AMS studies to investigate adduct formation in vivo by [^{14}C]-labelled compounds.

compound can be traced within an in vitro or in vivo system, as long as the isotope is incorporated in a metabolically stable position. For adduct analysis DNA is isolated from the tissue or cell type of interest using standard methods, such as solid phase extraction. All sample processing after the initial incubations or dosing must be conducted in a $^{14}C/^{3}H$-free laboratory specifically designated for AMS work and numerous precautions (described in refs. 70, 71) must be taken to prevent contamination of the samples. As with all organic material analysed by ^{14}C-AMS, the DNA is then converted to a form of carbon compatible with the ion source of the instrument, typically this is graphite which is produced in a two step process involving oxidation to CO_2 followed by reduction to filamentous carbon (71). Each individual graphite sample is then assayed by AMS to determine the radiocarbon content. This is calculated from the $^{14}C/^{13}C$ isotope ratio, by first subtracting the ratio measured for a control, such as a pre-dose or untreated sample, thereby taking into account the amount of ^{14}C present naturally plus any extraneous contamination introduced.

The essential features of an accelerator mass spectrometer are detailed in refs. 62, 71. In brief, the instrument itself is comprised of two mass spectrometers separated by an electrostatic accelerator, through which negative ions are accelerated to high energies. In the ion source graphite samples are bombarded by a beam of positively charged cesium ions, to generate negatively charged carbon ions, which then pass through an initial magnet that selects for ions with the desired mass-to-charge ratio (14 mass units, with a single negative charge for ^{14}C). The beam then enters the accelerator where ions are accelerated towards a positive terminal, here they are stripped of electrons and take on a positive charge, the magnitude of which depends on the size of the accelerator (most typically, $^{14}C^{4+}$ down to $^{1+}$). This process destroys any interfering molecular isobars. The positively charged ^{14}C ions are then accelerated out of the accelerator and further separated based on energy and charge upon passage through an electrostatic analyser and mass spectrometer. Finally, individual $^{14}C^{4+}$ ions are counted in a detector which identifies isotopes by their characteristic pattern of energy loss. Since AMS measurements are an isotope ratio, during the analysis of each sample a stable isotope of the same element is selected periodically (^{13}C or ^{12}C) and being vastly more abundant is measured as an electrical current in a Faraday cup.

Studies employing AMS to detect adducts most commonly involve the analysis of intact DNA and consequently, provide no information on the nature of the bound radiolabel. Theoretically, any increase in radiocarbon content measured could be due to non-covalent binding of the test compound or a related derivative, metabolic incorporation of the isotope into the DNA itself, or contamination with adducted protein. In order to unequivocally prove the formation of DNA adducts, it is therefore vital that the following confirmatory study (or analogous protocol) is performed (8). DNA should be hydrolyzed to nucleosides or nucleotides, alternatively the adducted bases could be liberated. The products are then separated by HPLC and fractions corresponding to the entire run collected and subject to AMS analysis, as illustrated for [^{14}C]-tamoxifen-adducted DNA in Fig. 6 (72). The presence of elevated ^{14}C concentrations in fractions with retention times equivalent to a synthetic DNA adduct standard and not the free parent compound, potential metabolites or unmodified nucleosides/nucleotides, confirms covalent DNA adduct formation, and may also enable putative structural identification of the actual adducts formed (65). HPLC-AMS is less sensitive than standard analysis of

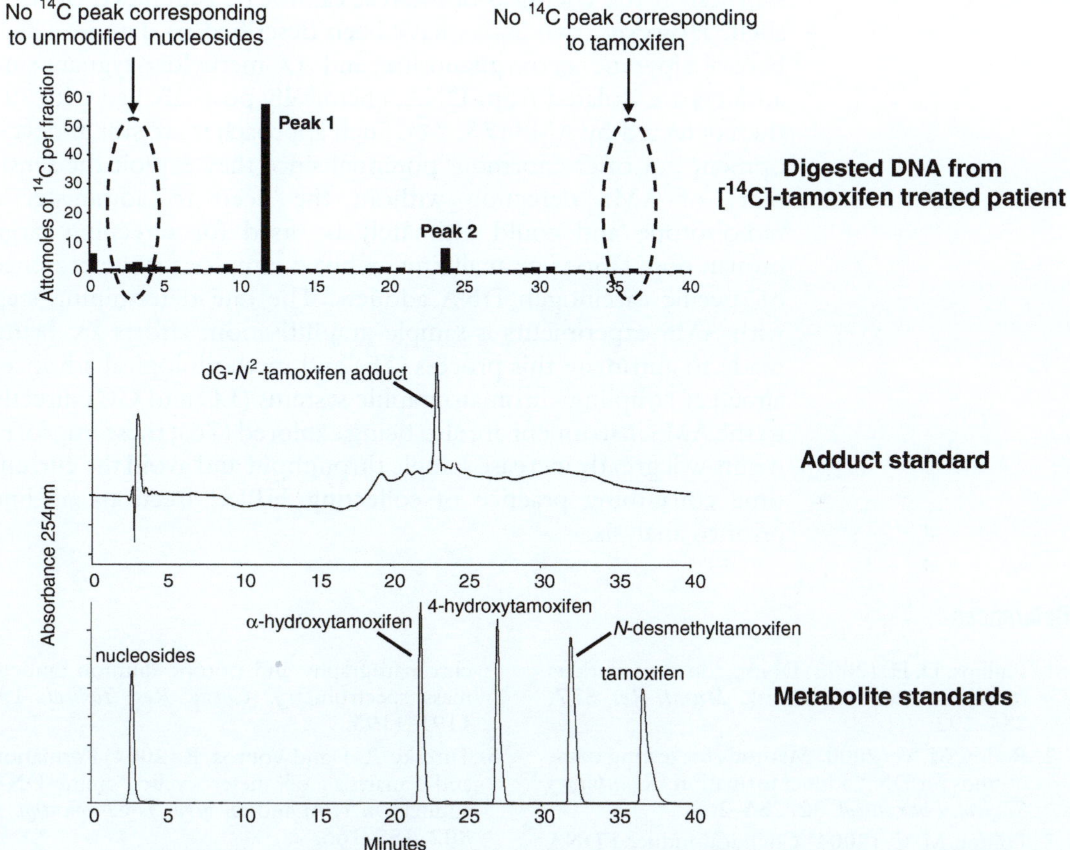

Fig. 6. Verification of adduct formation in radiolabel studies. HPLC-AMS analysis of digested colon DNA from a patient who received a single therapeutic (20 mg) dose of [^{14}C]-tamoxifen (adapted from ref. 72). Fractions were collected at 30 s intervals and the entire run analysed by AMS. Also shown for comparison are HPLC-UV chromatograms of a tamoxifen 2′-deoxyguanosine adduct standard and tamoxifen together with its major metabolites and a mixture of 2′-deoxynucleoside standards, which were run at the same time. The important points to note are that neither of the two peaks of radiocarbon detected corresponds in retention time to tamoxifen, any of its metabolites or the unmodified 2′-deoxynucleosides. Furthermore, peak 2 elutes at the same time as the adduct standard, providing evidence that tamoxifen is capable of forming adducts in human tissues. The identity of peak 1 is currently unknown but could be another 2′-deoxynucleoside adduct with a different structure or a di/trinucleotide adduct resistant to hydrolysis.

total DNA, therefore such verifications can only be performed with samples containing sufficiently high levels of adducts.

Potential disadvantages of using AMS as a tool for DNA adduct detection include the high instrumentation costs of commercially available systems and the need for qualified physicists to maintain the instrument. Accessibility is therefore currently limited and there are only a very small number of laboratories worldwide that routinely offer biomedical AMS analysis (62). The situation is however improving as the instruments become cheaper and more user-friendly to operate, this should in turn reduce the expenses associated with sample analysis. At the moment, administration of a [^{14}C] or [^{3}H]-labelled compound is a prerequisite for DNA adduct detection by AMS, which may necessitate a costly custom synthesis if the chemical of interest cannot be purchased off-the-shelf. However, two assays have been described to date in which benzo[*a*]pyrene deoxyguanosine and *O*^{6}-methyldeoxyguanosine adducts are isolated from DNA, chemically post-labelled with ^{14}C then detected by AMS (73, 74). Such approaches are still in development but offer enormous potential since they exploit the sensitivity of AMS detection without the need to administer a radioisotope and could ultimately be used for screening large human populations or analysing archived samples for the presence of specific carcinogen DNA adducts. The rate determining step with AMS experiments is sample graphitisation; efforts are being made to automate this process (75) and methodological advances aimed at coupling chromatographic systems (LC and GC) directly to the AMS instrument are also being explored (76); these improvements will greatly increase sample throughput and avoid the current time consuming practice of collecting HPLC fractions off-line prior to analysis.

References

1. Phillips, D. H. (2005) DNA adducts as markers of exposure and cancer risk. *Mutat. Res.* **577**, 284–292.
2. Reddy, M. V. (2000) Methods for testing compounds for DNA adduct formation. *Regulatory Toxicol. Pharmacol.* **32**, 256–263.
3. Poirier, M. C. (2004) Chemical-induced DNA damage and human cancer risk. *Nature Reviews Cancer* **4**, 630–636.
4. Sharma, R. A. and Farmer, P. B. (2004) Biological relevance of adduct detection to the chemoprevention of cancer. *Clin. Cancer Res.* **10**, 4901–4912.
5. Egner, P. A., Groopman, J. D., Wang, J. S., Kensler, T. W. and Friesen, M. D. (2006) Quantification of aflatoxin-B-1-N-7-guanine in human urine by high-performance liquid chromatography and isotope dilution tandem mass spectrometry. *Chem. Res. Toxicol.* **19**, 1191–1195.
6. Turesky, R. J. and Vouros, P. (2004) Formation and analysis of heterocyclic amine-DNA adducts *in vitro* and *in vivo*. *J. Chromatog. B* **802**, 155–166.
7. Tang DL, Phillips DH, Stampfer M, Mooney LA, Hsu YZ, Cho S, Tsai WY, Ma J, Cole KJ, Ni She M, Perera FP (2001) Association between carcinogen-DNA adducts in white blood cells and lung cancer risk in the physicians health study. *Cancer Res.* **62**, 6708–6712.
8. Phillips, D. H., Farmer, P. B., Beland, F. A., Nath, R. G., Poirier, M. C. Reddy, M. V. and Turteltaub, K. W. (2000) Methods of DNA

adduct determination and their application to testing compounds for genotoxicity. *Environ. Mol. Mutagenesis* **35**, 222–233.

9. Randerath, K., Reddy, M. V. and Gupta, R. C. (1981) ^{32}P-Labeling test for DNA damage. *Proc. Natl. Acad. Sci. USA.* **78**, 6126–6129.
10. Beach, A. C., Gupta, R. C. (1992) Human biomonitoring and the ^{32}P-postlabeling assay. *Carcinogenesis* **13**, 1053–1074.
11. Randerath, K., Randerath, E., Danna, T. F., van Golen, K. L., Putman, K. L. (1989) A new sensitive ^{32}P-postlabeling assay based on the specific enzymatic conversion of bulky DNA lesions to radiolabeled dinucleotides and nucleoside 5′-monophosphates. *Carcinogenesis* **10**, 1231–1239.
12. Guichard, Y., Jones, G. D. D., Farmer, P. B. (2000) Detection of DNA alkylphosphotriesters by ^{32}P-postlabelling: evidence for the non-random manifestation of phosphotriester lesions *in vivo*. *Cancer Res.* **60**, 1276–1282.
13. Phillips, D. H. and Arlt, V. M. (2007) The ^{32}P-postlabeling assay for DNA adducts. *Nature Protocols* **2**, 2772–2781.
14. Reddy, M. V. and Randerath, K. (1986) Nuclease P1-mediated enhancement of sensitivity of ^{32}P-postlabeling test for structurally diverse DNA adducts. *Carcinogenesis* **7**, 1543–1551.
15. Gupta, R. C. (1985) Enhanced sensitivity of ^{32}P-postlabeling analysis of aromatic carcinogen:DNA adducts. *Cancer Res.* **25**, 5656–5662.
16. Cooper, D. P., Griffin, K. A. and Povey, A. C. (1992) Immunoaffinity purification combined with ^{32}P-postlabelling for the detection of *O*^{6}-methylguanine in DNA from human tissues. *Carcinogenesis* **13**, 469–475.
17. Kumar, R., Staffas, J., Försti, A., Hemminki, K. (1995) ^{32}P-Postlabelling method for the detection of 7-alkyguanine adducts formed by reaction of different 1,2-alkyl epoxides with DNA. *Carcinogenesis* **16**, 483–489.
18. Arlt, V. M., Gingerich, J. Schmeiser, H. H., Phillips, D. H., Douglas, G. R. and White, P. A. (2008) Genotoxicity of 3-nitrobenzanthrone and 3-aminobenzanthrone in Muta (TM) Mouse and lung epithelial cells derived from Muta (TM) Mouse. *Mutagenesis* **23**, 483–490.
19. Martin, E. A.; Heydon, R. T.; Brown, K.; Brown, J. E.; Lim, C. K.; White, I. N. H. and Smith, L. L. (1998) Development of a novel automated on-line solid-phase extraction-High-Performance-Liquid-Chromatographic method for the analysis of ^{32}P-postlabelled tamoxifen-DNA adducts: a comparison with thin layer chromatography. *Carcinogenesis* **19**, 1061–1069.
20. Terashima, I., Suzuki, N. and Shibutani, S. (2002) ^{32}P-Postlabeling/polyacrylamide gel electrophoresis analysis: Application to the detection of DNA adducts. *Chem. Res. Toxicol.* **15**, 305–311.
21. Pavanello, S., Favretto, D., Brugnone, F., Mastrangelo, G., Dal Pra, G. and Clonfero, E. (1999) HPLC/fluorescence determination of anti-BPDE-DNA adducts in mononuclear white blood cells from PAH-exposed humans. *Carcinogenesis* **20**, 431–435.
22. Weaver, V. M. and Groopman, J. D. (1994) Fluorescence quantification of aflatoxin *N7*-guanine adducts. *Cancer Epidemiol. Biomarkers Prev.* **3**, 669–674.
23. Sharma, M. (2000) Analysis of tamoxifen-DNA adducts by high-performance liquid chromatography using postcolumn online photochemical activation. *Biochem. Biophys. Res. Commun.* **273**, 40–44.
24. Harris, C. C., LaVeck, G., Groopman, J., Wilson, V. L. and Mann, D. (1986) Measurement of aflatoxin B_1, its metabolites, and DNA adducts by synchronous fluorescence spectrophotometry. *Cancer Res.* **46**, 3249–3253.
25. Jang, H. G., Park, M., Wishnok, J. S., Tannenbaum, S. R. and Wogan G. N. (2006) Hydroxyl-specific fluorescence labeling of ABP-deoxyguanosine, PhIP-deoxyguanosine, and AFB1-formamidopyrimidine with BODIPY-FL. *Anal. Biochem.* **359**, 151–160.
26. Borthakur, G., Butryee, C., Stacewicz-Sapuntzakis, M. and Bowen, P. E. (2008) Exfoliated buccal mucosa cells as a source of DNA to study oxidative stress. *Cancer Epidemiol. Biomarkers Prev.* **17**, 212–219.
27. Loft, S., Svoboda, P., Kasai, H., Tjonneland, A., Vogel, U., Moller, P., Overvad, K. and Raaschou-Nielsen,O. (2006) Prospective study of 8-oxo-7,8-dihydro-2′-deoxyguanosine excretion and the risk of lung cancer. *Carcinogenesis* **27**, 1245–1250.
28. Sawa, T., Tatemichi, M., Akaike, T., Barbin, A. and Ohshima, H. (2006) Analysis of urinary 8-nitroguanine, a marker of nitrative nucleic acid damage, by high-performance liquid chromatography-electrochemical detection coupled with immunoaffinity purification: Association with cigarette smoking. *Free Radical Biology and Medicine* **40**, 711–720.
29. Poirier, M. C. and Weston, A. (1996) Human DNA adduct measurements: state of the art. *Environ. Health Perspectives* **104**, 883–893.
30. de Koc, T. M. C. M., Moonen, H. J. J., van Delft. J. and van Schooten F. J. (2002)

Methodologies for bulky DNA adduct analysis in biomonitoring of environmental and occupational exposures. *J. Chromatog. B.* **778**, 345–355.

31. Manchester, D. K., Weston, A., Choi, J. S., Trivers, G., Fennessey, P. V., Quintana, E., Farmer, P. B., Mann, D. L. and Harris, C. C. (1988) Detection of benzo(a)pyrene diol epoxide-DNA adducts in human placenta. *Proc. Natl. Acad. Sci. USA*, **85**, 9243–9247.
32. Shuker, D. E. G. and Farmer, P. B. (1992) Relevance of urinary DNA adducts as markers of carcinogen exposure. *Crit. Rev Toxicol.* **5**, 450–460.
33. Godschalk, R., Nair, J., Kliem, H. C., Wiessler, M., Bouvier, G. and Bertsch, H. (2002) Modified immunoenriched ^{32}P-HPLC assay for the detection of O^4-ethylthymidine in human biomonitoring studies. *Chem. Res. Toxicol.* **15**, 433–437.
34. Santella, R. M. (1999) Immunological methods for detection of carcinogen-DNA damage in humans. *Cancer Epidemiol. Biomarkers Prev.* **8**, 733–739.
35. Leuratti, C., Singh, R., Lagneau, C., Farmer, P. B., Plastaras, J. P., Marnett, L. J. and Shuker, D. E. G. (1998) Determination of malondialdehyde-induced DNA damage in human tissues using an immunoslot blot assay. *Carcinogenesis* **19**, 1919–1924.
36. Santella, R. M., Weston, A., Perera, F. P., Trivers, G. T., Harris, C. C., Young, T. L., Nguyen, D., Lee, B. M. and Poirier, M. C. (1988) Interlaboratory comparison of antisera and immunoassays for benzo[*a*]pyrene-diol-epoxide-I-modified DNA. *Carcinogenesis* **9**, 1265–1269.
37. Divi, R. L., Beland, F. A., Fu, P. F., Von Tungeln, L. S., Schoket, B., Camara, J. E., Ghei, M., Rothman, N., Sinha, R., Poirier, M. C. (2002) Highly sensitive chemiluminescence immunoassay for benzo[*a*]pyrene-DNA adducts: validation by comparison with other methods, and use in biomonitoring. *Carcinogenesis* **23**, 2043–2049.
38. Divi, R. L., Osborne, M. R., Hewer, A., Phillips, D. H. and Poirier, M. C. (1999) Tamoxifen-DNA adduct formation in rat liver determined by immunoassay and ^{32}P-postlabeling. *Cancer Res.* **59**, 4829–4833.
39. Santella, R. M. (1991) DNA adducts in humans as biomarkers of exposure to environmental and occupational carcinogens. *Environ. Cancer Rev.* **C9**, 57–81.
40. Mumford, J. L., Williams, K., Wilcosky, T. C., Everson, R. B., Young, T. L. and Santella, R. M. (1996) A sensitive color ELISA for detecting polycyclic aromatic hydrocarbon-DNA adducts in human tissues. *Mutat. Res.* **359**, 171–177.
41. Hsieh, L. L., Hsu, S. W., Chen, D. S. and Santella, R. M. (1988) Immunological detection of aflatoxin B_1-DNA adducts formed *in vivo*. *Cancer Res.* **48**, 6328–6331.
42. Wilson, V. L., Weston, A., Manchester, D. K., Trivers, G. E., Roberts, D. W., Kadlubar, F. F., Montesano, R., Willey, J. C., Mann, D. L. and Harris, C. C. (1989) Alkyl and aryl carcinogen adducts detected in human peripheral lung. *Carcinogenesis* **10**, 2149–2153.
43. Sharma, R. A., McLelland, H. R., Hill, K. A., Ireson, C. R., Euden, S. A., Manson, M. M., Pirmohamed, M., Marnett, L. J., Gescher, A. J. and Steward, W. P. (2001) Pharmacodynamic and pharmacokinetic study of oral Curcuma extract in patients with colorectal cancer. *Clin. Cancer Res.* **7**, 1894–1900.
44. Van Gijssel, H. E., Schild, L. J., Watt, D. L., Roth, M. J., Wang, G. Q., Dawsey, S. M., Albert, P. S., Qiao, Y. L., Taylor, P. R., Dong, Z. W. and Poirier, M. C. (2004) Polycyclic aromatic hydrocarbon-DNA adducts determined by semiquantitative immunohistochemistry in human esophageal biopsies taken in 1985. *Mutat. Res.* **547**, 55–62.
45. Rybicki, B. A., Rundle, A., Savera, A. T., Sankey, S. S. and Tang, D. (2004) Polycyclic aromatic hydrocarbon-DNA adducts in prostate cancer. *Cancer Res.* **64**, 8854–8859.
46. Zhang, Y. J., Hsu, T. M. and Santella, R. M. (1995) Immunoperoxidase detection of polycyclic aromatic hydrocarbon-DNA adducts in oral mucosa cells of smokers and non-smokers. *Cancer Epidemiol. Biomarkers Prev.* **4**, 133–138.
47. Rundle, A., Tang, D. L., Zhou, J. Z., Cho, S. and Perera, F. (2000) The association between glutathione S-transferase M1 genotype and polycyclic aromatic hydrocarbon-DNA adducts in breast tissue. *Cancer Epidemiol. Biomarkers Prev.* **9**, 1079–1085.
48. Blommaert, F. A., Michael, C., Terheggen, P. M. A. B., Muggia, F. M., Kortes, V., Schornagel, J. H., Hart, A. A. M. and Engelse, L. D. (1993) Drug-induced DNA modification in buccal cells of cancer patients receiving carboplatin and cisplatin combination chemotherapy, as determined by an immunocytochemical method: interindividual variation and correlation with disease response. *Cancer Res.* **53**, 5669–5675.
49. Farmer, P. B., Brown, K., Tompkins, E., Emms, V. L., Jones, D. J. L., Singh, R. and Phillips, D. H. (2005) DNA adducts: mass spectrometry methods and future prospects. *Toxicol. Appl. Pharmacol.* **207**, S293–S301.

50. Ziegel, R., Shallop, A., Jones, R., Tretyakova, N. (2003) *K-ras* gene sequence effects on the formation of 4-(methylnitrosamino)-1-(3-pyridyl)-1-butanone (NNK)-DNA adducts. *Chem. Res. Toxicol.* **16,** 541–550.
51. Brown, K.; Harvey, C. A.; Turteltaub, K. W.; Shields, S. J. (2003) Structural characterization of carcinogen-modified oligodeoxynucleotide adducts using matrix-assisted laser desorption/ionization mass spectrometry. *J. Mass Spectrom.***38**, 68–79.
52. Singh, R. and Farmer, P. B. (2006) Liquid chromatography-electrospray ionization-mass spectrometry: the future of DNA adduct detection. *Carcinogenesis* **27**, 178–196.
53. Koc, H. and Swenberg, J. A. (2002) Applications of mass spectrometry for quantitation of DNA adducts. *J. Chromatogr. B.* **778**, 323–343.
54. Tompkins, E. M., Jones, D. J. L., Lamb, J. H., Marsden, D. A., Farmer, P. B. and Brown, K. (2008) Simultaneous detection of five different 2-hydroxyethyl DNA adducts formed by ethylene oxide exposure, using a high-performance liquid chromatography/electrospray ionisation tandem mass spectrometry assay. *Rapid Commun. Mass Spectrom* **22**, 19–28.
55. Marsden, D. A.; Jones, D. J. L.; Lamb, J. H.; Tompkins, E.; Farmer, P. B.; Brown K. (2007) Determination of endogenous and exogenously derived *N7*-(2-hydroxyethyl)guanine adducts in ethylene oxide treated rats. *Chem. Res. Toxicol.* **20**, 290–299.
56. Gamboa da Costa, G., Marques, M. M., Beland, F. A., Freeman, J. P., Churchwell, M. I., Doerge, D. R. (2003) Quantification of tamoxifen DNA adducts using on-line sample preparation and HPLC-electrospray ionization tandem mass spectrometry. *Chem. Res. Toxicol.* **16**, 357–366.
57. Bessette, E. E., Goodenough, A. K., Langouët, S., Yasa, I., Kozekov, I. D., Spivack, S. D., Turesky, R. J. (2009) Screening for DNA adducts by data-dependent constant neutral loss-triple stage mass spectrometry with a linear quadruple ion trap mass spectrometer. *Anal. Chem.* **81**, 809–819.
58. Kanaly, R. A., Matsui, A., Hanaoka, T. and Matsuda, T. (2007) Application of the adductome approach to assess intertissue DNA damage variations in human lung and esophagus. *Mutat. Res.* **625**, 83–93.
59. Farmer, P. B. and Singh, R. (2008) Use of DNA adducts to identify human health risk from exposure to hazardous environmental pollutants: The increasing role of mass spectrometry in assessing biologically effective doses of genotoxic carcinogens. *Mutat. Res.* **659**, 68–76.
60. Phillips, D. H., Hewer A. and Osborne, M. R. (1992) Interaction of omeprazole with DNA in rat tissues. *Mutagenesis* 7, 277–283.
61. Vogel. J. S., Turteltaub, K. W., Finkel, R. and Nelson, D. E. (1995) Accelerator mass spectrometry. *Anal. Chem.* **67**, 353A–359A.
62. Brown, K., Tompkins, E. M. and White, I. N. H. (2006) Applications of accelerator mass spectrometry for pharmacological and toxicological research. *Mass Spec. Rev.* **25**, 127–145.
63. Turteltaub, K. W., Vogel, J. S., Frantz, C. E. and Fultz, E. (1993) Studies on DNA adduction with heterocyclic amines by accelerator mass spectrometry: a new technique for tracing isotope-labelled DNA adduction. *In Phillips, D. H., Castegno, M., Bartsch, H., editors. Postlabeling methods for detecting DNA adducts. Lyon: IARC*, p 293–391.
64. Turteltaub, K. W., Felton, J. S., Gledhill, B. L., Vogel, J. S., Southon, J. R., Caffee, M. W., Finkel, R. C., Nelson, D. E., Proctor, I. D. and Davis, J. C. (1990). Accelerator mass spectrometry in biomedical dosimetry: relationship between low-level exposure and covalent binding of heterocyclic amine carcinogens to DNA. *Proc Natl Acad Sci USA* **87**, 5288–5292.
65. Mauthe, R. J., Dingley, K. H., Leveson, S. H., Freeman, S. P., Turesky, R. J., Garner, R. C. and Turteltaub, K. W. (1999). Comparison of DNA-adduct and tissue-available dose levels of MeIQx in human and rodent colon following administration of a very low dose. *Int J Cancer* **80**, 539–545.
66. Marsden, D. A., Jones, D. J. L., Britton, R. G., Ognibene, T., Ubick, E., Farmer, P. B. and Brown, K. (2009) Dose response relationships for *N7*-(2-hydroxyethyl)guanine induced by low dose [^{14}C]-ethylene oxide: evidence for a novel mechanism of endogenous DNA adduct formation. *Cancer Res.* **69**, 3052–3059.
67. Martin, E. A., Brown, K., Gaskell, M., Al-Azzawi, F., Garner, R. C., Boocock, D. J., Mattock, E., Pring, D. W., Dingley, K., Turteltaub, K. W., Smith, L. L. and White, I. N. H. (2003) Tamoxifen DNA damage detected in human endometrium using accelerator mass spectrometry. *Cancer Research* **63**, 8461–8465.
68. Coldwell, K. E., Cutts, S. M., Ognibene, T. J., Henderson, P. T. and Phillips, D. R. (2008) Detection of adriamycin-DNA adducts by accelerator mass spectrometry at clinically relevant adriamycin concentrations. *Nucl. Acids Res.* **36**, e100.
69. Mally, A., Zepnik, H., Wanek, P., Eder, E., Dingley, K., Ihmels, H., Volkel, W. and Dekant, W. (2004) Ochratoxin A: lack of formation of

covalent DNA adducts. *Chem. Res. Toxicol.* **17**, 234–242.

70. Buchholz, B. A., Freeman, S. P. H. T., Haack, K. and Vogel, J. S. (2000). Tips and traps in the ^{14}C bio-AMS preparation laboratory. *Nucl Instr and Meth B* **172**, 404–408.
71. Brown, K., Dingley, K. H. and Turteltaub, K. W. (2005) Accelerator Mass Spectrometry for Biomedical Research, *in Biological Mass Spectrometry,* 402, *Methods in Enzymology,* 423–443 (Ed. Burlingame, A.). Academic Press. San Diego, California, U.S.
72. Brown, K., Tompkins, E. M., Boocock, D. J., Martin, E. A., Farmer, P. B., Turteltaub, K. W., Ubick, E., Hemingway, D., Horner-Glister, E. and White, I. N. H. (2007) Tamoxifen forms DNA adducts in human colon after administration of a single [^{14}C]-labeled therapeutic dose. *Cancer Res.* **67**, 6995–7002.
73. Goldman, R., Day, B. W., Carver, T. A., Mauthe, R. J., Turteltaub, K. W. and Shields, P. G. (2000) Quantitation of benzo[*a*]pyrene-DNA adducts by postlabeling with ^{14}C-acetic anhydride and accelerator mass spectrometry. *Chem. Biol. Interact.* **126**, 171–183.
74. Tompkins, E. M., Farmer, P. B., Lamb, J. H., Jukes, R., Dingley, K., Ubick, E., Turteltaub, K. W., Martin, E. A. and Brown, K. (2006) A novel ^{14}C-postlabelling assay using accelerator mass spectrometry for the detection of O^6-methyldeoxyguanosine adducts. *Rap. Commun. Mass Spectrom.* **20**, 883–891.
75. Ognibene, T. J., Bench, G., Vogel, J. S., Peaslee, G. F. and Murov, S. (2003) A high-throughput method for the conversion of CO_2 obtained from biochemical samples to graphite in septa-sealed vials for quantification of ^{14}C via accelerator mass spectrometry. *Anal. Chem.* **75**, 2192–2196.
76. Liberman, R. G., Tannenbaum, S. R., Hughey, B. J., Shefer, R. E., Klinkowstein, R. E., Prakash, C., Harriman, S. P. and Skipper, P. L. (2004) An interface for direct analysis of ^{14}C in nonvolatile samples by accelerator mass spectrometry. *Anal. Chem.* **76**, 328–334.

Chapter 12

The GADD45a-GFP GreenScreen HC Assay

Richard M. Walmsley and Matthew Tate

Abstract

Mutagens, clastogens, and aneugens cause increased expression of the human *GADD45a* gene. This has been exploited in the GreenScreen HC genotoxicity assay in which the gene's expression is linked to the expression of green fluorescent protein (GFP). The host for the reporter construct is the human lymphoblastoid cell line TK6. It was chosen for its growth as a cell suspension, which allows simple pipette transfers, and for its wild-type p53 competent status. P53 is required for proper *GADD45a* expression, and more generally for genome stability. TK6 is a karyotypically stable cell line.

The GreenScreen assays were designed to facilitate screening, and this is reflected in its microplate format and low compound requirement. Protocols are available for testing with and without S9 as a source of exogenous metabolic activation. Data is collected either spectrophotometrically or by flow cytometry, and a simple spreadsheet converts raw data into dose–response curves, and provides a statistically significant positive or negative result. Extensive validation has demonstrated that in contrast to other in vitro mammalian genotoxicity assays, the *GADD45a* assays have both high sensitivity and specificity – they very rarely produce misleading positive results.

Key words: GADD45a, GADD45a-GFP, GreenScreen HC, Screening, Genotoxicity

1.. Introduction

The *GADD45a-GFP* GreenScreen HC genotoxicity assay monitors genotoxin-induced transcription of the *GADD45a* gene (1). *GADD45a*, originally identified and named by the Fornace laboratory (2), has been implicated in the response to genome damage by genetic, biochemical, and genomic approaches (2–7). Mice lacking the gene are more prone to tumors induced by ionizing radiation and genotoxin exposure (3); their lymphobasts and fibroblasts have defective nucleotide excision repair; their fibroblasts show centrosome amplification, unequal segregation of chromosomes due to multiple spindle poles, and the induction of aneuploidy (4). The *GADD45a* protein modifies DNA accessibility in damaged chromatin and associates with nuclear factors associated

James M. Parry and Elizabeth M. Parry (eds.), *Genetic Toxicology: Principles and Methods*, Methods in Molecular Biology, vol. 817, DOI 10.1007/978-1-61779-421-6_12, © Springer Science+Business Media, LLC 2012

with cell cycle regulation (5, 6). In microarray studies, the gene is one of those most robustly induced by genotoxins (7). All these studies implicate *GADD45a* as a clear component of the pathway that contributes to the maintenance of genomic stability, and this is reflected in its induction by mutagens, clastogens, and aneugens (1).

The GreenScreen HC assay has been reviewed in ref. 8. It monitors the expression of the *GADD45a* gene, using an in-frame green fluorescent protein (GFP) reporter gene (see Fig. 1 for a representation of the reporter plasmid.) The reporter is hosted by the human lymphoblastoid cell line TK6, which is p53 competent – a necessary attribute for the proper genotoxic response in all mammals. For detailed information, readers are referred to the original validation paper (1), the subsequent transferability "ring trial" (9), the validation of a protocol for the assessment of S9 metabolites, which expanded the spectrum of genotoxins identified by the assay (10), and to some larger studies including the 1,266 compound "Sigma Library Of Pharmacologically Active Compounds" (11). Other material submitted for publication includes an assessment of the 320 compounds from the US ToxCast EPA programme, as

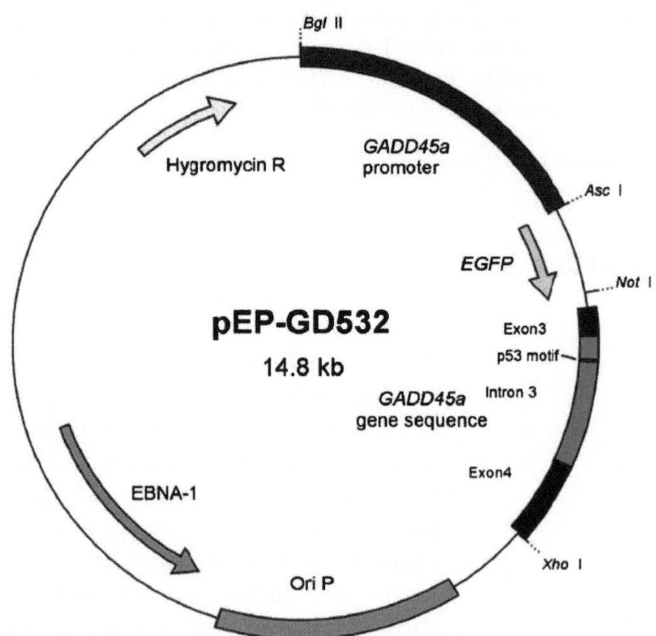

Fig. 1. Plasmid map representing the GADD45a-GFP reporter system expressed in the GenM-T01 cell strain. The plasmid contains the GFP sequence (eGFP), *GADD45a* promoter and regulatory elements contain within exons 3 and 4 as well as intron 3 which contains a p53 binding domain. The plasmid backbone contains the Epstein–Barr nuclear antigen and origin of replication and a hygromycin B resistance gene, all of which allow stable episomal replication of the reporter and specific strain selection in conditions containing hygromycin B. Reproduced from ref. 1.

well as data from 75 marketed pharmaceuticals and 61 compounds recommended by an ECVAM expert working group. Check http://www.gentronix.co.uk for new publications. Additional data has been generated from the 8,000+ compounds from 60 proprietary collections of pharmaceutical and biotechnology companies and laboratories that have used the assay routinely in the early identification of genotoxic hazard.

Flow cytometry provides the most effective method for the detection of genotoxic metabolites, generated by incubation with a source of exogenous metabolism, typically S9. The particulate, colored and fluorescent nature of the components of S9 can be effectively "gated" out in the data analysis. A microplate reader protocol for S9 allows the detection of the majority of promutagens, but its sensitivity is lower than that achieved when using flow cytometry. This chapter refers to the 96-well plate assays using spectrophotometric data collection both with and without S9. Details of the flow cytometry method are available from Gentronix Ltd.

Positive prevalence of GreenScreen HC data in pharmaceutical screening varies from around 7% (screened to 100 μM top dose) to 11% (300 μM top dose). This is a little lower than the figure of 12% of drug candidates that raise genotoxicity safety concerns following the battery of genotoxicity testing (12). In terms of overall performance, the sensitivity of this assay to genotoxic carcinogens is 87% and the specificity is 95%. This includes data from a study in which initial interpretation had suggested a very much reduced sensitivity (13), though this was based on the expectation that the assay would detect carcinogens with nongenotoxic modes of action. A secondary analysis, excluding nongenotoxic carcinogens, showed very much higher sensitivity for the assay (14]. The sensitivity and specificity of the assay to in vivo genotoxins are 78% and 94%, respectively.

There are four main procedures involved in the protocol: preparation of the cells; preparation of the test samples; preparation of the 96-well test microplate; data collection, processing, and analysis. The incubation time means that assays can be set up Monday to Wednesday, and data can be collected Wednesday to Friday. A single operator can test up to 72 compounds per week without the aid of robotic systems. Even higher throughput is possible when samples are supplied in solution by a dispensary, and where robotic liquid handling devices are employed. The assessment of a compound for the risk of genotoxic metabolites requires the addition of a source of exogenous metabolism, S9. An additional handling step is needed in which S9 is washed away after 3 h, to limit S9-induced toxicity. Cells are then incubated for a further 45 h and data is collected at the 48 h timepoint. A single operator can test 36 compounds in a week with S9. All these validated procedures are described below. Further information regarding assay preparation, instrumentation specifications, and settings are available from Gentronix Ltd., including detailed diagrams showing assay plate layouts.

2. Materials

2.1. Cell Culture and Reagents

1. RPMI 1640 + Glutamax/25 mM HEPES (e.g., Invitrogen Corp. catalog no. 72400-021).
2. Heat-inactivated horse serum (HIHS) (Recommended supplier – Sigma, catalog no. H1138).
3. Sodium pyruvate (e.g., Invitrogen Corp. catalog no. 11360-039).
4. Penicillin/streptomycin (e.g., Invitrogen Corp. catalog no. 15070-063).
5. Hygromycin B (e.g., Invitrogen Corp. 10687-010).
6. Cell culture flasks.
7. GreenScreen HC Assay Medium (Gentronix Ltd).
8. GreenScreen HC Assay Medium Supplement (Gentronix Ltd).
9. Dimethyl sulfoxide (e.g., Sigma-Aldrich, catalog no. 154938 – 100 ml).
10. Phosphate-buffered saline without calcium, magnesium, or phenol red (e.g., D-PBS Invitrogen Corp. catalog no. 14190-136).
11. Methyl methanesulfonate (MMS; e.g., Sigma-Aldrich, catalog no. M4016 – 1 g).
12. Sterile water.

Reagents required for GreenScreen HC with S9

1. Rat liver S9: (Rat liver S9, Aroclor 1254-induced male SD rat liver in 0.15 KCl; MolTox, Inc., catalog no. 11–101). Store at –80°C in ultra-low temperature freezer.
2. Cyclophosphamide monohydrate (CPA; Sigma, catalog no. C0768).
3. GreenScreen HC S9 DNA Binding Stain (Gentronix Ltd).
4. GreenScreen HC S9 Cell Lysis reagent (Gentronix Ltd).
5. GreenScreen HC S9 Co-factors: Each vial contains 1.8 ml of co-factor solution (Gentronix Ltd).
6. GreenScreen HC S9 Exposure Medium (Gentronix Ltd).
7. GreenScreen HC S9 Recovery Medium (Gentronix Ltd).
8. GreenScreen HC S9 Recovery Medium Supplement (Gentronix Ltd).

2.2. Specialist Equipment

1. Microplate reader capable of fluorescence and absorbance reading (suitable models are listed at http://www.genotronix.co.uk).

2. Standard materials for cell maintenance and passage: CO_2 incubator, Class II safety cabinet.
3. Black, clear flat-bottom, 96-well, sterile microplates (Matrix Technologies ScreenMates: catalog no. 4929 is recommended for optimum performance).
4. Other. Plastic troughs for reagent pipetting (e.g., Matrix Technologies Standard 25 ml Reagent Reservoir: catalog no. 8093 or 8094); Breathable membranes (e.g., Diversified Biotech, Breathe-Easy sealing membrane).

   The S9 version of the assay requires round/U-bottomed (*not conical*), solid, black, polystyrene, 96-well, sterile microplates (available from Gentronix Ltd and leading scientific consumable suppliers).

2.3. Preparation of Routine Culture and Assay Media

1. It is recommended that all reagents are purchased as sterile filtered, cell culture tested solutions. If reagents are bought as powders they should be dissolved in distilled, sterile water to the appropriate stock concentration, sterilized and stored appropriately (see Note 1).
2. To heat inactivate horse serum (if not inactivated when purchased) – heat to 56°C for 30 min. Hygromycin B can be purchased as a prefiltered, cell culture tested solution. For powder, resuspend in sterile distilled water at a concentration of 50 mg/ml. Filter sterilize and store at 2–8°C.

2.3.1. Routine Culture Medium (GS-HC-CM)

Combine the reagents listed in the table below: Reagents can be added to the recommended GIBCO 500 ml RPMI 1640 medium bottle. The prepared media should be labeled with its preparation date and a 2-week expiry date. It should be stored at 2–8°C, out of direct sunlight.

Reagent	Stock concentration	Final concentration	Volume (ml)
RPMI 1640 + Glutamax			500
Sodium pyruvate	100 mM	1.8 mM	10.4
Hygromycin B	50 mg/ml	200 μg/ml	2.3
Pen/strep	5,000 IU/ml /5,000 μg/ml	50 IU/ml/ 50 μg/ml	5.8
HIDHS	100%	10%	57

2.3.2. Assay Medium (Complete GS-HC-AM) for the Assay Without S9

GreenScreen HC Assay Medium is combined with GreenScreen HC Assay Medium Supplement to produce a cell culture medium with low autofluorescence, specially formulated by Gentronix Ltd for use in the GreenScreen HC assay without S9. To make the complete assay medium, combine the two reagents in the ratios listed in the table below. It is recommended that the volume of Medium Supplement required is added directly to the bottle of GS-HC-AM Stock Medium. When the reagents are combined, GS-HC-AM has a Medium Supplement concentration of 20% v/v. Complete GS-HC-AM should be labeled with its preparation date and has a shelf-life of 45 days after addition of Supplement. Media should be stored at 2–8°C, out of direct sunlight.

For the HC-20 kit (Gentronix Ltd), the aliquots of 11.25 ml of complete Assay Medium are sufficient for an individual microplate (four compound tests per plate). For HC-20B kits, up to 55 ml of complete Assay Medium is made, sufficient for five microplates (20 compound tests).

GreenScreen HC kit	Reagent	Volume (ml)
HC-20	GS-HC-AM stock medium	9
	GS-HC-AM medium supplement	2.25
HC-20B	GS-HC-AM stock medium	44
	GS-HC-AM medium supplement	11

2.3.3. Assay Medium for the Assay with S9

The S9 version of the GreenScreen HC assay has two stages, exposure and recovery, and this requires the preparation of two separate media. To ensure correct function of the assay with S9, it is very important that the following media are used – and only used – in the assay WITH S9.

1. GreenScreen HC S9 Exposure Medium (GS-HC S9-EM)

   GS-HC S9-EM is combined with HIHS to produce the medium used in the exposure period of the GreenScreen HC assay. To make the complete GS-HC S9-EM, combine the two reagents in the ratio listed in the table below. It is recommended that the volume of HIHS required is added directly to the bottle of GS-HC S9-EM. When the reagents are combined, the complete GS-HC S9-EM has an HIHS concentration of 10% v/v. Complete GS-HC S9-EM should be labeled with its preparation date and has a shelf-life of 14 days after addition of serum. Medium should be stored at 2–8°C, out of direct sunlight.

GreenScreen HC kit	Reagent	Volume (ml)
HC-20	GS-HC S9-EM stock medium	10.8
	Heat-inactivated horse serum (HIHS)	1.2

2. GreenScreen HC S9 Recovery Medium (GS-HC S9-RM)

   GS-HC S9-RM is combined with GreenScreen HC S9 Recovery Medium Supplement to produce a complete recovery medium used after the compound exposure and washing stages. To make the complete GS-HC S9-RM, combine the two reagents in the ratio listed in the table below. It is recommended that the GreenScreen HC S9 Recovery Medium Supplement is added directly to the bottle of GS-HC S9-RM. When the reagents are combined, the complete GS-HC S9-RM has a concentration of 10% v/v Recovery Medium Supplement. Complete GS-HC S9-RM should be labeled with its preparation date and has a shelf-life of 45 days after addition of Supplement. Medium should be stored at 2–8°C, out of direct sunlight.

GreenScreen HC kit	Reagent	Volume (ml)
HC-20	GS-HC S9-RM stock medium	12.6
	GreenScreen HC/S9 recovery medium supplement	1.4

3. Methods

3.1. Maintenance of Cell Lines

Two cell lines are provided; a control cell line (GenM-C01) and a test cell line (GenM-T01). These are both required to perform a GreenScreen HC assay and should be maintained as separate cultures.

3.1.1. Thawing Frozen Cells

1. Thaw vials of GenM-C01 and GenM-T01 quickly at 37°C, until the frozen core can be dislodged, then tip the entire contents into separate sterile 75 cm^2 culture flasks.
2. Over a period of 2 min, add 50 ml of prewarmed (37°C) culture medium (GS-HC-CM) to the cells.
3. Resuspended, cells should be placed in a CO_2 incubator for 2–3 days where they will reach a density of between 2×10^5 and 2×10^6 cells per milliliter. Harvest and resuspend in 50 ml prewarmed GS-HC-CM at an appropriate density for the intended passage time indicated below.

3.1.2. Routine Cell Culturing (Passaging)

Maintain cell lines in log phase by passaging every 1–4 days. Measure cell culture densities and dilute aliquots with fresh prewarmed GS-HC-CM to a final cell titre of $0.15–5 \times 10^5$ cells/ml, depending on requirements (see table below and Note 2).

Passage duration (days)	Seeding cell concentration (number of cells/ml)
1	5.0×10^5
2	1.5×10^5
3	5.0×10^4
4	1.5×10^4

The recommended passage dilutions will deliver the appropriate cell concentration of approximately 1×10^6 cells/ml on the day assays are commenced.

Cells are required to be in logarithmic growth phase, and must have achieved a density of between 5×10^5 cells/ml and 1.2×10^6 cells/ml before they can be used in the assay. Cell cultures with a density outside of these limits should not be used and instead should be passaged for a subsequent day using the guidance above to achieve the required density. Each assay plate will require approximately 5 ml of each cell line culture at a starting concentration of 2×10^6 cells/ml.

Cells should initially be passaged for 2 weeks in the presence of Hygromycin B before the cultures are used in assays, and can be maintained in log phase for up to 3 months.

3.2. Preparing Test and Control Compounds

All concentrations of standard and test compounds are stated as made up by the operator. When conducting the assay without S9, all concentrations described below for that assay are halved on the plate when a sample volume of 75 μl is combined with 75 μl of cell culture. For the assay with S9, all concentrations detailed below for that assay are 2.5 times greater than the final on plate concentration to take into account the dilution factor when 60 μl of sample volume is combined with 15 μl of S9 mix and 75 μl of cell culture. All standard and test chemicals should be prepared fresh shortly before the GreenScreen HC assay plate is set up.

3.2.1. Diluent: 2% DMSO in Sterile Water Is Required for Assays with and Without S9

100 ml of diluent should be prepared per 16 compound run (four microplates) and used to prepare and dilute all the standard and test compounds. To prepare 100 ml of 2% DMSO, combine 2 ml of DMSO with 98 ml of sterile water and mix thoroughly.

3.2.2. Control Compounds for Assays Without S9

The control compounds are prepared in diluent to the following concentrations

Standard 1 – MMS HIGH = 100 μg/ml
Standard 2 – MMS LOW = 20 μg/ml

A minimum of 300 μl of each MMS control is required per plate. The following suggested dilutions produce 2–2.5 ml for each standard: MMS should be freshly prepared from 100% stock solutions before use. When preparing MMS standards ensure thorough mixing at each dilution stage by repeated aspiration and dispensing of the measured aliquot in the diluent, and/or by vortexing the solution in a sealed vial (see Note 3).

Dilution from MMS stock

Stock dilution: Add 5 μl of 100% MMS to 1,300 μl diluent = 5 mg/ml MMS

MMS HIGH: Dilute 50 μl of 5 mg/ml MMS with 2,450 μl diluent = 100 μg/ml MMS

MMS LOW: Dilute 400 μl of MMS HIGH with 1,600 μl diluent = 20 μg/ml MMS

3.2.3. Control Compounds for Assays with S9

The control compounds are prepared in diluent to the following concentrations

Standard 1 – CPA HIGH = 62.5 μg/ml to give a final concentration of 25 μg/ml

Standard 2 – CPA LOW = 25 μg/ml to give a final concentration of 5 μg/ml

The following suggested dilutions produce sufficient solution for two microplates

When preparing CPA standards, ensure thorough mixing at each dilution stage by repeated aspiration and dispensing of the measured aliquot in the diluent and/or by vortexing the solution in a sealed vial (see Note 4).

Prepare a stock solution of 1 mg/ml CPA in diluent.

CPA HIGH: Dilute 75 μl of 1 mg/ml CPA with 1,125 μl diluent = 62.5 μg/ml CPA.

CPA LOW: Dilute 15 μl of 1 mg/ml CPA with 1,185 μl diluent = 12.5 μg/ml CPA.

Batch aliquots of these solutions can be prepared for use in the assay as the compound is stable if kept frozen at −20°C. Do not thaw and then re-freeze.

3.2.4. Test Compounds

The test compound must be dissolved in a solution that matches the diluent used, typically 2% v/v DMSO in sterile water. The diluent solvent itself is not diluted across the plate. Compounds prepared in water should be diluted in water.

A test concentration of 1 mM or 500 μg/ml (whichever is lowest) is recommended for screening. This equates to an initial stock concentration of test compound in 2% DMSO of either 2 mM or 1 mg/ml for the assay without S9 or a 2.5 mM or 1.25 mg/ml concentration for the S9 version to take into account the increased dilution factor during microplate set-up. It is desirable that the test compound is fully soluble at the top concentration tested. A minimum of 400 μl of each test compound is required per plate. The recommended method to prepare solutions of test compounds is as follows:

For compounds with high aqueous solubility – dissolve directly in aqueous diluent (i.e., 2% DMSO) and dilute, with diluent, as necessary.

For compounds of limited aqueous solubility – first dissolve in 100% DMSO at highest achievable concentration, and then add 20 μl of this DMSO stock standard to 980 μl sterile water to produce a test solution containing 2% v/v DMSO.

If the compound precipitates from solution when the DMSO standard is added to water, the original DMSO stock standard can be diluted further with 100% DMSO.

The 20 μl + 980 μl water dilution step is then repeated to produce a fresh test standard.

3.2.5. Preparation of the S9 Mix for Use in Assays Including Exogenous Metabolism

The S9 mix used in GreenScreen HC comprises of a standard Co-factor solution (supplied with the assay) which contains 5.56 mM β-NADP disodium salt and 27.8 mM disodium G-6-P and a final on-plate concentration of S9 of 1% v/v. To prepare this mix both the S9 vial and GreenScreen HC co-factor solution should be thawed on ice.

Add 200 μl S9 to 1.8 ml vial of GreenScreen HC S9 co-factor solution.

This provides sufficient S9 mix for 1 GreenScreen HC S9 assay.

3.3. Microplate Set-Up: General Comments

The microplate layout and data processing template is designed for the assay of four compounds per microplate. If fewer than four compounds are run on a microplate, the "missing" compounds must be substituted, i.e., wells must not be left empty. While operators might choose to complete the plate by using diluent as a test compound (effectively running a blank), it is more valuable to generate additional data by testing single compounds in duplicate or at differing top concentrations. Subheading 3.3.1 below refers to the set-up of the assay in the absence of S9, while Subheading 3.3.2 details the stages of preparation of an assay plate in the presence of S9. Following completion of either method, refer to Subheading 3.3.2 onwards to complete the assay set-up by addition of cell strains to the microplate.

3.3.1. Microplate Set-Up Without S9

1. Adding the diluent. Use an 8-channel pipette dispensing 75 μl.
   Excluding all wells in columns 1 and 11, and wells E12 to H12 inclusive, dispense 75 μl diluent into all other wells, column by column (see Note 5).
   Dispense an additional 75 μl of diluent into wells A12 and B12 (they now contain 150 μl).

2. Dispensing the test compounds. Use a single-channel pipette.
   (a) Dispense 150 μl of test compound 1 into wells A1 and E1, and 75 μl into E12.
   (b) Dispense 150 μl of test compound 2 into wells B1 and F1, and 75 μl into F12.
   (c) Dispense 150 μl of test compound 3 into wells C1 and G1, and 75 μl into G12.
   (d) Dispense 150 μl of test compound 4 into wells D1 and H1, and 75 μl into H12.
3. Serially diluting the test compounds. Use an 8-channel pipette dispensing 75 μl.

   Aspirate 75 μl from all wells in column 1 and dispense into column 2. When dispensing into these wells, mix the contents by repeated aspiration/dispense, or by using the mix feature common to many electronic auto-pipettes.
   (a) Using the same tips, repeat the entire process for the next column to the right, starting with aspiration of 75 μl from column 2 and transferring to column 3.
   (b) Repeat this procedure up to and including column 9. After mixing the well contents in column 9, aspirate 75 μl from column 9 and discard to waste.
4. Dispensing the standard control compounds. Single-channel pipette, 75 μl.
   (a) Dispense 75 μl of the MMS LOW control into wells A11, B11, E11, and F11.
   (b) Dispense 75 μl of the MMS HIGH control into wells C11, D11, G11, and H11.

3.3.2. Microplate Set-Up with S9 (See ***Note 5****)*

1. Adding the diluent. Use an 8-channel pipette dispensing 60 μl.
   (a) Omitting all wells in columns 1 and 11 and wells E12 to H12, dispense 60 μl diluent into all other wells column by column.
   (b) In addition, dispense 15 μl diluent to wells A12 and B12.
2. Dispensing the test compounds. Use a single-channel pipette.
   (a) Dispense 120 μl of test compound 1 into wells A1 and E1 and 60 μl into well E12.
   (b) Dispense 120 μl of test compound 2 into wells B1 and F1 and 60 μl into well F12.
   (c) Dispense 120 μl of test compound 3 into wells C1 and G1 and 60 μl into well G12.
   (d) Dispense 120 μl of test compound 4 into wells D1 and H1 and 60 μl into well H12.

3. Serially diluting the test compounds. Use an 8-channel pipette dispensing 60 μl.
   (a) Aspirate 60 μl from all wells in column 1 and dispense into column 2. When dispensing into these wells, mix the contents by repeated aspiration/dispense, or by using the mix feature common to many electronic auto-pipettes.
   (b) Using the same tips, repeat the entire process for the next column to the right, starting with aspiration of 60 μl from column 2 and transferring to column 3.
   (c) Repeat this procedure up to and including column 9.

   After mixing the well contents in column 9, aspirate 60 μl from column 9 and discard to waste.
4. Dispensing the standard control compounds. Single-channel pipette, 60 μl
   (a) Dispense 60 μl of the CPA LOW control into wells A11, B11, E11, and F11.
   (b) Dispense 60 μl of the CPA HIGH control into wells C11, D11, G11, and H11.
5. Addition of S9 mix. Use an 8-channel pipette dispensing 15 μl.

   Dispense 15 μl of the S9 mix to all wells in columns 1 through to 11 and wells C12 to H12.

3.4. Preparation of Cell Lines for the Microplate: For Assays with or Without S9

For each assay plate, prepare 5 ml suspensions of GenM-C01 and GenM-T01 cells at a density of 2×10^6 cells/ml in Complete Assay media (see Subheading 2.3.2 without S9 or Subheading 2.3.3 with S9). Before taking cell counts, shake cultures GENTLY to resuspend the cells (see Note 6).

1. Calculate the cell density from the routine cultures of GenM-C01 and GenM-T01 cells. Assay ready cell culture densities must be between 5×10^5 cells/ml and 1.2×10^6 cells/ml. If the cell culture density of either GenM-C01 or GenM-T01 lies outside of this range, the assay should not be run and the cells should be passaged for a subsequent day using the passage guidance (Subheading 3.1.2).

   5 ml at 2×10^6 cells per ml of each cell line are required per assay plate. Use the following equation to calculate the volume (V) of routine cell culture required to prepare cells for N number of assay plates:

$$V = \frac{(2 \times 10^6) \times (N \times 5)}{Y}$$

   where Y is the cell count per milliliter of the routine cultures.
2. Transfer volume V of GenM-C01 and GenM-T01 cell suspensions from the routine cultures to separate sterile centrifuge

tubes. Harvest cells at 1,400 rpm (~340 RCF) for 5 min. Decant the routine culture medium and resuspend cells in 5–10 ml prewarmed PBS (or D-PBS). Harvest cells a second time at 1,400 rpm for 5 min and decant the PBS. Resuspend cells in the appropriate volume of complete assay medium. Ensure that cells are fully suspended in media by repeated pipetting or gentle shaking just prior to dispensing.

3.5. Microplate Set-Up: Adding Cell Lines to the Microplate

When running the assay WITH S9, Gentronix Complete Assay Medium as described in Subheading 3.4 should be substituted for GreenScreen HC S9 Exposure Medium.

1. Dispensing the media contamination controls. Use a single-channel pipette, 75 μl.

   A small volume of Gentronix HC Complete Assay Medium is used as a standard on the plate in order to demonstrate that the media is clear of contamination. Assay medium for contamination controls should be taken from the same aliquot of complete assay medium used to prepare the cell cultures. Dispense 75 μl of Gentronix Complete Assay Medium into wells C12 to H12. For operators preparing the with S9 version of the assay, an additional 75 μl of Gentronix Complete Assay Medium should be added to wells A12 and B12.

2. Dispensing the GenM-C01 control cell line. 8- or 12-Channel pipette, 75 μl.

   Carefully pour the cells suspension into a reagent reservoir for dispensing.

   Pipette 75 μl of GenM-C01 culture (at 2×10^6 cells/ml) into rows A, B, C, and D from column 1 up to and including column 11.

3. Dispensing the GenM-T01 test-strain. 8- or 12-Channel pipette, 75 μl.

   Carefully pour the cells suspension into a reagent reservoir for dispensing.

   Pipette 75 μl of GenM-T01 culture (at 2×10^6 cells/ml) into rows E, F, G, and H from column 1 up to and including column 11.

3.6. Summary of Final Plate Layout

Each plate is divided into two halves. Rows A–D contain test cells and rows E–H contain control cells. In each half, four compounds are each present at nine dilutions. Columns 11 and 12 contain blanks and other controls. See Fig. 2 for an illustration of the final plate layout of (a) without S9 or (b) with S9.

3.7. Covering and Incubation

The plate is now complete. Ensure it is labeled. It is recommended that plates are covered with a breathable membrane. Be sure to remove both the protective layers, one on either side of the membrane

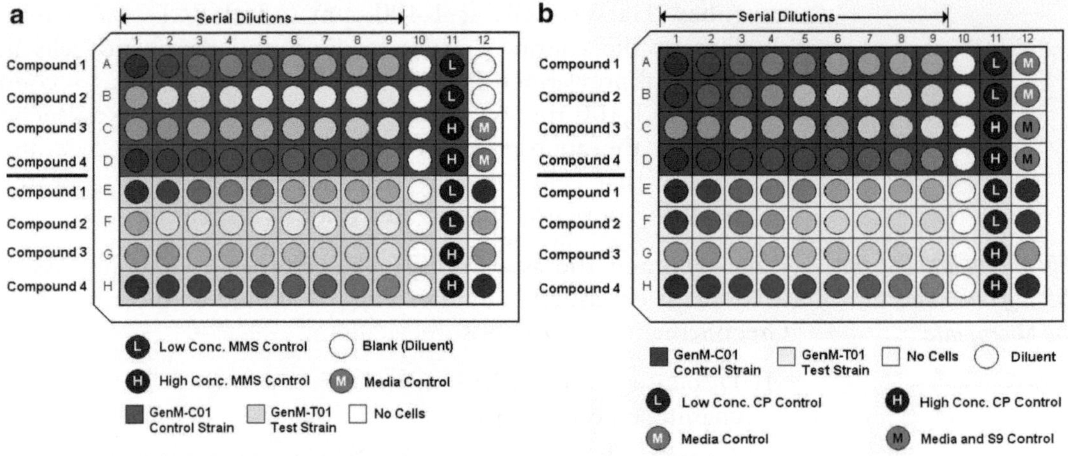

Fig. 2. Illustration of the final plate layouts of GreenScreen HC assays. (**a**) The without S9 version of the assay; (**b**) the assay with S9. Both layouts detail the location for test compound addition and serial dilution; addition of genotoxic standards and other controls and the addition of the control (GenM-C01) and test (GenM-T01) cell strains with a final well volume of 150 μl and a final cell concentration of 1.5 + E05 cells/well.

before applying to the plate. Shake the plate gently for 10–15 s on a microplate shaker (to fully mix the contents of each well).

3.8. S9 Wash Stages

Due to the inherent cytotoxicity of S9 to mammalian cells, the exposure of compound and S9 mix in this version of the GreenScreen HC assay is limited to 3 h. After the 3 h compound incubation period, several plate wash steps are carried out using prewarmed PBS (37°C). The cells are then resuspended in Complete GS-HC S9-RM and incubated for a further 45 h.

1. Incubate in CO_2 incubator without shaking for 3 h.
2. After incubation, centrifuge the microplates at 340 RCF for 5 min at room temperature.
3. Using a 8- or 12-channel pipette, aspirate 120 μl of supernatant from all wells on the microplate, taking care not to disturb the cell pellet.
4. Using a 8- or 12-channel pipette, dispense 120 μl of sterile PBS to all wells on the microplate.
5. Cover using a sterile membrane or plate lid and centrifuge at 340 RCF for 5 min.
6. Repeat steps 3,4 and 5.
7. After removing the supernatant from the second PBS wash, use an 8- or 12-channel pipette to dispense 120 μl of Complete GS-HC S9-RM to all wells on the microplate.

8. The wash stage is now complete. Cover the plate using a sterile breathable membrane and incubate at 37°C for 45 h.
9. Then follow step 2 of Subheading 3.9 for data collection.

3.9. Collection of Assay Data

There are many 96-well microplate readers on the market, and a list of instruments effective in delivering GreenScreen data is provided on the Gentronix Web site, together with step-by-step procedures for their set-up. They all deliver data in a Microsoft Excel compatible format which can be copied and pasted directly into the data processing template (supplied by Gentronix Ltd). As the GreenScreen HC assay is measuring the detection of accumulated GFP, standard methods for measuring fluorescein fluorescence intensity are used, as GFP has similar fluorescence excitation and emission spectra.

Note: the minus S9 version of the GreenScreen HC assay has two time points for data collection which occur 24 and 48 h after initial plate set-up. A positive result for genotoxicity at either of these timepoint indicates a genotoxic risk. The S9 version of the assay has a single 48 h data collection timepoint only.

1. Data collection for the assay without S9 treatment.
   (a) Incubate in a CO_2 incubator without shaking for 24 h.
   (b) Before collecting data, shake the microplate for 10–15 s on a microplate shaker to thoroughly resuspend the cells. Then carefully remove the breathable membrane.
   (c) Collect 24 h data set. Absorbance 620; GFP as 2(a) below.
   (d) After reading, recover the microplate with a fresh breathable membrane and incubate for a further 24 h. After the second period of incubation, again shake the plate for 10–15 s on a microplate shaker to thoroughly resuspend the cells and carefully remove the breathable membrane, before taking a final set of microplate readings at the 48 h time point.
2. Data collection for the assay with S9 treatment.

   At the end of the 48 h incubation period, gently shake plate on an orbital microplate shaker to resuspend cells.
   (a) Collect GFP data using a plate reader capable of detecting fluorescence. Standard settings for fluorescein ($485_{ex}/535_{em}$) are suitable for GFP measurement.
   (b) During the GFP data collection, bring the GreenScreen HC S9 Cell Lysis reagent and GreenScreen HC S9 DNA Binding Stain to room temperature.
   (c) For each microplate prepared, combine 5 ml of Cell Lysis Reagent with 24 μl of DNA Binding Stain.

(d) Following GFP data collection, pipette 50 µl of the combined Cell Lysis Reagent and DNA Binding Stain into each well of the microplate.

(e) Incubate for 20 min at 37°C, 5% CO_2 in a humidified atmosphere protected from light exposure.

(f) Collect fluorescence data. Standard filters for fluorescein ($485_{ex}/535_{em}$) are also suitable for this fluorescence measurement.

3.10. Disposal

When all results processing is completed, contain the used microplates within a plastic bag containing a suitable absorbent material. Seal and dispose according to local regulations for handling GM organisms and genotoxic compounds.

3.11. Data Interpretation

The GreenScreen data processing software package performs simple arithmetic tasks in an Excel spread sheet. It is supplied free of charge with the assay components. The expression of GADD45a is inferred from the fluorescence intensity of GFP and in the processing software a "brightness" value is calculated. "Brightness" is the fluorescence signal normalized to the cell density in the sample, which is measured as either absorbance (without S9 version) or a DNA fluorescence signal (with S9 version). Brightness distinguishes between the fluorescence signal from a small highly fluorescent population, and a large, weakly fluorescent population. Brightness is calculated for both test and control wells and scaled with reference to the average brightness of the GenM-T01 untreated control only, which is set to 1, to give a fold GFP induction value. The genotoxicity dose–response graphs results reflect the degree of induction in GFP expression and can be readily compared between different data sets.

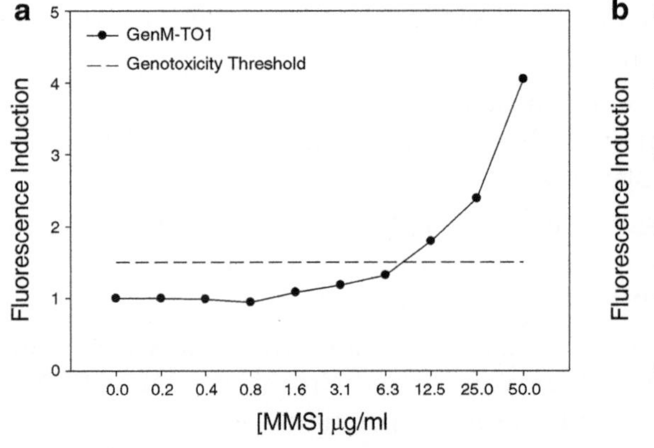

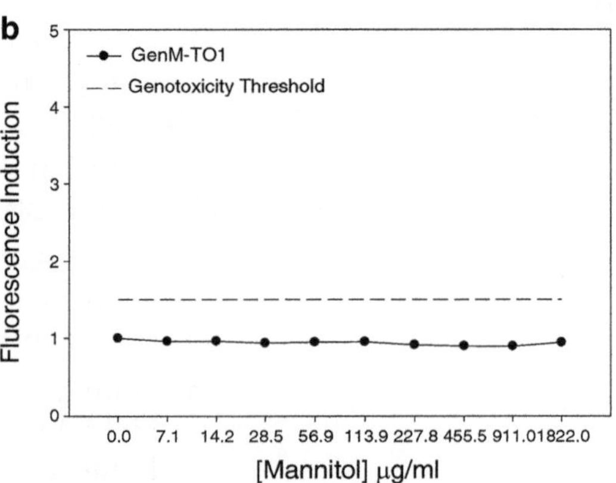

Fig. 3. Example data from the GreenScreen HC assay without S9. (**a**) Increase in GFP induction for the known genotoxic carcinogen MMS. (**b**) The negative assay response for the nongenotoxic, noncarcinogen mannitol. Data shown is the corrected and normalized brightness signal from GenM-T01 test cells. A positive genotoxicity result is achieved when the fluorescence induction at particular compound dose exceeds the genotoxicity threshold of 1.5-fold (without S9 assay).

A positive genotoxicity result is signaled if the GFP induction for a test compound increases beyond the threshold of a statistically significant increase in brightness. This corresponds to a 1.5-fold induction compared to the untreated control for the without S9 version of the assay and a 1.3-fold induction in the S9 assay reflecting the shorter (3 h) compound exposure time. The lowest effective concentration is recorded as the first concentration at which this threshold is exceeded.

See Fig. 3 for an example of the dose–response to (a) MMS, a genotoxic agent and (b) Mannitol, a nongenotoxic noncarcinogenic compound.

Light absorbance (without S9) and DNA fluorescence (S9 assay) provide measures of cell growth. Cytotoxicity results are scaled such that the average final cell density of the untreated controls equals 100% and values for test compound doses are scaled to those achieved in the untreated controls giving a value for relative cell density (RCD). The specified inoculation cell density and the use of the supplied assay medium ensure that data are in the linear range of correspondence between light absorbance (DNA stain fluorescence for S9 assays) and cell density for most spectrophotometric plate readers. In a well where no growth has occurred, the RSG is around 30% and anything below figure leads to data rejection, since it indicates cell lysis. This is an automated step in the data processing software. Almost any compound at a high enough dose will interfere with cell growth (remember Paracelsus!). However, colored compounds lead to increasing light absorbance with dose. This is obvious from graphs and alerted by the software. The subtraction of control data from test data provides a simple correction, but graphs should be inspected in case of a high absorbance alert. If the test compound itself is fluorescent, there will be a dose-dependent increase in fluorescence from the control strain, and the software produces a fluorescence alert which advises the operator to read the plate again using fluorescence polarization. It is important to look at data from individual strains as well as this summary data to check for compound fluorescence interference.

The software produces the following additional alerts

1. Growth in uninnoculated medium. This indicates that other wells might also be contaminated and the plate should be rejected.
2. Control failure. There are wells containing high and low concentrations of MMS (CPA for S9 assay), and their brightness data provide performance criteria: that the genotoxic standard is causing induction of the reporter and that there is an appropriate increase in induction at the higher dose.

4. Notes

The best way to learn is to learn with experts. The assay developers from Gentronix Ltd will provide free training to commercial users anywhere in the world when the first two assay kits are purchased. You are also welcome to visit our laboratory for training in Manchester. The training generally spans 4 days, but it is not 4 full days of work. The plate set-up is demonstrated, the new user(s) runs the assay on successive days, and the data processing and interpretation protocols are explained and demonstrated. Gentronix has written detailed Standard Operating Procedures and data interpretation notes for these demonstrations, and they can be obtained directly from the company at the following address: Gentronix Ltd, CTF Building, 46 Grafton Street, Manchester, M13 9NT. Tel: 0161 603 7662.

General

Read experimental protocols in advance and ensure that you have all materials to hand. Make sure that you have a properly set-up for spectrophotometric microplate reader, and that you will have access to it when you need to collect data. Follow the protocol precisely. Run through the protocol using water in place of reagents and practice. This will familiarize you with the plate layout, and allow you to have the appropriate pipettes in place, and set to the correct volumes.

Specific

1. You probably already have experience in cell passage, but to ensure success, only use the prescriptive protocols provided. Ensure that you follow media preparation protocols precisely. If you do not, and the cells are diluted from higher cell densities, the assay will fail. Only use the assay medium provided. Before passaging or using cells in an assay, examine cells microscopically for the presence of any contamination. Before taking cell counts for passaging or assays the cultures should be gently agitated to achieve a homogeneous mix. Operators should avoid excessive frothing of the cell suspension as this can have a detrimental effect on cell viability.
2. The modified TK6 cell lines used in the GreenScreen HC assay may not share the growth rate of cell strains you are familiar with, therefore it is essential the passage routine detailed above is followed to ensure assay performance.
3. MMS is unstable in aqueous solution. Make fresh control sample every time. If you do not, you risk positive control failures for an assay which might otherwise be fine. MMS has a higher density than water: take care to take a precise volume from the 100% stock and ensure the solution is thoroughly mixed using

a vortex mixer to ensure the correct 5 mg/ml stock concentration is achieved. Failure to do this will lead to genotoxic control failures on the assay plate due to technical error.

4. The results template is useful – not only for data processing. Use it. Fill in the date, the user, and the test concentration (in mg/ml, mM, etc.).
5. You must use all the wells, or the software will produce meaningless results.
6. TK6 cells are less robust than yeast or bacteria: they are larger and have no cell wall. They are easily affected by liquid shear forces. Froth is bad. Viscosity indicates lysis! Take care when passaging and handling cells during assay preparation.

References

1. Hastwell,P.W., Chai,L., Roberts,K.J., Webster,T.W., Harvey,J.S., Rees,R.W. and Walmsley,R.M. (2006) High-specificity and high-sensitivity genotoxicity assessment in a human cell line: Validation of the GreenScreen HC *GADD45a-GFP* genotoxicity screening assay. *Mutat. Res.* **607**, 160–175.
2. Fornace Jr.,A.J., Jackman,J., Hollander,M.C., Hoffman-Liebermann,B. and Liebermann,D.A. (1992) Genotoxic-stress-response genes and growth-arrest genes. GADD, MyD and other genes induced by treatment eliciting growth arrest. *Ann. N. Y. Acad. Sci.* **663**, 139–153.
3. Hildesheim,J., Bulavin,D., Anver,M.R., Alvord,W.G., Hollander,M.C., Vardanian, L., and Fornace, Jr, A. (2002) Gadd45a protects against UV irradiation-induced skin tumors and promotes apoptosis and stress signaling via MAP kinases and p53. *Cancer Res.* **62,** 7305–7315.
4. Hollander,M.C., Sheikh,M.S., Bulavin,D., Lundren,K., Augeri-Henmueller,L., Shehee,R., Molinaro,T., Kim,K., Tolosa,E., Ashwell,J.D., Rosenberg,M.D., Zhan,Q., Fernández-Salguero,P.M., Morgan,W.F., Deng,C.X., and Fornace,A.J., Jr. (1999) Genomic instability in Gadd45a-deficient mice. *Nature Genetics* **23,** 176–184.
5. Carrier,F., Georgel,P.T., Pourquier,P., Blake,M., Kontny,H.U., Antinore,N.J., Gariboldi,M., Myers,T.G., Weinstein,J.N., Pommier,Y., Fornace Jr.,A.J. (1999) Gadd45, a p53-responsive stress protein, modifies DNA accessibility on damaged chromatin. *Mol. Cell. Biology,* **19**, 1673–1685.
6. Hildesheim,J. and Fornace Jr,A.J. (2002) Gadd45a: An Elusive Yet Attractive Candidate Gene in Pancreatic Cancer. *Clin. Cancer Res.* **8**, 2475–2479.
7. Newton,R., Aardema,M., and Aubrecht,J. (2004) The utility of DNA microarrays for characterizing genotoxicity. *Env. Health. Perspect.* **112**, 420–422
8. Walmsley,R.M. (2008) *GADD45a-GFP* GreenScreen HC genotoxicity screening assay. *Expert Opin. Drug Metabol. Toxicol.* **4**, 827–835.
9. Billinton,N., Hastwell,P.W., Beerens,D., Birrell,L., Ellis,P., Maskell,S., Webster,T.W., Windebank,S., Woestenborghs,F., Lynch,A.M., Scott,A.D., Tweats,D.J., van Gompel,J., Rees,R.W. and Walmsley,R.M. (2008) Interlaboratory assessment of the GreenScreen HC *GADD45a-GFP* genotoxicity screening assay: an enabling study for independent validation as an alternative method. *Mutat. Res.* **653**, 22–33.
10. Jagger,C., Tate,M., Cahill,P.A., Hughes,C., Knight,A.W., Billinton,N. and Walmsley,R.W. (2008) Assessment of the genotoxicity of S9-generated metabolites using the GreenScreen HC *GADD45a-GFP* assay. *Mutagenesis,* doi:10.1093/mutagen/050
11. Knight,A.W., Birrell,L. and Walmsley,R.M. (2008) Development and validation of a higher throughput screening approach to genotoxicity testing using the *GADD45a-GFP* GreenScreen HC assay. *J. Biomol. Screening,* in press.
12. Aubrecht J, Osowski JJ, Persaud P, Cheung JR, Ackerman J, Lopes SH, Ku WW (2007) Bioluminescent Salmonella reverse mutation assay: a screen for detecting mutagenicity with high throughput attributes. *Mutagenesis* **22**: 335–342.
13. Olaharski, A., Albertini, S., Kirchner, S., Platz, S., Uppal, H., Lin, H. and Kolaja, K.(2007).

Evaluation of the GreenScreen GADD45a-GFP indicator assay with non-proprietary and proprietary compounds. *Mutat. Res.:Genet. Toxicol. Environ. Mutagen.* doi:10.1016/j.mrgentox.2008.08.019

14. Walmsley, R.M. and Billinton, N. (2009) Genotoxic Carcinogen or not Genotoxic Carcinogen? That is the question. *Mutat, Res: Genet. Toxicol. Environ. Mutagen.* **672:** 17–19

Chapter 13

Real-Time Reverse-Transcription Polymerase Chain Reaction: Technical Considerations for Gene Expression Analysis

Shareen H. Doak and Zoulikha M. Zaïr

Abstract

The reverse transcription – polymerase chain reaction (RT-PCR) is a sensitive technique for the quantification of steady-state mRNA levels, particularly in samples with limited quantities of extracted RNA, or for analysis of low level transcripts. The procedure amplifies defined mRNA transcripts by taking advantage of retroviral enzymes with reverse transcriptase (RT) activity, coupled to PCR. The resultant PCR product concentration is directly proportional to the initial starting quantity of mRNA, therefore allowing quantification of gene expression by incorporation of a fluorescence detector for the appropriate amplicons. In this chapter, we describe a number of the most popular techniques for performing RT-PCR and detail the subsequent analysis methodologies required to interpret the resultant data in either a relative manner or through absolute quantification of gene expression levels.

Key words: Gene expression, Relative quantification, Absolute quantification, Real-time RT-PCR, Primer validation, SYBR Green, TaqMan

1. Introduction

1.1. Concept of RT-PCR

Only a small proportion of the genes in a cell are expressed at any given time, appropriate to the cell types' function and degree of growth or differentiation. The pattern of gene expression within a cell therefore changes over time in response to micro-environmental adaptations and the stage of cellular differentiation, responsible for specialising the various cell types that make up multi-cellular organisms. Both the external and internal cellular environments provide signals that co-ordinate the pattern of gene expression, thus investigations into the steady-state mRNA levels within a cell under differing conditions can provide an understanding of how a cell copes and

James M. Parry and Elizabeth M. Parry (eds.), *Genetic Toxicology: Principles and Methods*, Methods in Molecular Biology, vol. 817, DOI 10.1007/978-1-61779-421-6_13, © Springer Science+Business Media, LLC 2012

adapts to a continually shifting environment. The success of such analyses relies on accurate quantification of mRNA concentration as a measure of gene expression. Traditionally, gene expression analysis involved the use of northern blots. However, this technique requires large amounts of RNA (~10 μg total RNA) and is also time-consuming generating semi-quantitative data. The most popular method currently utilised is the reverse transcription – polymerase chain reaction (RT-PCR), whereby reverse transcriptase enzymes catalyse the synthesis of cDNA from mRNA, which subsequently acts as a template for the ensuing PCR reaction using primers specific to the gene of interest. RT-PCR is highly sensitive (detecting the presence of very rare transcripts) and only requires a small amount of starting RNA (>10 pg total RNA – SuperScript One-step RT-PCR kit from Invitrogen, Paisley, UK), due to the exponential amplification during PCR. Furthermore, the coupling of RT-PCR to real-time thermal cycling technology has also greatly enhanced the quantitative power of the technique.

1.2. Detection and Quantification of mRNA

RT-PCR has been used for mRNA quantification since 1989 (1, 2) but such studies were initially considered as unreliable. Experimental results were found to vary considerably both between different laboratories and upon repetition within the same facilities largely due to slight inconsistencies in post-PCR processing stages. Furthermore, the RT reaction has long been considered the source of most variability due to slightly differing efficiencies between reactions (3). Despite these initial differences being minute, exponential amplification by PCR results in significant accentuated differences in the final measurements, thereby confounding expression quantification.

The automation of real-time PCR technology has now omitted the need for post-PCR processing, thereby dramatically improving the reliability and reproducibility of RT-PCR. Real-time PCR relies on the detection and quantification of a fluorescent reporter that accumulates during the course of the PCR reaction in a directly proportional manner to amplicon generation. The technique therefore eliminates post-PCR processing for quantification of the amount of PCR product produced. The technology was first developed in 1992 when Higuchi et al. (4) followed a PCR reaction by measuring the amount of fluorescence released by ethidium bromide (a DNA intercalator) as the DNA was amplified. Now there are four main fluorescence systems utilised for automated real-time PCR using thermal cyclers with fluorescence detection capabilities (5):

1. Hybridisation Probes – this involves the use of two probes, one labelled with a fluorescein donor typically at the 3′ end, while the other has an acceptor fluorophore at its 5′ end. The two probes are designed to bind adjacent to one-another, on a specific PCR product. A fluorescent signal is only released

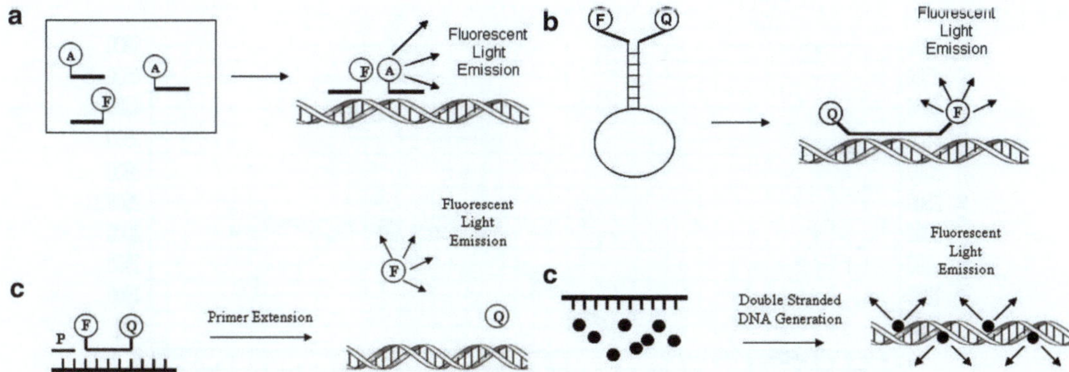

Fig. 1. (**a**) Hybridisation probes – a fluorescent signal is only released when the fluorescein donor (F) and acceptor (A) molecules are held in close proximity to each other. (**b**) Molecular Beacons – in the hairpin structure the fluorescence dye (F) is attenuated by the quencher molecule (Q) due to their close proximity, but upon hybridisation to the PCR product this structure is disrupted thus allowing the release of a fluorescent signal. (**c**) Hydrolysis probes – the 5′–3′ exonuclease activity of Taq polymerase (upon primer extension – P) separates the fluorescent dye (F) from the quencher molecule (Q), therefore permitting the release of subsequent signal. (**d**) SYBR Green – although the unbound dye releases a minimal amount of fluorescence, this emission is dramatically increased upon the dye intercalating with double stranded DNA.

when the two probes have hybridised, as at this time they are close enough to allow fluorescence resonance energy transfer (FRET) from the fluorescein donor to the acceptor, subsequently resulting in light emission (Fig. 1a).

2. Molecular Beacons – these are a form of hybridisation probe that are designed to form a hairpin structure with a stem and loop when in solution (Fig. 1b). These probes have a fluorescein molecule attached to one end, with a quencher at the other. In the hairpin structure, the quencher suppresses the fluorescent signal. However, upon binding to the specific, complementary PCR product, the two molecules are separated, permitting fluorescence release.
3. Hydrolysis Probes (also known as TaqMan probes) – these probes rely on the 5′–3′ exonuclease activity of Taq polymerase. They are oligonucleotides with a fluorescent dye at their 5′ end and a quencher dye attached to their 3′ base. Due to their close proximity, fluorescence is not emitted when the probe is intact. The probe is designed to anneal to an internal region within the specific PCR product along with the PCR primers, hence upon priming of the PCR reaction, the TaqMan probe is hydrolysed. The subsequent separation of the fluorescent dye and quencher allows the emission of a fluorescent signal (Fig. 1c).
4. SYBR Green – an intercalating dye that binds to double stranded DNA via the minor grooves and as a result of this interaction, releases a fluorescent signal (Fig. 1d). The distinct advantage in the use of this particular reporter is that it binds to all double-stranded DNA, therefore eliminating the need to

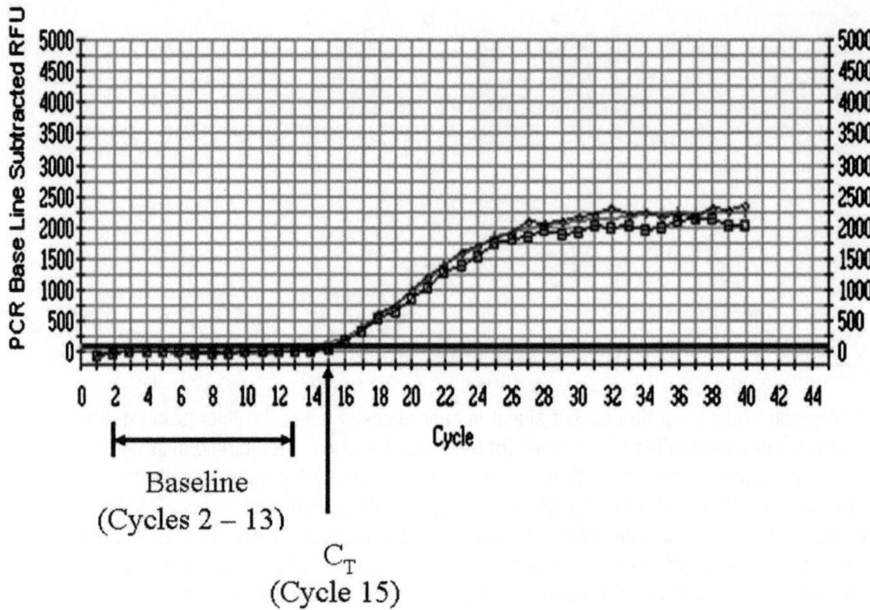

Fig. 2. Amplification plot for a sample in triplicate illustrating its C_T (threshold cycle), the plots *baseline range* and the position of the fixed threshold level (*horizontal line*).

design, produce, and optimise specific probes. However, this benefit is also its downfall as all non-specific double stranded DNA (including primer dimer) will be detected. To avoid false positive results, the PCR reaction must therefore be optimised to minimise primer dimer formation and avoid the generation of non-specific PCR products.

As the PCR product accumulates with each cycle, the fluorescent signal emitted from the chosen probe or dye increases in direct proportion and therefore can be detected and quantified by the automated real-time thermal cyclers. The increase in fluorescence is plotted against the cycle number (as shown in Fig. 2) to determine the threshold cycle (C_T) of the sample i.e., the point at which the fluorescent signal first significantly increases above background fluorescence. The higher the amount of PCR template in a reaction, the lower the number of cycles required before the fluorescent signal emitted first appreciably rises above the background (i.e. the lower the C_T value). Thus, C_T values can be used to infer the starting quantity of the template in the reaction, and thereby the degree of expression of the test gene. The sensitivity of fluorescence detection coupled to the high specificity of PCR in amplification of a single template therefore allows the detection of mRNA transcripts from a single, laser micro-dissected cell (6).

Real-time RT-PCR can be used to establish the patterns of gene expression either in an absolute or relative quantitative fashion. Absolute quantification measures the number of mRNA molecules in a sample by comparing the C_T value against a standard curve, while relative quantification ascertains the level of test mRNA expression in arbitrary units relative to a calibrator or control sample. In this chapter, we will provide the methodology required to enable analysis by both means.

Real-time RT-PCR technology is consequently a powerful tool and by removing the need for post-PCR processing, the technique has greatly enhanced the reliability, reproducibility, and accuracy of gene expression analysis, in addition to increasing sample throughput.

2. Materials

2.1. cDNA Synthesis

1. The user will need to supply the extracted RNA for study. RNA is prone to degradation, thus for long-term storage following extraction, it should be maintained at –80°C in aliquots that must not be freeze-thawed more than twice.
2. First strand cDNA synthesis kit (there are many commercial companies that now supply these kits (e.g. RETROscript kit, Ambion Ltd, Cambridgeshire, UK; SuperScript First-Strand Synthesis System, Invitrogen; DyNAmo cDNA Synthesis kit, Labtech Int Ltd, Sussex, UK; Reverse Transcription System, Promega Ltd, Southampton, UK).

2.2. Real-Time PCR

1. Custom PCR primers optimised for the amplification of the specific test gene under investigation will be needed – store in aliquots at –20°C to avoid repeated freeze-thawing.
2. iQ SYBR Green Supermix (BioRad, Hertfordshire, UK) – store at –20°C and thaw on ice immediately prior to use, additionally it is light sensitive so it must be stored in the dark and excessive lighting avoided when being used.
3. Sterile 96-well 0.2 ml PCR plates (BioRad).
4. Optical Quality Sealing Tape (BioRad).
5. TaqMan master mix (Applied Biosystems, Cheshire, UK) – store at 4°C.
6. TaqMan probe mix (Applied Biosystems) – store at –20°C and thaw on ice immediately prior to use.

3. Method

Several fluorescent detection methods exist for real-time PCR (as detailed in Subheading 1.2) all of which require generation and optimisation of probes. The exception, however, is SYBR Green dye, which due to its simplicity, ease of use and cost, is often utilised – but as it is not gene specific, users must be aware that it will bind to any double-stranded DNA present within the reaction including non-target amplicons and primer dimer. In this chapter, we will therefore detail both the use of SYBR Green and TaqMan probes. Additionally, we will highlight the validation steps necessary to avoid spurious or false results.

3.1. RNA Sample Preparation

3.1.1. RNA Integrity

Isolation of intact RNA is essential in generating good quality cDNA for RT-PCR detection. Several RNA extraction kits are commercially available, each one favourable to the cell or tissue type under analysis. Validation of your RNA integrity is highly recommended, especially when using a new RNA isolation methodology.

The most common method used to assess the integrity of total RNA is to run an aliquot of the RNA sample on a denaturing agarose gel stained with ethidium bromide (EtBr). Details on how to carry out gel electrophoresis can be found at www.protocol-online.org. Alternative nucleic acid stains include SYBRI Green II RNA gel stain (Molecular Probes, Invitrogen). Intact total RNA run on a denaturing gel will have sharp 28S and 18S ribosomal RNA (rRNA) bands for eukaryotic samples. Partially to completely degraded RNA will have a smeared appearance on an agarose gel; however, it is important to note that Poly(A) selected samples will not contain strong rRNA bands and will also appear as a smear from approximately 6 to 0.5 kb.

Automated systems also exist as an alternative to the agarose gel approach to determine RNA integrity. Various automated systems, such as the RNA 6000 LabChip® (Caliper Tech. Co., Chesire, UK) or Experion Automated Electrophoresis system (BioRad) are provided by a number of gene expression companies, all typically providing a good estimate of RNA concentration and purity in a sample (i.e. rRNA contamination in mRNA preparations). Alternatively, a spectrophotometric approach can also be used to ascertain the purity of an RNA sample, where their absorbance ratio at 260/280 nm, must lie within the limits of 1.7–2.2.

3.1.2. One-Step Versus Two-Step RT-PCR

Following total RNA extraction from the test sample, mRNA will need to be reverse transcribed into cDNA and this can either be performed as a one- or two-step RT-PCR reaction. With one-step RT-PCR, no separate cDNA synthesis stage is required. It therefore allows you to go straight from the reverse transcription (cDNA synthesis) incubation into PCR cycling without opening tubes or adding reagents. In contrast, two-step RT-PCR involves separate steps for reverse transcription, followed by the PCR reaction as an independent stage.

When working with multiple RNA samples, examining a small number of transcripts and where the quantity of RNA is not a limiting factor, one-step RT-PCR is likely to be the more suitable method to increase through-put. Additionally, one-step real-time RT-PCR is quicker to set up, less expensive to use, and involves reduced handling of samples, thereby minimising pipetting errors and other sources of error such as sample crossover or contamination. However, if limited RNA is available for a test sample and particularly if several genes need to be analysed and thus amplified from a single RNA source, two-step RT-PCR can be far more efficient because only a single reverse transcription reaction mix is required to facilitate multiple subsequent PCR reactions. Furthermore, two-step RT-PCR allows you to be flexible with the amount of reverse transcriptase you add to the reaction thereby providing the possibility of increasing cDNA synthesis, thus improving the flexibility and scope of further analysis.

Ultimately, the use of one-step versus two-step RT-PCR is dependent upon the end user and the samples they have to analyse, with the primary difference being the RT-PCR reaction kit that is purchased. For the most part of this chapter, however, we will refer to the two-step RT-PCR methodology.

3.1.3. cDNA Synthesis for Two-Step RT-PCR

Many commercial companies have cDNA synthesis kits available for this reaction (see Subheading 2.1). Manufacturers' instructions can simply be followed with these reverse transcription kits, but one consideration is the choice of primers to initiate this reaction. Oligo (dT) primers, random hexamers/decamers, or a mixture of all may be supplied. The most suitable primers to use at this stage will largely depend on the gene-specific primers needed to amplify the target sequence during the PCR phase of the technique and therefore needs to be optimised accordingly (see Notes 1 and 2). There are several pros and cons to consider when deciding on the primers to use at this stage. Gene-specific primers are generally preferred when analysing very rare messages, particularly as they considerably reduce background noise. These primers also allow one-step RT-PCR to be performed, i.e. the RT and PCR steps are performed together in the same tube, which minimises the risk of carry over contamination. However, the main disadvantage with this technique is that a separate RT reaction needs to be performed for each gene of interest, which might be problematic if only limited RNA is available. The alternative is use of non-specific primers such as oligo (dT) (anneals to the mRNA poly A tail) or random hexamers/decamers (oligonucleotide pools of 6–10 nucleotides, with all base combinations present). The use of oligo (dT) primers reduces background noise as only mRNA is reverse transcribed, but this cDNA generally displays 3′ end bias, with the transcription of large mRNA molecules rarely being completed. This can be overcome by using random hexamer/decamer primers; however, as they anneal to all RNA types, the background noise may be higher. Although the most suitable RT primer needed to satisfy an assays' requirements must be ascertained, a non-specific primed RT reaction allows the analysis of multiple mRNA transcripts from a single cDNA pool. This is particularly advantageous when the RNA sample available is limited. It is worth noting, however, that different methods of priming have been shown to provide different sensitivities and efficiencies (7). For instance, there is more linearity in a real-time run with template dilutions using gene-specific primers than using random hexamers (8). Therefore, choosing priming methods in two-step RT-PCR such as oligo dT or random hexamers that would allow for ease of use, may lead to altered real-time results (7).

Once synthesised, all cDNA preparations should either be used immediately or stored at –20°C for no more than 2-weeks. Repeated freeze-thawing must also be avoided; cDNA should not be used if it has been freeze-thawed more than twice to minimise artefacts caused by degradation.

3.2. Assay Design

3.2.1. Endogenous Controls

A prerequisite for accurate and reliable quantification of gene copy number is the normalisation of gene expression against an internal standard to compensate for slight variations in reaction efficiency, small sample-to-sample pipetting errors, or differences in RNA template starting quantities. Such internal standards include cellular RNA that simultaneously undergoes RT-PCR along with the gene of interest, and is subsequently used as a reference to which the target gene expression is normalised (9). Housekeeping genes are typically used for normalisation on the assumption that their expression levels of are not significantly altered in response to different experimental conditions. Recent data, nevertheless, are now emerging to show this not always the case (10). As a consequence, it is recommended that a range of endogenous control genes be tested for their suitability. This will involve quantifying the level of housekeeping gene expression by RT-PCR under intended experimental conditions and subsequently utilising bioinformatics software programs, such as geNorm and BestKeeper, to provide a statistical means of determining the most stable set of endogenous controls for your experimental assay. Where possible, multiple housekeeping genes should be used in parallel as internal standards.

3.2.2. PCR Primer Design

The specificity and efficiency of the PCR primers is a critical consideration. Various software programs can be used to help design your primers or alternatively "validated primers" are commercially available for most genes of interest. Below is a set of guidelines for primer design (11):

1. Forward and reverse PCR primers should have an equal melting temperature (T_m) of 58–60°C, and where necessary (i.e. fluorescence detection methods other than SYBR Green), probes should be designed with a T_m value of 10°C higher.
2. Primers should be 15–30 bases in length.

3. The G+C content should ideally be 30–80%.
4. The total number of Gs and Cs in the last five nucleotides at the 3′ end of the primer should not exceed two. This minimises relative instability at the 3′ end of primers to reduce non-specific priming.
5. PCR primers should be designed to amplify a product of 100–150 bp in length as small PCR products amplify with optimal efficiency (maximum amplicon size should not exceed 400 bp).
6. To avoid false-positive results due to amplification of contaminating genomic DNA in the cDNA preparation, it is preferable to have primers spanning exon–exon junctions in the cDNA sequence. This way, genomic DNA amplification will be minimised.
7. Where SYBR Green is being used, melt curve analysis should be included in the real-time PCR program at the end of the amplification stages to verify the purity of the resultant PCR products (see Note 3).

When considering whether to use SYBR Green or TaqMan, it should be noted that the SYBR Green assay only requires a validated gene-specific PCR primer pair in addition to the regular PCR components. The TaqMan chemistry utilizes FRET technology

Table 1
The advantages and disadvantages of SYBR Green and TaqMan amplicon detection methods

SYBR Green		TaqMan	
Pros	**Cons**	**Pros**	**Cons**
Preparation only requires a few days for primer design and validation	The double-stranded DNA binding property means that non-specific products and mRNAs with high sequence identity may be detected	Considered to be more sensitive when detecting low copy numbers (<10 copies) because of its ability to resolve the signal of a single copy of template (11)	TaqMan requires the additional synthesis of the dual-labeled probe after the validation of the potential primer set. Validation of primers and probes usually takes 2–3 weeks
Detection of the PCR product will occur at earlier cycles, which, is especially important in the case of low-abundance transcripts (>10 copies)		The probe offers additional specificity since the probe sequence exactly matches the target sequence	

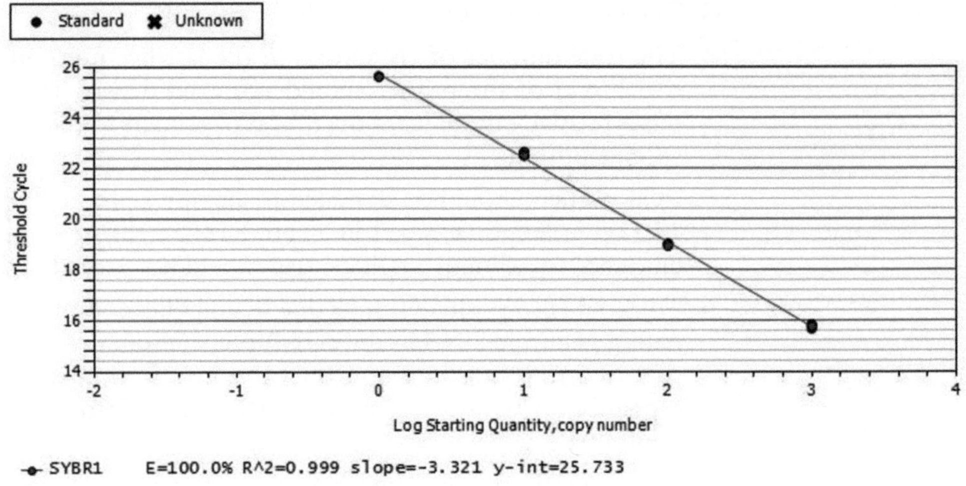

Fig. 3. Example of an ideal real-time PCR standard curve used to assess the amplification efficiency of a test primer set.

and requires the additional design of a probe (Fig. 1c). Table 1 outlines the advantages and disadvantages to using both types of detection methods with regards to specificity of subsequent PCR amplification.

3.2.3. PCR Efficiency

The individual PCR amplification efficiencies for each gene-specific primer set needs to be determined to ensure optimal reaction conditions prior to using them for gene expression analysis by RT-PCR, as differences in efficiency can dramatically confound the interpretation of consequent results. This is evaluated by performing real-time PCR on a tenfold dilution series of a template (10^0–10^{-4}) and subsequently plotting the dilution factor (or log starting quantity) against the threshold cycle (C_T) recorded for each sample (such outputs are usually accessible on the analysis software accompanying the real-time thermal cycler as shown in Fig. 3). The efficiency of each PCR reaction is determined by the following equation:

$$E = \left[10^{(-1/a)}\right] - 1,$$

a = slope of the standard curve.

An optimum amplification efficiency of 1.0 is obtained when the standard curve slope is −3.3. Only PCR efficiency values of E = 0.9–1.1 (i.e. >90% efficiency) are therefore considered suitable.

Furthermore, it is vital that the primer sets for the test genes have the same PCR efficiency as those specific to the internal standards (housekeeping genes), particularly when using the relative quantification method for data analysis (Subheading 3.4.1) as this equation assumes equal efficiency. If in practice this is not the case,

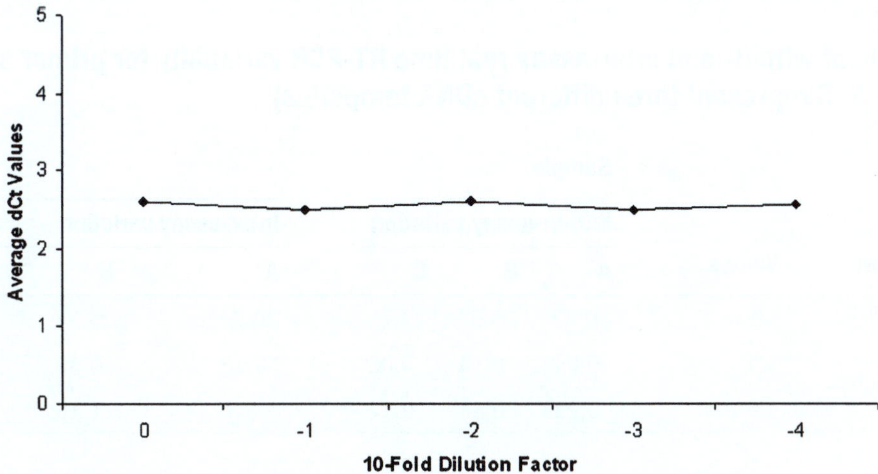

Fig. 4. Validation experiment demonstrating the relative efficiency plot comparing a test to internal standard primer set. The variation in ΔC_T values with template dilution was investigated and the resultant *graph slope* is 0.006. Hence, the primer efficiencies were approximately equal and can be used for relative quantification using the $2^{-\Delta\Delta C_T}$ method (Subheading 3.4.1).

then substantial errors in interpreting gene expression patterns can arise. If the efficiencies of two primers sets are approximately equal then there should be no difference in their ΔC_T value with template dilution (12), where:

$$\Delta C_T = C_{T(\text{test gene})} - C_{T(\text{internal standard})}.$$

Thus, to test the comparative efficiency of two primer sets, a tenfold dilution series (10^0–10^{-4}) of three individually synthesised cDNA templates from different RNA extractions need to be produced. Real-time PCR must then be performed for each of the three dilution series, with both the test and internal standard primer sets. Based on the resultant C_T values, the ΔC_T at each dilution can be calculated using the above equation and then a graph of the average ΔC_T (from the three cDNA templates) against log template dilution input needs to be plotted as shown in Fig. 4. Hence, for a test and internal standard gene primer set to be considered equally efficient at amplifying their respective target sites and therefore suitable for relative quantification using real-time RT-PCR and the $2^{-\Delta\Delta C_T}$ method (as described in Subheading 3.4.1), the slope of the resultant graph must be <0.1.

3.2.4. Establishing Assay Variation

To establish experimental reproducibility, within-assay and inter-assay variation in the real-time RT-PCR data generated will need to be assessed with each primer set utilised. This is evaluated by performing five repeats of the real-time PCR assay (as described in Subheading 3.3) using cDNA template originating from three different RNA extractions.

Table 2
Example of within- and inter-assay real time RT-PCR variability for primer set *X* (where A–C represent three different cDNA templates)

		Sample					
		Within-assay variation			Inter-assay variation		
Primer set	Values	A	B	C	A	B	C
X	Mean C_T	20.07	25.55	22.75	20.94	23.77	24.79
	SD	0.04	0.14	0.05	0.46	0.35	0.29
	CV (%)	0.22	0.55	0.22	2.18	1.46	1.16

Within-assay variability is determined by running all five repetitions of each cDNA sample on the same real-time PCR plate, while inter-assay variation is based on the results generated by performing the five repeats of the three cDNA templates over an 8-day period, with each replicate on a different day (see Note 4).

The within- and inter-assay variability is described by their coefficient of variation (CV), where CV = (Standard Deviation/Mean C_T) × 100. The mean C_T and corresponding standard deviation are calculated from the raw C_T data from the five replicates per sample. Hence, as three cDNA templates are assessed, CV is usually described as a range. For example, using the data displayed in Table 2, the CV for within-assay and inter-assay variation for real time RT-PCR with primer set *X* was 0.22–0.55% and 1.16–2.18% respectively.

3.3. Real-Time RT-PCR Reaction Set-up (SYBR Green and TaqMan Probes)

All the reactions must be set up under aseptic conditions in laminar flow fume hoods using sterile, nuclease-free, filtered pipette tips. The fume hood and all equipment to go into it (e.g. pipettes, filter tip boxes, etc.) should be cleaned with 70% ethanol prior to use and avoid opening tip boxes outside of the hood. Additionally, due to the light-sensitive nature of the SYBR Green and fluorescence-based probes, the reaction set-up should be performed with minimal lighting. The components of each individual real-time PCR reaction are outlined in Table 3.

A master mix containing either the SYBR Green Supermix or TaqMan mix, primers and water should first be made for all reactions requiring the same components within the 96-well plate, as depicted in Fig. 5. These master mixes then need to be sub-divided into aliquots for three reactions as each individual sample must be run in triplicate. The appropriate volume of sample cDNA to serve three replicate reactions must then be added to each appropriate sub-master mix and 25 μl of each resultant mix is aliquoted into the wells of a sterile 96-well 0.2 ml PCR plate ensuring that the exact same component quantities are present in each triplicate (Fig. 6). Negative

Table 3
Real time PCR reaction components for two-step RT-PCR

SYBR Green	TaqMan
12.5 μl iQ SYBR Green Supermix	12.5 μl TaqMan master mix (final dilution of 1×)
1 μl Forward primer (0.2 μM final concentration)	1.25 μl TaqMan probe mix (final dilution of 1×)
1 μl Reverse primer (0.2 μM final concentration)	
1–3 μl Sample cDNA	1–3 μl sample cDNA
Water to a final 25 μl volume	Water to a final 25 μl volume

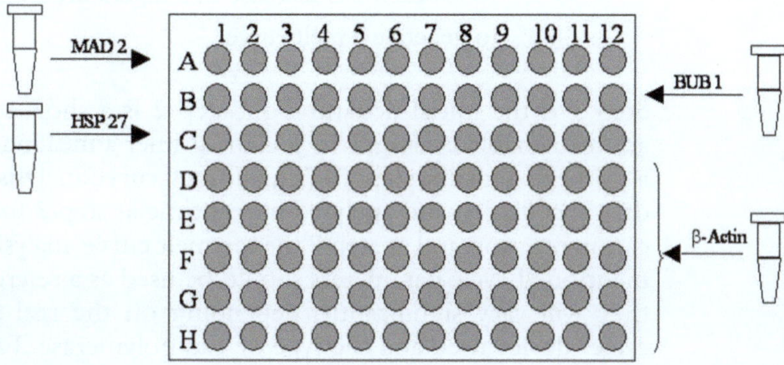

Fig. 5. A master mix for all samples containing the same reagents is generated to ensure all samples contain equal quantities of all required components. In this example, MAD2, BUB1 and HSP27 are the test genes, while β-actin is the housekeeping gene.

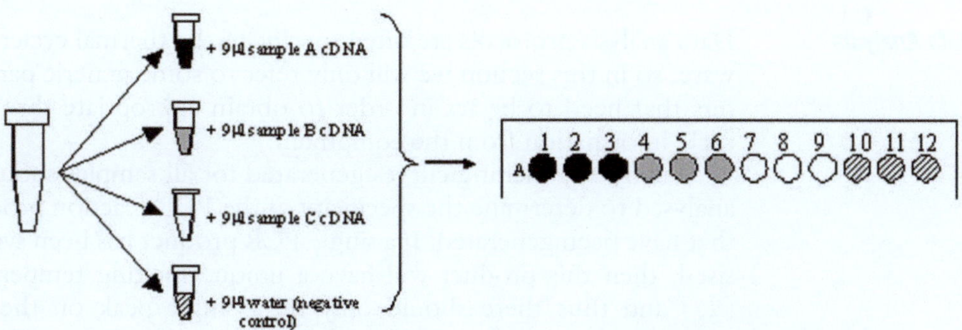

Fig. 6. A sub-master mix containing the components for three reactions per sample plus 9 μl cDNA (3 μl per reaction) is split between three wells, thus ensuring each reaction triplicate contained the same reagent quantities.

controls must also be included on the 96-well plate for each primer pair, where the cDNA is replaced with water (see Note 5).

Once all 25 μl reactions are loaded into the wells, the plate is sealed with Optical Quality Sealing Tape and briefly centrifuged to collect all the contents in the bottom of the wells. The sample plate is then inserted into the real-time thermal cycler and run on the appropriate program optimised for the gene-specific primers (see Note 6). An example of such a program is:

1. 95°C for 3 min.
2. 94°C for 30s.
3. 60°C for 30s.
4. 72°C for 30s.

(steps 2–4: × 40 cycles.)

5. 55°C for 30s.
6. 95°C for 30s.
7. 10s at each 1°C increase in temperature from 55 to 95°C to generate a melt curve.

Step 1 is the initial denaturation; step 2 is a short-term denaturation to enable cycling; step 3 is the primer annealing stage; step 4 is primer extension; steps 5–7 enable melt curve analysis. Fluorescent data is collected and analysed in real time at step 3 for the amplification reaction and at step 7 for the melt curve analysis. The aforementioned cycle parameters should be used as a reference guide as they will vary significantly, depending on the real-time thermal cycler device used and the type of Taq polymerase. Fast-PCR runs can be performed using specially engineered Taq polymerase that can reduce the total reaction time to 45 min. However, other items such as extra thin walled 96-well plates and quick-temperature changing ramping blocks also must be used in order to achieve this reduced reaction time.

3.4. Data Analysis

Data analysis protocols are often specific to the thermal cycler software, so in this section we will only refer to some generic parameters that need to be set in order to obtain appropriate threshold cycle information from the equipment.

Firstly, the melting curves generated for all samples should be analysed to determine the specificity of the PCR reaction products that have been generated. If a single PCR product has been synthesised, then this product will have a unique melting temperature (T_m) and thus there should only be a single peak on the plot (Fig. 7). However, if inefficient primer annealing has resulted in the generation of several PCR products or primer dimer, then multiple peaks may be seen (Fig. 8) and such samples will need to be removed from subsequent analysis.

PCR baseline subtraction then needs to be selected; the baseline for the amplification plot is defined as the range of cycles during which the detection system measured no target amplification above the background signal (illustrated in Fig. 2). This range is

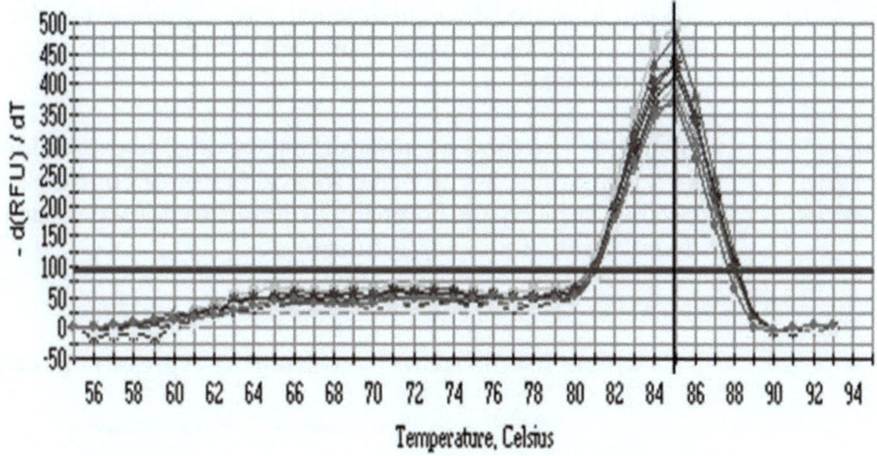

Fig. 7. Example of a melt curve for a PCR product where all reactions have the correct T_m.

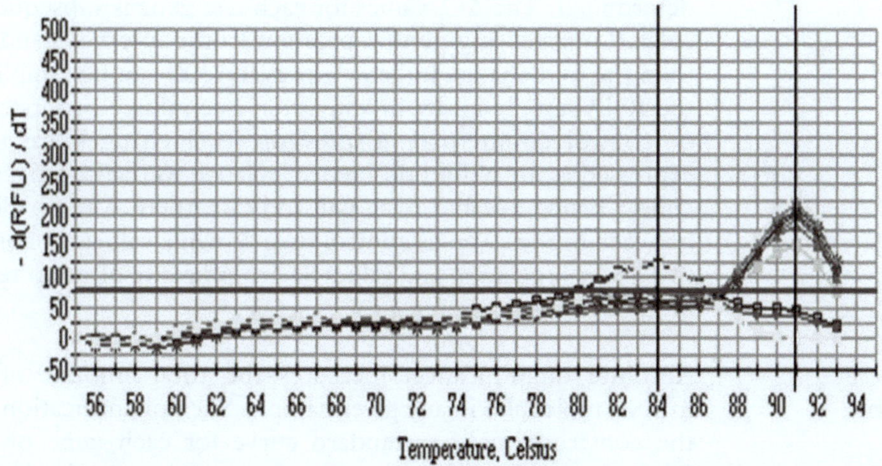

Fig. 8. Example of a melt curve demonstrating the presence of non-specific amplicons where different PCR products that have a T_m 84 or 91°C are synthesised.

set from the second cycle to the point two cycles earlier than the C_T value of the most abundant sample. The software then sets a fixed fluorescence threshold level at a statistically significant point above the baseline. This threshold is calculated by averaging the standard deviation of fluorescence signals over the baseline cycle range and then multiplying by an adjustable factor of 10 (i.e. the threshold is 10 SD from the baseline).

The next stage in the data analysis requires the manual review of all resultant C_T values. Samples should be discarded if the C_T is lower than 10 or >35 cycles. Additionally, it is important to compare the C_T values in each triplicate reaction to identify any outliers that will need to be removed from the analysis as the invalid data (possibly caused by experimental error) could skew the results – there should be no more than one C_T difference between the sample triplicates.

3.4.1. Relative Quantification Using the $2^{-\Delta\Delta C_T}$ Method

The $2^{-\Delta\Delta C_T}$ method quantifies the level of gene expression changes in a test sample as compared to a calibrator or control sample, in a relative fashion, e.g. comparing tumour to normal tissue (12). It is based on the following formula:

$$X = 2^{-\Delta\Delta C_T},$$

where: X = the factor by which gene expression has altered.

$$\Delta\Delta C_T = \Delta C_{T(\text{Diseased Tissue})} - \Delta C_{T(\text{Normal Tissue})}.$$

$$\Delta C_T = C_{T(\text{Test Gene})} - C_{T(\text{Internal Standard Gene})}.$$

Hence, for each sample (run in triplicate) the mean of the three C_T values obtained for the test and internal standard genes are determined. The ΔC_T values for each test gene is subsequently calculated normalising gene expression to the internal standard, and then the fold change in gene expression (X) can be established for the RNA sample under investigation as compared to its calibrator or control counterpart. Expression level changes are therefore expressed as an N-fold difference (relative to the internal standard gene); hence a value of 1 will indicate there was no difference between the test versus control tissues, while values <1 correspond to a down-regulation and values >1 are indicative of an up-regulation in gene expression.

3.4.2. Absolute Quantification

Absolute quantification measures the total amount of specific mRNA molecules in any given sample. Such quantification requires the construction of a standard curve for each gene of interest. Different sources of starting template may be used in the production of a standard curve from which to extrapolate absolute quantification. Several sources of DNA standards are used, including recombinant DNA, genomic DNA and commercially synthesized oligonucleotides. Here we shall describe the three most common approaches.

cRNA Standard

Synthetic construction of a DNA fragment, made via in vitro transcription, provides a means to labelling the newly formed cRNA with a fluorescent probe. In preparing cRNA templates, total RNA must first be transcribed to cDNA, as described in Subheading 3.1.3. cDNA amplification is performed with target specific primers modified to contain a T7-promoter sequence at the 5′ region of the 5′ primer and an oligo-dT sequence at the 5′ end of the 3′ primer (Sigma-Genosys, St. Louis, USA). Incorporation of the T7 promoter is required for in vitro transcription while oligo-dT, at the 3′ primer end generates a cRNA with a poly(dA) tail. In vitro transcription

(Promega Ltd) generate cRNA standards that should be aliquoted and stored at −80°C when not in use. When setting up the RT-PCR reaction, quantification of cRNAs should be performed in triplicate and converted to the molecule number using the following formula:

$$N(\text{molecules}/\mu\text{l}) = \frac{C(\text{cRNA}\mu\text{g}/\mu\text{l})}{K(\text{fragmentsize/bp})} 182.5\times1013.$$

The formula shown above gives the molecules per microliter (N), if the concentration of the cRNA (C) is known in relation to the fragment size (K) multiplied by a factor derived from the molecular mass and the Avogadro constant (13).

recDNA Standard

Generation of cloned recDNA provides an alternative template for the production of a standard curve. In this instance, the gene of interest is PCR amplified as a standard and cloned into an expression vector (Promega Ltd). With the molecular weight of the plasmid and insert known, it is possible to calculate the copy number as follows (14):

Weight in daltons (g/mol) = bp size of ds product (330 Da × 2 nt/bp).

Hence: (g/mol) = Avogadro's number = g/molecule = copy number (where: bp = base pairs, ds = double-stranded and nt = nucleotides).

Generation of a Standard Curve

Either standard described above may be utilised to generate a standard curve from which to calculate absolute levels of your gene of interest. recDNA standards are highly stable, generating reproducible standard curves in comparison cRNA. Furthermore, the longer templates derived from recDNA and genomic DNA mimic the average native mRNA. While cRNA standards are more prone to degradation or cleavage, they, nevertheless, are more relevant in their use since cRNAs take into account the conversion efficiency of mRNA to cDNA. Irrespectively, the calibration curve may be created by plotting the threshold cycle (C_T) corresponding to each standard, versus the value of their corresponding log number of test concentration. This curve is used as a reference standard for extrapolating quantitative information for mRNA targets of unknown concentrations by amplifying a dilution series corresponding to the target gene C_T values. An example of this is shown in Fig. 9.

4. Notes

1. If oligo(dT) primers are preferred for the cDNA synthesis step, then the gene-specific primers for subsequent PCR should by

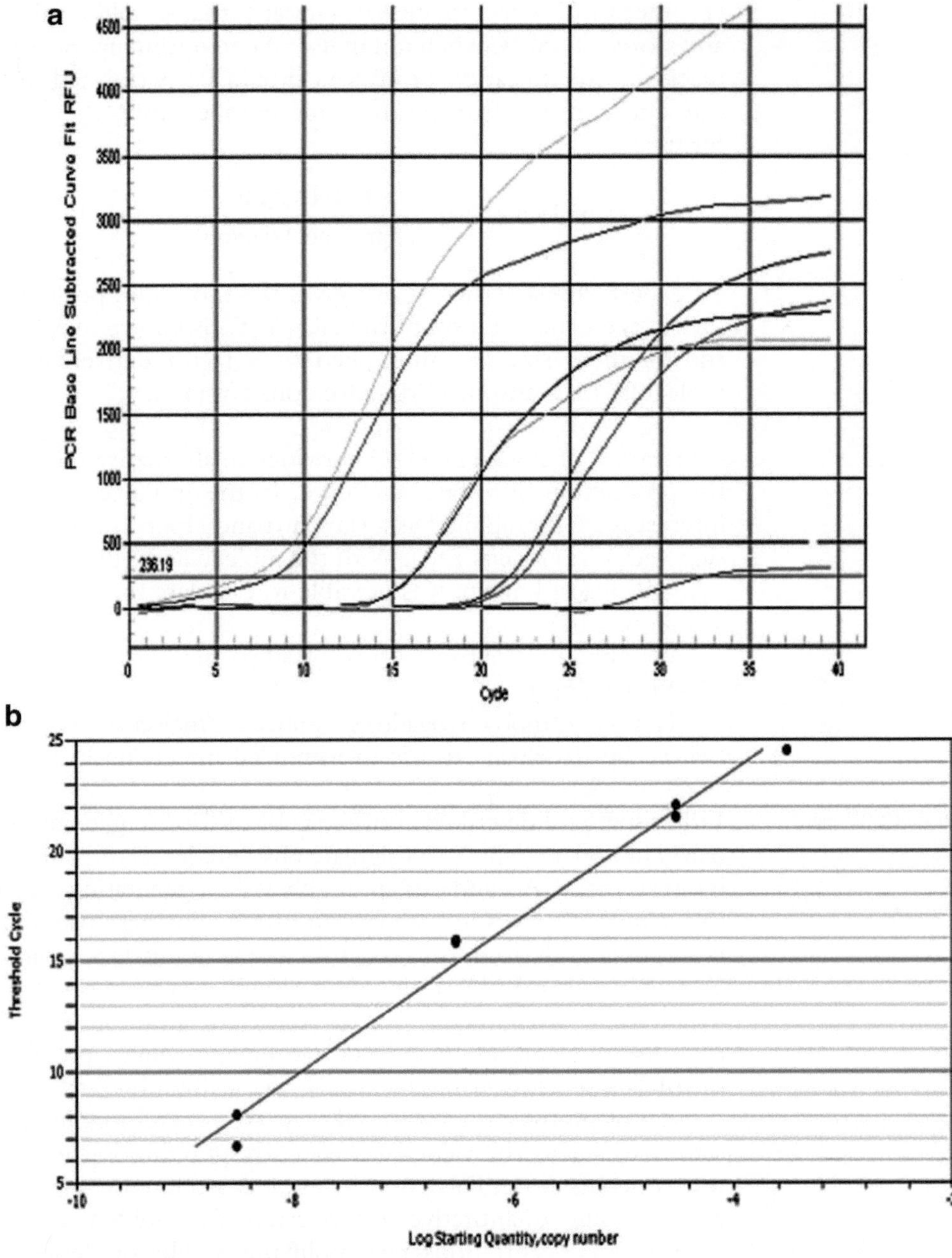

Fig. 9. Generation of standard curves for absolute quantification. (**a**) Amplification plot showing the threshold cycle of each standard used. (**b**) Standard curve generated by plotting the threshold cycle values against the known concentration of starting material. Threshold cycles obtained from unknown samples can be plotted along the standard curve and their concentration extrapolated.

selected at positions no more than 2 kb upstream from the polyA tail, as this region is the most efficiently reverse transcribed when using oligo (dT) primers. Additionally, the forward and reverse primers should be located in separate, adjacent exons so as to differentiate between PCR products generated from cDNA and contaminating genomic DNA.

2. To determine the most suitable primers for the cDNA synthesis stage, it is recommended that separate cDNA reactions are performed using all primers provided individually. Each of the resultant cDNA products should then be used as the template in a PCR reaction with the gene-specific primers. The subsequent PCR products will then need to be separated on a 6% polyacrylamide gel for visualisation of the banding pattern – the cDNA template which results in a single strong band of the appropriate size (with no evidence of multiple bands or smears) is the most suitable.
3. Melt curve analysis involves gradually increasing the temperature to determine the melting temperature (T_m) of the double-stranded DNA molecules present in the reaction. SYBR Green only binds to double-stranded DNA, so when the T_m is reached, there is a dramatic decrease in fluorescent signal as the double-stranded DNA denatures. The T_m of a molecule is characteristic as it is dependent upon variables such as the GC content, sequence and length, hence there will only be a single peak at one temperature if a single PCR product is present in the reaction. This analysis therefore allows the identification of any reactions containing non-specific products.
4. It is imperative that sufficient cDNA is synthesised to serve the entire variation experiment to ensure that exactly the same cDNA pool is used for each of the ten replicates (five each for the within- and inter-assay variability determinations). It is therefore recommended that an appropriate quantity of cDNA is synthesised and stored in aliquots that are only freeze-thawed once. This will minimise variation associated with different reverse transcription reactions, thereby ensuring that the resultant CV is directly related to the reproducibility of the real-time PCR procedure alone.
5. All reaction components including the 96-well plate should be kept on ice during the entire set-up process when loading a real-time reaction.
6. When using multiple primer sets within the same plate, it is important to bear in mind that they must all perform at their optimal efficiency when using the same PCR program; in particular, they must require the same annealing temperature.

References

1. Becker-Andre, M. and Hahlbrock, K. (1989) Absolute mRNA quantitation using the polymerase chain reaction (PCR). A novel approach by a PCR aided transcriptional titration assay (PATTY). *Nucleic Acid. Res.* **17**, 9437–9446.
2. Wang, A.M., Doyle, M.V. and Mark, D.F. (1989) Quantitation of mRNA by the polymerase chain reaction. *PNAS* **86**, 9717–9721.
3. Freeman, W.M., Walker, S.J. and Vrana, K.E. (1999) Quantitative RT-PCR: Pitfalls and potential. *BioTechniques* **26**, 112–125.
4. Higuchi, R., Dollinger, G., Walsh, P.S. and Griffith, R. (1992) Simultaneous amplification and detection of specific DNA sequences. *BioTechnology* **10**, 413–417.
5. Bustin, S.A. (2000) Absolute quantification of mRNA using real-time reverse transcription polymerase chain reaction assays. *J. Mol. Endocrinol.* **25**, 169–193.
6. Bustin, S.A. (2002) Quantification of mRNA using real-time reverse transcription PCR (RT-PCR): trends and problems. *J. Mol. Endocrinol.* **29**, 23–39.
7. Wacker, M.J. and M.P. (2005) Godard, Analysis of one-step and two-step real-time RT-PCR using SuperScript III. *J. Biomol. Tech.* **16**, 266–271.
8. Bustin, S.A., Benes, V., Nolan, T. and Pfaffl, M. W. (2005) Quantitative real-time RT-PCR – a perspective. *J. Mol. Endocrinol.* **34**, 597–601.
9. Karge, W.H. Schaefer, E.J. and Ordovas J.M. (1998) Quantification of mRNA by polymerase chain reaction (PCR) using an internal standard and a nonradioactive detection method. *Methods Mol. Biol.* **110**, 43–61.
10. Morse, D.L., Carroll, D., Weberg, L., Borgstrom, M. C., Ranger-Moore, J. and Gillies, R. J. (2005) Determining suitable internal standards for mRNA quantification of increasing cancer progression in human breast cells by real-time reverse transcriptase polymerase chain reaction. *Anal. Biochem.* **342**, 69–77.
11. Dorak MT (Ed) (2006) *Real-Time PCR (Advanced Methods Series)*. Oxford: Taylor & Francis.
12. Livak, K.J. and Schmittgen, T.D. (2001) Analysis of relative gene expression data using real time quantitative PCR and the 2^{-66Ct} method. *Methods* **25**, 402–408.
13. Fronhoffs, S., Totzke, G., Stier, S., Wernert, N., Rothe, M., Bruning, T., Koch, B., Sachinidis, A., Vetter, H. and Ko, Y. (2002) A method for the rapid construction of cRNA standard curves in quantitative real-time reverse transcription polymerase chain reaction. *Mol. Cell. Probes* **16**, 99–110.
14. Whelan, J.A., Russell N.B. and Whelan M.A. (2003) A method for the absolute quantification of cDNA using real-time PCR. *J. Immunol. Methods* **278**, 261–269.

Chapter 14

Cytogenetic In Vivo Assays in Somatic Cells

Ann T. Doherty, Adi Baumgartner, and Diana Anderson

Abstract

Chromosome aberration assays are employed to detect the induction of chromosome breakage (clastogenesis) in somatic and germ cells by direct observation of the chromosomal damage during metaphase analysis, or by indirect observation of chromosomal fragments. Thus, various types of cytogenetic change can be detected such as structural chromosome aberrations (CA), sister chromatid exchanges (SCE), ploidy changes, and micronuclei. Following the induction of the chromosomal damage, most of the aberrations and abnormalities detected by these assays can be detrimental or even lethal to the cell. Their presence, however, indicates a potential to also induce more subtle and therefore transmissible chromosomal damage which survives cell division to produce heritable cytogenetic changes. Usually, induced cytogenetic damage is accompanied by other genotoxic damage such as gene mutations.

Key words: Cytogenetics, In vivo assays, Somatic cells, Bone marrow micronucleus test, Chromosome aberration test

1. Introduction

There were attempts in the late nineteenth century (1) to identify human chromosomes, but the correct number for diploid cells in man was established as 46 chromosomes in 1956 (2, 3). Within the next few years numerical abnormalities in human syndromes were cytogenetically identified (4–6) and the well known Philadelphia chromosome, a tumour marker for myeloid leukaemia, was characterised (7). Using better and more specific DNA stains, the next step in cytogenetic analysis was the determination of the full human karyotype in 1970 (8). Then, in the 1970s, Giemsa banding replaced earlier techniques (9, 10), leading to a significantly improved resolution of chromosomal bands which identified the Philadelphia chromosome as a translocation t(9, 22) (11). Chemical mutagenesis and effects of radiation were often

James M. Parry and Elizabeth M. Parry (eds.), *Genetic Toxicology: Principles and Methods*, Methods in Molecular Biology, vol. 817,
DOI 10.1007/978-1-61779-421-6_14, © Springer Science+Business Media, LLC 2012

studied in the early 1970s via the in vivo induction of micronuclei in the bone marrow (12, 13), which resulted in the development of the cytochalasin B micronucleus (CBMN) test (14). More than half a century after the correct chromosome number in humans was announced, Giemsa staining of chromosomes and micronuclei has remained essential in cytogenetic procedures. Despite modern techniques based on fluorescence in situ hybridisation (15), these earlier techniques are rapid and still very useful for the screening of chromosomal damage.

1.1. Types and Consequences of Cytogenetic Damage at the Individual Chromosome and Chromosome Set Level

Chromosome defects can arise at the level of the individual chromosome or at the level of the chromosomal set affecting the number of chromosomes. In humans for instance, over 80% of all structural aberrations occur de novo and are of paternal origin (16). Numerical abnormalities as well as structural chromosomal aberrations are involved in the aetiology of neoplasia in somatic cells (17); while in germ cells they can lead to perinatal mortality, dominant lethality, or congenital malformations in the offspring (18–20) and to congenital tumours (19, 21). Individual chromosome damage consists of breakage of chromatids due to the direct or indirect action of intrinsic or exogenous genotoxins. Reactive oxygen and/or nitrogen species are important for the initial DNA lesion by directly producing oxidative DNA damage or indirectly by forming DNA adducts via oxidised lipids (22). In contrast to eukaryotes, DNA double-strand breaks in bacteria or other haploid organisms are usually fatal as they do not have the capability to reconstitute the breaks via non-homologous end-joining or direct repair of breaks, resulting in no apparent cytogenetic damage (23). However, unrepaired or mis-repaired breaks may very well result in chromosomal damage (24) and consequently most of these aberrations may contribute to mutagenesis and carcinogenesis (25). DNA damage may even result in cell death (apoptosis) at the next or following mitoses. If, for example, unrepaired chromosomal fragments are introduced into the zygote via an affected germ cell, the embryo may die at a very early stage from a dominant lethal mutation (26), although up to 8% sperm DNA damage can be tolerated by the oocytes due to their repair capacity (27).

Open DNA ends at chromosome breaks are exposed to nucleolytic attacks (28) and prone to loss of genetic material, hence the necessity to efficiently repair at any cost. When chromosome breaks are then rejoined in a different order from the original one, chromosomal rearrangements are the consequence (29). Hence, chromosome-type and chromatid-type aberrations and micronuclei but not sister chromatid exchanges can predict cancer risk (30, 31). There are various types of chromosomal rearrangements (32) depending on the rearrangement and the position of the centromeres (see also Fig. 1):

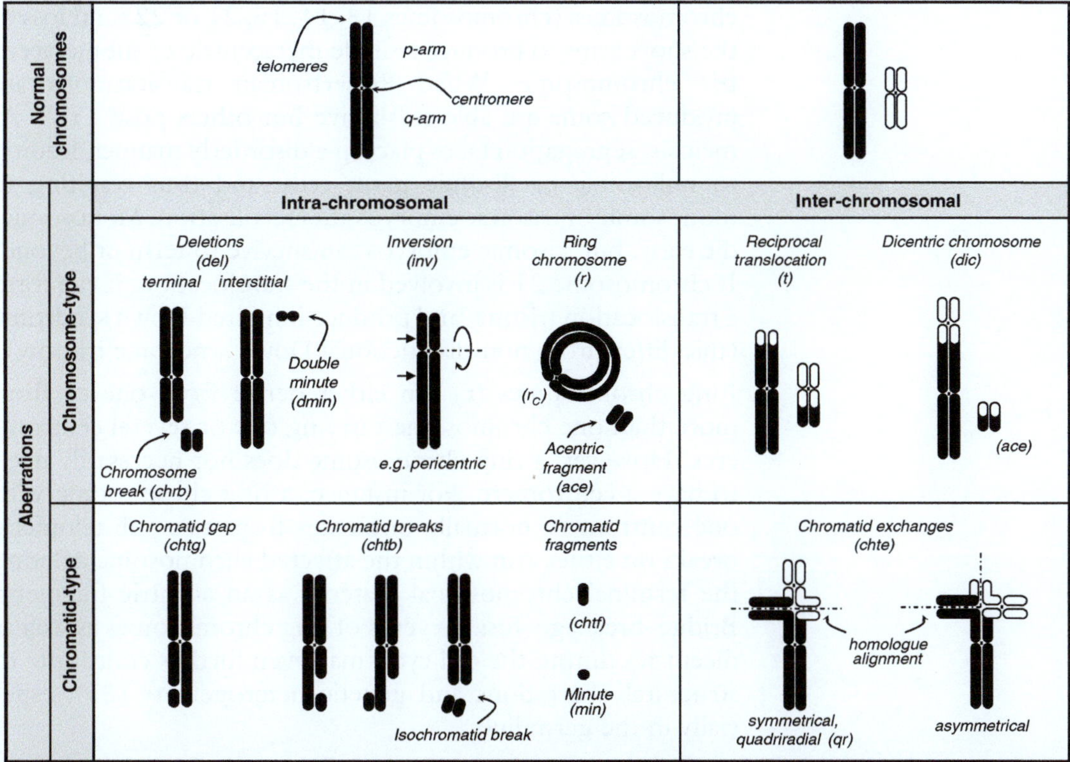

Fig. 1. Overview of frequent chromosome-type and chromatid-type aberrations, which can occur within a single chromosome (intra-chromosomal) or between two or even more chromosomes (inter-chromosomal) after chromosome breakage (adapted from Bauchinger (109)). A more detailed description of various other aberrations and their abbreviations can be found in An International System for Human Cytogenetic Nomenclature (ISCN) (32).

1. Reciprocal translocations (t) are symmetrical exchanges which occur when each rearranged chromosome carries just one centromere. This allows the zygote to develop normally, but when such heterozygotes form germ cells at meiosis, about half of their gametes will be genetically unbalanced, since they are partly mono- or trisomic (33). The unbalanced gametes which survive produce unbalanced zygotes, which result in death shortly before and after birth, or congenital malformations (34).
2. Asymmetrical exchanges arise when one of the rearranged chromosomes carries both centromeres leaving in most cases an acentric (ace) fragment. Somatic or germ cells carrying such a dicentric (dic) chromosome usually die due to segregation difficulties at cell division contributing to dominant lethality. However, cells carrying a dicentric could in ≤50% of the cases progress past mitosis causing various other disruptions via bridge–breakage–fusion events (35, 36).
3. Robertsonian translocations (rob), very common in humans (37), originate through centric fusion of two acrocentric

chromosomes (chromosomes 13, 14, 15, 21 or 22) and loss of the short arms to produce a single metacentric or submetacentric chromosome. When Robertsonian translocations are produced some are able to survive but others pose a risk. At meiosis, segregation takes place in a disorderly manner, leading to nullisomic or disomic germ cells, and thus resulting in monosomic or trisomic embryos after fertilisation. Monosomics die early, but trisomic embryos can survive to term or beyond. If chromosome 21 is involved in the translocation, it can form a translocation trisomy and produce inherited Down syndrome (this differs from non-disjunctional Down syndrome trisomy).

4. Ring chromosomes (r) can either derive from one or from more than one chromosome carrying one or several centromeres. However, a ring chromosome does not necessarily need to have a centromere. For instance, a ring chromosome with one centromere normally originates from two sub-telomeric breaks on either arm within the affected chromosome deleting the terminal chromosomal material as an acentric fragment. Bridge–breakage–fusion events of ring chromosomes as well as dicentrics during the cell cycle may even further contribute to structural aberrations and genetic heterogeneity (35), especially in the germ line.
5. Inversions (inv) arise when two breaks occur in the same chromosome. For paracentric inversions these breaks are located on the same arm as for pericentric inversions the breaks occur on either side of the centromere. The portion between the two breaks is detached and becomes reinserted in the opposite way to its original position, i.e. the gene order is reversed. This need not cause a genetic problem in somatic cells, but imbalanced gametes could result in congenital malformation or foetal death.
6. Deletions (del) and deficiencies (loss of chromosomal material) are produced interstitially when two breaks arise close together in the same chromosome. The two ends of the chromosome join when the fragment between the breaks becomes detached. Terminal deletions, on the contrary, can cause the loss of a chromosome end at the next cell division. Large deletions may contribute to dominant lethality, while small cryptic deletions are difficult to distinguish from point mutations. Deletions may uncover pre-existing recessive gene.

Different types of cytogenetic events can lead to neoplasia in humans, such as reciprocal translocations or non-reciprocal rearrangements with chromatin loss, deletions or duplication of whole chromosomes or chromosome segments (38–40). Cytogenetic aberrations have been found in over 45,000 human neoplasms and tumours linked to specific structural balanced rearrangements (41, 42). Such changes can cause the elimination of tumour suppressor genes, resulting in malignancy, or activation of proto-oncogenes (43, 44). The mechanisms by which proto-oncogenes become activated to their oncogenic forms in tumour cells include single-point mutations, deletions, inversions, and translocations of gene material between chromosomes, creating for instance fusion genes as well as gene amplification (39, 42, 45, 46). The consequence of this genetic change may be the altered production of an otherwise normal gene product (47). This can occur either by increasing the expression rate of the gene (transcriptional activation) or by post-transcriptional stabilisation of the messenger RNA or the final protein product (48–50). Initiating dysregulation of the proto-oncogene expression (51) also causes elevated levels of protein or inability to switch off the gene at the appropriate time and finally uncontrolled cell proliferation (52). Thus, chromosome rearrangements may induce neoplasia by activating a potential oncogene that is a proto-oncogene (53). Chronic myeloid leukaemia cases may carry a marker called the "Philadelphia chromosome", which is a derivative chromosome 22 originating from a reciprocal t(9, 22) (q34;q11) translocation (32, 54). As a consequence the ABL1 (= c-ABL) proto-oncogene on chromosome 9 (9q34.1) becomes joined with the BCR region on chromosome 22 (22q11.23) resulting in the expression of a fusion protein encoded by both DNA sequences (54, 55). The exact mechanism of this fusion protein causing cell transformation was poorly understood (56); however, in recent years the molecular underlying mechanism became clear and lead to the development of specific BCR-ABL tyrosine kinase inhibitors, which allow a targeted approach in treating CML (57–59).

The tumour state can also be inherited. Most hereditary predispositions affect only one particular cell type; however, the genes with the relevant germ-line mutation are not cell-type specific and thus increase the risk to unrelated cell types (60). Retinoblastomas and osteosarcomas, for instance, develop in children who inherit a defective chromosome 13 from one parent. The defect has been shown to involve a deletion or mutation in the RB1 gene at the 13q14.2 locus (61, 62). Tumours arise when the normal copy of the same gene on the other chromosomes is lost or mutated in early childhood. This demonstrates that the gene when present in a functional state has suppressive effects on the development of tumours. Genes on chromosome 11 (e.g. WT1 at 11p13 or H19 at 11p15.5) are also frequently lost or epigenetically modified and thus silenced in tumours or in children with the Tay-Sachs familial variant of Wilms' tumour (63, 64). The predisposition per se is due to the inheritance of a defective allele of a tumour suppressor (65).

Cytogenetic damage can also arise at the level of the chromosome set. Accuracy of chromosome replication and segregation of chromosomes to daughter cells require accurate maintenance of the chromosome complement of a eukaryotic cell (66). Chromosome segregation in meiosis and mitosis is dependent upon the synthesis and functioning of the proteins of the spindle apparatus and upon the attachment and movement of chromosomes on the spindle. The kinetochores attach the chromosomes to the spindle and the centrioles are responsible for the polar orientation of the division apparatus (67, 68). Sometimes such segregation events proceed incorrectly and homologous chromosomes separate, with deviations from the normal number (aneuploidy) into daughter cells or as a multiple of the complete karyotype (polyploidy). When both copies of a particular chromosome move into a daughter cell and the other cell receives none, the event is known as non-disjunction (69, 70). Aneuploidy in live births and abortions arises from aneuploid gametes during germ cell meiosis. Trisomy or monosomy of large chromosomes leads to early embryonic death. Trisomy of the smaller chromosomes allows survival but is detrimental to the health of an affected person (71), e.g. Down syndrome (trisomy 21), Patau syndrome (trisomy 13), and Edward syndrome (trisomy 18) (72). Sex chromosome trisomies (Klinefelter's and XXX syndromes) and the sex chromosome monosomy (XO), known as the Turner syndrome, are also compatible with survival. Aneuploidy in somatic cells is involved in the formation of human tumours (60, 73), The "Mitelman Database of Chromosome Aberrations in Cancer" gives a detailed overview how chromosomal aberrations relate to tumour characteristics (74, 75). An abnormal number of chromosomes, seems to be the primary cause of the genomic instability in neoplasms due to destabilisation of the karyotype and the genes (76, 77). Up to 10% of tumours are monosomic or trisomic for a specific chromosome as the single observable cytogenetic change. Most common among such tumours are trisomies 8, 9, 12, and 21 and monosomy for chromosomes 7, 22, and Y (78).

1.2. Regulatory Guidelines

At the national level, various regulatory guidelines have been developed to measure cytogenetic damage, e.g. US Environmental Protection Agency (EPA), United Kingdom Environmental Mutagen Society (UKEMS), at the European level – the Official Journal of the European Union (OJEC), but most importantly at the international level of the Organisation for Economic Co-operation and Development (OECD). The OECD currently has 30 member states and the membership includes many of the world's major industrial nations. However, it is important to note that a number of significant states are not members, including the BRIC countries (Brazil, Russia, India, and China). Of course, this does not prevent contract houses in India and Brazil from being able to perform studies to OECD guidelines. The pharmaceutical industry has agreed a harmonised approach through the acceptance and implementation of the ICH (International Conference on Harmonisation of Technical Requirements for Registration of Pharmaceuticals for Human Use) and ICH issue guidelines for pharmaceuticals in three major economic regions, Europe,

Japan and the USA. The veterinary equivalent to the ICH is the International Cooperation on Harmonisation of Technical Requirements for Registration of Veterinary Medicinal Products (VICH) and the objective of VICH is the issuance of guidelines for veterinary pharmaceuticals. The OECD and the ICH/VICH guidelines constitute the two major sets of internationally harmonised genotoxicity guidelines in regulatory use.

1.3. Cytogenetic In Vivo Assays

Chromosome changes can have severe consequences in cells and whole organisms, and the present chapter concerns with cytogenetic assays which measure gross chromosome changes in rodent somatic cells in vivo. It primarily concerns with how to carry out the bone marrow micronucleus and peripheral blood micronucleus assays and bone marrow metaphase assay.

The methods described in this chapter have all come from a pharmaceutical laboratory and as such the methods have all been conducted accordingly to GLP requirements. Good laboratory practice generally refers to a system of management controls for laboratories and research organisations to ensure the consistency and reliability of results as outlined in the OECD Principles of GLP and national regulations.

1.4. Bone Marrow Micronucleus Test

The micronucleus test is a short-term mutagenicity test for the detection of chromosome damaging agents and/or spindle poisons (79). The OECD guideline 474 for mammalian erythrocyte micronucleus test (80) was adopted in July 2007. Micronuclei (induced or spontaneous) can arise in any dividing cell population and are the result of structural chromosome damage and/or whole chromosome(s) lagging at anaphase. One cell-type particularly suitable for observing this phenomenon is the newly formed erythrocyte (immature erythrocyte – IE) in the bone marrow. During the last maturation division in the erythrocyte precursor cell (the erythroblast) the nucleus is extruded to form an anucleated cell (this occurs approximately 6–8 h after the final mitosis (81)). Chromosome fragments (or whole chromosomes) that have not been incorporated into the main nucleus remain in the cytoplasm and form small nuclei termed micronuclei, and are conspicuous in spite of their small size. The IEs (because they do not lose their RNA for approximately 24 h) stain differently from the mature erythrocytes (E) during this period and are easily distinguished. Thus, anucleation and differential staining combine to make IE particularly useful cells for examining micronuclei (Fig. 2).

Mice or rats are recommended and any appropriate mammalian species may be used (80).

For the purpose of this chapter immature erythrocytes (IE) are the same as polychromatic erythrocytes. As a chromatic stain is not used routinely anymore, IE is the correct term in the bone marrow. When the cells are released into the peripheral blood they are then called reticulocytes (Also see Note 4).

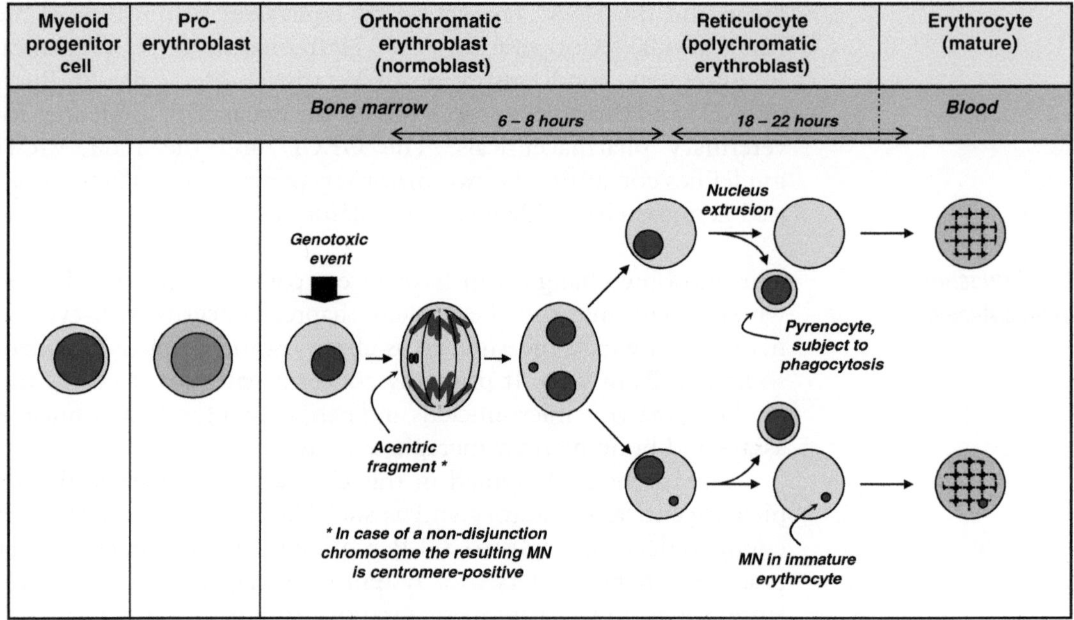

Fig. 2. Schematic overview of the bone marrow MN test. The genetic damage can be observed in immature erythrocytes (IE) carrying one or more MN resulting from chromosomal breakage or non-disjunction.

Flow-cytometric methods for both peripheral blood and bone marrow are available and these methods are discussed later in this chapter.

1.4.1. Factors Influencing the Micronucleus Test

Both hypothermia and hyperthermia have been shown to be associated with increases in micronucleated erythrocytes in the bone marrow of mice (82). Hyperthermia induced micronuclei would be removed if the animal's body temperature is regulated by warming (Fig. 3).

The sensitivity of the in vivo micronucleus test is increased by inducing erythropoiesis which has been demonstrated by bleeding mice to induce erythropoiesis (83, 84). When the compound induces more micronuclei at 48 h than that seen at 24 h with no genotoxic potential in the in vitro tests, it is suggested that haematological measurements be taken to look at erythropoietin levels (Fig. 4).

1.4.2. Limitations of Micronucleus Test In Vivo

The test compound of interest must reach the target cells of the bone marrow in sufficient concentration to induce damage. The damage may result from clastogenicity or aneugenicity. Analysis for centromere presence either by kinetochore protein of the centromere in the rat or pan-centromeric probe in the mouse is needed to determine this (85, 86). This centromeric labelling technique is much clearer and less subjective than kinetochore labelling. In addition, kinetochore labelling is prone to false negatives, due to

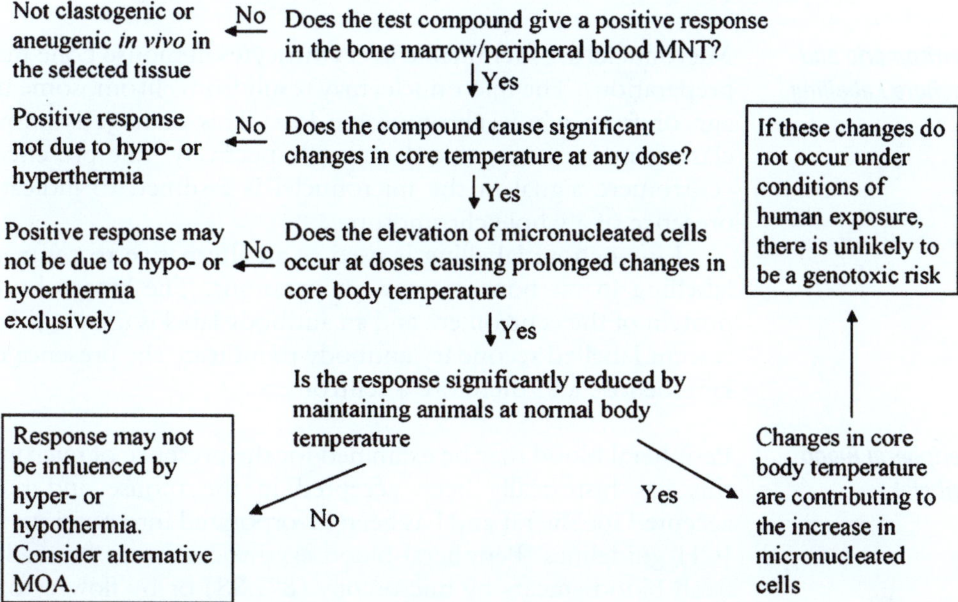

Fig. 3. Flow chart showing the approaches to be taken when a hypothermic or hyperthermic response occurs in rodent bone marrow micronucleus tests. MOA = mode of action (reproduced from Tweats et al. (82) with authors' permission).

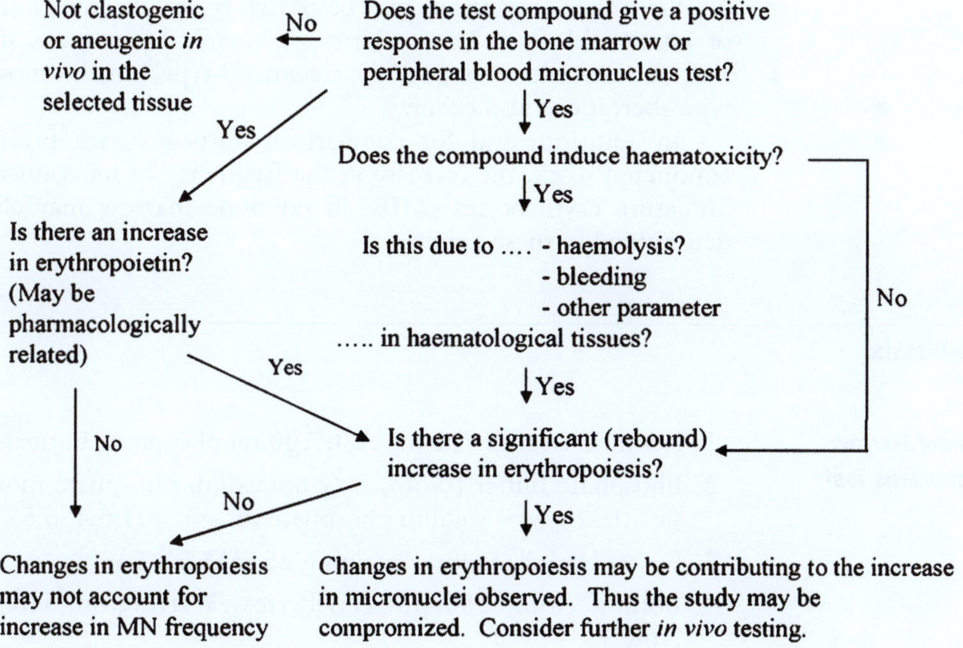

Fig. 4. Flow chart showing the approaches to be taken when an effect on erythropoiesis occurs in rodent bone marrow micronucleus tests (reproduced from Tweats et al. (82) with authors' permission).

the possible interaction of the test compound with the epitope for the kinetochore antibody.

1.5. Centromeric and Kinetochore Labelling

Micronuclei are identifiable in erythrocytes in mouse bone marrow preparations. The micronuclei may result from chromosome breakage or from whole chromosome loss. This damage is known as clastogenic or aneugenic damage, respectively. The presence of a centromere signal in the micronuclei is assumed to indicate the presence of a whole chromosome.

Kinetochore labelling is used as an alternative to centromeric labelling in rat bone marrow preparations. The kinetochore is a protein of the centromere and an antibody label is used with a fluorescent labelled secondary antibody to indicate the presence of the kinetochore and therefore a centromere.

1.6. Peripheral Blood Micronuclei

Peripheral blood may be examined for the presence of micronuclei, this has historically been accepted in the mouse and recently accepted for the rat and has been incorporated into revisions of the ICH guidelines. Peripheral blood erythrocytes may be read from fresh blood smears by microscopy (87, 88) or by flow cytometry (89, 90).

1.7. Chromosome Aberration Test

The ex vivo rat lymphocyte cytogenetic assay is used to identify agents that induce damage in circulating peripheral blood lymphocytes resulting in the expression of structural chromosome aberrations in the cultured cells.

Structural aberrations may be of two types, chromosome-type or chromatid-type. The majority of chemical mutagens induce aberrations predominantly of the chromatid-type, but chromosome-type aberrations also occur.

In addition, and for comparison purposes with previously conducted work, the increase in the frequency of micronucleated immature erythrocytes (MIE) in rat bone marrow may also be determined in these animals.

2. Materials

2.1. Bone Marrow Micronucleus Test

1. Acridine Orange (12 mg AO/100 ml phosphate buffer).
2. Phosphate buffer (0.66% w/v potassium phosphate monobasic + 0.32% w/v sodium phosphate dibasic, pH 6.4–6.5).
3. Foetal bovine serum containing 25 mM EDTA.
4. Bolting cloth, 150 μm (Lockertex, Warrington, UK). See Note 1.
5. Methanol (HPLC Gradient Grade).

2.2. Centromeric and Kinetochore Labelling

2.2.1. Centromeric Labelling for Micronucleated Erythrocytes in the Mouse

1. 20× SSC stock solution (3 M NaCl – 17.53 g per 100 ml, 0.3 M trisodium citrate – 8.82 g per 100 ml).
2. 2× SSC (1 in 10 dilution of stock).
3. 2× SSC with 0.1% Tween-20 (100 μl of Tween-20 in 100 ml of 2× SSC).
4. 0.4× SSC with 0.3% Tween-20 (20 ml 2× SSC+80 ml water+300 μl Tween-20).
5. All solutions are prepared by dilution from stock of 20× SSC, which is stable for at least 1 year and stored at room temperature.

2.2.2. Kinetochore Labelling for Micronucleated Erythrocytes in the Rat

1. Anti-kinetochore antibody (Antibodies Inc.).
2. PBS (phosphate buffered saline).
3. 1% Tween-20 in PBS.
4. VECTASHIELD mounting medium with propidium iodide.
5. Fluorescein isothiocyanate (FITC) – conjugated goat anti-human IgG (Sigma).

2.3. Peripheral Blood Micronuclei

1. RNaseZap (Sigma R2020-250ML).
2. DEPC (Sigma D5758-5ML).
3. HBSS (Invitrogen 14065-500ML).
4. Heparin sodium salt (Sigma H3149-50KU).
5. Guanidinium thiocyanate (GTC).
6. Heat inactivated, filtered foetal bovine serum containing 25 mM EDTA.
7. Acridine orange (1 mg/ml).
8. HBSS is freshly prepared for each test using DEPC-treated water (HBSS/DEPC).
9. Heparinised HBSS/DEPC is prepared by adding 14.1 mg Heparin sodium salt (Sigma H3149-50KU) to 150 ml of 1× HBSS.
10. A 2% GTC solution is prepared by adding 2 ml of GTC to 98 ml of heparinised HBSS (GTC/NaHep/HBSS).
11. GTC must not be added to Virkon or acid, as this will produce poisonous gas.

2.4. Chromosome Aberration Test

1. PBS/NaHep: PBS tablet is dissolved in sterile H_2O, autoclaved then 14.1 mg of sodium heparin – NaHep (Sigma Aldrich) is added to 150 ml of PBS.
2. 125 μl Colcemid™.
3. 0.075 M KCl.
4. Methanol/glacial acetic acid fixative (3:1, v/v).

5. Filtered 6% solution Giemsa at pH 6.8.
6. Permanent mounting medium, e.g. DPX (Fluka 44581).

2.4.1. Media Preparation

1. Pen–strep media
   RPMI 1640 (Dutch modification) obtained from Invitrogen supplemented with the following: 100 ml foetal bovine serum, heat-inactivated (FBSHI) 12 ml Pen–strep 6 ml l-glutamine (100×), 5 ml concanavalin A (1%), 5 ml phytohaemagglutinin (PHA) (1%).
2. Gentamycin media
   RPMI 1640 (Dutch modification) obtained from Invitrogen supplemented with: 100 ml foetal bovine serum, heat inactivated (FBSHI) 1 ml gentamycin 6 ml l-Glutamine (100×) 5 ml concanavalin A (1%) 5 ml PHA (1%).
3. Add 5 ml of sterile PBS to 5 mg concanavalin A to make a 1 mg/ml solution. (Mix on a roller to prevent foaming).
4. Add 5 ml of sterile water to 5 mg pre-weighted bottle of PHA to make a 1 mg/ml solution.

3. Methods

3.1. Bone Marrow Micronucleus Test

To fulfil current "regulatory" guidelines, two alternative protocols can be used:

3.1.1. Treatment Regimens: Single or Double Dose

1. The test article is given as two doses 24 h apart and the animals killed and sampled 24 h after the final dose.

   or

2. The test article is given once and the animals killed and sampled 24 and 48 h after dosing.

3.1.2. Integration into Repeat Dose Studies

At the time of going to press ICH S2 A and B revisions (91) are on going and at step 2 and when signed off two testing cascades will be allowed. Both of these options will include integration of the micronucleus study into the repeat dose toxicology studies (14 or 28-day studies) (92).

ICH guidance version ICH S2 (R1) recommended for adoption at Step 4 of the ICH Process on June 11 2009.

Option 1

1. A test for gene mutation in bacteria.
2. A cytogenetic test for chromosomal damage (the *in vitro* metaphase chromosome aberration test or *in vitro* micronucleus test), or an *in vitro* mouse lymphoma tk gene mutation assay.

3. An *in vivo* test for genotoxicity, generally a test for chromosomal damage using rodent hematopoietic cells, either for micronuclei or for chromosomal aberrations in metaphase cells.

Option 2

1. A test for gene mutation in bacteria.
2. An *in vivo* assessment of genotoxicity with two different tissues, usually an assay for micronuclei using rodent hematopoietic cells and a second *in vivo* assay. Typically, this would be a DNA strand breakage assay (93, 94).

There is more historical experience with Option 1, partly because it is based on S2A and B. Nevertheless, the reasoning behind considering Options 1 and 2 equally acceptable is as follows: When a positive result occurs in an *in vitro* mammalian cell assay, clearly negative results in two well conducted in vivo assays, in appropriate tissues and with demonstrated adequate exposure, are considered sufficient evidence for lack of genotoxic potential *in vivo* (see section 5.4.1.1 below). Thus a test strategy in which two *in vivo* assays are conducted is the same strategy that would be used to follow up a positive result in vitro.

Under both standard battery options, either acute or repeat dose study designs *in vivo* can be used. In case of repeated administrations, attempts should be made to incorporate the genotoxicity endpoints into toxicity studies, if scientifically justified. When more than one endpoint is evaluated *in vivo* it is preferable that they are incorporated into a single study. Often sufficient information on the likely suitability of the doses for the repeat-dose toxicology study is available before the study begins and can be used to determine whether an acute or an integrated test will be suitable.

For compounds that give negative results, the completion of either option of the standard test battery, performed and evaluated in accordance with current recommendations, will usually provide sufficient assurance of the absence of genotoxic activity and no additional tests are warranted. Compounds that give positive results in the standard test battery might, depending on their therapeutic use, need to be tested more extensively.

The integration of the micronucleus test into the repeat dose toxicity studies is not a new one. Early studies showed no differences for aneugens or clastogens when performing the micronucleus test following a single or a five-dose treatment schedule. This schedule was found to be more sensitive for anti-metabolites 5-fluorouracil and methotrexate (95). A comparison of acute micronucleus data for 15 chemicals with micronucleus frequencies in the 90-day sub-chronic bioassay revealed 12 of the results correlated. However, in the ones that did not, Riddelline was positive in the acute test and not in the repeat dose, and phenolphthalein was negative in acute dose and positive at the end of 3 months dosing (96). A collaborative trial was undertaken comparing 15 chemicals positive in the acute dose protocols with that of the 28-day study. Thirteen of the 15 chemicals were positive in the repeat dose testing and suggesting that integration was possible for the rat studies (97).

3.1.3. Dose Selection

Dose level selection follows the recommendation of OECD Guideline 474 (80). Three dose levels are used in both protocols. The highest dose must satisfy one or more of the following criteria:

1. The limit dose.
2. The maximum tolerated dose.
3. The highest achievable dose (i.e. restricted by formulation).
4. A dose which produces a significant depression in the percentage of immature erythrocytes at either 24 or 48 h post-dose.

The other two dose levels are set at 50% and 10% of the highest dose, unless otherwise stated as in a repeat experiment to determine a "no effect" dose level. In this case, the log interval between the highest dose and 10% of the highest dose would be used.

The test compound is usually administered via the intended clinical route (usually oral or intravenous).

1. Limit dose
   dose level will be used when a single administration of the test compound is well tolerated. Tolerance of the test compound will have been assessed in a dose finding study. The limit dose for the micronucleus test is 2,000 mg/kg (80).
2. The maximum tolerated dose (MTD)
   If there are insufficient data to support this choice of dose level a preliminary test to the micronucleus assay must be performed to confirm survival of the main study.

   This is described as the highest dose level where there are signs of evident toxicity but no deaths, as previously recommended (98). The overt signs of toxicity include subdued behaviour, piloerection, reduced body temperature, and hunched posture. The severity of which can be categorised as mild, moderate, or severe and is also dependent on the duration from time of onset of the clinical observation to recovery.

   The MTD is usually determined in the dose range finding study, but if the data are unavailable then a preliminary test to the main micronucleus study may be performed. This may be required to establish the MTD if higher than a dose level previously tested.

   It is recommended that observation of mild toxic effects will be used to define the MTD in the main micronucleus test since the severity banding for this is moderate. This MTD is judged to comprise of mild/moderate category observations of short duration (usually only a few hours).

3. The highest achievable dose
This is determined by the solubility and formulation of the test compound. The test compound is usually presented in HPMC or other relevant solvent.

4. Depression of immature erythrocytes
Assessment of the ratio of immature erythrocytes (IE) to mature erythrocytes (E) can provide evidence that a sufficiently toxic dose has been administered, and that bone marrow exposure has taken place. Reduction in the number of IE:E is either due to inhibition of the division of maturation of the nucleated erythropoietic cells or to the replenishment of the marrow with peripheral blood.

A significant depression in this ratio at either 24 or 48 h post-dose can be used to justify a maximum dose.

3.1.4. Animals and Sex

Each treatment group and control must include at least five analysable animals per sex (99). If sufficient data exist to show no differences in toxicicity exist between the sexes then a single sex may be used (80).

A standard treatment protocol includes the micronucleus assay conducted in rats, using a single sex (males), unless the intended human exposure is sex specific or there are substantial differences (>5-fold) in toxicity between the sexes. Strain is dependent on the laboratories and should include a historical control for that strain. The animals are used for the micronucleus test in the range of 9–10 weeks at intake.

On arrival, the rats are housed up to 4 per cage (mice are housed singly) and given, pelletted rodent diet, supplied by an appropriate supplier (e.g. Special Diet Services Ltd, England) and water from the site drinking water supply ad libitum.

The temperature and relative humidity of the animal rooms are monitored by an environmental monitoring system, excursions outside of the pre-set limits are recorded and any excursions other than those resulting from routine animal husbandry are reported. The animal room is controlled to provide a 12 h dark and a 12 h artificial light cycle. An acclimatisation period of at least 6 days is required.

Animals are randomly allocated to cages within a standard rack plan for micronucleus assays and are identified by ear and/or tail marks and cage cards. Animals are dosed according to the study plan.

3.1.5. Preparation and Staining of Bone Marrow Smears

For rats, at least four slides are prepared from the bone marrow for each animal according to the method described by Tinwell and Ashby in 1989 (100). Acridine orange fluorescent staining has advantages over Giemsa stain, as it does not stain RNA and other acidic materials that Giemsa does stain, and it allows the discrimination of DNA from RNA. The acridine orange stain has been found to give more reliable scoring between scorers (101).

For mice the same procedure as the rat is used to prepare the slides with the exception that slides maybe stained with Giemsa or acridine orange and scored for the presence of micronuclei using the criteria of Schmid (1975) (79), details will be recorded in the study file.

1. Animals are killed 24 h after the final dose (Protocol 1) or either 24 or 48 h after dosing (Protocol 2) by halothane inhalation, and death is confirmed by cervical dislocation unless a blood sample is required. In this case, an alternative confirmation of death is used, e.g. opening of the thorax. The ventral surface and rear leg(s) are swabbed and one femur is removed and the bone marrow exposed at both ends. Both femurs from cyclophosphamide dosed animals may be removed to prepare additional positive control slides to be used in other studies.
2. The femoral cells are flushed out with foetal bovine serum containing 25 mM EDTA. The cellular suspension is filtered through bolting cloth and spun at approximately 200×*g* for 5 min. See Note 1.
3. Supernatant is drawn off with a Pasteur pipette. The amount left will vary according to the pellet size. As a guideline, the amount of supernatant left should be approximately twice the volume of the pellet size. The cells in the sediment are carefully mixed by repeated aspiration and a small drop is pipetted onto the middle of a clean microscope slide.
4. The droplet is spread by placing another slide on top of the slide with the droplet and pulling the two slides apart in a sliding motion.
5. After air drying the cells are fixed in methanol (see Note 2) for at least 15 min and allowed to air dry. At least four slides are prepared from each animal and two are stained and the rest stored unstained in a –20°C freezer in the event that further analysis is required.
6. Slides are dipped in fresh phosphate buffer (0.66% w/v potassium phosphate monobasic + 0.32% w/v sodium phosphate dibasic, pH 6.4–6.5) and stained in a solution of acridine orange (AO), (12 mg AO/100 ml buffer), for 1 min. Slides are then placed in buffer for 10 min followed by a further 15 min in a fresh batch of buffer.
7. After staining slides are air-dried and stored protected from light.

3.1.6. Coding of Slides

All slides are coded prior to being scored for micronuclei.

1. A code sheet is produced using a macro in an excel spreadsheet. All animals from the main test are allocated a slide code.
2. The slide codes are written or printed on adhesive labels together with the study number.
3. The code labels are applied to the appropriate slides according to the code sheet. A blank label is also placed on the reverse of the frosted end of the slide to cover all identification marks.
4. Ideally, someone not involved in the micronucleus analysis should perform the coding but when this is not possible it will not invalidate the study.
5. The code sheet must be signed and dated by the individual who has performed the coding procedure and the code sheet then placed in a sealed envelope. The envelope is only opened after all analysis is complete for de-coding. The code sheet is archived as raw data.

3.1.7. Analysis of Slides

1. Slides are "wet" mounted (carefully avoiding air bubbles) prior to scoring with phosphate buffer and a glass coverslip. Microscopic analysis is performed using a fluorescence microscope with BG-12 excitation filter and 0–530 barrier filter.
2. The cells are identified by the following staining properties of acridine orange:
   (a) Immature erythrocytes (IE) – bright orange fluorescence
   (b) Micronuclei - bright yellow/green fluorescence within the IE
   (c) Mature erythrocytes (E) – dull khaki green colour.
3. Scoring is performed on electronic digital counters and data recorded on paper.
4. Relative proportions of immature and mature erythrocytes (IE:E ratio) will be determined in a total of 1,000 erythrocytes. See Note 3.
5. 2,000 IE are scored for each animal (from either of the two slides initially stained) and the number of micronucleated immature erythrocytes (MIE) counted.
6. In the event of an equivocal result, analysis may be extended up to 6,000 IE per animal (93).

3.1.8. Decoding of Micronucleus Slides

Once all the slide analysis has been completed the code sheet is opened and the signature and date of who opened the envelope recorded on the code sheet itself. The data are de-coded and tabulated manually into an Excel table according to treatment group and documented by a signature and date. The tabulated data are then 100% checked and documented with a signature and date by a second person and archived as raw data.

3.1.9. Number of Cells Scored

The OECD guideline requires that the proportion of immature and mature erythrocytes is determined for each animal at least 200 erythrocytes for bone marrow, however, 1,000 is standard practice. At least 2,000 immature erythrocytes per animal are required for the incidence of micronucleated immature erythrocytes.

Similar statistical power can be achieved by scoring 2,000 IE from seven rats or 4,000 IE from five rats, respectively. However, this is based only on control animals and does not consider possible differences in responses between animals to treatment with a potential genotoxin. In order to minimise the possible influence of responders and non-responders, the preferred study design is to score 2,000 IE from groups of seven rats.

Power analyses show that if an equivocal result is obtained after scoring 2,000 immature erythrocytes (IE), it is appropriate to re-code the slides and score an additional 4,000 IE, i.e. analysing a total of 6,000 IE. No meaningful increase in statistical power is gained by scoring more than 6,000 IE. This is consistent with the variability observed between separate counts on the same slide (102).

In addition increased power can be achieved by scoring more cells up to 20,000 with flow cytometric methods in both rats and mice (103).

Systems for automated analysis are acceptable alternatives to manual evaluation if appropriately justified and validated.

3.1.10. Controls

Concurrent negative solvent/vehicle controls should be included for each sex in each test (80). The solvent/vehicle should not produce toxic effects or react with the test chemical. Treatment regimes should be identical to that used in the test compound treated groups. A negative (vehicle) control should be presented in the same form as the test compound, e.g. pH.

Positive controls should be included or positive slides coded into a study to control for both staining and scoring procedures (104). The positive control should produce micronuclei at exposure levels expected to give a detectable increase over background.

Examples of positive control substances in the OECD guideline include the following: ethyl methane sulphonate, ethyl nitrosourea, mitomycin C, cyclophosphamide (monohydrate), and triethylenemelamine.

Cyclophosphamide, when considered a positive control, is dosed orally (or IP) at 20 mg/kg body weight to rats or at 65 mg/kg body weight to mice. Numerous slides can also be prepared from a positive control dosed animal and then coded into future studies.

3.1.11. Evaluation Criteria

1. Criteria for a valid assay
   For a test to be considered valid, the following criteria should be fulfilled:
   (a) The mean concurrent vehicle control values fall within the acceptable limits as defined in the Laboratory historical control data. The positive control group data clearly demonstrates a statistically and biologically significant increase when compared with the concurrent vehicle control group.
   (b) The test compound should be tested at a dose level equivalent to the maximum tolerated dose (MTD) (as determined in a preliminary study or based on the results of preliminary dose setting studies). Alternatively, a reduction in the ratio of IE to E is indicative of bone marrow exposure. If neither of these conditions are fulfilled the test compound should be administered at a dose of 2,000 mg/kg/day (the limit dose for the testing laboratory).
2. Evaluation of data
   The result of the test will be assessed using the following criteria:
   (a) The test will be regarded as clearly negative if there are no increases of either statistical or biological significance in the number of MIE at any dose compared with concurrent vehicle control animals.
   (b) The test will be regarded as clearly positive if there is an increase in MIE that is of statistical and biological significance and is above the higher acceptance limit (see historical control data in testing laboratory) and that clearly demonstrates a dose-response relationship.
   (c) If an increase is seen that is statistically different from the concurrent control but does not fulfil the criteria for a positive result as defined in (b), further statistical and/or microscopic analyses may be performed as detailed in (1), (2), and (3). However biological relevance will remain the primary consideration.
      (1) Combining the 24 and 48 h concurrent control values and reapplying the statistical analysis to determine if the increases remain statistically significant. Combining the control data from the males and females are also acceptable in the event of an unusually low control in one sex.
      (2) If the concurrent control value(s) are below the mean historical control value, comparing the increase(s) with the Laboratory historical control may be deemed appropriate. See Note 4.

(3) Recoding the slides and analysing additional IE up to a maximum of 6,000 IE per animal to increase the statistical power of the test.

(d) If the result of the test remains equivocal or uninterpretable after the additional analyses described in (c), then a repeat test will be considered.

3.1.12. Interpretation of Data (Statistics)

The application of formal statistics may be deemed unnecessary if no increases over the concurrent vehicle control values are observed.

Many parametric and non-parametric tests are considered suitable for statistical analyses of micronucleus data, these have been reviewed by Lovell et al., 1989 (105).

One parametric approach of particular note is detailed:

Before analysis the number of micronucleated immature erythrocytes (MIE) is subjected to an average square root transformation ((sqrt(MIE) + sqrt(MIE + 1))/2); this is to achieve an approximate normal distribution (Lovell et al., p197; Snedecor and Cochran, p287) (105, 106). The variance of the transformed count should be around 0.25 (105). A box-plot of the untransformed count should be produced to provide a visual summary that will reveal gross features.

In the statistical analysis each expression time (usually 24 h, and occasionally 48 h) is considered separately. The SAS GLM program is used to perform an analysis of variance. Appropriate linear contrasts (e.g. −3, −1, 1, 3 for a control and three test group study) are used to test for dose-related trends in MIE. Where there is only one test group, this reduces to a single two-sample *t*-test between the test and the control group (−1, 1).

In a dose response study, the No Observable Effect Level is determined in the following way.

If a significant increase is observed across all test doses then the trend test will be reapplied in a closed testing "cascade" – first excluding the top dose, then the intermediate dose(s) and finally with only the control and the lowest dose. This closed testing procedure stops when either a non-significant effect is detected or all dose levels have been tested. The above tests are all one sided at the 5% level. Positive controls are excluded from this analysis.

If the tests for trend fail to reach statistical significance, and there is some suggestion of increased levels of MIE in the intermediate dose groups ('umbrella effect'), pair-wise tests with control (one-sided at the 1% level) are performed.

They may also compare positive controls with vehicle controls using Fisher's exact test (Lovell et al., p200) (105).

In cases where counts of more than 2,000 IE are available (e.g. 6,000 from 3 counts of 2,000 IE from the same slide, or from different slides), the average of the separate counts will be analysed. An appropriate plot should be used to assess any differences between slides.

3.2. Centromeric and Kinetochore Labelling

Care should be taken at all stages of using the chromosome paints to minimise light exposure, as they are photo-degraded.

3.2.1. Procedure for Centromeric Labelling for Micronucleated Erythrocytes in the Mouse

Day 1

The slides are to be aged for at least 24 h at room temperature prior to centromeric labelling.

A programmable hotplate (Hybrite) is switched on prior to use to allow the hotplate to reach 42°C (approx. 10 min). The slides to be painted are warmed by placing them on the hotplate surface prior to use.

The Centromeric paint is taken from the freezer to thaw prior to use (at least 15 min).

15–20 μl of mouse centromeric chromosome paint is added to each slide with a small coverslip. The coverslip is then sealed by adding rubber cement glue around the edges of the coverslip. When the glue has dried the slides are placed on the hotplate. The programmable hotplate denatures the slide and paint at 69°C for 5 min then the slide is left for between 16 and 40 h at 42°C on the hotplate in a humidified atmosphere.

Day 3

Make up 0.4× SSC with 0.3% Tween-20 in glass Coplin jar and place in the water bath with water covering approx. ¾ of the height of the Coplin jar.

Heat to 73±1°C and allow approximately 1 h getting to temperature. Place 2× SSC/0.1% Tween-20 in a Coplin jar at room temperature.

Check the temperature of the water bath.

Remove slides from the hotplate, remove glue and gently slide off the coverslips. Place two slides only at one time (as each slide reduces the temperature of the Coplin jar by 1°C) in the Coplin jar in the water bath, agitate for 3–4 s and leave in the Coplin jar for 1 min.

Transfer slides to a Coplin jar containing 2× SSC, 0.1% Tween-20 and incubate for 2 min. Drain the slides (N.B. do not allow to dry) and add antifade containing DAPI counterstain, add large (22×40 mm^2, or 22×50 mm^2) coverslip and place in card tray store flat and in the dark prior to scoring on a suitable fluorescent microscope.

Allow the temperature in the Coplin jar to return to previous level (approx. 5–10 min) before washing the next two slides.

Slide checking
Check the slides after counterstain and antifade is added to determine the presence of centromeric signals (either red or green) under the microscope in any nucleated cells (blue from DAPI counterstain), to determine that the hybridisation has taken place.

Slide coding
The slides may be coded at the discretion of the Study Director. When coded the Study Director produces a code sheet using an Excel spreadsheet. All slides are allocated a slide code. The slide codes are written or printed on adhesive labels together with the study number. The code labels are applied to the appropriate slides according to the code sheet. The code sheet must be signed and dated by the individual who has performed the coding procedure and the code sheet then placed in a sealed envelope. The envelope is only opened after all analysis is complete for de-coding. The code sheet is archived as raw data.

Slide scoring
The slides are scored 'blind' using an appropriate fluorescence microscope and scoring is performed on electronic digital counters. Data are recorded on paper. The aim is to score 100 micronuclei. In control cultures or low levels of micronuclei induction this may not be possible – scoring is stopped at 20,000 cells.

3.2.2. Procedure for Kinetochore Labelling for Micronucleated Erythrocytes in the Rat

Care should be taken at all stages when using the antihuman IgG antibody with fluorescent CY3 or FITC labels to minimise light exposure as they are photo-degraded. The procedure is therefore split into two distinct phases. Phase 1, which can be conducted in the light, and phase 2, which must be conducted under minimal light to preserve the fluorescent label.

The indirect immuno-fluorescence labelling of kinetochore proteins is performed as described by Ellard et al. (107). To preserve the integrity of the kinetochore proteins, cells are fixed in analytical reagent grade methanol at approximately –20°C for 15 min and then air dried at room temperature. As soon as the fixative has evaporated, the slides are stored in slide boxes in a –20°C freezer until analysis (to preserve the kinetochore epitope).

The slide preparations are immediately hydrated with PBS on removal from the freezer. While the slides are in the buffer, 100 μl of anti-kinetochore antibody (Antibodies Inc.) is diluted 1:1 with PBS. The slides are removed from the PBS after about 10 min, allowing any excess PBS to drain off. Next 100 μl of the diluted antibody is applied to the slides. A flexible plastic coverslip is gently placed on the slides, spreading the antibody evenly across the surfaces of the slides. The slides are then incubated in a humidified chamber at 37°C for approximately 1 h.

After incubation, unbound antibody is removed from the slides by washing in a Coplin jar as detailed in steps 1–5 below (care is taken not to disturb the cells):

Step 1: Place in PBS for 1 min.

Step 2: Place in PBS containing 1% Tween-20 for 3 min, with gentle agitation.

Step 3: Rinse in PBS.

Step 4: Rinse in PBS.

Step 5: Place in PBS for 5 min.

While the slides are in the final wash, 5 μl of fluorescein isothiocyanate (FITC)-conjugated goat anti human IgG (Sigma) is diluted 1:50 (10 μl in 490 μl of PBS). Once the slides have been removed from their final wash and drained, approximately 120 μl of the fluorescent probe is placed onto the slides and spread evenly with a plastic coverslip as in phase 1. The slides are then incubated in a dark humidified chamber at 37°C for approximately 1 h. After incubation the slides are washed in a Coplin jar as detailed in steps 1–5 above. Once the slides have been removed from their final wash and drained, two to three drops of VECTASHIELD mounting medium containing propidium iodide are placed on the slides with a glass coverslip, and placed in a card slide tray for 20 min to develop prior to scoring.

Evaluation of data – Analysis of slides

The slides are checked after counter stain and antifade is added, to determine the presence of kinetochore signals (green) under the microscope in any nucleated cells (red from counter stain), to determine that the hybridisation has taken place.

The slides may be coded by producing a code sheet using an excel spreadsheet. All slides are allocated a slide code. The slide codes are written or printed on adhesive labels together with the study number. The code labels are applied to the appropriate slides according to the code sheet. The code sheet must be signed and dated by the individual who has performed the coding procedure and the code sheet then placed in a sealed envelope. The envelope is only opened after all analysis is complete for de-coding. The code sheet is archived as raw data.

1. Immature Erythrocytes – orange/red.
   Mature Erythrocytes – dark grey almost invisible.
2. Propidium iodide – stains RNA and DNA red.
3. Nucleated cells – appear bright red/almost yellow.
4. Micronuclei – fluorescent green/yellow.

Scoring takes place on the red filter to identify the immature erythrocytes containing micronuclei and then swapped to green and triple band pass filters to ascertain if a kinetochore signal is present.

One hundred micronucleated erythrocytes are scored and the number of erythrocytes is recorded. If 20,000 erythrocytes are reached and 100 micronucleated erythrocytes are not seen, scoring is stopped.

3.3. Peripheral Blood Micronuclei

3.3.1. Fresh Blood Smears Method to Preserve RNA from Deterioration

To reduce exogenous RNases in the class 2 cabinet, spray RNaseZap onto cabinet surfaces. Wipe thoroughly with clean blue roll. Rinse with dH_2O and dry with clean blue roll.

To reduce RNases in solutions DEPC can be added to the water used to prepare the solutions.

Add 1.0 ml DEPC to 1 L dH_2O in a glass bottle and shake vigorously to bring the DEPC into solution. Incubate solution for 12 h at 37°C. Then, autoclave the solution for 15 min to remove any trace of DEPC.

The rat blood is washed thoroughly in heparinised HBSS/DEPC containing 2% guanidinium thiocyanate (GTC) to remove factors in the plasma that cause cell deterioration and to preserve RNA in the immature reticulocytes.

Approximately 2 ml of GTC/NaHep/HBSS solution is added to each tube containing 1 ml of blood and gently mixed. The tubes are then centrifuged at 690×*g* for 5 min. The supernatant is removed with a stripette and discarded into a disposable bottle (see Note 5). A further 2 ml of GTC/NaHep/HBSS solution is added to blood pellets, mixed gently and centrifuged at 690×*g* for 5 min. The supernatant is removed with a stripette and discarded into a disposable bottle. The blood is re-suspended to 3 ml with heat inactivated, filtered foetal bovine serum containing 25 mM ethylenediamine tetraacetic acid (FBSHI/EDTA). Blood (120 μl) is placed into an Eppendorf tube containing 1 ml of serum and allowed to sediment for at least 24 h and no more than 2 weeks. The fresh blood smear is produced by mixing 2 μl of sedimented blood and 2 μl of serum on a slide with 1 μl of acridine orange 1 mg/ml and a 22×22 mm coverslip placed over. The slide should rest for 10 min prior to scoring to prevent movement on the fresh slide.

Flow cytometry methods are available from Torous et al. and Abramsson-Zetterberg (89, 108). The method given is for the preparation of samples to be analysed by Litron laboratories; however, there are other protocols for use on your own Flow Cytometer.

1. Blood collection and fixation
   Samples are prepared according to the protocol of Rat Micro Flow Plus Micronucleus Kit (Litron, Rochester, USA).
2. Preparing the fixative tubes and anticoagulant/diluent vials.
   Preparing the fixative tubes (at least 1 day prior to blood collection):
   (a) Duplicate samples are prepared as a precaution against potential problems.

(b) Label two 15-ml polypropylene centrifuge tubes per sample. Add 2 ml of solution A (methanol at least 99.8%) to each tube, replace lids and put in a rack.

(c) Store rack of tubes at −75°C to −85°C overnight (or longer) to ensure sufficient cooling of the fixative.

Preparing the anticoagulant/diluent vials:

(a) For each sample, aseptically aliquot 350 μl anticoagulant/diluent into a labelled vial.

(b) Refrigerate vials until required.

3. Collecting blood samples

(a) Collect peripheral blood using an approved method into K_2EDTA-coated tubes.

(b) Add 60 μl to 120 μl to 350 μl anticoagulant/diluent.

(c) Samples are stable for 2 days in K_2EDTA-coated tubes when stored at 2–8°C.

4. Fixing blood samples

The tubes containing Fixative must remain ultra-cold (−75°C to −85°C) and must not be stored in a freezer containing dry ice and the CO_2 vapours can cause cell aggregation.

(a) If one person is fixing the blood only, remove one tube of fixative from freezer at a time

(b) Invert diluted blood sample prior to fixing.

(c) Draw 180 μl of diluted blood sample into the tip of a micro-pipettor and remove corresponding 15 ml tube from freezer.

(d) Un-cap fixative tube, position pipette tip approximately 1 ml above the ultra-cold fixative and forcibly eject diluted blood into fixative.

(e) Cap the tube and vortex briefly (~3–5 s) or strike the bottom of the tube several times before returning it to the freezer.

(f) Change pipette tip and repeat with the rest of samples as above.

(g) If freezer temperature rises by 5°C stop processing the sample and wait until freezer returns to correct temperature. The diluted blood is stable for 6 h at room temperature and up to 24 h if stored at 2–8°C.

(h) Store samples at −75°C to −85°C for at least 48 h before analysis.

(i) Samples are stable for 1–2 years as long as the temperature is maintained at −75°C to −85°C.

Samples are then shipped to Litron Laboratories for analysis.

3.4. Chromosome Aberration Test

3.4.1. Animal Husbandry and Dosing

Rats aged 9–10 weeks at intake are used for this study type. Groups of seven male rats can be used unless there is a specific requirement to use females.

On arrival, the rats are housed up to 4 per cage and given appropriate animal food, e.g. pelletted rodent diet, supplied by for example Special Diet Services Ltd, England, and water supply ad libitum.

The temperature and relative humidity of the animal rooms are monitored by an environmental monitoring system, excursions outside of the pre-set limits are recorded and any excursions other than those resulting from routine animal husbandry are reported. The animal room is controlled to provide a 12 h dark–12 h artificial light cycle. An acclimatisation period of at least 6 days is required prior to the start of procedures.

Animals are randomly allocated to cages and are identified by ear or tail marks and cage cards. Animals are dosed according to the study plan.

The animals are weighed and then administered a dose by the route specified in the study plan. After dosing the animals are observed at the time slots indicated in the study plan for clinical signs and then at least once on day of termination.

3.4.2. Sample Collection and Lymphocyte Culture

1. Blood sample collection
   Blood samples are taken from each animal pre-dose and at termination (Day 29). Each pre-dose blood sample acts as a control for each animal. Positive control animals will only have pre-dose and terminal samples.

   Other blood samples may be taken at intervals during the study period (e.g. Day 2 or Day 15 etc.).

   The appropriate volume of whole blood will be taken from each animal into labelled lithium heparin tubes. The blood is mixed to ensure adequate contact with the anticoagulant and transferred to the laboratory in a timely manner to begin culture initiation.

2. Washing of blood
   Pre-terminal blood samples will be taken from the tail vein and terminal blood samples will be taken by cardiac puncture after administration of halothane.

   The rat blood is washed thoroughly in phosphate buffered saline (PBS) containing 9.4%, w/v, sodium heparin (NaHep), to remove factors in the plasma, which cause cell deterioration.

   Approximately 2 ml of PBS/NaHep solution is added to each tube containing 2 ml of blood and gently mixed. The tubes are then centrifuged for approximately 690 × g, for 5 min and the supernatant aspirated and discarded. A further 4 ml PBS/NaHep is added to each tube, the pellet gently mixed and the procedure repeated twice. After the final spin, the supernatant is aspirated and PBS/NaHep added to give approximately 2 ml final volume. Volumes of blood will be adjusted to the appropriate starting volume.

3. Lymphocyte culture initiation
   At least duplicate cultures are initiated for each animal at each sample time. If more blood is available then additional cultures will be included.

   Aliquots (0.5 ml) of washed blood/PBS/NaHep are dispensed into appropriately labelled culture tubes containing 9.5 ml of medium.

   For each duplicate culture initiated one will have media containing penicillin–streptomycin antibiotic (pen–strep) and the other with media containing gentamycin (gen) (see 2.4.1). This is to safeguard against the study failing due to contamination of the cultures; however, only the pen–strep cultures will be analysed for chromosome aberrations

   Cultures may be stored at 4°C until time of initiation which is the time at which cultures are placed in the incubator at 37°C. Cultures are incubated for approximately 68–70 h and are shaken daily.

3.4.3. Procedure

At least 24 h prior to cell harvest the media is changed. Cultures are shaken and poured into corresponding centrifuge tubes, then centrifuged at 300 × *g* for 5 min, the supernatant removed and 9.5 ml of appropriate fresh media added which has been warmed to 37°C. Cultures are then transferred into new corresponding culture flasks and are then placed back in the incubator at 37°C until harvesting.

Harvesting of lymphocyte cultures

Approximately 2 h prior to cell harvest, 125 μl Colcemid™ is added to (final concentration of approximately 0.125 μg/ml) to arrest the dividing cells in metaphase. At 68–70 h the cultures are transferred into appropriately labelled centrifuge tubes and centrifuged at 300 × *g* for 5 min then the supernatant carefully removed. The cells then re-suspended in 9.5 ml 0.075 M KCl (hypotonic) at 37°C for 12 min to allow swelling of the cells to occur.

The cultures are centrifuged at approximately 300 × *g* for 5 min and the supernatant removed and discarded.

The cells are fixed in freshly prepared methanol /glacial acetic acid fixative (3:1, v/v) at room temperature by adding the first 1 ml of fixative drop-wise whilst vortexing and then making up to approximately 10 ml. After this first fixation the cells are centrifuged at 300 × *g* for 7 min and re-suspended in approximately 10 ml fresh fixative and vortexed to mix thoroughly. This step is repeated and the cells are then re-suspended in third fixative, spun down and the majority of the supernatant removed leaving the pellet in remaining fixative. Sufficient fixative should be left to give a milky suspension once mixed with the cell pellet (approximately 0.5 ml).

After slide preparation the remaining cell pellet can be kept indefinitely in final fixative if fixative is filled to the top of the tube excluding all air. If cells are left in final fix they must be spun down and re-suspended in fresh fixative before slide preparation.

1. Slide preparation
   Metaphase preparations are made by dropping a concentrated cell suspension on to clean moist slides (labelled with the study number, animal number and date) and allowed to air dry. When completely dry, the slides are stained in filtered 6% solution Giemsa at pH 6.8 for 5 min or for a time that produces optimal staining (staining intensity may vary due to the effects of the test compound or the degree of chromosomal condensation). Excess Giemsa is removed by rinsing in tap water. Slides are then coverslipped using a permanent mounting medium, e.g. DePX.
2. Slide analysis
   All slides are checked for suitable quality prior to aberration analysis.

   The slides may be scanned using the automated metaphase finder or scanned manually. All analysis is performed manually and data recorded on paper.

   One hundred metaphases from each pen–strep culture are analysed for the incidence of chromosomal aberrations. If there is failure of the pen–strep culture then the gentamycin cultures will be analysed. This will be at the discretion of the Study Director and will be documented in a file note. From the positive control, only sufficient cells to confirm a positive response are analysed.

   Suitable metaphases are selected at a magnification of 100× and analysed at a magnification of least 1,000×. Each metaphase is analysed for the presence of structural chromosomal abnormalities.
3. Coding of slides
   The slide codes are written or printed on to adhesive labels together with the study number and the code labels applied to the appropriate slides according to the code sheet as in previous slide coding sections.
4. Statistical analysis
   The Fisher's Exact probability test is employed to evaluate the incidence of metaphases showing aberrations. This is a useful non-parametric technique for analysing data when comparing two small independent samples, and is useful when scores for the sample all fall into one or other of two mutually exclusive

classes. The test determines whether the two groups differ in the proportions with which they fall into the two classifications.

The data from cultures in each group are pooled and the number of aberrant cells with and without gap-type damage calculated. Statistical analysis is performed on the pooled data from each time point relative to the pre-dose incidence of aberrations.

The test is considered acceptable if the mean solvent control aberrations are within the Laboratory historical control range and the positive control cultures indicate a clear clastogenic response to the reference compounds.

5. Evaluation of results

Data are assessed from each animal and dose group separately. Data will be interpreted using both concurrent and historical control data as appropriate and will follow the scheme listed below:

- No statistically significant increase in aberrations at any dose level above concurrent solvent control values. NEGATIVE.
- A statistically significant increase in the number of aberrant cells above concurrent control levels which is unrelated to dose and which falls within the Laboratory solvent control range. NEGATIVE.
- An increase in the number of aberrant cells, at least at one concentration, which is substantially greater than the Laboratory historical solvent control range and is reproducible (excluding gap-type aberrations). POSITIVE.

Data sets that do not fulfil the above criteria will be evaluated further on a case-by-case basis. Further Evaluation may comprise of the following.

(i). Extended scoring of affected cultures

(ii). Repeating the appropriate part of the study, possibly with a narrower range of doses.

Once (i) or (ii) of the above have been carried out, interpretation is based on the previously stated criteria.

4. Notes

1. An alternative to bolting cloth for filtering bone marrow is a 100-μm Luer-Lok™ filter on a syringe.
2. The slides are fixed in methanol at –20°C to preserve the kinetochore epitope; duplicate slides are then stored at –20°C.

3. The terms immature erythrocyte (IE) and mature erythrocyte (E) mean the same population of cells as polychromatic erythrocyte (PCE) and normochromatic erythrocyte (NCE), respectively. The term IE and E are used, as we do not stain with a chromatic stain.
4. The presence of a very low control value can affect the statistical result; therefore, in this case it may be appropriate to compare to the laboratory historical control value.
5. GTC must not be added to Virkon or acid, as this produces poisonous gas.

References

1. Arnason, U. (2006) 50 years after – examination of some circumstances around the establishment of the correct chromosome number of man. *Hereditas* **143**, 202–211.
2. Tjio, J. H., and Levan, A. (1956) The chromosome number of man. *Hereditas* **42**, 1–6.
3. Ford, C. E., and Hamerton, J. L. (1956) The chromosomes of man. *Nature* **178**, 1020–1023.
4. Lejeune, J. M., Gautier, M., and Turpin, R. (1958) Etude des chromosomes somatiques de neuf enfants mongoliens. *CR Acad. Sci. Paris* **248**, 1721–1722.
5. Jacobs, P. A., and Strong, J. A. (1959) A case of human intersexuality having a possible XXY sex-determining mechanism. *Nature* **183**, 302–303.
6. Ford, C. E., Jones, K. W., Polani, P. E., De Almeida, J. C., and Briggs, J. H. (1959) A sex-chromosome anomaly in a case of gonadal dysgenesis (Turner's syndrome). *Lancet* **1**, 711–713.
7. Nowell, P. C., and Hungerford, D. A. (1960) A minute chromosome in human chronic granulocytic leukemia. *Science* **132**, 1497
8. Caspersson, T., Zech, L., and Johansson, C. (1970) Analysis of human metaphase chromosome set by aid of DNA-binding fluorescent agents. *Exp Cell Res* **62,** 490–492.
9. Drets, M. E., and Shaw, M. W. (1971) Specific banding patterns of human chromosomes. *Proc Natl Acad Sci U S A* **68**, 2073–2077.
10. Seabright, M. (1971) A rapid banding technique for human chromosomes. *Lancet* **2**, 971–972.
11. Rowley, J. D. (1973) Letter: A new consistent chromosomal abnormality in chronic myelogenous leukaemia identified by quinacrine fluorescence and Giemsa staining. *Nature* **243**, 290–293.
12. Boller, K., and Schmid, W. (1970) (Chemical mutagenesis in mammals. The Chinese hamster bone marrow as an in vivo test system. Hematological findings after treatment with trenimon). *Humangenetik* **11**, 35–54.
13. Heddle, J. A., and Carrano, A. V. (1977) The DNA content of micronuclei induced in mouse bone marrow by gamma-irradiation: evidence that micronuclei arise from acentric chromosomal fragments. *Mutat Res* **44**, 63–69.
14. Fenech, M., and Morley, A. A. (1985) Measurement of micronuclei in lymphocytes. *Mutat Res* **147**, 29–36.
15. Swiger, R. R., and Tucker, J. D. (1996) Fluorescence in situ hybridization: a brief review. *Environ Mol Mutagen* **27**, 245–254.
16. Thomas, N. S., Durkie, M., Van Zyl, B., Sanford, R., Potts, G., Youings, S., Dennis, N., and Jacobs, P. (2006) Parental and chromosomal origin of unbalanced de novo structural chromosome abnormalities in man. *Hum Genet* **119**, 444–450.
17. Beckmann, M. W., Niederacher, D., Schnurch, H. G., Gusterson, B. A., and Bender, H. G. (1997) Multistep carcinogenesis of breast cancer and tumour heterogeneity. *J Mol Med* **75**, 429–439.
18. Chandley, A. C. (1981) The origin of chromosomal aberrations in man and their potential for survival and reproduction in the adult human population. *Ann Genet* **24**, 5–11.
19. Anderson, D., Jenkinson, P. C., Edwards, A. J., Hughes, J. A., and Brinkworth, M. H. (1999) Commentary: paternal legacies. *Teratog Carcinog Mutagen* **19**, 105–108.
20. Marchetti, F., and Wyrobek, A. J. (2005) Mechanisms and consequences of paternally-transmitted chromosomal abnormalities. *Birth Defects Res C Embryo Today* **75,** 112–129.

21. Nomura, T., Nakajima, H., Ryo, H., Li, L. Y., Fukudome, Y., Adachi, S., Gotoh, H., and Tanaka, H. (2004) Transgenerational transmission of radiation- and chemically induced tumors and congenital anomalies in mice: studies of their possible relationship to induced chromosomal and molecular changes. *Cytogenet Genome Res* **104**, 252–260.
22. Martinez, G. R., Loureiro, A. P., Marques, S. A., Miyamoto, S., Yamaguchi, L. F., Onuki, J., Almeida, E. A., Garcia, C. C., Barbosa, L. F., Medeiros, M. H., and Di Mascio, P. (2003) Oxidative and alkylating damage in DNA. *Mutat Res* **544**, 115–127.
23. van Gent, D. C., and van der Burg, M. (2007) Non-homologous end-joining, a sticky affair. *Oncogene* **26**, 7731–7740.
24. Iliakis, G., Wang, H., Perrault, A. R., Boecker, W., Rosidi, B., Windhofer, F., Wu, W., Guan, J., Terzoudi, G., and Pantelias, G. (2004) Mechanisms of DNA double strand break repair and chromosome aberration formation. *Cytogenet Genome Res* **104**, 14–20.
25. Hanawalt, P. C., Ford, J. M., and Lloyd, D. R. (2003) Functional characterization of global genomic DNA repair and its implications for cancer. *Mutat Res* **544,** 107–114.
26. Fukamachi, S., Shimada, A., Naruse, K., and Shima, A. (2001) Genomic analysis of gamma-ray-induced germ-cell mutations at the b locus recovered from the medaka specific-locus test. *Mutat Res* **458**, 19–29.
27. Ahmadi, A., and Ng, S. C. (1999) Fertilizing ability of DNA-damaged spermatozoa. *J Exp Zool* **284**, 696–704.
28. Fisher, T. S., and Zakian, V. A. (2005) Ku: a multifunctional protein involved in telomere maintenance. *DNA Repair (Amst)* **4**, 1215–1226.
29. Bryant, P. E. (2004) Repair and chromosomal damage. *Radiother Oncol* **72**, 251–256.
30. Norppa, H., Bonassi, S., Hansteen, I. L., Hagmar, L., Stromberg, U., Rossner, P., Boffetta, P., Lindholm, C., Gundy, S., Lazutka, J., Cebulska-Wasilewska, A., Fabianova, E., Sram, R. J., Knudsen, L. E., Barale, R., and Fucic, A. (2006) Chromosomal aberrations and SCEs as biomarkers of cancer risk. *Mutat Res* **600**, 37–45.
31. Bonassi, S., Znaor, A., Ceppi, M., Lando, C., Chang, W. P., Holland, N., Kirsch-Volders, M., Zeiger, E., Ban, S., Barale, R., Bigatti, M. P., Bolognesi, C., Cebulska-Wasilewska, A., Fabianova, E., Fucic, A., Hagmar, L., Joksic, G., Martelli, A., Migliore, L., Mirkova, E., Scarfi, M. R., Zijno, A., Norppa, H., and Fenech, M. (2007) An increased micronucleus frequency in peripheral blood lymphocytes predicts the risk of cancer in humans. *Carcinogenesis* **28**, 625–631.
32. Shaffer, L. G., and Tommerup, N. (eds.) (2005) *An international system for human cytogenetic nomenclature (2005): recommendations of the International Standing Committee on Human Cytogenetic Nomenclature.* Karger, Karger, Basel.
33. Benet, J., Oliver-Bonet, M., Cifuentes, P., Templado, C., and Navarro, J. (2005) Segregation of chromosomes in sperm of reciprocal translocation carriers: a review. *Cytogenet Genome Res* **111**, 281–290.
34. Munne, S. (2002) Preimplantation genetic diagnosis of numerical and structural chromosome abnormalities. *Reprod Biomed Online* **4**, 183–196.
35. Gisselsson, D., Pettersson, L., Hoglund, M., Heidenblad, M., Gorunova, L., Wiegant, J., Mertens, F., Dal Cin, P., Mitelman, F., and Mandahl, N. (2000) Chromosomal breakage-fusion-bridge events cause genetic intratumor heterogeneity. *Proc Natl Acad Sci U S A* **97**, 5357–5362.
36. Krishnaja, A. P., and Sharma, N. K. (2004) Transmission of gamma-ray-induced unstable chromosomal aberrations through successive mitotic divisions in human lymphocytes in vitro. *Mutagenesis* **19**, 299–305.
37. Bandyopadhyay, R., Heller, A., Knox-DuBois, C., McCaskill, C., Berend, S. A., Page, S. L., and Shaffer, L. G. (2002) Parental origin and timing of de novo Robertsonian translocation formation. *Am J Hum Genet* **71**, 1456–1462.
38. Gilbert, F. (1983) Chromosomes, genes, and cancer: a classification of chromosome abnormalities in cancer. *J Natl Cancer Inst* **71**, 1107–1114.
39. Albertson, D. G. (2006) Gene amplification in cancer. *Trends Genet* **22**, 447–455.
40. Solomon, E., Borrow, J., and Goddard, A. D. (1991) Chromosome aberrations and cancer. *Science* **254**, 1153–1160.
41. Croce, C. M. (1987) Role of chromosome translocations in human neoplasia. *Cell* **49**, 155–156.
42. Mitelman, F., Johansson, B., and Mertens, F. (2004) Fusion genes and rearranged genes as a linear function of chromosome aberrations in cancer. *Nat Genet* **36**, 331–334.
43. Suhardja, A., Kovacs, K., and Rutka, J. (2001) Genetic basis of pituitary adenoma invasiveness: a review. *J Neurooncol* **52**, 195–204.
44. Yang, L., Han, Y., Suarez Saiz, F., and Minden, M. D. (2007) A tumor suppressor

and oncogene: the WT1 story. *Leukemia* **21**, 868–876.

45. O'Connell, P. (2003) Genetic and cytogenetic analyses of breast cancer yield different perspectives of a complex disease. *Breast Cancer Res Treat* **78**, 347–357.
46. Peter, M. E., Legembre, P., and Barnhart, B. C. (2005) Does CD95 have tumor promoting activities? *Biochim Biophys Acta* **1755**, 25–36.
47. Cox, P. M., and Goding, C. R. (1991) Transcription and cancer. *Br J Cancer* **63**, 651–662.
48. Boxer, L. M., and Dang, C. V. (2001) Translocations involving c-myc and c-myc function. *Oncogene* **20**, 5595–5610.
49. Hecht, J. L., and Aster, J. C. (2000) Molecular biology of Burkitt's lymphoma. *J Clin Oncol* **18**, 3707–3721.
50. Janz, S. (2006) Myc translocations in B cell and plasma cell neoplasms. *DNA Repair (Amst)* **5**, 1213–1224.
51. Rui, L., and Goodnow, C. C. (2006) Lymphoma and the control of B cell growth and differentiation. *Curr Mol Med* **6**, 291–308.
52. Gartel, A. L., and Shchors, K. (2003) Mechanisms of c-myc-mediated transcriptional repression of growth arrest genes. *Exp Cell Res* **283**, 17–21.
53. Bos, T. J. (1992) Oncogenes and cell growth. *Adv Exp Med Biol* **321**, 45–49; discussion 51.
54. Jorgensen, H. G., and Holyoake, T. L. (2007) Characterization of cancer stem cells in chronic myeloid leukaemia. *Biochem Soc Trans* **35,** 1347–1351.
55. Shtivelman, E., Lifshitz, B., Gale, R. P., and Canaani, E. (1985) Fused transcript of abl and bcr genes in chronic myelogenous leukaemia. *Nature* **315**, 550–554.
56. Maru, Y. (2001) Molecular biology of chronic myeloid leukemia. *Int J Hematol* **73**, 308–322.
57. Melo, J. V., Hughes, T. P., and Apperley, J. F. (2003) Chronic myeloid leukemia. *Hematology Am Soc Hematol Educ Program*, 132–152.
58. Panigrahi, I., and Naithani, R. (2006) Imatinib mesylate: A designer drug. *J Assoc Physicians India* **54**, 203–206.
59. Steinberg, M. (2007) Dasatinib: a tyrosine kinase inhibitor for the treatment of chronic myelogenous leukemia and philadelphia chromosome-positive acute lymphoblastic leukemia. *Clin Ther* **29**, 2289–2308.
60. Bignold, L. P. (2007) Mutation, replicative infidelity of DNA and aneuploidy sequentially in the formation of malignant pleomorphic tumors. *Histol Histopathol* **22**, 321–326.
61. Houdayer, C., Gauthier-Villars, M., Lauge, A., Pages-Berhouet, S., Dehainault, C., Caux-Moncoutier, V., Karczynski, P., Tosi, M., Doz, F., Desjardins, L., Couturier, J., and Stoppa-Lyonnet, D. (2004) Comprehensive screening for constitutional RB1 mutations by DHPLC and QMPSF. *Hum Mutat* **23**, 193–202.
62. Issing, W. J., Wustrow, T. P., Oeckler, R., Mezger, J., and Nerlich, A. (1993) An association of the RB gene with osteosarcoma: molecular genetic evaluation of a case of hereditary retinoblastoma. *Eur Arch Otorhinolaryngol* **250**, 277–280.
63. Fukuzawa, R., Umezawa, A., Ochi, K., Urano, F., Ikeda, H., and Hata, J. (1999) High frequency of inactivation of the imprinted H19 gene in "sporadic" hepatoblastoma. *Int J Cancer* **82**, 490–497.
64. Koufos, A., Hansen, M. F., Copeland, N. G., Jenkins, N. A., Lampkin, B. C., and Cavenee, W. K. (1985) Loss of heterozygosity in three embryonal tumours suggests a common pathogenetic mechanism. *Nature* **316**, 330–334.
65. Rohde, K., Teare, M. D., and Santibanez Koref, M. (1997) Analysis of genetic linkage and somatic loss of heterozygosity in affected pairs of first-degree relatives. *Am J Hum Genet* **61**, 418–422.
66. Ghosh, S. K., Hajra, S., Paek, A., and Jayaram, M. (2006) Mechanisms for chromosome and plasmid segregation. *Annu Rev Biochem* **75,** 211–241.
67. Bloom, K. (2005) Chromosome segregation: seeing is believing. *Curr Biol* **15**, R500-503.
68. Taylor, S. S., Scott, M. I., and Holland, A. J. (2004) The spindle checkpoint: a quality control mechanism which ensures accurate chromosome segregation. *Chromosome Res* **12**, 599–616.
69. Duesberg, P., Rasnick, D., Li, R., Winters, L., Rausch, C., and Hehlmann, R. (1999) How aneuploidy may cause cancer and genetic instability. *Anticancer Res* **19**, 4887–4906.
70. Hassold, T., Hall, H., and Hunt, P. (2007) The origin of human aneuploidy: where we have been, where we are going. *Hum Mol Genet* **16 Spec No. 2**, R203-208.
71. Nagaishi, M., Yamamoto, T., Iinuma, K., Shimomura, K., Berend, S. A., and Knops, J. (2004) Chromosome abnormalities identified in 347 spontaneous abortions collected in Japan. *J Obstet Gynaecol Res* **30**, 237–241.

72. Altug-Teber, O., Bonin, M., Walter, M., Mau-Holzmann, U. A., Dufke, A., Stappert, H., Tekesin, I., Heilbronner, H., Nieselt, K., and Riess, O. (2007) Specific transcriptional changes in human fetuses with autosomal trisomies. *Cytogenet Genome Res* **119**, 171–184.
73. Duesberg, P., and Rasnick, D. (2000) Aneuploidy, the somatic mutation that makes cancer a species of its own. *Cell Motil Cytoskeleton* **47**, 81–107.
74. Schaefer, C., Grouse, L., Buetow, K., and Strausberg, R. L. (2001) A new cancer genome anatomy project web resource for the community. *Cancer J* **7**, 52–60.
75. Mitelman Database of Chromosome Aberrations in Cancer (2009). Mitelman, F., Johansson, B., and Mertens, F. (eds.), http://cgap.nci.nih.gov/Chromosomes/Mitelman, accessed: March 12, 2009.
76. Atkin, N. B. (2003) Aneuploidy in carcinomas may be initiated by the acquisition of a single trisomy. *Cytogenet Genome Res* **101**, 99–102.
77. Duesberg, P., Fabarius, A., and Hehlmann, R. (2004) Aneuploidy, the primary cause of the multilateral genomic instability of neoplastic and preneoplastic cells. *IUBMB Life* **56**, 65–81.
78. Dey, P. (2004) Aneuploidy and malignancy: an unsolved equation. *J Clin Pathol* **57**, 1245–1249.
79. Schmid, W. (1975) The micronucleus test. *Mutat Res* **31**, 9–15.
80. Organisation for Economic Co-Operation and Development, OECD Guideline for Testing Chemicals 474: Mammalian Erythrocyte Micronucleus Test (1997). OECD (ed.), http://titania.sourceoecd.org/vl=354845/cl=18/nw=1/rpsv/ij/oecdjournals/1607310x/v1n4/s41/p1, accessed: March 12, 2009.
81. Hart, J. W., and Hartley-Asp, B. (1983) Induction of micronuclei in the mouse. Revised timing of the final stage of erythropoiesis. *Mutat Res* **120**, 127–132.
82. Tweats, D. J., Blakey, D., Heflich, R. H., Jacobs, A., Jacobsen, S. D., Morita, T., Nohmi, T., O'Donovan, M. R., Sasaki, Y. F., Sofuni, T., Tice, R., and Group, I. W. (2007) Report of the IWGT working group on strategies and interpretation of regulatory in vivo tests I. Increases in micronucleated bone marrow cells in rodents that do not indicate genotoxic hazards. *Mutat Res* **627**, 78–91.
83. Hirai, O., Miyamae, Y., Fujino, Y., Izumi, H., Miyamoto, A., and Noguchi, H. (1991) Prior bleeding enhances the sensitivity of the in vivo micronucleus test. *Mutat Res* **264**, 109–114.
84. Suzuki, Y., Nagae, Y., Li, J., Sakaba, H., Mozawa, K., Takahashi, A., and Shimizu, H. (1989) The micronucleus test and erythropoiesis. Effects of erythropoietin and a mutagen on the ratio of polychromatic to normochromatic erythrocytes (P/N ratio). *Mutagenesis* **4**, 420–424.
85. Attia, S. M., Kliesch, U., Schriever-Schwemmer, G., Badary, O. A., Hamada, F. M., and Adler, I. D. (2003) Etoposide and merbarone are clastogenic and aneugenic in the mouse bone marrow micronucleus test complemented by fluorescence in situ hybridization with the mouse minor satellite DNA probe. *Environ Mol Mutagen* **41**, 99–103.
86. Degrassi, F., Tanzarella, C., Ieradi, L. A., Zima, J., Cappai, A., Lascialfari, A., Allegra, F., and Cristaldi, M. (1999) CREST staining of micronuclei from free-living rodents to detect environmental contamination in situ. *Mutagenesis* **14**, 391–396.
87. Holden, H. E., Majeska, J. B., and Studwell, D. (1997) A direct comparison of mouse and rat bone marrow and blood as target tissues in the micronucleus assay. *Mutat Res* **391**, 87–89.
88. Proudlock, R. J., Statham, J., and Howard, W. (1997) Evaluation of the rat bone marrow and peripheral blood micronucleus test using monocrotaline. *Mutat Res* **392**, 243–249.
89. Abramsson-Zetterberg, L. (2003) The dose-response relationship at very low doses of acrylamide is linear in the flow cytometer-based mouse micronucleus assay. *Mutat Res* **535**, 215–222.
90. Cammerer, Z., Elhajouji, A., and Suter, W. (2007) In vivo micronucleus test with flow cytometry after acute and chronic exposures of rats to chemicals. *Mutat Res* **626**, 26–33.
91. Tweats, D. J., Blakey, D., Heflich, R. H., Jacobs, A., Jacobsen, S. D., Morita, T., Nohmi, T., O'Donovan, M. R., Sasaki, Y. F., Sofuni, T., Tice, R., and Group, I. W. (2007) Report of the IWGT working group on strategies and interpretation of regulatory in vivo tests I. Increases in micronucleated bone marrow cells in rodents that do not indicate genotoxic hazards. *Mutat Res* **627**, 78–91. M3(R2): Guidance on Nonclinical Safety Studies for the Conduct of Human Clinical Trials and Marketing Authorization for Pharmaceuticals (2009). ICH (ed.), http://www.ich.org/LOB/media/MEDIA4744.pdf, accessed: March 12, 2009.
92. MacGregor, J. T., Wehr, C. M., Henika, P. R., and Shelby, M. D. (1990) The in vivo eryth-

rocyte micronucleus test: measurement at steady state increases assay efficiency and permits integration with toxicity studies. *Fundam Appl Toxicol* **14**, 513–522.

93. Hartmann, A., Agurell, E., Beevers, C., Brendler-Schwaab, S., Burlinson, B., Clay, P., Collins, A., Smith, A., Speit, G., Thybaud, V., Tice, R. R., and th International Comet Assay, W. (2003) Recommendations for conducting the in vivo alkaline Comet assay. 4th International Comet Assay Workshop. *Mutagenesis* **18**, 45–51.
94. Tice, R. R., Agurell, E., Anderson, D., Burlinson, B., Hartmann, A., Kobayashi, H., Miyamae, Y., Rojas, E., Ryu, J. C., and Sasaki, Y. F. (2000) Single cell gel/comet assay: guidelines for in vitro and in vivo genetic toxicology testing. *Environ Mol Mutagen* **35**, 206–221.
95. Yamamoto, K. I., and Kikuchi, Y. (1981) Studies on micronuclei time response and on the effects of multiple treatments of mutagens on induction of micronuclei. *Mutat Res* **90**, 163–173.
96. Heddle, J. A., Cimino, M. C., Hayashi, M., Romagna, F., Shelby, M. D., Tucker, J. D., Vanparys, P., and MacGregor, J. T. (1991) Micronuclei as an index of cytogenetic damage: past, present, and future. *Environ Mol Mutagen* **18**, 277–291.
97. Hamada, S., Sutou, S., Morita, T., Wakata, A., Asanami, S., Hosoya, S., Ozawa, S., Kondo, K., Nakajima, M., Shimada, H., Osawa, K., Kondo, Y., Asano, N., Sato, S., Tamura, H., Yajima, N., Marshall, R., Moore, C., Blakey, D. H., Schechtman, L. M., Weaver, J. L., Torous, D. K., Proudlock, R., Ito, S., Namiki, C., and Hayashi, M. (2001) Evaluation of the rodent micronucleus assay by a 28-day treatment protocol: Summary of the 13th Collaborative Study by the Collaborative Study Group for the Micronucleus Test (CSGMT)/Environmental Mutagen Society of Japan (JEMS)-Mammalian Mutagenicity Study Group (MMS). *Environ Mol Mutagen* **37**, 93–110.
98. van den Heuvel, M. J., Clark, D. G., Fielder, R. J., Koundakjian, P. P., Oliver, G. J., Pelling, D., Tomlinson, N. J., and Walker, A. P. (1990) The international validation of a fixed-dose procedure as an alternative to the classical LD50 test. *Food Chem Toxicol* **28**, 469–482.
99. Hayashi, M., Tice, R. R., MacGregor, J. T., Anderson, D., Blakey, D. H., Kirsh-Volders, M., Oleson, F. B., Jr., Pacchierotti, F., Romagna, F., Shimada, H., and et al. (1994) In vivo rodent erythrocyte micronucleus assay. *Mutat Res* **312**, 293–304.
100. Tinwell, H., and Ashby, J. (1989) Comparison of acridine orange and Giemsa stains in several mouse bone marrow micronucleus assays – including a triple dose study. *Mutagenesis* **4**, 476–481.
101. Hayashi, M., Sofuni, T., and Ishidate, M., Jr. (1983) An application of Acridine Orange fluorescent staining to the micronucleus test. *Mutat Res* **120**, 241–247.
102. Hayes, J., Doherty, A. T., Adkins, D. J., Oldman, K., and O'Donovan, M. R. (2009) The rat bone-marrow micronucleus test – study design and statistical power. *Mutagenesis* **24**, 419–424.
103. Kissling, G. E., Dertinger, S. D., Hayashi, M., and MacGregor, J. T. (2007) Sensitivity of the erythrocyte micronucleus assay: dependence on number of cells scored and inter-animal variability. *Mutat Res* **634**, 235–240.
104. Hayashi, M., MacGregor, J. T., Gatehouse, D. G., Adler, I. D., Blakey, D. H., Dertinger, S. D., Krishna, G., Morita, T., Russo, A., and Sutou, S. (2000) In vivo rodent erythrocyte micronucleus assay. II. Some aspects of protocol design including repeated treatments, integration with toxicity testing, and automated scoring. *Environ Mol Mutagen* **35**, 234–252.
105. Lovell, D. P., Anderson, D., Albanese, R., Amphlett, G. E., Clare, G., Ferguson, R., Richold, M., Papworth, D. G., and Savage, J. K. R. (1989) *Statistical analysis of in vivo cytogenetic assays.* Cambridge University Press, Cambridge.
106. Snedecor, G. W., and Cochran, W. G. (1989) *Statistical methods.* Iowa State University Press, Ames.
107. Ellard, S., Mohammed, Y., Dogra, S., Wolfel, C., Doehmer, J., and Parry, J. M. (1991) The use of genetically engineered V79 Chinese hamster cultures expressing rat liver CYP1A1, 1A2 and 2B1 cDNAs in micronucleus assays. *Mutagenesis* **6**, 461–470.
108. Torous, D. K., Dertinger, S. D., Hall, N. E., and Tometsko, C. R. (2000) Enumeration of micronucleated reticulocytes in rat peripheral blood: a flow cytometric study. *Mutat Res* **465**, 91–99.
109. Bauchinger, M. (1974) Zytogenetische Strahlenwirkungen, in *Medizinische Strahlenkunde – Biophysikalische Einführung für Studierende und Ärzte* (Hug, O., ed.), Springer, Berlin, Germany, p. 136.

Chapter 15

Cytogenetic Methods in Human Biomonitoring: Principles and Uses

Raluca A. Mateuca, Ilse Decordier, and Micheline Kirsch-Volders

Abstract

Cellular phenotypes can be applied as biomarkers to differentiate normal from abnormal biological conditions. Several cytogenetic methods have been developed and allow the accurate detection of such phenotypic changes.

Based on their mechanisms of formation, cellular phenotypes may be used either as biomarkers of exposure or as biomarkers of effect. Therefore, it is important that cytogenetic methods implemented in human biomonitoring should be based on a good knowledge of these mechanisms.

In this chapter, we aim to review the mechanistic basis, the methodology, and the use in human biomonitoring studies of four major cytogenetic endpoints: sister chromatid exchanges (SCEs), high frequency cells (HFCs), chromosomal aberrations (CAs), and micronuclei (MN). In addition, an overview of potential confounding factors on the induction of these cytogenetic makers is presented. Furthermore, the combination of cytogenetics with molecular methods, which allows chromosome and gene identification on metaphase as well as in interphase cells with high resolution, is discussed. Finally, practical recommendations for an efficient application of these cytogenetic assays and a correct interpretation of the results on the basis of cellular phenotype(s) assessment in human biomonitoring are highlighted.

Key words: Cellular phenotypes, Sister chromatid exchanges, High frequency cells, Chromosomal aberrations, Micronuclei, Cytokinesis-block micronucleus assay

1. Cellular Phenotypes as Biomarkers of Exposure and Effect in Human Populations

Cellular phenotypes, as defined by the observable traits of a cell, can be used as biomarkers to differentiate normal from abnormal biological conditions. For example, abnormal biological conditions resulting from endogenous and/or exogenous exposures can trigger reversible (e.g. recombination events) or irreversible (e.g. chromosomal translocations) phenotypic changes which can easily be detected at the cellular level. Several cytogenetic methods have been developed up to date, which allow the accurate detection of such phenotypic changes. Their implementation in human

James M. Parry and Elizabeth M. Parry (eds.), *Genetic Toxicology: Principles and Methods*, Methods in Molecular Biology, vol. 817, DOI 10.1007/978-1-61779-421-6_15, © Springer Science+Business Media, LLC 2012

biomonitoring studies allows the understanding of the complex mechanisms underlying genotoxic responses to endogenous and/or exogenous agents. Detection of abnormal cellular phenotypes in human populations may reflect either recent or long-term exposure to various endogenous and/or exogenous sources. As a result of recent exposure, transient cellular phenotypes (e.g. recombination events) can be detected, while long-term accumulated exposures may result in stable cellular phenotypes (e.g. chromosomal translocations) with greater impact on cell fate. Based on their mechanisms of formation, cellular phenotypes may, therefore, be used either as biomarkers of exposure or as biomarkers of effect. Therefore, a sound selection of cytogenetic methods for use in human biomonitoring should be based on a good knowledge of these mechanisms.

Critical for using cellular phenotypes as biomarkers in human populations is the selection of biological specimens to be investigated. Assessment of cytogenetic changes in readily available surrogate cells has been frequently used to estimate events occurring at the target organs and to provide early warning signals for health risk. The most frequently used surrogate cells in human studies are the peripheral blood lymphocytes (PBL), although the usefulness of non-blood cells (e.g. buccal, nasal, scalp, sputum, urothelial) in the biomonitoring of mutagens/carcinogens exposure is also documented (for review (1)). The major reason for using lymphocytes is that these cells circulate throughout the body and that they have reasonably long life-span if a suitable cell type is considered (e.g. T-lymphocytes); therefore, they can be damaged in any tissue/organ-specific toxic environment (2). However, the relevance of cytogenetic changes measured in surrogate cells to the corresponding phenotypic changes in target tissues/organs is often unknown.

The objective of this chapter is to review the mechanistic basis, the methodology and the uses of four major cytogenetic endpoints applied for biomonitoring purposes: sister chromatid exchanges (SCEs), high frequency cells (HFCs), chromosomal aberrations (CAs), and micronuclei (MN).

2. SCEs/HFCs and Their Use in Human Biomonitoring

Sister chromatid exchanges (SCEs) are reciprocal DNA exchanges occurring during the S-phase of the cell cycle between the two sister chromatids of a duplicated chromosome (for review (3)). SCEs are induced by a large number of S-phase-dependent clastogens, including UV-light and some metals (e.g. Cr, Cd, As, Co, Ni) (4, 5). Although little is known about their molecular basis, SCEs appear to be the consequence of DNA replication on a damaged template, possibly at the replication fork. The simplest pathway by which SCEs are likely occur, involves the initial collapse of a replication fork when it encounters a pre-existing nick or gap in one parental strand (6). The subsequent generation of a replication-associated double-strand break (DSB) with one free end, will initiate the RAD51-mediated invasion of the intact strand by conservative homologous recombination (HR) repair. Resolution of the resulting Holliday junction by non-crossing over will trigger SCEs formation (Fig. 1). Another possible mechanism of SCEs induction involves the initial stall of a replication fork owing to obstacles on the DNA template (7). A stalled replication fork may reverse due to positive tortional strain in the DNA or due to enzymatic action, generating a half chicken foot structure.

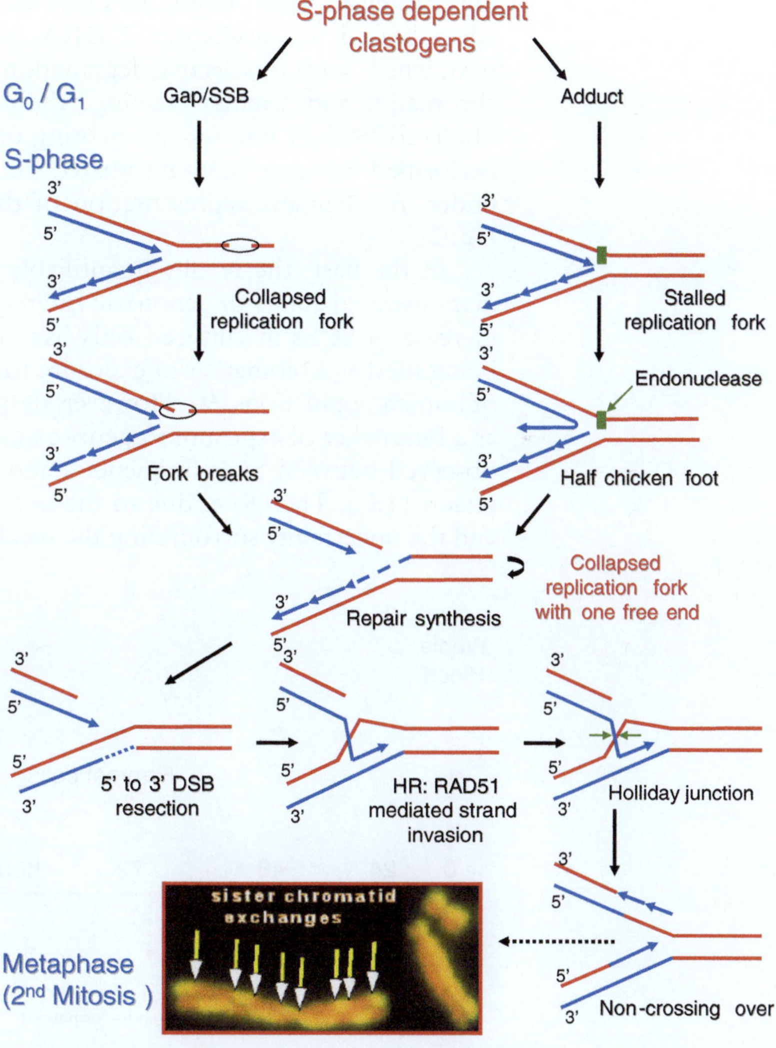

Fig. 1. Mechanisms of SCEs formation (Adapted from refs. 6, 7).

Cleavage of this intermediary structure by endonucleases will give rise to a collapsed replication fork with one free end, which can be further repaired by conservative HR, resulting in SCEs (Fig. 1).

The standard fluorescence plus Giemsa (FPG) assay used to visualize SCEs is based on the differential staining of the sister chromatids, after two rounds of replication in the presence of bromodeoxyuridine (BrdU) (5). BrdU is an analogue of thymidine which is efficiently incorporated into the elongating DNA strands during replication (6). When cells are cultured through a single replication cycle in the presence of this thymidine analogue, one DNA strand in each daughter chromatid is substituted with BrdU (8). After a second round of replication, one chromatid contains one substituted DNA strand, while both strands of its sister chromatid are substituted. The chromatids can be further differentiated by treatment with the 33258 Hoechst dye, which fluoresces at a lower intensity when bound to DNA substituted with BrdU than when bound to unsubstituted DNA. Following photosensitization, which leads to selective degradation of the highly substituted chromatid, and Giemsa staining, the sister chromatids can be observed by light microscopy. Scoring of SCE frequencies can be performed using a semi-automated computer-based metaphase finder. A schematic representation of the SCE assay is given in Fig. 2.

In the past, the readily quantifiable nature of SCEs and the demonstrated ability of genotoxic chemicals to induce a significant increase in SCEs in cultured cells have resulted in this endpoint being used as a biomarker of exposure to genotoxic agents in PBL of human populations (9). However, despite its good performance as a biomarker of exposure to genotoxins, no association has been observed between SCE frequencies and cancer risk ((10–12) for review (13)). Therefore, due to the lack of predictivity for cancer and the uncertainty surrounding the mechanism and the biological

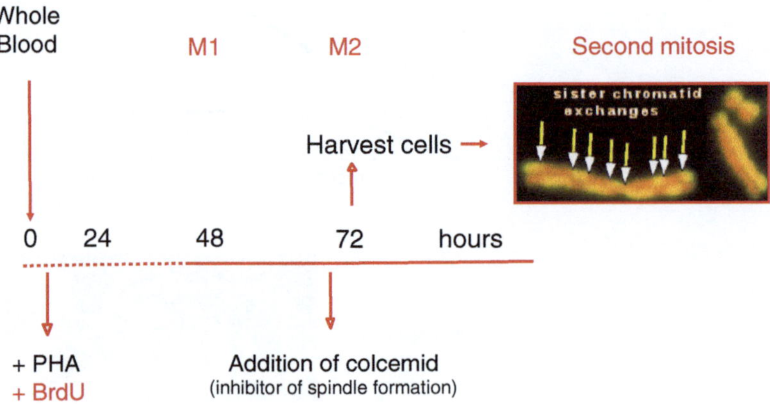

Fig. 2. Schematic illustration of the SCE assay.

significance of SCEs, this biomarker is not frequently used as a routine test in human biomonitoring at the present moment.

In the early 1980, the concept of high frequency cells (HFCs) was introduced as a way to increase the sensitivity of the SCE assay for detecting the effect of genotoxic exposures (9, 14). HFCs were defined as cells whose SCE frequency exceeded the 95th percentile of the SCE distribution in a pooled dataset from control individuals. In many instances, the frequency of HFCs was shown to be more sensitive than the mean SCE level for discriminating exposed groups from the baseline (15–17). Therefore, the HFC frequency was increasingly used as an additional endpoint in the SCE assay, although the real nature of HFCs was still uncertain. Some authors had postulated that HFCs could represent either a subpopulation of cells particularly susceptible to genotoxic stress or a subpopulation of DNA repair-deficient lymphocytes (18–20). More recently, it has been shown that HFCs more likely represent a subpopulation of long-lived lymphocytes which accumulated a large number of SCE-inducing lesions resulting from exposure to natural or artificial genotoxins (21). The accumulation of this damage could be due to the fact that many of these agents produce lesions that are efficiently repaired in mitogen-stimulated lymphocytes in vitro, but not in circulating lymphocytes. However, despite the sensitivity of HFCs for exposure assessment, no association between increased HFCs and cancer risk was observed in a recently analyzed cohort of healthy individuals (12).

Potential confounders which may influence the baseline SCEs and HFCs induction in PBL are shown in Table 1.

Table 1
Influence of potential confounders on baseline SCEs and HFCs induction in PBL

Cytogenetic endpoints: SCEs and HFCs
Age: ↑SCE (22) No significant influence of age on SCE induction (23, 24)
Gender: ↑SCE (**females**[a]) (23–25)
Micronutrient status: ↑SCE and HFC positively correlated with plasma vitamin B12 levels in smokers (26)
Smoking: ↑SCE [**smokers**[b], all levels combined; **heavy smokers**[c] (>10 cigarettes/day)] (25) ↑SCE [**smokers**[b], all levels combined] (23, 24, 27) ↑SCE [**smokers**[d] (≥ 7 years)] (28)

(continued)

Table 1 (continued)

Cytogenetic endpoints: SCEs and HFCs

Genetic polymorphisms: Xenobiotic metabolizing enzymes
GSTs: No significant influence of *GSTM1/GSTT1/GSTP1*^{105} on SCE/HFC induction (25, 27, 29–32) ↑SCE (*GSTT1* **null**)[e] (33–36); ↓SCE (*GSTT1* **null**)[e] (28) ↑SCE (smokers with *GSTM1* **null**)[f] (37); ↑SCE (*GSTM1* **null**)[g] (28, 38, 39) ↑SCE [variant (**Ile/Val or Val/Val**) *GSTP1*^{105}] [h] (28)
EPHX: No significant influence of *EPHX* on SCE induction (30, 31)
CYPs: ↑SCE and HFC [**wild type** *CYP2E1(Pst)*][i] (40) No significant influence of *CYP2E1* on SCE induction (27, 31)
NAT2: No significant influence of NAT2 on SCE induction (25, 31) Folate metabolism enzymes
MTHFR^{222}*:* No influence of *MTHFR*^{222} polymorphism on SCE/HFC induction (26, 32) DNA repair enzymes
hOGG1^{326}*:* No significant influence of *hOGG1*^{326} on SCE/HFC induction (32)
XRCC3^{241}*:* No significant influence of *XRCC1*^{241} on SCE/HFC induction (25, 32)
XRCC1^{194}*:* No significant influence of *XRCC1*^{194} on SCE induction (25)
XRCC1^{280}*:* No significant influence of *XRCC1*^{280} on SCE induction (25)
XRCC1^{399}*:* No significant influence of *XRCC1*^{399} on SCE/HFC induction (25, 27, 32) ↑SCE (smokers with **Gln/Gln** *XRCC1*^{399}) [j] (39) ↑SCE [**smokers** (>10 cigarettes/day) with variant(**Arg/Gln or Gln/Gln**) *XRCC1*^{399}] [j] (30)
XPD^{751}*:* No significant influence of *XPD*^{751} on SCE induction (39)

[a] Reference category: **males**
[b] Reference category: **non-smokers**
[c] Reference category: **light smokers**(<10 cigarettes/day)
[d] Reference category: **smokers** (< 7 years)
[e] Reference category: *GSTT1* **positive**
[f] Reference category: smokers with *GSTM1* **positive**
[g] Reference category: *GSTM1* **positive**
[h] Reference category: *GSTP1*^{105} **wild type (Ile/Ile)**
[i] Reference category: **heterozygote** *CYP2E1*(Pst)
[j] Reference category: smokers with **wild type** (**Arg/Arg**) *XRCC1*^{399}

3. CAs and Their Use in Human Biomonitoring (Reviewed in (41))

Chromosomal aberrations (CAs) are changes in normal chromosome structure or number that can occur spontaneously or as a result of chemical/radiation treatment (42). Structural CAs in PBL, as assessed by the chromosome aberration (CA) assay, have been used for over 30 years in occupational and environmental settings as a biomarker of early effects of genotoxic carcinogens (43). Structural CAs are most commonly scored in metaphase-arrested cells that have been fixed, spread on microscope slides, and Giemsa stained (8). However, this method is not suitable for estimation of numerical CAs as artefactual chromosome loss may occur. Therefore, this section will exclusively focus on the mechanistic basis of structural CAs formation and on their use as biomarkers of early genotoxic effects in the human population.

Structural CAs may be induced by direct DNA breakage, by replication on a damaged DNA template, by inhibition of DNA synthesis, and by other mechanisms (e.g. topoisomerase II inhibitors) (for review (3)). Based on morphological criteria, structural CAs can be divided into two main classes: chromosome-type aberrations (CSAs), involving both chromatids of one or multiple chromosomes, and chromatid-type aberrations (CTAs), involving only one of the two chromatids of a chromosome or several *chromosomes* (3, 43). Generation of structural CAs requires one or several DNA DSBs, but the mechanisms of CSAs and CTAs formation appear to differ with the mutagen (ionizing radiation versus chemicals) and involve specific DNA repair mechanisms (43). CSAs are mostly generated in vivo in G_0/G_1 lymphocytes by S-phase-independent clastogens (e.g. ionizing radiation), and reflect DSBs which are incompletely repaired or unrepaired by the non homologous end joining (NHEJ) and single strand annealing (SSA) mechanisms. After DNA synthesis and chromosome duplication, the aberrations formed in G_0/G_1 are doubled and chromosome-type breaks and exchanges (e.g. dicentric and ring chromosomes, translocations) are seen in metaphase. CTAs (e.g. chromatid type breaks and exchanges) arise predominantly in vitro during the S-phase of the cultured lymphocytes, in response to base modifications and single-strand breaks induced in vivo by S-phase- dependent clastogens (e.g. chemicals); incomplete or failed repair of these lesions by conservative HR will trigger CTAs formation in the subsequent metaphase (3, 43). Figure 3 shows some examples of CSAs and CTAs formation in response to S-phase independent and S-phase dependent clastogens, respectively.

Since structural CAs may be induced via DNA breakage, their survival depends on the fate of the DNA breaks. DNA breaks may either rejoin such that the chromosome is restored to its original state, rejoin incorrectly or not rejoin at all. These last two cases

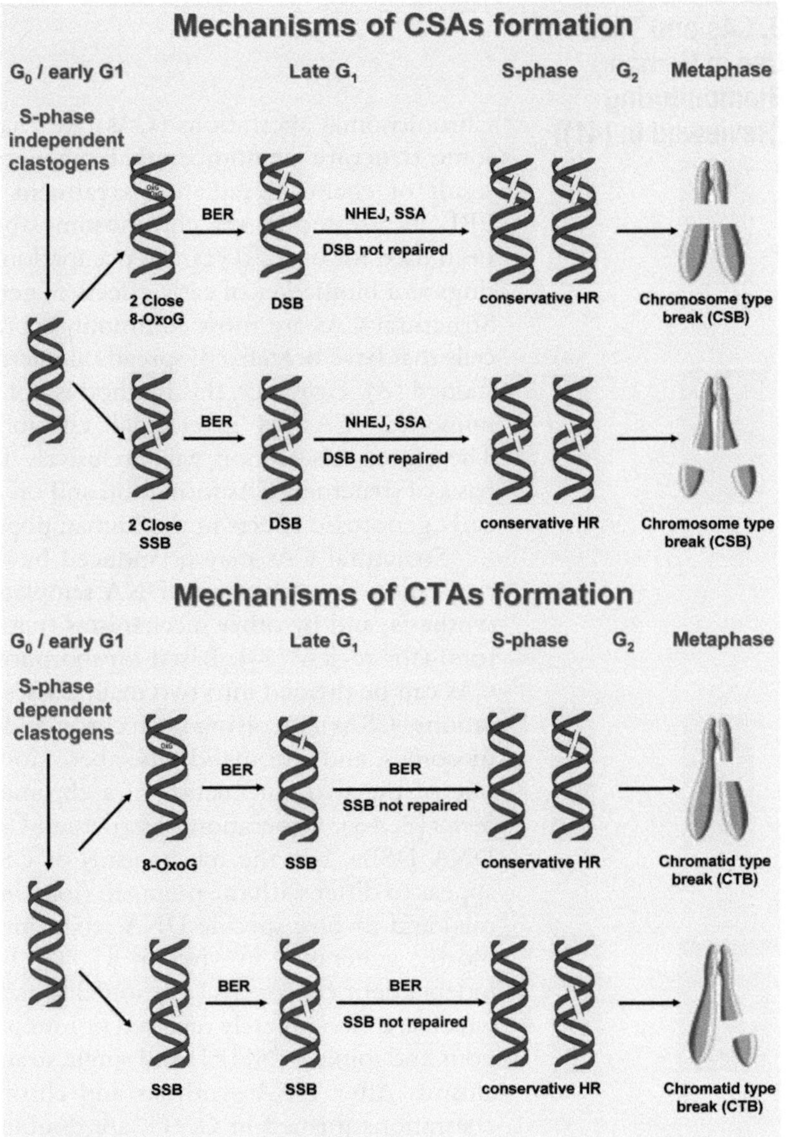

Fig. 3. Examples of CSAs and CTAs formation.

may be observable on microscopic preparations of metaphase cells. The type of chromosomal aberration will be decisive for the fate of the cell. Cells bearing unstable aberrations such as dicentrics, rings, and chromosome fragments can be eliminated by apoptosis in a p53-dependent way (44). Stable aberrations, such as balanced translocations, on the other hand may have deleterious consequences for the organism since they are much less effective in causing apoptotic cell death.

In human biomonitoring studies, detection of structural CAs is most commonly performed in PBL by use of the ex vivo/in vitro

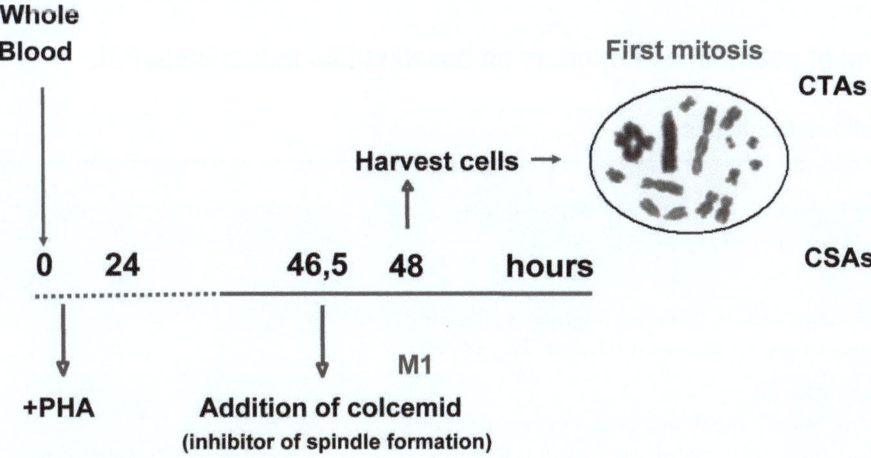

Fig. 4. Schematic illustration of the CA assay.

CA assay. A schematic representation of the classical in vitro CA test is given in Fig. 4. As peripheral lymphocytes are in the resting G_0 stage of the cell cycle, they are stimulated to divide by an aspecific antigen [e.g. phytohaemagglutinin (PHA)]. After 46.5 h, a spindle inhibitor (e.g. colcemid, colchicine) is added to block the cells in the (pro)metaphase of the first mitosis. After a subsequent hypotonic treatment, fixation (at 48 h), even spreading of the chromosomes in a single plane on the microscope slides, and classically Giemsa staining, the metaphases can be analyzed for structural CAs under the microscope (8). No full automation of the CA assay has been developed up to the present time, but interactive scoring is possible using a semi-automated metaphase finder. The sensitivity (lowest detectable dose/concentration) of the CA assay depends on the solubility, reactivity, uptake and metabolism of the mutagen/carcinogen (45). The specificity of the CA assay lies in its ability to identify a specific type of mutagen and is rather limited. Indeed the CA assay detects mutagens which are capable of inducing DNA-strand breaks, but does not allow identification of the clastogen class. However, information on the types of aberrations induced [S-phase-independent (CSAs) versus S-phase-dependent (CTAs)] following occupational and/or environmental exposure, gives some indication on the nature of the clastogenic damage produced (i.e. strand breaks versus base damage, respectively). The recently introduced fluorescence in situ hybridization (FISH) chromosome painting methods (see Subheading 5), increase the efficiency and specificity of the CA assay by allowing the detection of chromosome-specific rearrangements and/or loss. Potential confounders which may influence the baseline CAs induction in PBL are shown in Table 2.

At the time the CA test was included in the regular medical check-ups of workers exposed to mutagens/carcinogens, extensive

Table 2
Influence of potential confounders on baseline CAs induction in PBL

Cytogenetic endpoint: CAs

Age: ↑ AB.C (22); ↑CSAs (46); ↑Tr. (47–50), ↑ insertions (47), ↑ exchanges (49), ↑dicentrics (47), ↑acentric fragments (23, 47, 51)
Gender: ↑CTGs (**females**[a]) (25); ↑acentric fragments (**females**[a]) (23) No evidence of gender effect on baseline Tr. (48, 50)
Micronutrient status: No effect of vitamin supplementation on spontaneous AB.C (52) ↓AB.C after antioxidant supplementation (vitamin C, E, carotene, selenium) in smokers (53, 54)
Smoking: ↑CTGs (**smokers**[b], all levels combined) (25); ↑CTAs (**smokers**[b], all levels combined) (55); ↑Tr. and ↑ exchanges (**smokers**[b], all levels combined) (49) ↑dicentrics in **heavy smokers**[b] (>30 cigarettes/day) (51) No significant influence of smoking (all levels combined) on CAs levels (23, 56)
Genetic polymorphisms: Xenobiotic metabolizing enzymes
GSTs: ↑ AB.C (*GSTT1* **null**)[c] (29, 57); ↑ CSBs (*GSTT1* **null**[c]) (25) ↑ total CAs (*GSTM1* **positive**)[d] (58); ↑ CTBs (*GSTM1* **positive** non-smokers)[e] (25) ↑ Tr. [**wild type (Ile/Ile)** $GSTP1^{105}$][f] (59)
EPHX: ↓ total CAs (**fast** *EPHX*)[g] (60); ↑ total CAs (**slow** *EPHX*)[h] (61, 62)
CYPs: ↑ Tr. (new-borns *CYP1A1*MspI **heterozygotes**)[i] (63) ↑ AB.C [**wild type (Ile/Ile)***CYP1A1*2C*][j] (59)
NAT2: ↑ AB.C (*NAT2* **slow acetylators**)[k] (64); ↑ CTBs (*NAT2* **slow acetylators**)[k] (35) ↑ CSAs (*NAT2* **slow acetylators**)[k] (65) ↑ CSBs (non-smokers *NAT2* **slow acetylators**)[l] (25); ↑CSAs (exchanges) (*NAT2* **fast acetylators)**[m] (55) Folate metabolism enzymes
$MTHFR^{222}$*:* ↑Tr. (**Val/Val** $MTHFR^{222}$)[n] (59); ↑CTBs and CTAs [variant (**Ala/Val** or **Val/Val)** $MTHFR^{222}$][o] (55) DNA repair enzymes
$hOGG1^{326}$*:* No significant influence of $hOGG1^{326}$ on baseline CAs induction (55, 59)
$XRCC3^{241}$*:* ↑ CTBs (**Thr/Met** $XRCC3^{241}$)[p] (25); ↑ CSAs (**Met/Met** $XRCC3^{241}$)[p] (55)
$XRCC1^{194}$*:* ↓ CSBs (**Arg/Trp** $XRCC1^{194}$)[q] (25)

(continued)

Table 2 (continued)

Cytogenetic endpoint: CAs

XRCC1^{280}: ↓CSBs [**variant (Arg/His or His/His)** *XRCC1*^{280}] [r] (25)
XRCC1^{399}: ↓CTGs (non-smokers with **Gln/Gln** *XRCC1*^{399}) [s] (25) ↑AB.C and total CAs [**variant (Arg/Gln or Gln/Gln)** *XRCC1*^{399}] [t](32)
XPD^{312}: ↑ total CAs (breaks) [**variant**[u] **(Asp/Asn or Asn/Asn)** *XPD*^{312} **smokers; old (>48y) variant**[v] *XPD*^{312} carriers] (66)
XPD^{751}: ↓ total CAs (**Gln/Gln**[w] *XPD*^{751}; **Gln/Gln**[x] *XPD*^{751} **smokers**) (67)

AB.C aberrant cells, *CSAs* chromosome-type aberrations, *CTAs* chromatid-type aberrations, *CSBs* chromosome-type breaks, *CTGs* chromatid-type gaps, *CTBs* chromatid-type breaks, *Tr* translocations

[a] Reference category: **males;**
[b] Reference category: **non-smokers**
[c] Reference category: *GSTT1* **positive**
[d] Reference category: *GSTM1* **null**
[e] Reference category: non-smokers with *GSTM1* **null**
[f] Reference category: **heterozygote (Ile/Val)** *GSTP1*^{105}
[g] Reference category: **slow** *EPHX* activity
[h] Reference category: **fast** *EPHX* activity
[i] Reference category: new-borns with **wild type** *CYP1A1*MspI
j Reference category: **heterozygote (Ile/Val)** *CYP1A1*2C*
[k] Reference category: *NAT2* **fast acetylators**
[l] Reference category: non-smokers *NAT2* **fast acetylators**
[m] Reference category: *NAT2* **slow acetylators**
[n] Reference category: **Ala/Ala** *MTHFR*^{222} and **Ala/Val** *MTHFR*^{222}
[o] Reference category: **Ala/Ala** *MTHFR*^{222}
[p] Reference category: **wild type (Thr/Thr)** *XRCC3*^{241}
[q] Reference category: **wild type (Arg/Arg)** *XRCC1*^{194}
[r] Reference category: **wild type (Arg/Arg)** *XRCC1*^{280}
[s] Reference category: non-smokers with **wild type (Arg/Arg)** *XRCC1*^{399}
[t] Reference category: **wild type (Arg/Arg)** *XRCC1*^{399}
[u] Reference category: **wild type (Asp/Asp)** *XPD*^{312} carriers
[v] Reference category: young (< 48y) **wild type (Asp/Asp)** *XPD*^{312} carriers
[w] Reference category: **wild type (Lys/Lys)** and **heterozygote (Lys/Gln)** *XPD*^{751} carriers;
[x] Reference category: **wild type (Lys/Lys)** *XPD*^{751} carriers.

coordinated validation studies were not required. However, the CAs test was widely accepted and considered as validated through its intensive application in many laboratories. The extensive use of the CA assay over the last 30 years has resulted in the accumulation of analytical data and has enabled the examination of the potential association between previously measured structural CA frequency and subsequent cancer outcome (for review (68)). The first epidemiological data showing that the frequency of CAs in PBL may predict cancer incidence in human populations was provided in the

1990s by Nordic and Italian cohort studies (10, 11, 69–71). Later analyses nested within those cohorts, suggested that the cancer risk predictivity of CAs is not modified by occupational exposure to carcinogens or tobacco smoking, supporting the hypothesis that the association between structural CAs and cancer might be independent of exposure (72). A more recent examination of the Nordic-Italian cohorts aimed at evaluating whether CSAs and CTAs have different cancer risk predicitivity (43). A significantly elevated cancer risk was observed in the Nordic cohorts for subjects with both high CSAs and high CTAs at test, while the results of the Italian cohort did not indicate any clear-cut difference in cancer predictivity between the CSA and CTA biomarkers. Aside from the Nordic-Italian studies, the cancer risk predictivity of CAs sub-classes has been addressed in Taiwanese, Czech and Central European cohorts (73–75). The results of these studies indicated that the association between CAs frequency and cancer risk might be limited to CSAs. Moreover, some discrepancies in the strength of the association between CAs and cancer risk and in the independence from exposure to carcinogens were observed in the Czech and Central European cohorts compared to the Nordic-Italian cohorts. To clarify these issues, a large pooled analysis of 22,358 cancer free individuals was recently conducted, including CAs data from all European cohorts published so far (76). To standardize for inter-laboratory variation, subjects were classified within each laboratory according to tertiles of CA frequency. An increased relative risk (RR) of cancer was observed for subjects in the medium [RR = 1.31; 95% confidence interval (CI) = 1.07–1.60] and in the high [RR = 1.41; 95% CI = 1.16–1.72] tertiles of the CAs distribution, when compared with the low tertile. This increase was mostly driven by CSAs. Moreover, the presence of ring chromosomes increased the RR to 2.22 (95% CI = 1.34–3.68). When the association of CAs frequency with specific cancer sites was investigated, a significant increase in stomach cancer was observed for subjects in the high tertile of the CAs distribution [RR = 3.13; 95% CI = 1.17–8.39]. The effect of CAs levels on the overall cancer risk was not modified by occupational exposure to carcinogens or tobacco smoking. The results of this pooled analysis reinforce the evidence of a link between CAs frequency and cancer risk, highlighting the importance of structural CAs as biomarkers of health risk prediction in human populations.

4. MN and Their Use in Human Biomonitoring (Reviewed in (41))

Micronuclei (MN) are small, extra-nuclear bodies that arise in dividing cells from acentric chromosome/chromatid fragments or whole chromosomes/chromatids that lag behind in anaphase and are not included in the daughter nuclei in telophase (77). MN harbouring chromosomal fragments may result from direct double-strand DNA breakage, conversion of SSBs into DSBs after cell replication, or inhibition of DNA synthesis. Misrepair of two chromosome breaks may lead to an asymmetrical chromosome rearrangement producing a dicentric chromosome and an acentric fragment. Frequently, the centromeres of the dicentric chromosomes are pulled to opposite poles of the cells at anaphase resulting in the formation of a nucleoplasmic bridge (NPB) between the daughter nuclei and an acentric fragment that lags behind to form a micronucleus (78) (for review (79)). MN harbouring whole chromosomes are primarily formed from defects in the chromosome segregation machinery such as deficiencies in the cell cycle controlling genes, failure of the mitotic spindle, kinetochore, or other parts of the mitotic apparatus or by damage to chromosomal substructures, mechanical disruption (for review (3)), and hypomethylation of centromeric DNA (80). MN can also arise by gene amplification via breakage–fusion–bridge (BFB) cycles when amplified DNA is selectively localized to specific sites at the periphery of the nucleus and eliminated via nuclear budding (NBUD) during the S-phase of the cell cycle (for review (81)). Figure 5 shows the most common mechanisms of MN formation.

The fate of MN after their formation in the micronucleated cell is poorly understood. Their post-mitotic fate includes:

- Elimination of the micronucleated cell as a consequence of apoptosis (82);
- Expulsion from the cell (when the DNA within the MN is not expected to be functional or capable of replication owing to the absence of the necessary cytoplasmic components) (83);
- Reincorporation into the main nucleus (when the reincorporated chromosome may be indistinguishable from those of the main nucleus and might resume normal biological activity) (83);
- Retention within the cell's cytoplasm as an extra-nuclear entity (when MN may complete one or more rounds of DNA/chromosome replication) (83).

The use of MN as a measure of early genotoxic effects has become a standard assay in human biomonitoring studies. Micronuclei can be scored in exfoliated epithelial cells (from buccal or nasal mucosa, or urine) and in erythrocytes, but the standard ex vivo/in vitro MN test is usually performed in lymphocytes. The cytokinesis-block micronucleus (CBMN) assay is the most extensively used method for measuring MN in cultured human lymphocytes because scoring is specifically restricted to cells that have divided once after mitogen stimulation (77, 84) (for review (79)). A schematic representation of the classical ex vivo/in vitro CBMN test is given in Fig. 6. In a classical ex vivo/in vitro CBMN test,

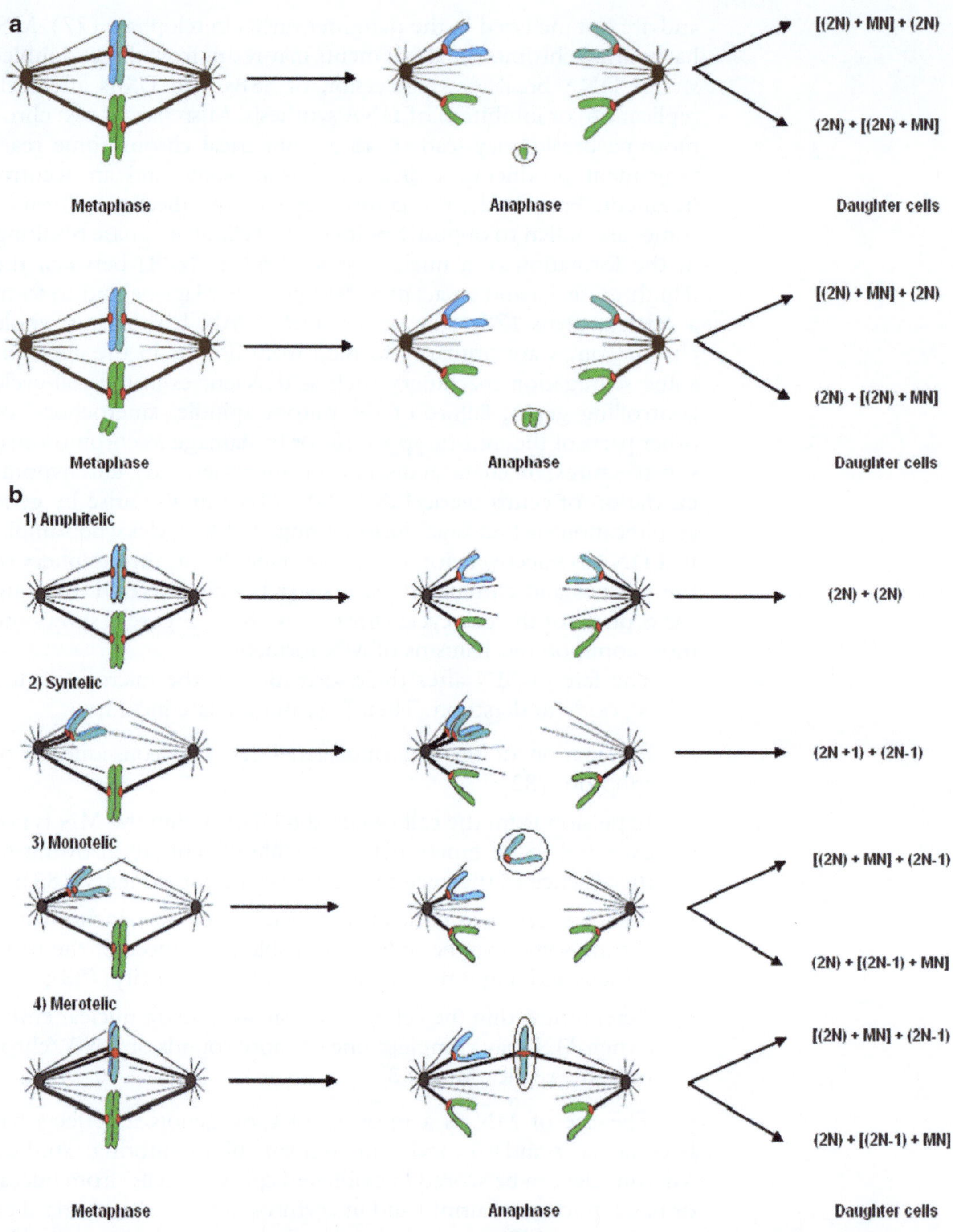

Fig. 5. Most common mechanisms of MN formation: (**a**) acentric chromosome/chromatid fragments; (**b**) misattachment of tubulin fibres on kinetochore; (**c**) tubulin depolymerization; (**d**) defects in centromeric DNA, in kinetochore proteins or in kinetochore assembly; (**e**) late replication, peripheral location in the nucleus and epigenetic modifications of histones (for review ref. 41).

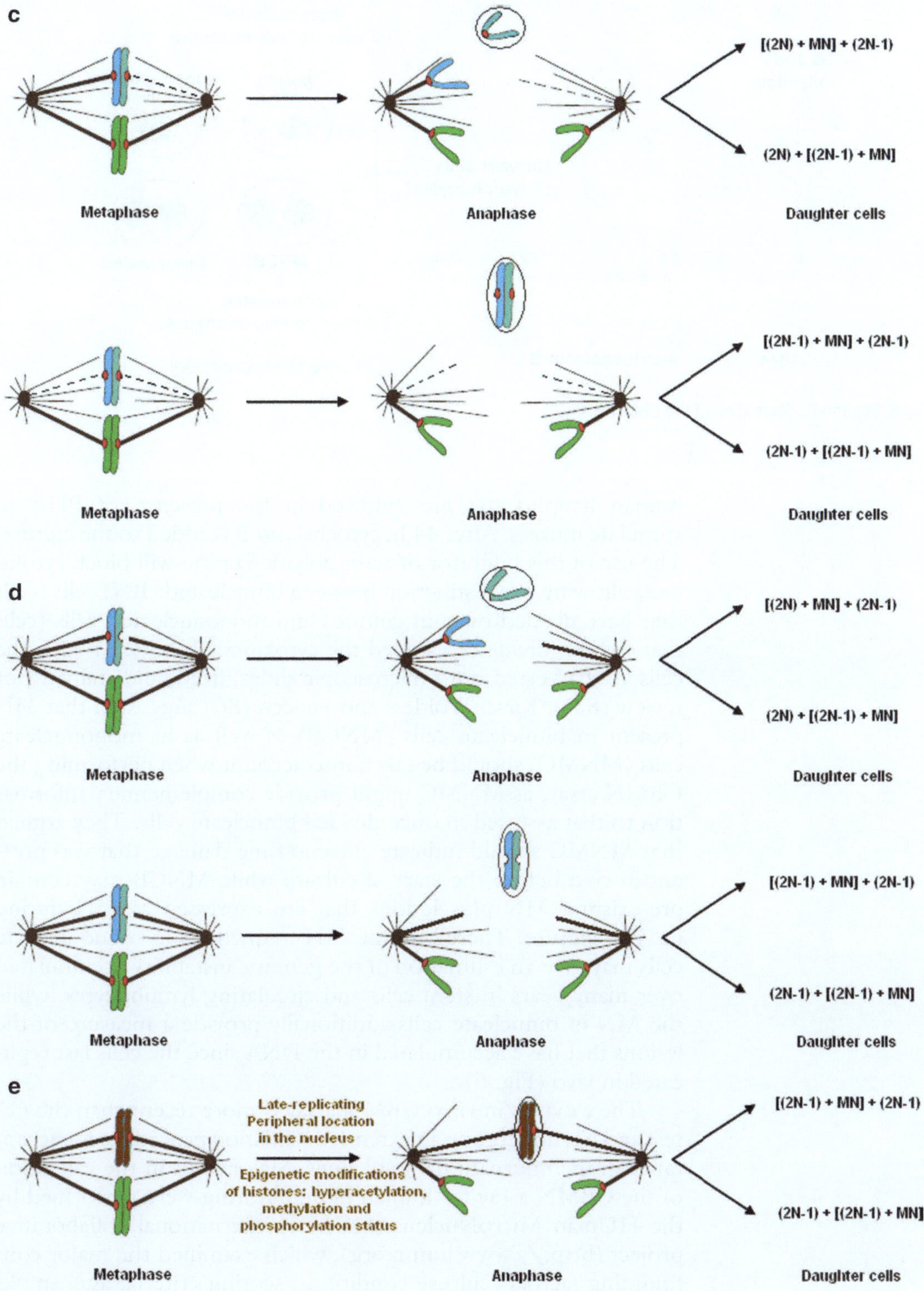

Fig. 5. (continued)

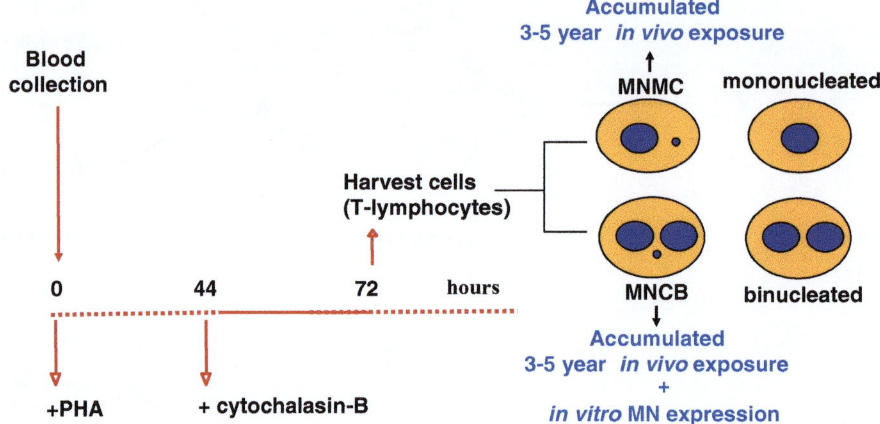

Fig. 6. Schematic illustration of the CBMN assay.

human lymphocytes are cultured in the presence of PHA to stimulate mitosis. After 44 h, cytochalasin B is added to the culture. The use of this inhibitor of actin polymerization will block cytokinesis allowing the distinction between binucleated (BN) cells (cells that have divided once in culture) and mononucleated cells (cells that did not divide or escaped the cytokinesis-block). At 72 h the cells are harvested onto microscopic slides, fixed, and stained (for review (85)). Kirsch-Volders and Fenech (86) suggested that MN present in binucleate cells (MNCB) as well as in mononucleate cells (MNMC) should be taken into account when performing the CBMN assay, as MNMC might provide complementary information to that assessed in once-divided binucleate cells. They argued that MNMC should indicate chromosome damage that was present in vivo before the start of culture while MNCB may contain pre-existing MN plus lesions that are expressed as MN during in vitro culture. Therefore, the MN frequencies in mononucleate cells may give an estimation of the genome instability accumulated over many years in stem cells and circulating lymphocytes, while the MN in binucleate cells additionally provide a measure of the lesions that have accumulated in the DNA since the cells last replicated in vivo (Fig. 6).

The ex vivo/in vitro CBMN assay is more recent than the CA test and has undergone an extensive validation procedure for acceptance in the international guidelines. Major steps in the validation of the CBMN assay for human biomonitoring were performed by the HUman MicroNucleus (HUMN) international collaborative project (http://www.humn.org), which examined the major confounding factors (culture conditions, scoring criteria, age, smoking, genotype, exposure) influencing MN induction. The major advantage of the CBMN assay over the traditional CA test lies in its ability to detect both clastogenic and aneugenic events, leading to structural and numerical CAs, respectively (for review (87, 88)). The distinction between the two phenomena, which can be achieved even at low doses of mutagen/carcinogen exposure by centromere (FISH) and kinetochore [Calcinosis, Raynaud's phenomenon, Esophageal dysmotility, Sclerodactyly, and Telangiectasia (CREST)] identification, contributes to the high sensitivity and specificity of the method. However, since MN formed from entire chromosomes with disrupted or detached kinetochore may not be identified by CREST-derived anti-kinetochore antibodies (for review (89)), the use of FISH with probes labelling the pan (peri) centromeric region of chromosomes is recommended to distinguish between micronuclei containing a whole chromosome (centromere positive micronucleus) and an acentric chromosome fragment (centromere negative micronucleus) (*see* Subheading 5). Besides its capacity to detect clastogenic and aneugenic events, the CBMN assay can provide additional measures of genotoxicity and cytotoxicity: nucleoplasmic bridges (NPB, a marker of chromosome rearrangement), nuclear buds (NBUD, a marker of gene amplification) (90), cell division inhibition (by estimation of the nuclear division index) (for review (88)), necrosis and apoptosis (91) (for review (79)). For this reason, the CBMN test can be considered as a "cytome" assay covering chromosome instability, mitotic dysfunction, cell proliferation and cell death (for review (92)). Another advantage of the CBMN assay lies in its recent automation (93). The implementation of automated methods for MN detection allows the analysis of a large number of cells and the exclusion of subjective judgment and individual scoring skills. However, at this point, automated systems for MN detection require further validation for large scale applicability in human biomonitoring studies. Potential confounders which may influence the baseline MN induction in PBL are shown in Table 3.

Considering the cancer risk predictivity of CAs and the mechanistic similarities between CAs and MN formation, an association between MN frequency and cancer risk was also expected. The possibility of a link between MN induction and cancer development was first addressed by Nordic and Italian cohort studies (10, 11, 70, 71), which found that high MN frequencies in PBLs were not predictive of an increased cancer risk. However, these studies did not have sufficient power and/or follow up time to allow conclusions to be drawn concerning the cancer predictivity of MN. Moreover, most of the data had not been obtained by using the sensitive ex vivo/in vitro cytokinesis block methodology. A more recent analysis performed within the framework of the HUMN project indicates that an increased frequency of MN in peripheral blood lymphocytes predicts the cancer risk in humans (119). The analysis was performed on a total of 6,718 disease-free subjects from 10 countries (20 laboratories), who were screened for MN frequency between 1980 and 2002. To standardize for the inter-laboratory

Table 3
Influence of potential confounders on baseline MN induction in PBL

Cytogenetic endpoint: MN

Age: ↑MNCB (22, 24, 80, 94–102)
Gender: ↑MNCB (**females**[a])(for review (89, 96)) (101, 103)
Micronutrient status: ↑MNCB negatively correlated with plasma folate and B12 and positively correlated with homocysteine and vitamin C (79, 104–108)
Smoking: ↓ MNCB (**smokers**[b], all levels combined) but ↑MNCB in **heavy smokers**[b] (> 30 cigarettes/day) (109) ↓ MNCB (**smokers**[b], all levels combined) (101) No significant influence of smoking (all levels combined) on MNCB levels (102)
Genetic polymorphisms: Xenobiotic metabolizing enzymes
GSTs: ↑C- MNCB (*GSTM1* **positive**[c]) (110); ↑MNMC (*GSTM1* **positive**[c]) (32) No significant influence of *GSTM1/GSTT1/GSTP1*^{105}on MNCB induction (32, 101)
EPHX: ↓ MNCB (**fast** *EPHX*)[d] (60)
CYPs: ↓ MNCB [CYP2E1 (*)3][e] (111)
NAT2: No influence of NAT2 on MNCB induction (112) Folate metabolism enzymes
MTHFR^{222}*:* ↑ MNCB and ↓ NBUD (**Val/Val** *MTHFR*^{222})[f] (90) No significant influence of *MTHFR*^{222} polymorphism on MNCB/MNMC induction (32, 113) DNA repair enzymes
hOGG1^{326}*:* ↓ MNCB (Ser/Ser *hOGG1*^{326} **smokers**)[g] (102); ↓ MNCB (**Ser/Ser** *hOGG1*^{326})[h] (114) No significant influence of *hOGG1*^{326} on MNCB/ MNMC induction (32, 115)
XRCC3^{241}*:* No significant influence of *XRCC3*^{241} on MNCB/MNMC induction (32, 110, 114–117)
hOGG1^{326}*-XRCC3*^{241}*:* ↑ MNCB [**Met/Met** *XRCC3*^{241}-variant (Ser/Cys or Cys/Cys) *hOGG1*^{326}][i] (102)
XRCC1^{194}*:* No significant influence of *XRCC1*^{194} on MNCB/MNMC induction (114, 115)
XRCC1^{280}*:* No significant influence of *XRCC1*^{280} on MNCB/MNMC induction (114, 115)

(continued)

Table 3 (continued)

Cytogenetic endpoint: MN

XRCC1^{399}:
No significant influence of *XRCC1*^{399} on MNCB/MNMC induction (32, 110, 114, 115, 117); ↑ MNCB (**Arg/Gln** *XRCC1*^{399})[j] (118)
XPD^{312}:
↑ MNCB [**wild type (Asp/Asp)** *XPD*^{312}][k] (118); No significant influence of *XPD*^{312} on MNCB induction (116)

[a] Reference category: **males**
[b] Reference category: **non-smokers**
[c] Reference category: *GSTM1* **null**
[d] Reference category: **slow** *EPHX* activity
[e] Reference category: **wild type** CYP2E1 (*)1/(*)1
[f] Reference category: **wild type (Ala/Ala)** *MTHFR*^{222}
[g] Reference category: **non-smokers** with wild-type (Ser/Ser) *hOGG1*^{326} genotype
[h] Reference category: **variant** (Ser/Cys or Cys/Cys) *hOGG1*^{326}
[i] Reference category: **wild type** (**Thr/Thr**) *XRCC3*^{241}-variant (Ser/Cys or Cys/Cys) *hOGG1*^{326}
[j] Reference category: **wild type (Arg/Arg)** *XRCC1*^{399}
[k] Reference category: **heterozygote (Asp/Asn)** *XPD*^{312}

variability, subjects were classified according to the percentiles of MN distribution within each laboratory as low, medium, or high frequency. A significant overall increase in cancer incidence in subjects with medium (RR = 1.84; 95% CI = 1.28–2.66) and high MN frequency (RR = 1.53; 95% CI = 1.04–2.25) was observed. Moreover, the same groups showed a decreased cancer-free survival ($p = 0.001$ and $p = 0.025$, respectively) which was present in all national cohorts and for all major cancer sites, especially urogenital (RR = 2.80; 95% CI = 1.17–6.73) and gastro-intestinal cancers (RR = 1.74; 95% CI = 1.01–4.71). The predictive value of MN frequency as a biomarker for cancer risk in the general population was recently confirmed by a case-control study nested within a longitudinal cohort of 1,650 disease-free individuals (120). A significantly higher MN frequency was found in PBL of subjects who developed cancer within 14 years after blood sampling (cases) as compared to those who were still cancer free at the end of the follow-up period (controls) (4.7 ± 3.4 versus 1.5 ± 1.7 MN/1,000 BN cells). Moreover, an increased risk of cancer death was found in individuals with high MN frequency (>2.5/1,000 BN cells) (OR = 10.7; 95% CI = 4.6–24.9) when compared to individuals with low MN frequency (∴2.5/1,000 BN cells). The existing evidence linking MN frequencies with cancer risk was also substantiated by a recent meta-analysis of 37 publications, which clearly showed a 28–64% increase in the baseline MN level of untreated cancer patients compared to cancer-free referents (121). Besides the cancer risk predictivity of

MN, other biomarkers of the CBMN assay (i.e. NPB and NBUD) were also shown to be strongly associated with cancer risk, indicating that the integration of various cytogenetic biomarkers within one assay may improve cancer risk prediction (122). All the recently accumulated evidence on the cancer predictive value of increased MN frequencies make the ex vivo/in vitro CBMN assay a good candidate for wide usage in human biomonitoring.

5. FISH-Cytogenetics

In recent years, cytogenetics in combination with molecular methods has made rapid progress, resulting in new molecular cytogenetic methodologies such as fluorescence in situ hybridization (FISH). These new methodologies bridge classic cytogenetics with molecular approach and allow chromosome and gene identification on metaphase as well as in interphase with high resolution.

FISH uses fluorescently labelled DNA probes complementary to regions of individual chromosomes. These labelled DNA segments hybridize with the cytological targets in the sample and can be visualized by fluorescence microscopy in interphase nuclei or on metaphase chromosomes (123).

FISH with whole chromosome paints has been the primary method to quantify and characterize chromosome damage from environmental or occupational exposures. The advantages of chromosome painting are the speed of the assay and the ability to identify relatively stable events such as translocations in parallel with the enumeration of unstable dicentrics. Currently, most painting is performed with just one colour of paint, but sometimes with two or three. Each additional probe in the cocktail increases the proportion of the genome in which aberrations can be observed and also increases the fraction of all exchanges that can be detected. Some years ago, spectral karyotyping (SKY) (124) and multiplex FISH (mFISH) (125) made it possible to paint each of the 24 human chromosomes in a unique colour. These approaches have a wide range of uses, including the characterization of structural interchromosomal aberrations and complex chromosomal rearrangements. However, this approach requires expensive probes and the analysis time per cell is substantially longer than when only a few chromosomes are painted. Besides interchromosomal exchanges commonly detected by chromosome painting, intrachromosomal exchanges such as pericentric and paracentric inversions occur and may form an important component of risk evaluation. These are not detectable by chromosome painting and require the use of chromosome bands. The bands may be natural, e.g. G-bands, or synthetic, i.e. based on region-specific partial chromosome paints that are hybridized simultaneously and labelled

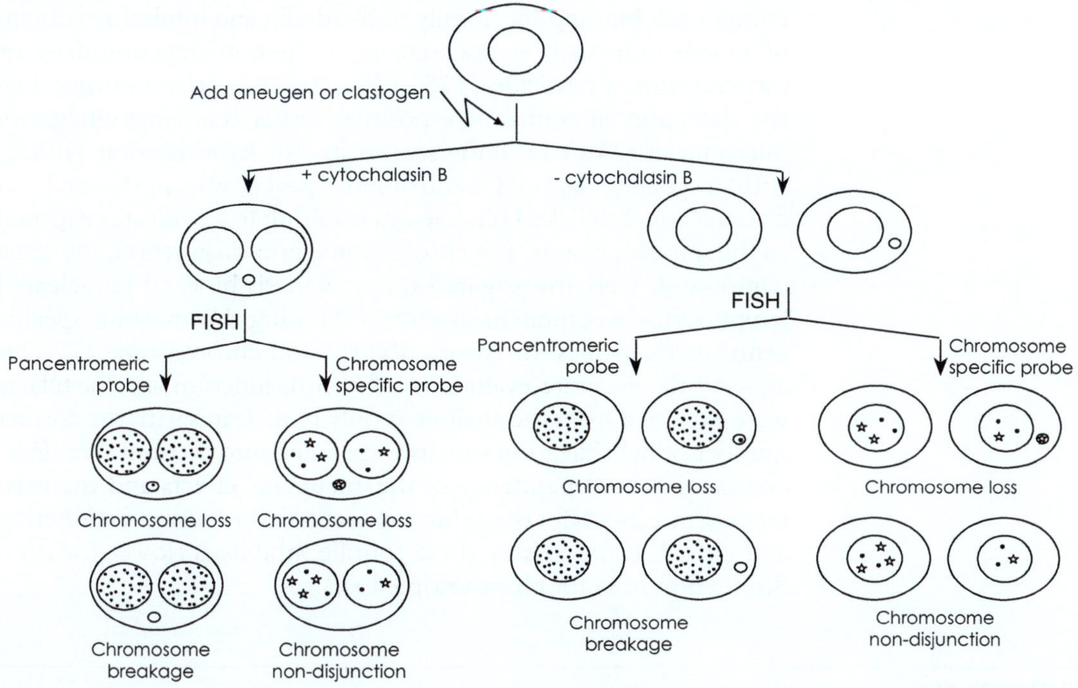

Fig. 7. Detection of chromosome loss and chromosome non-disjunction in the cytokinesis-block micronucleus assay combined with FISH (for review ref. 85).

in multiple colours with the multicolour banding technique (mBANDs) (126, 127). This technology is very accurate and well validated, but labour- and cost-demanding.

Combination of fluorescence in situ hybridization (FISH) using probes for pancentromeric, regions with the cytokinesis-block micronucleus assay allows discrimination between clastogenic (inducing micronuclei containing chromosome fragments) and aneugenic agents (inducing micronuclei containing whole chromosomes) (91) (Fig. 7). Discriminating between these two phenomena is important in studies of human genotoxic effects in vivo. The specific analysis of the induced type of micronuclei may considerably improve the sensitivity of detecting the exposure effect.

Detection of FISH signals for chromosome specific sequences (centromeric chromosome-specific probes) in both macronuclei and micronuclei also allows discrimination between aneuploidy due to chromosome non-disjunction or to chromosome loss and provides an accurate analysis of non-disjunction (unequal distribution of unique homologous chromosome pairs in the daughter nuclei) (Fig. 7). This is very helpful to perform risk assessment for compounds with threshold type of dose-responses. Our laboratory used the MN assay in combination with FISH for the in vitro demonstration of thresholds for microtubule inhibitors aneugenic compounds binding specifically to β-tubulin and inhibiting tubulin polymerization such as nocodazole, a chemotherapeutic drug or carbendazim, a pesticide (128, 129). To assess chromosome loss the detection of centromere-positive versus centromere-negative micronuclei (MN) by fluorescence in situ hybridization (FISH) with a general alphoid centromeric probe was performed on cytochalasin-B blocked binucleates resulting from cultures exposed to the spindle poisons. For chromosome non-disjunction, the same compounds were investigated on cytokinesis-blocked binucleated lymphocytes in combination with FISH using chromosome specific centromeric probes for chromosome 1 and chromosome 17. This allowed the accurate evaluation of non-disjunction since artefacts were excluded from the analysis as only binucleates with the correct number of hybridization signals were taken into account. We demonstrated dose dependency of the aneugenic effects and the existence of thresholds for the induction of chromosome non-disjunction and chromosome loss by these spindle inhibitors (lower for non-disjunction than for chromosome loss).

6. Practical Recommendations

Cellular phenotypes, as assessed by SCEs, CAs, and MN assays, can be used in human biomonitoring studies to determine the impact of environmental, occupational, or medical factors on genome stability. However, the efficient application of these cytogenetic assays in human biomonitoring requires:

1. Validated biomarkers;
2. Good protocols and sufficient expertise in the laboratories conducting the tests;
3. A sound selection of endpoints and assays based on the good knowledge of the exposure source, the exposure route and the exposure-cellular phenotype(s) relationship; if the exposure source is difficult to characterize (e.g. complex mixtures), the complementary assessment of several cellular phenotypes is recommended;
4. Knowledge of the potential confounding factors influencing the baseline induction of SCEs/HFCs, CAs, and MN in human populations (from our expertise, we classify their importance as shown in Table 4);
5. Availability of well-matched controls for each study design;
6. Adequate statistics, sample sizes and data analysis.

These conditions are a prerequisite for the sound interpretation of the biomonitoring results on the basis of cellular phenotype(s) assessment.

Table 4
Overview of the main confounding factors influencing the baseline SCEs/HFCs, CAs, and MN induction in PBL

Cytogenetic endpoints	Confounding factors				Genetic polymorphisms		
	Age	Gender	Smoking	Micronutrient status	Xenobiotic metabolism	Folate metabolism	DNA repair
SCEs/HFCs	±	+	++	i	+	i	+
CAs	+	±	++	±	+	+	+
MN	++	++	+	++	+	+	+

++: strong/sufficient evidence; +: weak evidence or more research required; ±: conflicting data; i: insufficient data and more research required

Application of the SCEs/HFCs, CAs, and MN cytogenetic endpoints in human biomonitoring has been extensively performed through single studies (25, 39, 48, 55, 59, 110, 114, 115, 117, 130), which are usually small sized due to the costs and working-time required by the technical procedures; moreover, in single biomonitoring studies of occupational exposure, small sample sizes are inherent to the modest workforce in the industrial settings. Therefore, a common obstacle in reaching definite conclusions based on the measurement of SCEs/HFC, CAs, and MN endpoints in human populations has been the lack of statistical power of such single study approaches. To overcome this problem, several pooled/meta-analyses (12, 76, 99, 101, 102, 109, 119, 121) have been undertaken over the last years, allowing a better assessment of the questions raised by the preliminary single biomonitoring studies (e.g. association with cancer risk, impact of potential confounders on the induction of cellular phenotypes). However, while the association between SCEs/HFC, CAs, MN, and cancer risk has been addressed in very large pooled/meta-analyses (12, 76, 119), few similar approaches have been undertaken, on a much smaller scale, to assess the link between cellular phenotypes (CAs, MN) and genetic polymorphisms (101, 102, 131). The future investigation of such biomarker-genotype associations in larger scale pooled/meta-analyses, possibly focused on only one type of exposure, could greatly improve our understanding of the mechanistic basis underlying the formation of altered cellular phenotype(s).

Finally, the investigation of cytogenetic endpoints in human biomonitoring studies has so far involved risk assessment at group level, while little research has been focused on the evaluation of risk for individual people (132). Therefore, another crucial issue in the future application of SCEs/HFC, CAs, and MN in biomonitoring studies will be the meaning of altered cellular phenotypes at individual level.

Acknowledgements

Financial support for this review paper was provided by the ECNIS (Environmental Cancer, Nutrition and Individual Susceptibility) (contract number FOOD-CT-2005-513943) and the NewGeneris (contract number FOOD-CT-2005-016320-2) European Union projects.

References

1. Salama SA, Serrana M and Au WW (1999). Biomonitoring using accessible human cells for exposure and health risk assessment. Mutat Res 436(1):99–112.
2. Au WW (2007). Usefulness of biomarkers in population studies: from exposure to susceptibility and to prediction of cancer. Int J Hyg Environ Health 210(3–4):239–46.
3. Albertini RJ, Anderson D, Douglas GR, Hagmar L, Hemminki K, Merlo F, Natarajan AT, Norppa H, Shuker DE, Tice R and others (2000). IPCS guidelines for the monitoring of genotoxic effects of carcinogens in humans. International Programme on Chemical Safety. Mutat Res 463(2):111–72.
4. Helleday T, Nilsson R and Jenssen D (2000). Arsenic[III] and heavy metal ions induce intrachromosomal homologous recombination in the hprt gene of V79 Chinese hamster cells. Environ Mol Mutagen 35(2):114–22.
5. Garcia-Sagredo JM (2008). Fifty years of cytogenetics: A parallel view of the evolution of cytogenetics and genotoxicology. Biochim Biophys Acta 1779(6–7):363–75.
6. Wilson DM, 3rd and Thompson LH (2007). Molecular mechanisms of sister-chromatid exchange. Mutat Res 616(1–2):11–23.
7. Helleday T (2003). Pathways for mitotic homologous recombination in mammalian cells. Mutat Res 532(1–2):103–15.
8. Dean BJ and Danford, N (1984). Assays for the detection of chemically-induced chromosome damage in cultured mammalian cells. In: S. Venitt, Parry JM, editors. Mutagenicity Testing – a Practical Approach. Oxford: IRL Press. p 187–232.
9. Carrano A and Moore D (1982). The Rationale and Methodology for quantifying Sister Chromatid Exchange in Humans. In: Heddle JA, editor. Mutagenicity: New Horizons in Genetic Toxicology. New York: Academic Press Inc. p 267–304.
10. Hagmar L, Brogger A, Hansteen IL, Heim S, Hogstedt B, Knudsen L, Lambert B, Linnainmaa K, Mitelman F, Nordenson I and others (1994). Cancer risk in humans predicted by increased levels of chromosomal aberrations in lymphocytes: Nordic study group on the health risk of chromosome damage. Cancer Res 54(11):2919–22.
11. Hagmar L, Bonassi S, Stromberg U, Brogger A, Knudsen LE, Norppa H and Reuterwall C (1998). Chromosomal aberrations in lymphocytes predict human cancer: a report from the European Study Group on Cytogenetic Biomarkers and Health (ESCH). Cancer Res 58(18):4117–21.
12. Bonassi S, Lando C, Ceppi M, Landi S, Rossi AM and Barale R (2004). No association between increased levels of high-frequency sister chromatid exchange cells (HFCs) and the risk of cancer in healthy individuals. Environ Mol Mutagen 43(2):134–6.
13. Norppa H, Bonassi S, Hansteen IL, Hagmar L, Stromberg U, Rossner P, Boffetta P, Lindholm C, Gundy S, Lazutka J and others (2006). Chromosomal aberrations and SCEs as biomarkers of cancer risk. Mutat Res 600(1–2):37–45.
14. Moore D and Carrano A (1984). Statistical Analysis of High SCE Frequency. In: Tice R, Hollaender A, editors. Sister Chromatid Exchanges. New York and London Plenum Press. p 469–479.
15. Marez T, Shirali P, Hildebrand HF and Haguenoer JM (1991). Increased frequency of sister chromatid exchange in workers exposed to high doses of methylmethacrylate. Mutagenesis 6(2):127–9.
16. Motykiewicz G, Michalska J, Pendzich J, Perera FP and Chorazy M (1992). A cytogenetic study of men environmentally and occupationally exposed to airborne pollutants. Mutat Res 280(4):253–9.
17. Van Hummelen P, Gennart JP, Buchet JP, Lauwerys R and Kirsch-Volders M (1993). Biological markers in PAH exposed workers and controls. Mutat Res 300(3–4):231–9.

18. Yager JW (1987). Effect of concentration-time parameters on sister-chromatid exchanges induced in rabbit lymphocytes by ethylene oxide inhalation. Mutat Res 182(6):343–52.
19. Kelsey KT, Wiencke JK, Eisen EA, Lynch DW, Lewis TR and, Little JB (1988). Persistently elevated sister chromatid exchanges in ethylene oxide-exposed primates: the role of a subpopulation of high frequency cells. Cancer Res 48(17):5045–50.
20. Kelsey KT, Wiencke JK, Eisen EA, Lynch DW, Lewis TR and Little JB (1989). Sister chromatid exchange in chronic ethylene oxide-exposed primates: unexpected effects of in vitro culture duration, incubation temperature, and serum supplementation. Cancer Res 49(7):1727–31.
21. Ponzanelli I, Landi S, Bernacchi F and Barale R (1997). The nature of high frequency sister chromatid exchange cells (HFCs). Mutagenesis 12(5):329–33.
22. Bolognesi C, Abbondandolo A, Barale R, Casalone R, Dalpra L, De Ferrari M, Degrassi F, Forni A, Lamberti L, Lando C and others (1997). Age-related increase of baseline frequencies of sister chromatid exchanges, chromosome aberrations, and micronuclei in human lymphocytes. Cancer Epidemiol Biomarkers Prev 6(4):249–56.
23. Bender MA, Preston RJ, Leonard RC, Pyatt BE, Gooch PC and Shelby MD (1988). Chromosomal aberration and sister-chromatid exchange frequencies in peripheral blood lymphocytes of a large human population sample. Mutat Res 204(3):421–33.
24. Bukvic N, Gentile M, Susca F, Fanelli M, Serio G, Buonadonna L, Capurso A and Guanti G (2001). Sex chromosome loss, micronuclei, sister chromatid exchange and aging: a study including 16 centenarians. Mutat Res 498(1–2):159–67.
25. Tuimala J, Szekely G, Wikman H, Jarventaus H, Hirvonen A, Gundy S and Norppa H (2004). Genetic polymorphisms of DNA repair and xenobiotic-metabolizing enzymes: effects on levels of sister chromatid exchanges and chromosomal aberrations. Mutat Res 554(1–2):319–33.
26. Zijno A, Andreoli C, Leopardi P, Marcon F, Rossi S, Caiola S, Verdina A, Galati R, Cafolla A and Crebelli R (2003). Folate status, metabolic genotype, and biomarkers of genotoxicity in healthy subjects. Carcinogenesis 24(6):1097–103.
27. Wong RH, Wang JD, Hsieh LL and Cheng TJ (2003). XRCC1, CYP2E1 and ALDH2 genetic polymorphisms and sister chromatid exchange frequency alterations amongst vinyl chloride monomer-exposed polyvinyl chloride workers. Arch Toxicol 77(8):433–40.
28. Perez-Cadahia B, Lafuente A, Cabaleiro T, Pasaro E, Mendez J and Laffon B (2007). Initial study on the effects of Prestige oil on human health. Environ Int 33(2):176–85.
29. Sram RJ, Rossner P, Peltonen K, Podrazilova K, Mrackova G, Demopoulos NA, Stephanou G, Vlachodimitropoulos D, Darroudi F and Tates AD (1998). Chromosomal aberrations, sister-chromatid exchanges, cells with high frequency of SCE, micronuclei and comet assay parameters in 1, 3-butadiene-exposed workers. Mutat Res 419(1–3):145–54.
30. Lei YC, Hwang SJ, Chang CC, Kuo HW, Luo JC, Chang MJ and Cheng TJ (2002). Effects on sister chromatid exchange frequency of polymorphisms in DNA repair gene XRCC1 in smokers. Mutat Res 519(1–2):93–101.
31. Laczmanska I, Gil J, Karpinski P, Stembalska A, Kozlowska J, Busza H, Trusewicz A, Pesz K, Ramsey D, Schlade-Bartusiak K and others (2006). Influence of polymorphisms in xenobiotic-metabolizing genes and DNA-repair genes on diepoxybutane-induced SCE frequency. Environ Mol Mutagen 47(9):666–73.
32. Mateuca RA, Carton C, Roesems S, Roelants M, Lison D and Kirsch-Volders M. (2010) Genotoxicity surveillance program in workers dismantling arsenic trichloride-containing ammunition. Int Arch Occup Environ Health 83: 483–495.
33. Schroder KR, Wiebel FA, Reich S, Dannappel D, Bolt HM and Hallier E (1995). Glutathione-S-transferase (GST) theta polymorphism influences background SCE rate. Arch Toxicol 69(7):505–7.
34. Wiencke JK, Pemble S, Ketterer B and Kelsey KT (1995). Gene deletion of glutathione S-transferase theta: correlation with induced genetic damage and potential role in endogenous mutagenesis. Cancer Epidemiol Biomarkers Prev 4(3):253–9.
35. Norppa H, Bemardini S, Wikman H, Jarventaus H, Rosenberg C, Bolognesi C and Hirvonen A (2000). Cytogenetic biomarkers in occupational exposure to diisocyanates: influence of genetic polymorphisms of metabolic enzymes. Environ Mol Mutagen 35(Suppl. 31):45.
36. Laffon B, Pasaro E and Mendez J (2002). Evaluation of genotoxic effects in a group of workers exposed to low levels of styrene. Toxicology 171(2–3):175–86.
37. van Poppel G, de Vogel N, van Balderen PJ and Kok FJ (1992). Increased cytogenetic damage in smokers deficient in glutathione

S-transferase isozyme mu. Carcinogenesis 13(2):303–5.

38. Cheng TJ, Christiani DC, Wiencke JK, Wain JC, Xu X and Kelsey KT (1995). Comparison of sister chromatid exchange frequency in peripheral lymphocytes in lung cancer cases and controls. Mutat Res 348(2):75–82.
39. Duell EJ, Wiencke JK, Cheng TJ, Varkonyi A, Zuo ZF, Ashok TD, Mark EJ, Wain JC, Christiani DC and Kelsey KT (2000). Polymorphisms in the DNA repair genes XRCC1 and ERCC2 and biomarkers of DNA damage in human blood mononuclear cells. Carcinogenesis 21(5):965–71.
40. Carere A, Andreoli C, Galati R, Leopardi P, Marcon F, Rosati MV, Rossi S, Tomei F, Verdina A, Zijno A and others (2002). Biomonitoring of exposure to urban air pollutants: analysis of sister chromatid exchanges and DNA lesions in peripheral lymphocytes of traffic policemen. Mutat Res 518(2):215–24.
41. Mateuca R, Lombaert N, Aka PV, Decordier I and Kirsch-Volders M (2006). Chromosomal changes: induction, detection methods and applicability in human biomonitoring. Biochimie 88(11):1515–31.
42. Russell PJ (2002). Chromosomal mutations. In: Cummings B, editor. Genetics. San Francisco: Pearson Education Inc. p 595–621.
43. Hagmar L, Stromberg U, Bonassi S, Hansteen IL, Knudsen LE, Lindholm C and Norppa H (2004). Impact of types of lymphocyte chromosomal aberrations on human cancer risk: results from Nordic and Italian cohorts. Cancer Res 64(6):2258–63.
44. Schwartz JL and Jordan R (1997). Selective elimination of human lymphoid cells with unstable chromosome aberrations by p53-dependent apoptosis. Carcinogenesis 18(1):201–5.
45. Mateuca R and Kirsch-Volders, M (2007). Chromosomal damage. In: P.B. Farmer, S.A. Kyrtopoulos, Emmeny JM, editors. State of validation of biomarkers of carcinogen exposure and early effects and their applicability to molecular epidemiology. Lodz, Poland: Nofer Institute of Occupational medicine. p 24–32.
46. Carbonell E, Peris F, Xamena N, Creus A and Marcos R (1996). Chromosomal aberration analysis in 85 control individuals. Mutat Res 370(1):29–37.
47. Ramsey MJ, Moore DH, 2nd, Briner JF, Lee DA, Olsen L, Senft JR and Tucker JD (1995). The effects of age and lifestyle factors on the accumulation of cytogenetic damage as measured by chromosome painting. Mutat Res 338(1–6):95–106.
48. Pressl S, Edwards A and Stephan G (1999). The influence of age, sex and smoking habits on the background level of fish-detected translocations. Mutat Res 442(2):89–95.
49. Tawn EJ and Whitehouse CA (2001). Frequencies of chromosome aberrations in a control population determined by G banding. Mutat Res 490(2):171–7.
50. Whitehouse CA, Edwards AA, Tawn EJ, Stephan G, Oestreicher U, Moquet JE, Lloyd DC, Roy L, Voisin P, Lindholm C and others (2005). Translocation yields in peripheral blood lymphocytes from control populations. Int J Radiat Biol 81(2):139–45.
51. Stephan G and Pressl S (1999). Chromosomal aberrations in peripheral lymphocytes from healthy subjects as detected in first cell division. Mutat Res 446(2):231–7.
52. Sidneva ES, Katosova LD, Platonova VI, Beketova NA, Kodentsova VM, Chebotarev AN, Durnev AD and Bochkov NP (2005). Effects of vitamin supply on spontaneous and chemically induced mutagenesis in human cells. Bull Exp Biol Med 139(2):230–4.
53. Dusinska M, Kazimirova A, Barancokova M, Beno M, Smolkova B, Horska A, Raslova K, Wsolova L and Collins AR (2003). Nutritional supplementation with antioxidants decreases chromosomal damage in humans. Mutagenesis 18(4):371–6.
54. Volkovova K, Barancokova M, Kazimirova A, Collins A, Raslova K, Smolkova B, Horska A, Wsolova L and Dusinska M (2005). Antioxidant supplementation reduces inter-individual variation in markers of oxidative damage. Free Radic Res 39(6):659–66.
55. Catalan J, Heilimo I, Falck GC, Jarventaus H, Rosenstrom P, Nykyri E, Kallas-Tarpila T, Pitkamaki L, Hirvonen A and Norppa H (2009). Chromosomal aberrations in railroad transit workers: Effect of genetic polymorphisms. Environ Mol Mutagen. 50: 304–316.
56. van Diemen PC, Maasdam D, Vermeulen S, Darroudi F and Natarajan AT (1995). Influence of smoking habits on the frequencies of structural and numerical chromosomal aberrations in human peripheral blood lymphocytes using the fluorescence in situ hybridization (FISH) technique. Mutagenesis 10(6):487–95.
57. Landi S, Norppa H, Frenzilli G, Cipollini G, Ponzanelli I, Barale R and Hirvonen A (1998). Individual sensitivity to cytogenetic effects of 1,2:3,4-diepoxybutane in cultured human lymphocytes: influence of glutathione S-transferase M1, P1 and T1 genotypes. Pharmacogenetics 8(6):461–71.

58. Sorsa M, Osterman-Golkar S, Peltonen K, Saarikoski ST and Sram R (1996). Assessment of exposure to butadiene in the process industry. Toxicology 113(1–3):77–83.
59. Sram RJ, Beskid O, Binkova B, Chvatalova I, Lnenickova Z, Milcova A, Solansky I, Tulupova E, Bavorova H, Ocadlikova D and others (2007). Chromosomal aberrations in environmentally exposed population in relation to metabolic and DNA repair genes polymorphisms. Mutat Res 620(1–2):22–33.
60. Migliore L, Naccarati A, Coppede F, Bergamaschi E, De Palma G, Voho A, Manini P, Jarventaus H, Mutti A, Norppa H and others (2006). Cytogenetic biomarkers, urinary metabolites and metabolic gene polymorphisms in workers exposed to styrene. Pharmacogenet Genomics 16(2):87–99.
61. Vodicka P, Soucek P, Tates AD, Dusinska M, Sarmanova J, Zamecnikova M, Vodickova L, Koskinen M, de Zwart FA, Natarajan AT and others (2001). Association between genetic polymorphisms and biomarkers in styrene-exposed workers. Mutat Res 482(1–2):89–103.
62. Vodicka P, Kumar R, Soucek P, Haufroid V, Stetina R, Dusinska M, Kuricova M, Zamecnikova M, Buchanova J, Norppa H and others (2003). Genetic polymorphisms of DNA repair and biotransformation genes and possible links with chromosomal aberrations and single-strand breaks in DNA. Toxicol Lett 144(Suppl. 1):s154.
63. Pluth JM, Ramsey MJ and Tucker JD (2000). Role of maternal exposures and newborn genotypes on newborn chromosome aberration frequencies. Mutat Res 465(1–2):101–11.
64. Knudsen LE, Norppa H, Gamborg MO, Nielsen PS, Okkels H, Soll-Johanning H, Raffn E, Jarventaus H and Autrup H (1999). Chromosomal aberrations in humans induced by urban air pollution: influence of DNA repair and polymorphisms of glutathione S-transferase M1 and N-acetyltransferase 2. Cancer Epidemiol Biomarkers Prev 8(4 Pt 1):303–10.
65. Norppa H, Lindholm C, Ojajarvi A, Salomaa S and Hirvonen A (2001). Metabolic polymorphisms and frequency of chromosomal aberrations. Environ Mol Mutagen 37(Suppl. 32):58.
66. Affatato AA, Wolfe KJ, Lopez MS, Hallberg C, Ammenheuser MM and Abdel-Rahman SZ (2004). Effect of XPD/ERCC2 polymorphisms on chromosome aberration frequencies in smokers and on sensitivity to the mutagenic tobacco-specific nitrosamine NNK. Environ Mol Mutagen 44(1):65–73.
67. Vodicka P, Kumar R, Stetina R, Sanyal S, Soucek P, Haufroid V, Dusinska M, Kuricova M, Zamecnikova M, Musak L and others (2004). Genetic polymorphisms in DNA repair genes and possible links with DNA repair rates, chromosomal aberrations and single-strand breaks in DNA. Carcinogenesis 25(5):757–63.
68. Bonassi S and Au WW (2002). Biomarkers in molecular epidemiology studies for health risk prediction. Mutat Res 511(1):73–86.
69. Bonassi S, Abbondandolo A, Camurri L, Dal Pra L, De Ferrari M, Degrassi F, Forni A, Lamberti L, Lando C, Padovani P and others (1995). Are chromosome aberrations in circulating lymphocytes predictive of future cancer onset in humans? Preliminary results of an Italian cohort study. Cancer Genet Cytogenet 79(2):133–5.
70. Hagmar L, Bonassi S, Stromberg U, Mikoczy Z, Lando C, Hansteen IL, Montagud AH, Knudsen L, Norppa H, Reuterwall C and others (1998). Cancer predictive value of cytogenetic markers used in occupational health surveillance programs. Recent Results Cancer Res 154:177–84.
71. Hagmar L, Bonassi S, Stromberg U, Mikoczy Z, Lando C, Hansteen IL, Montagud AH, Knudsen L, Norppa H, Reuterwall C and others (1998). Cancer predictive value of cytogenetic markers used in occupational health surveillance programs: a report from an ongoing study by the European Study Group on Cytogenetic Biomarkers and Health. Mutat Res 405(2):171–8.
72. Bonassi S, Hagmar L, Stromberg U, Montagud AH, Tinnerberg H, Forni A, Heikkila P, Wanders S, Wilhardt P, Hansteen IL and others (2000). Chromosomal aberrations in lymphocytes predict human cancer independently of exposure to carcinogens. European Study Group on Cytogenetic Biomarkers and Health. Cancer Res 60(6):1619–25.
73. Liou SH, Lung JC, Chen YH, Yang T, Hsieh LL, Chen CJ and Wu TN (1999). Increased chromosome-type chromosome aberration frequencies as biomarkers of cancer risk in a blackfoot endemic area. Cancer Res 59(7):1481–4.
74. Rossner P, Boffetta P, Ceppi M, Bonassi S, Smerhovsky Z, Landa K, Juzova D and Sram RJ (2005). Chromosomal aberrations in lymphocytes of healthy subjects and risk of cancer. Environ Health Perspect 113(5):517–20.
75. Boffetta P, van der Hel O, Norppa H, Fabianova E, Fucic A, Gundy S, Lazutka J,

Cebulska-Wasilewska A, Puskailerova D, Znaor A and others (2007). Chromosomal aberrations and cancer risk: results of a cohort study from Central Europe. Am J Epidemiol 165(1):36–43.

76. Bonassi S, Norppa H, Ceppi M, Stromberg U, Vermeulen R, Znaor A, Cebulska-Wasilewska A, Fabianova E, Fucic A, Gundy S and others (2008). Chromosomal aberration frequency in lymphocytes predicts the risk of cancer: results from a pooled cohort study of 22 358 subjects in 11 countries. Carcinogenesis 29(6):1178–83.
77. Fenech M and Morley AA (1985). Measurement of micronuclei in lymphocytes. Mutat Res 147(1–2):29–36.
78. Thomas P, Umegaki K and Fenech M (2003). Nucleoplasmic bridges are a sensitive measure of chromosome rearrangement in the cytokinesis-block micronucleus assay. Mutagenesis 18(2):187–94.
79. Fenech M (2005). The Genome Health Clinic and Genome Health Nutrigenomics concepts: diagnosis and nutritional treatment of genome and epigenome damage on an individual basis. Mutagenesis 20(4):255–69.
80. Fenech M, Baghurst P, Luderer W, Turner J, Record S, Ceppi M and Bonassi S (2005). Low intake of calcium, folate, nicotinic acid, vitamin E, retinol, beta-carotene and high intake of pantothenic acid, biotin and riboflavin are significantly associated with increased genome instability – results from a dietary intake and micronucleus index survey in South Australia. Carcinogenesis 26(5):991–9.
81. Fenech M (2002). Chromosomal biomarkers of genomic instability relevant to cancer. Drug Discov Today 7(22):1128–37.
82. Decordier I, Dillen L, Cundari E and Kirsch-Volders M (2002). Elimination of micronucleated cells by apoptosis after treatment with inhibitors of microtubules. Mutagenesis 17(4):337–44.
83. Leach NT and Jackson-Cook C (2004). Micronuclei with multiple copies of the X chromosome: do chromosomes replicate in micronuclei? Mutat Res 554(1–2):89–94.
84. Van Hummelen P and Kirsch-Volders M (1990). An improved method for the 'in vitro' micronucleus test using human lymphocytes. Mutagenesis 5(2):203–4.
85. Kirsch-Volders M, Vanhauwaert A, De Boeck M and Decordier I (2002). Importance of detecting numerical versus structural chromosome aberrations. Mutat Res 504(1–2):137–48.
86. Kirsch-Volders M and Fenech M (2001). Inclusion of micronuclei in non-divided mononuclear lymphocytes and necrosis/apoptosis may provide a more comprehensive cytokinesis block micronucleus assay for biomonitoring purposes. Mutagenesis 16(1):51–8.
87. Fenech M, Holland N, Chang WP, Zeiger E and Bonassi S (1999). The HUman MicroNucleus Project – An international collaborative study on the use of the micronucleus technique for measuring DNA damage in humans. Mutat Res 428(1–2):271–83.
88. Fenech M (2000). The in vitro micronucleus technique. Mutat Res 455(1–2):81–95.
89. Norppa H and Falck GC (2003). What do human micronuclei contain? Mutagenesis 18(3):221–33.
90. Kimura M, Umegaki K, Higuchi M, Thomas P and Fenech M (2004). Methylenetetrahydrofolate reductase C677T polymorphism, folic acid and riboflavin are important determinants of genome stability in cultured human lymphocytes. J Nutr 134(1):48–56.
91. Kirsch-Volders M, Elhajouji A, Cundari E and Van Hummelen P (1997). The in vitro micronucleus test: a multi-endpoint assay to detect simultaneously mitotic delay, apoptosis, chromosome breakage, chromosome loss and non-disjunction. Mutat Res 392(1–2):19–30.
92. Fenech M (2006). Cytokinesis-block micronucleus assay evolves into a "cytome" assay of chromosomal instability, mitotic dysfunction and cell death. Mutat Res 600(1–2):58–66.
93. Decordier I, Papine A, Plas G, Roesems S, Vande Loock K, Moreno-Palomo J, Cemeli E, Anderson D, Fucic A, Marcos R and others (2009). Automated image analysis of cytokinesis-blocked micronuclei: an adapted protocol and a validated scoring procedure for biomonitoring. Mutagenesis 24(1):85–93.
94. Fenech M and Rinaldi J (1994). The relationship between micronuclei in human lymphocytes and plasma levels of vitamin C, vitamin E, vitamin B12 and folic acid. Carcinogenesis 15(7):1405–11.
95. Gaziev AI, Sologub GR, Fomenko LA, Zaichkina SI, Kosyakova NI and Bradbury RJ (1996). Effect of vitamin-antioxidant micronutrients on the frequency of spontaneous and in vitro {gamma}-ray-induced micronuclei in lymphocytes of donors: the age factor. Carcinogenesis 17(3):493–499.
96. Fenech M (1998). Important variables that influence base-line micronucleus frequency in cytokinesis-blocked lymphocytes-a biomarker for DNA damage in human populations. Mutat Res 404(1–2):155–65.

97. Fenech M (1998). Chromosomal damage rate, aging, and diet. Ann N Y Acad Sci 854:23–36.
98. Bolognesi C, Lando C, Forni A, Landini E, Scarpato R, Migliore L and Bonassi S (1999). Chromosomal damage and ageing: effect on micronuclei frequency in peripheral blood lymphocytes. Age Ageing 28(4):393–7.
99. Bonassi S, Fenech M, Lando C, Lin YP, Ceppi M, Chang WP, Holland N, Kirsch-Volders M, Zeiger E, Ban S and others (2001). HUman MicroNucleus project: international database comparison for results with the cytokinesis-block micronucleus assay in human lymphocytes: I. Effect of laboratory protocol, scoring criteria, and host factors on the frequency of micronuclei. Environ Mol Mutagen 37(1):31–45.
100. Neri M, Ceppi M, Knudsen LE, Merlo DF, Barale R, Puntoni R and Bonassi S (2005). Baseline micronuclei frequency in children: estimates from meta- and pooled analyses. Environ Health Perspect 113(9):1226–9.
101. Kirsch-Volders M, Mateuca RA, Roelants M, Tremp A, Zeiger E, Bonassi S, Holland N, Chang WP, Aka PV, Deboeck M and others (2006). The effects of GSTM1 and GSTT1 polymorphisms on micronucleus frequencies in human lymphocytes in vivo. Cancer Epidemiol Biomarkers Prev 15(5):1038–42.
102. Mateuca RA, Roelants M, Iarmarcovai G, Aka PV, Godderis L, Tremp A, Bonassi S, Fenech M, Berge-Lefranc JL and Kirsch-Volders M (2008). hOGG1(326), XRCC1(399) and XRCC3(241) polymorphisms influence micronucleus frequencies in human lymphocytes in vivo. Mutagenesis 23(1):35–41.
103. Sari-Minodier I, Orsiere T, Auquier P, Martin F and Botta A (2007). Cytogenetic monitoring by use of the micronucleus assay among hospital workers exposed to low doses of ionizing radiation. Mutat Res 629(2):111–21.
104. Fenech M and Rinaldi J (1995). A comparison of lymphocyte micronuclei and plasma micronutrients in vegetarians and non-vegetarians. Carcinogenesis 16(2):223–30.
105. Fenech MF, Dreosti IE and Rinaldi JR (1997). Folate, vitamin B12, homocysteine status and chromosome damage rate in lymphocytes of older men. Carcinogenesis 18(7):1329–36.
106. Fenech M, Dreosti I and Aitken C (1997). Vitamin-E supplements and their effect on vitamin-E status in blood and genetic damage rate in peripheral blood lymphocytes. Carcinogenesis 18(2):359–64.
107. Fenech M, Aitken C and Rinaldi J (1998). Folate, vitamin B12, homocysteine status and DNA damage in young Australian adults. Carcinogenesis 19(7):1163–71.
108. Odagiri Y and Uchida H. (1998). Influence of serum micronutrients on the incidence of kinetochore-positive or -negative micronuclei in human peripheral blood lymphocytes. Mutat Res 415(1–2):35–45.
109. Bonassi S, Neri M, Lando C, Ceppi M, Lin YP, Chang WP, Holland N, Kirsch-Volders M, Zeiger E and Fenech M (2003). Effect of smoking habit on the frequency of micronuclei in human lymphocytes: results from the Human MicroNucleus project. Mutat Res 543(2):155–66.
110. Iarmarcovai G, Sari-Minodier I, Orsiere T, De Meo M, Gallice P, Bideau C, Iniesta D, Pompili J, Berge-Lefranc JL and Botta A (2006). A combined analysis of XRCC1, XRCC3, GSTM1 and GSTT1 polymorphisms and centromere content of micronuclei in welders. Mutagenesis 21(2): 159–65.
111. Ishikawa H, Yamamoto H, Tian Y, Kawano M, Yamauchi T and Yokoyama K (2004). Evidence of association of the CYP2E1 genetic polymorphism with micronuclei frequency in human peripheral blood. Mutat Res 546(1–2):45–53.
112. Scarpato R, Migliore L, Hirvonen A, Falck G and Norppa H (1996). Cytogenetic monitoring of occupational exposure to pesticides: characterization of GSTM1, GSTT1, and NAT2 genotypes. Environ Mol Mutagen 27(4):263–9.
113. Crott JW, Mashiyama ST, Ames BN and Fenech M (2001). The effect of folic acid deficiency and MTHFR C677T polymorphism on chromosome damage in human lymphocytes in vitro. Cancer Epidemiol Biomarkers Prev 10(10):1089–96.
114. Aka P, Mateuca R, Buchet JP, Thierens H and Kirsch-Volders M (2004). Are genetic polymorphisms in OGG1, XRCC1 and XRCC3 genes predictive for the DNA strand break repair phenotype and genotoxicity in workers exposed to low dose ionising radiations? Mutat Res 556(1–2):169–81.
115. Mateuca R, Aka PV, De Boeck M, Hauspie R, Kirsch-Volders M and Lison D (2005). Influence of hOGG1, XRCC1 and XRCC3 genotypes on biomarkers of genotoxicity in workers exposed to cobalt or hard metal dusts. Toxicol Lett 156(2):277–88.
116. Angelini S, Kumar R, Carbone F, Maffei F, Forti GC, Violante FS, Lodi V, Curti S, Hemminki K and Hrelia P (2005). Micronuclei in humans induced by exposure to low level of ionizing radiation: influence of polymorphisms in DNA repair genes. Mutat Res 570(1):105–17.

117. Iarmarcovai G, Sari-Minodier I, Chaspoul F, Botta C, De Meo M, Orsiere T, Berge-Lefranc JL, Gallice P and Botta A (2005). Risk assessment of welders using analysis of eight metals by ICP-MS in blood and urine and DNA damage evaluation by the comet and micronucleus assays; influence of XRCC1 and XRCC3 polymorphisms. Mutagenesis 20(6):425–32.
118. Rzeszowska-Wolny J, Polanska J, Pietrowska M, Palyvoda O, Jaworska J, Butkiewicz D and Hancock R (2005). Influence of polymorphisms in DNA repair genes XPD, XRCC1 and MGMT on DNA damage induced by gamma radiation and its repair in lymphocytes in vitro. Radiat Res 164(2):132–40.
119. Bonassi S, Znaor A, Ceppi M, Lando C, Chang WP, Holland N, Kirsch-Volders M, Zeiger E, Ban S, Barale R and others (2007). An increased micronucleus frequency in peripheral blood lymphocytes predicts the risk of cancer in humans. Carcinogenesis 28(3):625–31.
120. Murgia E, Ballardin M, Bonassi S, Rossi AM and Barale R (2008). Validation of micronuclei frequency in peripheral blood lymphocytes as early cancer risk biomarker in a nested case-control study. Mutat Res 639(1–2):27–34.
121. Iarmarcovai G, Ceppi M, Botta A, Orsiere T and Bonassi S (2008). Micronuclei frequency in peripheral blood lymphocytes of cancer patients: a meta-analysis. Mutat Res 659(3):274–83.
122. El-Zein RA, Schabath MB, Etzel CJ, Lopez MS, Franklin JD and Spitz MR (2006). Cytokinesis-blocked micronucleus assay as a novel biomarker for lung cancer risk. Cancer Res 66(12):6449–56.
123. Dudarewicz L, Holzgreve W, Jeziorowska A, Jakubowski L and Zimmermann B (2005). Molecular methods for rapid detection of aneuploidy. J Appl Genet 46(2):207–15.
124. Schrock E, du Manoir S, Veldman T, Schoell B, Wienberg J, Ferguson-Smith MA, Ning Y, Ledbetter DH, Bar-Am I, Soenksen D and others (1996). Multicolor spectral karyotyping of human chromosomes. Science 273(5274):494–7.
125. Speicher MR and Ward DC (1996). The coloring of cytogenetics. Nature Med 2(9):1046–8.
126. Horstmann M and Obe G (2003). CABAND: Classification of aberrations in multicolor banded chromosomes. Cytogenet Genome Res 103(1–2):24–7.
127. Chudoba I, Hickmann G, Friedrich T, Jauch A, Kozlowski P and Senger G (2004). mBAND: a high resolution multicolor banding technique for the detection of complex intrachromosomal aberrations. Cytogenet Genome Res 104(1–4):390–3.
128. Elhajouji A, Van Hummelen P and Kirsch-Volders M (1995). Indications for a threshold of chemically-induced aneuploidy in vitro in human lymphocytes. Environ Mol Mutagen 26(4):292–304.
129. Elhajouji A, Tibaldi F and Kirsch-Volders M (1997). Indication for thresholds of chromosome non-disjunction versus chromosome lagging induced by spindle inhibitors in vitro in human lymphocytes. Mutagenesis 12(3):133–40.
130. Godderis L, De Boeck M, Haufroid V, Emmery M, Mateuca R, Gardinal S, Kirsch-Volders M, Veulemans H and Lison D (2004). Influence of genetic polymorphisms on biomarkers of exposure and genotoxic effects in styrene-exposed workers. Environ Mol Mutagen 44(4):293–303.
131. Skjelbred CF, Svendsen M, Haugan V, Eek AK, Clausen KO, Svendsen MV and Hansteen IL (2006). Influence of DNA repair gene polymorphisms of hOGG1, XRCC1, XRCC3, ERCC2 and the folate metabolism gene MTHFR on chromosomal aberration frequencies. Mutat Res 602(1–2):151–62.
132. McCanlies E, Landsittel DP, Yucesoy B, Vallyathan V, Luster ML and Sharp DS (2002). Significance of genetic information in risk assessment and individual classification using silicosis as a case model. Ann Occup Hyg 46(4):375–81.

Chapter 16

The Measurement of Induced Genetic Change in Mammalian Germ Cells

Ilse-Dore Adler, Francesca Pacchierotti, and Antonella Russo

Abstract

In vivo methods are described to detect clastogenic and aneugenic effects of chemical agents in male and female germ cells in vivo. The knowledge of stages of germ cell development and their duration for a given test animal is essential for these experiments. Commonly, mice or rats are employed. Structural chromosome aberrations can be analyzed microscopically in mitotic cell divisions of differentiating spermatogonia, zygotes, or early embryos as well as in first meiotic cell divisions of spermatocytes and oocytes. Numerical chromosome aberrations are scorable during second meiotic divisions of spermatocytes and oocytes. The micronucleus test is applicable to early round spermatids and to first cleavage embryos, and as in somatic cells, it assesses structural as well as numerical chromosome aberrations. In contrast to the somatic micronucleus assay, the timing of cell sampling determines whether the micronuclei scored in round spermatids were formed from structural or numerical aberrations, i.e. with short treatment-sampling intervals the micronuclei are formed by exposed meiotic divisions and represent induced non-disjunction. On the contrary, after longer intervals of 12–14 days micronuclei are formed from induced unstable structural aberrations in differentiating spermatogonia or during the last round of DNA-synthesis in early spermatocytes. Furthermore, labelling with fluorescent DNA-probes can be used to confirm these theoretical expectations. The mouse sperm-FISH assay is totally based on scoring colour spots from individual chromosomes (e.g. X, Y, and 8) hybridized with specific DNA-probes. The most animal demanding assay described here is the dominant lethal test. It is commonly performed with treated male laboratory rodents and allows the determination of the most sensitive developmental stage of spermatogenesis to a particular chemical under test. Theoretically, unstable structural chromosome aberrations in sperm will lead to foetal deaths after fertilization at around the time of implantation in the uterus wall. These can be scored as deciduomata or early dead foetuses in the uterus wall of the females at mid-pregnancy. None of the tests described in this chapter provide data for a quantitative estimate of the genetic risk to progeny from exposed germ cells. The only tests on which such calculations can be based, the heritable translocation assay and the specific locus test, are so animal and time-consuming that they can no more be performed anywhere in the world and thus are not described here.

Key words: Aneugenicity, Clastogenicity, Dominant lethal effects, Germ cell tests, Mice, Rats

James M. Parry and Elizabeth M. Parry (eds.), *Genetic Toxicology: Principles and Methods*, Methods in Molecular Biology, vol. 817, DOI 10.1007/978-1-61779-421-6_16, © Springer Science+Business Media, LLC 2012

1. Introduction

Animal experiments for mutagenicity testing are performed for two purposes:

1. To detect genotoxic effects in somatic cells indicative of a carcinogenic potential of a test chemical and
2. To detect genotoxic effects in germ cells indicating a genetic hazard for progeny of exposed individuals. The latter is dealt with in this chapter.

Basically, all three classes of genetic alterations can be determined in germ cells: chromosomal aberrations, i.e. structural alterations of chromosomes which entail gain, loss, or translocation of chromosomal sections, gene mutations, i.e. changes in the genetic code, and genome mutations, i.e. changes in chromosome number. In germ cells, cytogenetic analyses and micronucleus studies are performed to detect structural and numerical chromosome aberrations. The consequences of these chromosomal effects in germ cells for the survival of resulting embryos are detected in the dominant lethal assay. The consequences of gene mutations are determined in the specific locus assay. However, since this assay requires specific mouse strains, thousands of animals, and consequently large animal facilities, it is not commonly used and will not be described here.

The analysis of genotoxic effects of environmental agents in germ cells is usually performed with mice. However, cytogenetic tests can be performed also with rats or Chinese hamsters.

Animals should be kept in the animal quarters for at least 1 week to acclimatize. Animals should be ear-marked for later identification (1). The animal quarters should be air-conditioned with a regulated room temperature of 27 ± 2°C, relative humidity of 57 ± 2% and a photo cycle of 12:12 h light and dark. The animals should be kept on standard species-specific pellet diet and water ad libitum. The animals may be housed in Macrolon cages Type 2 (for three rats) or Type 3 (for 5 mice) filled with saw dust. Cages with mesh-wire bottoms are not adequate for the dominant lethal test but may be used for the other tests where animals are only kept in the experiment for up to 2 weeks. Cage covers of gauze will prevent bedding material falling into cages below. It may be contaminated with either the test chemical or germs. All cages have to be clearly labelled with experiment number, animal number and sex, animal age, test compound, and other experimental details, e.g. plug date etc. Cage bedding has to be changed at least once per week. If water is supplied in bottles, these have to be changed along with the bedding. The floor of the animal rooms and the racks have to be washed every week using a disinfectant solution. Face mask, cotton gloves, clean lab coats, and shoe covers should be worn when entering the animal room. Specific-pathogen-free (SPF) animal houses require additional precautions.

Animals are treated with the test substances by intraperitoneal injection (i.p.), sometimes by gavage (p.o.) or by inhalation. Exposure to the test substance in food or drinking water is rarely used because the uptake of the test substance cannot be determined exactly. Likewise, intravenous, intramuscular, or subcutaneous injections are seldom used. Skin painting is only used when human exposure requires this route. In most cases, acute exposure is preferred over sub-acute or chronic exposure. However, the comparison between results of acute and chronic dosing for individual chemicals has given surprising results that question the paradigm which says that acute exposure to a high dose gives the optimum of response (2). Dosing is performed in mg per kg body weight (mg/kg) with a maximum tolerated dose (MTD) to optimize the effect and to avoid systemic toxicity. For dose–response studies, two additional lower doses are employed (e.g. MTD/2 and MTD/5). Solvents or suspension media should be applied by the same route as to the exposed animals to a control group of animals (solvent or negative controls). To demonstrate the efficiency of the test performance, a group of animals should be treated with a known clastogen or aneugen, respectively (positive control). Ideally, the route of exposure and the solvent for the positive control compound should be the same as for the test compound. Positive controls from experienced laboratories must not necessarily be included in every experiment while the concurrent negative control group is absolutely mandatory. Completely untreated negative controls are only sensible if the solvent or suspension medium itself is suspected to cause genotoxic effects.

For statistical evaluations of significant differences between results from treated and negative control groups, it is mandatory that the animals are distributed randomly to all experimental groups and that the individual groups are of the same size, i.e. for most cytogenetic tests five animals per group are recommended. Adsorption, metabolism, detoxification, and excretion of the test compound but also repair processes within the repair competent cells of an organ contribute to inter-animal variability in in vivo studies. Thus, in contrast to in vitro studies, the statistical comparison should be based on the response of the animals and not on the total number of cells evaluated per experimental group. Therefore, the correct statistical evaluation procedure would be a non-parametric test such as the Mann–Whitney test. If inter-animal variabilities within treated and control groups are non-significant, tests such as the chi-square test may be applied.

All mutagenicity tests in germ cells require the exact knowledge of the duration of individual developmental stages of germ cells in the experimental animals of the study (3). The timing of spermatogenesis stages in the mouse is illustrated in Fig. 1.

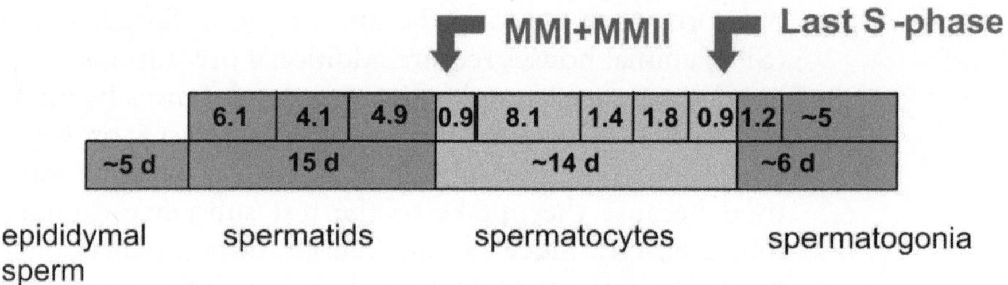

Fig. 1. Timing (in days) of different stages of spermatogenesis in the mouse. MMI and MMII (first and second meiotic division) follow each other without interkinesis within only 22 h. Drawn by Adi Baumgartner in his thesis, based on Oakberg (3).

Abundant numbers of male germ cells in dividing stages (spermatogonial mitoses or meiotic divisions in spermatocytes) are available for cytogenetic studies. The spermatogonial test is best used for the assessment of induced structural chromosome aberrations in male germ cells. It demonstrates whether or not a chemical or its reactive metabolites reach the germ cells and cause clastogenic effects. Spermatogonial stem cells of the mouse divide slowly with a cell cycle of 6–8 days (3). The resulting differentiating spermatogonia divide every 26–32 h (type A, intermediate, and type B spermatogonia). The majority of analyzable mitoses represent type B spermatogonia. Analyses of chromosomal aberrations at mitosis in differentiating spermatogonia resemble in most parts the cytogenetic analyses of mitosis in somatic cells, e.g. bone marrow. The same types of aberrations can be scored.

Cytogenetic analyses in first meiotic metaphase chromosomes is more difficult because paired homologous chromosomes form bivalents with less clearly defined structures. Depending on the interval between treatment and cell sampling, chromosome-type aberrations (reciprocal translocations) or chromatid-type aberrations (gaps, breaks, fragments, and exchange configurations) can be scored. Often, unpaired homologous chromosomes, either autosomes or sex chromosomes, are also noted. They are not the result of clastogenicity but represent failures of homologous pairing or premature separation of homologous chromosomes.

Chromosomal aberrations induced in post-meiotic male germ cell stages – spermatids and spermatozoa – can only be analyzed during first cleavage division after fertilization. The preparation technique for the isolation of first cleavage division zygotes is very tedious, which prevents routine use of this methodology. It is only used when scientifically indicated.

The spermatid micronucleus assay determines micronuclei in early round spermatids as the result of chromosomal breakage induced in differentiating spermatogonia or malsegregation induced during the meiotic divisions. Similar to the somatic micronucleus test, the origin of the micronuclei observed in spermatids can be determined by centromere-specific staining procedures.

Even though it is very useful, the spermatid micronucleus test has not been included in the international guidelines for chemical testing.

In females, germ cells are limited in number and mitotic divisions of oocyte propagation and the initial stages of meiosis occur during foetal development. At birth, the numbers of oocytes are fixed and they are arrested in a certain prophase stage of meiosis, the dictyate stage in the mouse. Oocytes are released from the arrest in small numbers during each oestrus cycle of the adult female. They undergo meiosis I (MMI) in the ovary and are ovulated during the second meiotic division (MMII) which is only completed after sperm entry in case of fertilization. For cytogenetic analyses female germ cells are difficult to collect from the ovarium (MMI) or oviduct (MMII) and are limited in number so that their use cannot be recommended for routine cytogenetic analyses. However, it is particularly important to include female germ cells in studies where sex specific effects are expected.

2. Materials

2.1. Chromosome Aberration Analysis in Mouse Spermatogonia

1. A shaking water bath.
2. A good set of dissecting instruments, including straight and curved forceps and scissors.
3. 100-ml glass flasks.
4. 60-mm diameter Petri dishes.
5. 10-ml round-bottom centrifuge tubes.
6. Glass Pasteur pipettes and rubber bulbs.
7. TIM (Testis Isolation Medium; (4)). For 5 L of medium prepare two separate solutions: (a) 30.25 g NaCl, 16.90 g KCl, 4.25 g Na_2HPO_4, 0.45 g KH_2PO_4, and 5 g glucose are dissolved in a final volume of 4 L of purified distilled water. (b) 0.9 g $CaCl_2 \cdot 2\ H_2O$ and 1.5 g $MgSO_4 \cdot 7\ H_2O$ are dissolved in 0.9 L of purified distilled water. Dropwise add solution (b) to solution (a) under slow agitation; then add 0.025 g phenol red, adjust pH to 7.2–7.3, and bring to final volume (5 L) (see Note 1).
8. Collagenase: Type I (Sigma) or collagenase A (Roche Applied Science, Mannheim, Germany).
9. Isotonic solution (2.2% w/v trisodium citrate dihydrate) (see Note 1).
10. Hypotonic solution (0.9% w/v trisodium citrate dihydrate) (see Note 1).
11. Fixative: 3:1 ethanol–glacial acetic acid (see Note 2).

12. Colchicine: to obtain a 10^{-3} M solution ready to be injected, dilute 1:25 (in purified distilled water) an aliquot of the stock solution (see Note 3).
13. Pre-cleaned microscope slides (see Note 4).
14. Giemsa solution.
15. 24 × 35 mm cover glasses.
16. Mounting medium.
17. Laboratory centrifuge.
18. Research microscope with 20× and 100× objectives.

2.2. Chromosome Aberration Analysis in Mouse Spermatocytes

1. A good set of dissecting instruments, including straight and curved forceps and scissors.
2. 60-mm diameter Petri dishes.
3. 10-ml round-bottom centrifuge tubes.
4. Glass Pasteur pipettes and rubber bulbs.
5. Isotonic solution (2.2% w/v trisodium citrate dihydrate) (see Note 1).
6. Hypotonic solution (0.9% w/v trisodium citrate dihydrate) (see Note 1).
7. Fixative: 3:1 ethanol–glacial acetic acid (see Note 2).
8. Colchicine 0.5% (optional) (see Note 3).
9. Pre-cleaned microscope slides (see Note 4).
10. Giemsa solution.
11. 2% acetic Orcein solution.
12. 24 × 35 mm cover glasses.
13. Mounting medium.
14. Research microscope with 20× and 100× objectives (with phase contrast for Orcein staining).

2.3. Spermatid Micronucleus Assay

1. A shaking water bath.
2. A cytocentrifuge (Cytospin, Shandon) (see Note 5).
3. A good set of dissecting instruments, including straight and curved forceps and scissors.
4. 100-ml glass flasks.
5. 60-mm diameter Petri dishes.
6. 10-ml round-bottom centrifuge tubes.
7. Glass Pasteur pipettes and rubber bulbs.
8. TIM (Testis Isolation Medium, (4)). The detailed recipe is reported in Subheading 2.1 (see Note 1).
9. Collagenase: Type I (Sigma) or collagenase A (Roche Applied Science, Mannheim, Germany).

10. Phosphate buffered saline (PBS), Ca^{+2} to Mg^{+2} free.
11. Percoll solution.
12. Helly's fixative: 1 L contains 50 g $HgCl_2$ and 25 g $K_2Cr_2O_7$. Immediately before use, add 50 µl formaldehyde per ml of saline solution (see Note 6).
13. Pre-cleaned microscope slides (see Note 4).
14. Mayer's hemallume.
15. Coplin jars containing 70, 90, and 100% ethanol.
16. Cover glasses (24 × 50 mm).
17. Mounting medium.
18. Research microscope with 20× and 100× objectives.

2.4. Sperm FISH Assay

1. A good set of dissecting instruments, including straight and curved forceps and scissors.
2. Petri dishes of 30 mm diameter.
3. Glass Pasteur pipettes and rubber bulbs.
4. Pre-cleaned microscope slides (see Note 4).
5. Cover glasses (24 × 35 mm).
6. Coplin jars.
7. Diamond pen.
8. Hot plate (70°C).
9. Slide warmer (37°C).
10. Incubator.
11. Freezers (–20 and –80°C).
12. Micropipettes and tips (300 µl and 50 µl).
13. Eppendorf cups (500 µl and 50 µl).
14. Vortex centrifuge.
15. Water bath (78°C).
16. Cooled centrifuge (4°C).
17. Rubber cement and wax.
18. Moisture chamber.
19. Fluorescence Photo Microscope equipped with various colour filters.
20. Foetal calf serum.
21. DTT (Dithiothreitol; Sigma) 1 mM in 0.1 M Tris–HCl pH 8.0.
22. LIS-Solution: 4 mM LIS (3,5-Diiodosalicylic acid, Sigma) in 0.1 M Tris–HCl, pH 8.0.
23. DAPI. Working solution 0.6 µg/ml 4,6-diamidino-2-phenylindole in 2× SSC, keep in a brown bottle and refrigerate.

24. PNBR buffer: Solve 5% Blocking Reagent (Boehringer Mannheim) at 4°C in PN buffer at 50–70°C. Stir while increasing the heat. Add 0.02% Na azide (bacteriostatic). Let cool for 30–60 min and filter-sterilize. Freeze portions of 1 ml at −4°C.
25. PN-buffer: Add 500 ml of a solution containing 6.9 g NaH_2PO_4 to 5 L of a solution containing 88.95 g Na_2HPO_4. Adjust pH with NaH_2PO_4 to 8.0 and then add 0.1% NP-40.
26. Formamide solutions (can be reused for 1–2 months). 70% FA: 70 ml formamide + 10 ml 20× SSC + 15 ml Millipore water. 50% FA: 50 ml formamide + 10 ml 20× SSC + 35 ml Millipore water. 30% FA: 30 ml formamide + 10 ml 20× SSC + 55 ml Millipore water. Adjust pH to 7.0 with 50% HCl in 1 N NaOH and then fill up to 100 ml with Millipore water.
27. 2×PBS. For 1 L: 16.0 g NACl + 0.4 g KCl + 2.88 g $Na_2HPO_4 \times 2\ H_2O$ + 3.0 g KH_2PO_4. Adjust to pH 7.5 with HCl and fill 300 ml into a 500-ml flask and autoclave.
28. 20× SSC. In a final volume of 500 ml Millipore water, prepare a solution with 87.66 g NaCl and 42.11 g trisodium citrate dihydrate (3.0 M NaCl, 0.3 M TriNa citrate). 2× SSC is obtained with nine parts of distilled H_2O + 1 part of 20× SSC, adjusting to pH 7.0.
29. Paraformaldehyde 4% (can be reused for 1 month). Attention: wear gloves and mask! Add 4 g paraformaldehyde to 50 ml 100 mM $MgCl_2$ plus ~12 drops of 1 N NaOH and heat to 70°C. Add another few drops of NaOH to clear the solution. Cool down to 35°C, filter and add 50 ml of 2× PBS (pH 7.5). Keeps refrigerated for weeks. Only use under a hood.
30. Anti-DIG-FITC. 1 mg/ml (Roche Applied Science, Mannheim, Germany), freeze at −20°C in Eppendorf cups in portions of 50 μl. For use, dilute 1:100 in PNBR buffer.
31. Biotinylated anti-streptavidin. 5 mg/ml in sterile H_2O (Vector Laboratories, California, USA), freeze at −20°C in Eppendorf cups in portions of 50 μl. For use, dilute 1:100 in PNBR buffer.
32. RAS. 1.5 mg/1.5 ml (Vector Laboratories, California, USA), freeze at −20°C in Eppendorf cups in portions of 100 μl. For use dilute 1:100 in PNBR buffer.
33. Streptavidin-CY3. 1 mg/0.6 ml (Dianova, Hamburg, Germany), freeze at −20°C in Eppendorf cups in portions of 10 μl. For use dilute 1:330 in PNBR buffer.
34. Vectashield (Vector Laboratories, California, USA).
35. DNA Probes. Plasmid DNAs for:
    (a) chromosome X (clone DXWas70)
    (b) chromosome Y (clone pY353/B)

(c) chromosome 8 (clone 8-4a and 8-5e)

are transformed in *E. coli* XL1-blue and extracted using the Qiagen Plasmid Maxi Kits (Qiagen, Hilden, Germany). The probes for chromosomes 8 and Y are labelled with biotin-16-dUTP and digoxigenin-11-dUTP (Roche Applied Science, Mannheim), respectively. The X-chromosome probe is labelled with a combination of biotin and digoxigenin-dUTP. For labelling, the nick translation system (GIBCO BRL, Life Technologies, USA) is used.

2.5. Dominant Lethal Assay

1. A good set of dissecting instruments, including straight and curved forceps, preparation needles and scissors.
2. A waxen or wooden board and pins.
3. A dissection microscope.

2.6. Cytogenetic Analysis of Oocytes

1. Follicle Stimulating Hormone (FSH) and Luteinizing Hormone (LH). These are the hormones needed to synchronize the oestrous cycle and induce oocyte maturation and ovulation. Pregnant Mare Serum (PMS) and Human Chronic Gonadotrophin (HCG) are used as FSH and LH, respectively (see Note 7). Working solutions of PMS and HCG are prepared in physiologic saline and frozen, ready to inject each animal with 0.2 ml containing 7.5 and 5 IU respectively. The frozen aliquots can be maintained at –20°C for up to 2 months.
2. Dissecting instruments: fine scissors, iris scissors, straight and curved forceps, micro-dissecting needle.
3. Glass Pasteur pipettes and rubber bulbs.
4. Custom-made capillary pipettes (obtained by pulling Pasteur pipettes through a flame).
5. Petri dishes, 30 mm diameter.
6. Multiwell plates (4–6 wells).
7. Microcentrifuge tubes. These tubes are obtained from glass Pasteur pipettes: the thin portion of the pipette is cut with the help of a diamond pen to a total length of approximately 6 cm, sealed by flaming to obtain the bottom of the microtube and graduated in a cm-scale. The latter will help to gradually replace hypotonic solution with fixative according to a standardized reproducible protocol. The wide end of the microcentrifuge tube is inserted into a rubber stopper to fix the tube within a normal 10-ml centrifuge tube and so obtain sedimentation of a pellet with a standard centrifuge.
8. Sterile Hank's Balanced Salt Solution (HBSS).
9. Hyaluronidase solution in HBSS (150 IU/ml). It can be stored at –20°C in 1–2-ml vials up to 3 months.

10. 0.3% (w/v) trisodium citrate dihydrate (hypotonic solution), prepared freshly or autoclaved for storing.
11. Fixative: 3:1 ethanol–glacial acetic acid (methanol can replace ethanol) to be prepared immediately before use.
12. Microscope slides (see Note 4).
13. Chromosome staining solutions. For the analysis of numerical chromosome changes in metaphase II oocytes it is advisable to apply a c-banding method to stain the chromosome preparations. Thereby, the centromere is darker than the rest of the chromosome, which allows better discrimination of single unpaired chromatids from chromosome fragments. A detailed protocol for c-banding, applied with good reproducible results in mouse oocytes, is described by Salamanca and Armendares (5). In principle, molecular cytogenetic techniques can be also applied to metaphase II oocytes, e.g. DNA painting probes may be used to recognize specific chromosome aneuploidies. This approach has been mostly applied in farmyard animals, i.e. for optimal breeding of cows (6) or pigs (7).

2.7. Cytogenetic Analysis of Zygotes

In addition to all the materials listed in Subheading 2.6, 2×10^{-3} M colchicine solution in physiological saline is required to arrest first cleavage metaphases.

3. Methods

3.1. Metaphase Analyses in Male Germ Cells

The mammalian male gametogenesis is a continuous process which allows the collection of adequate numbers of chromosome preparations from mitotically dividing cells (spermatogonia) or from meiotically dividing cells (primary and secondary spermatocytes).

The methods described here refer to the mouse but they can be applied with small modifications to other laboratory rodent species. General criteria for housing and handling animals, for the dosage and the route of exposure, and for the use of negative and positive controls, are discussed in Subheading 1.

3.1.1. Chromosome Aberration Analysis in Mouse Spermatogonia

Cytogenetic analysis of spermatogonial cells gives evidence for the ability of the test compound to reach the testis. The contribution of this assay is of particular relevance in the case of weak genotoxicity of the chemical because spermatogonial cells represent a sensitive cell stage with respect to chemical mutagens. Structural chromosome aberrations (8–11) and sister chromatid exchanges (12) can be evaluated in spermatogonial mitoses. Both end points are included in international regulatory guidelines (see for example the recent European Regulations: 13, 14). An example of a symmetrical chromatid exchange in a mouse spermatogonial cell at mitosis is shown in Fig. 2.

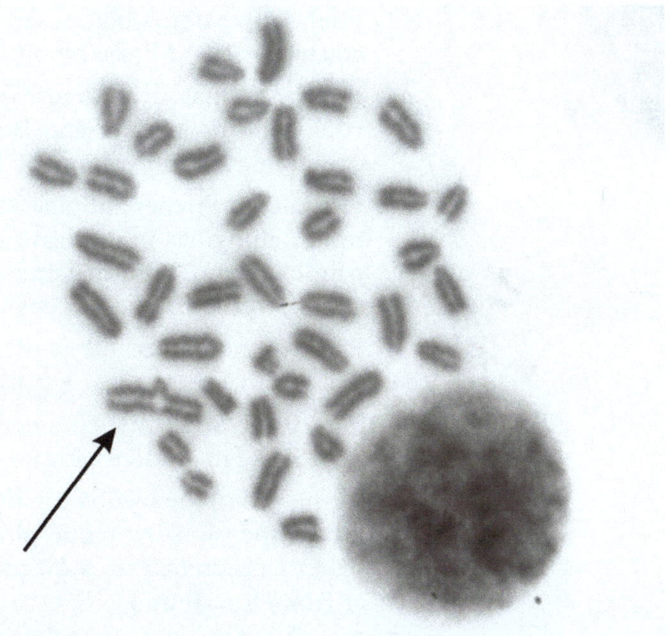

Fig. 2. Chromatid exchange (*arrow*) in a spermatogonial mouse cell at mitotic metaphase.

According to the cell cycle duration of the diverse developmental stages of differentiating spermatogonia (3, 15), the optimal time interval to evaluate structural chromosome aberrations in spermatogonial cells corresponds to 24–30 h after treatment. Five animals per dose should be employed and 100 cells at mitosis should be scored.

For sister chromatid exchange (SCE) evaluation, a rather prolonged exposure to 5-bromodeoxyuridine (BrdU) is required because of the length of the spermatogonial cell cycle. This can cause cytotoxicity and systemic toxicity. As a general rule, the intraperitoneal BrdU administration is not an efficient labelling method for in vivo proliferating cell compartments, because of the fast metabolization rate of BrdU. To achieve sister chromatid differentiation in spermatogonia, subcutaneous implantation of agar-coated BrdU tablets is necessary (16). The number of papers reporting SCE data in mouse spermatogonia in the last 5–8 years is negligible. Protocol details can be found elsewhere (16, 17) and will not be further described here.

Spermatogonial cells reside in the basal layer of the seminiferous tubules, and therefore, a mild enzymatic digestion step is necessary to isolate these cells from testis. Alternatively, 50% acetic acid can be used after fixation of the whole mass of testicular tubules to enrich the cell preparations with spermatogonial mitoses (18).

1. Four to five hours before sacrifice inject males with 0.3 ml of colchicine 10^{-3} M (see Notes 8 and 9).
2. Kill the animals according to regulatory guidelines applied in your country. Dissect and isolate both testes in a Petri dish containing a small volume of TIM (37°C) (see Note 10).
3. Use curved forceps and fine scissors to make an incision of the tunica albuginea and release the seminiferous tubules with a gentle pressure on the surface of the testis (you can use for this operation the curved forceps).
4. Remove the tunica and gently tease apart the tubules.
5. Transfer the seminiferous tubules into a 100-ml flask containing 10 ml of TIM supplemented with collagenase (0.5 mg/ml). Shake in a water bath for 15 min at 37°C (see Note 11).
6. At the end of the incubation time, let the germ cells be released into suspension by gentle pipetting. Filter through a 90-μm nylon membrane into a 10-ml centrifuge tube and centrifuge at 72×*g* for 10 min.
7. Resuspend dropwise the pellet in 2.2% sodium citrate solution and centrifuge for 10 min at 72×*g*.
8. Discard the supernatant and resuspend dropwise the pellet in 3 ml of hypotonic citrate solution. The hypotonic treatment should last at least 15 min at room temperature (invert the tubes once during this step).
9. Centrifuge for 10 min at 72×*g*.
10. Discard the supernatant and resuspend the pellet in 5 ml freshly prepared, cold fixative. Incubate for 10 min on ice.
11. Centrifuge for 10 min at 135×*g*.
12. Repeat twice steps 10 and 11.
13. Finally, discard the supernatant and resuspend the pellet in an appropriate volume of fresh fixative (see Note 12). Spot few drops of the cell suspension on perfectly clean slides and air-dry (see Note 4). Adjust the cell density if necessary.
14. Slides can be stained in Giemsa (8%, 10 min) and permanently mounted.
15. Analyze the slides according to the international guidelines for chromosome aberrations analysis. Briefly, select only well-spread metaphases but exclude those showing excessive scattering or isolated chromosomes in the proximity. Record chromosome and chromatid type of aberrations separately, and for each class distinguish breaks from exchanges. Score gaps but do not include gaps in the calculation of chromosomal aberrations. At least 100 metaphase spreads per male should be evaluated to calculate chromosome aberration frequencies (see Note 13).

16. Pool together individual data to make statistical comparisons. You can apply chi-square analysis to exclude the presence of significant inter-individual variability. Control and treated experimental groups are compared by chi-square analysis or G statistics. A regression test or a non-parametric trend test can be applied to verify dose–effect relationships (19).

3.1.2. Chromosome Aberration Analysis in Mouse Spermatocytes

To prepare meiotic chromosomes the reference protocol is that described by Evans and coworkers (20) which consists in the mechanical disruption of the seminiferous tubules. In the meiotic divisions of primary spermatocytes, the presence of bivalent chromosomes allows the immediate detection of reciprocal translocations as tri- or quadrivalent configurations, i.e. chains of 3 + 1, chains of 4 or rings of 4 chromosomes. Among other structural aberrations, only chromosome-type fragments can be identified easily according to standard criteria, while chromatid-type aberrations cannot be detected without intensive training. An isochromatid fragment (left) and a ring of four chromosomes at first meiotic metaphase (right) in mouse spermatocytes are shown in Fig. 3.

This method is not often used to study the effects of S-dependent clastogens in germ cells. It can be applied to evaluate reciprocal translocations induced in stem cell spermatogonia, but the data with known clastogens are mostly negative in contrast to data with ionizing radiation. This preparation procedure is applied predominantly to identify carriers of reciprocal translocations in the heritable translocation assay (8, 21).

Metaphase spreads from secondary spermatocytes cannot be analyzed for the presence of structural chromosome aberrations due to their poor chromosome morphology. However, this cell type contributes to aneuploidy studies because numerical anomalies in

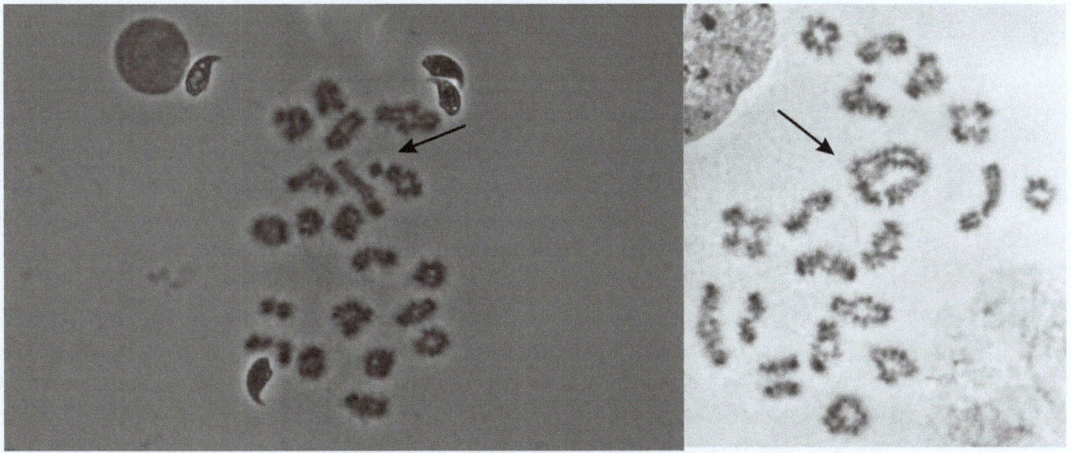

Fig. 3. First meiotic division chromosomes of mouse spermatocytes. *Left*: Bivalent with isochromatid fragment (*arrow*). *Right*: reciprocal translocation (ring of four chromosomes, R IV, *arrow*).

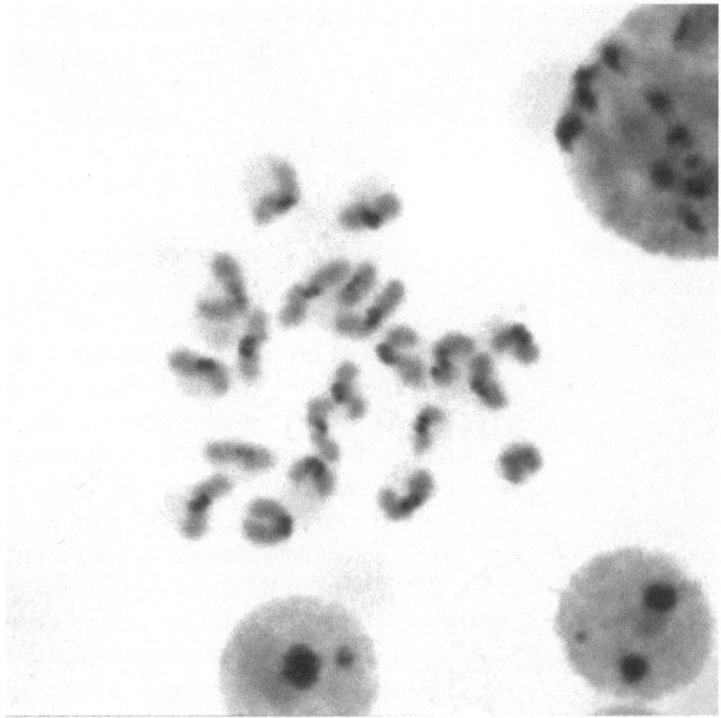

Fig. 4. Second meiotic division chromosomes ($n+1=21$) of a mouse spermatocyte (c-banded).

secondary spermatocytes provide estimations of non-disjunction events at the first meiotic division (17, 22–24). A mouse spermatocyte at second meiotic metaphase with 21 chromosomes is shown in Fig. 4.

The time interval between treatment of males and sampling of spermatocytes is determined not only by the duration of the developmental stages of the germ cells (Fig. 1) but also by the type of meiotic metaphase under study. For the analysis of translocation induction in spermatogonial cells, the interval has to be at least 3 weeks. For studies of induced chromatid aberrations in first meiotic metaphase spreads, the interval has to be between 11 and 13 days so that the cells are exposed shortly before or during the last S-phase in spermatogenesis. If non-disjunction during meiosis I is the end point of interest, the treatment-sampling interval has to be as short as 24 h because the first and the second meiotic divisions follow each other without interkinesis in a time span of less than 24 h (3). To determine induced rates of reciprocal translocations, the sample sizes have to be considerably higher than for the other two end points because reciprocal translocations are extremely rare events. Thus, at least 100 meiotic metaphase spreads from each of ten males have to be analyzed per experimental group. The other two studies can be performed with the usual sample size of 100 cells from each of five males.

Interestingly, the in vivo anti-mitotic treatment is not necessary to obtain analyzable spermatocyte metaphases: primary and secondary spermatocytes are relatively frequent in the testicular cell population even though it has been demonstrated that colchicine treatment increases the MI/MII yield (25). In primary spermatocytes, the morphology of the bivalents is good enough for scoring without colchicine treatment. Instead, the use of colchicine substantially improves the morphology of secondary spermatocyte metaphases. The presence of shortened MII chromosomes favours the accuracy of chromosome counting procedures.

1. Three hours before sacrifice of the animals, inject the males with 0.3 ml of 0.5% colchicine (see Notes 3 and 8). However, this step is optional for MI preparation.
2. Kill the animals according to the regulatory guidelines of your country. Dissect and isolate both testes in a 60-mm Petri dish containing isotonic solution (2.2% sodium citrate) at room temperature. Keep testes as free as possible from fat tissue (see Note 10).
3. Quickly transfer the testes in a second Petri dish containing fresh isotonic solution. Isolate the seminiferous tubules from the tunica albuginea as described in Subheading 3.1.1 (steps 3 and 4).
4. Cut through the tubular mass and repeatedly squeeze out the tubules with curved forceps to achieve their mechanical disruption. The increasing turbidity of the medium is an indicator of the dissociation of meiotic cells from the tubules and their release into suspension.
5. Transfer the suspension into a 10-ml centrifuge tube, together with the tubular debris. The latter will settle in a few minutes at the bottom of the tube. Now transfer the supernatant (enriched of meiotic germ cells) into a new tube.
6. Centrifuge for 10 min at 40–72 × *g* (see Note 14).
7. Discard the supernatant and resuspend dropwise the pellet in 3 ml of hypotonic solution (0.9% trisodium citrate). The cell suspension must be incubated for 12 min at room temperature (invert the tubes once during this step).
8. Centrifuge as in step 6 and then go through steps 10–13 of Subheading 3.1.1.
9. The slides can be stained in Giemsa (8%, 10 min) and permanently mounted. To microscopically detect reciprocal translocations, the slides should be stained with 2% acetic Orcein solution and viewed under phase contrast.
10. Scoring criteria for primary spermatocyte metaphases (for secondary spermatocytes skip to step 11): metaphase spreads should be selected at low magnification (20×) on the basis

of standard morphological criteria (see also step 15, Subheading 3.1.1). The presence of reciprocal translocations must be verified at high magnification (100×). Discard from the analysis primary spermatocyte metaphases with less than 20 bivalents. Classify reciprocal translocations as chain or ring multivalent. Record the presence of anomalies such as univalents, chromosome breaks, and chromosome fragments.

11. Scoring criteria for secondary spermatocyte metaphases (for primary spermatocytes skip this step): on the basis of conventional morphological criteria (see also step 15, Subheading 3.1.1), record the metaphases with $18 < n < 22$ chromosomes. The frequency of aneuploidy is the ratio of hyperploid MII spermatocytes ($n > 20$) to the total MII spermatocytes scored. Hypoploid MII spermatocytes must not be included in the calculation of aneuploidy frequency because one cannot distinguish true hypoploid from chromosome loss due to technical artefacts. Also, do not consider polyploid MII spermatocytes in your analysis: the existence of cytoplasm bridges between synchronous meiotic cells makes very common this type of technical artefacts.
12. From slides obtained in the absence of anti-mitotic treatment, the frequency of MI and MII figures can be determined with respect to 1,000 mid-pachytene nuclei, to verify the kinetics of the meiotic divisions. Normal MII/MI ratios should be equal to 2, while lower values indicate meiotic delay (23). This information must be considered to explain negative results or weak increases induced by the treatment: in particular, a meiotic block could prevent the formation of aneuploid MII spermatocytes.
13. Chromosome aberration frequencies or aneuploidy frequencies in treated and control animals should be compared by chi-square analysis or G statistics after pooling individual data. Preliminary chi-square analysis should be done to verify possible inter-individual variability. A regression test or a non-parametric trend test should be applied to verify dose–effect relationships. Ratios between MII and MI spermatocytes can be compared by applying the Student's *t* test on transformed mean values (e.g. square root transformation) or a non-parametric test (19).

3.2. Spermatid Micronucleus Assay

The micronucleus is a widely accepted indicator of chromosome damage, and protocol variants were proposed to be applied for germ cell studies. The eligible cell type in the testis is represented by early spermatids. These cells correspond to the stage immediately beyond the two consecutive meiotic divisions. The spermatid morphology allows a clear identification with respect to other cell types of the heterogeneous testis population. Furthermore, on the basis of the well known duration of spermatogenesis in rodents (3, 26), it is possible to define the adequate intervals between treatment and cell preparations, to evaluate the response of different premeiotic and meiotic stages to the chemical under investigation.

The spermatid micronucleus assay has been performed either in the rat or in the mouse. An approach known as the dissection method (27) is based on the isolation of short fragments of tubules, specifically selected (by using the dissection microscope) among those carrying early spermatids. In fact, in rodents a peculiar organization of seminiferous tubules (the so-called seminiferous epithelial wave) exists, consisting of specific spatial associations of diverse cell stages which occur in an ordered fashion along segments of the tubules. The segments containing the correct association of cell stages can be identified microscopically and squash preparations can be made. The dissection method was especially designed for analysis of rat spermatids, but application to mouse cells was reported (28, 29). An alternative approach was proposed (30) based on the preparation of a cell suspension from testicular tubules. Because the preparation includes a sample from the whole germ cell population, it is necessary to distinguish round (early) spermatids from other cell types. This is achieved by morphological criteria coupled with the Periodic Acid-Schiff (PAS) reaction, which makes evident the acrosomal structure under development. Both approaches were validated in collaborative studies (31, 32). The sensitivity of the spermatid micronucleus assay is probably lower with the suspension method than with the dissection one, since in the latter approach highly homogeneous cell populations can be isolated, which represent cells immediately deriving from the last meiotic division; on the contrary, the dissection method is skill demanding and hardly reproducible without direct training.

The results coming from the spermatid micronucleus assay in rats and mice were reviewed in the course of the second International Workshop on Genotoxicity Test Procedures (33). It appears that either clastogenic or aneugenic compounds can be detected by this assay and that very good agreement exists between the two methods and the response of the two species.

In planning the spermatid micronucleus assay, one must take into account that clastogenic compounds are expected to give a peak of effect after treatment of preleptotene (premeiotic S-phase). By contrast, aneugenic compounds should be more efficient when administered immediately before the meiotic divisions. Mouse early spermatids exposed during differentiation divisions of spermatogonia/preleptotene can be sampled 14–16 days after an acute treatment; this is a rather long time interval separating the sensitive S-phase stage from the analyzable cell type. A repeated treatment at preleptotene (e.g. 4 i.p. injections at 24 h intervals, harvesting 16 days later) may increase the sensitivity of the assay for weak

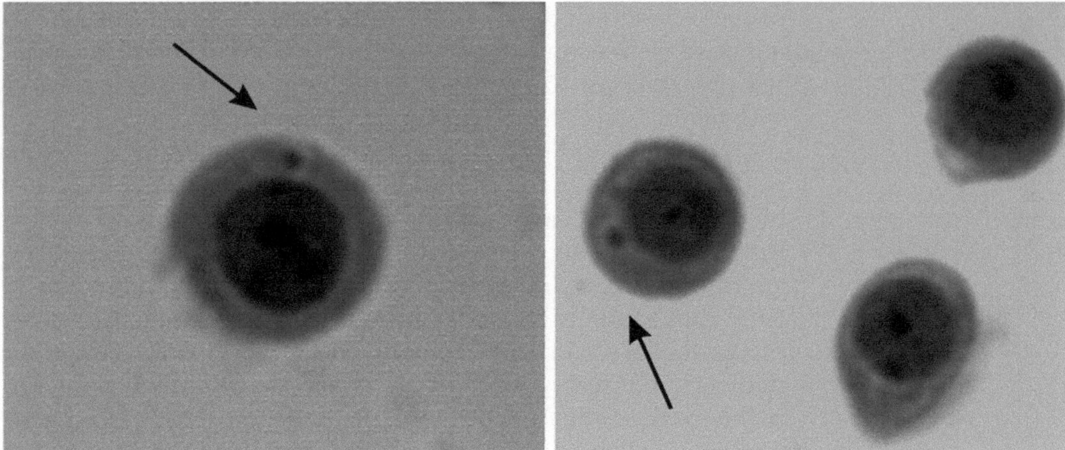

Fig. 5. Early spermatids with micronuclei (*arrows*).

clastogens (reviewed in ref. 31). The average time interval allowing to evaluate effects induced at diakinesis/MI/MII, which are expected to be caused by chromosome segregation errors, is 48 h (reviewed in ref. 31). In conclusion, to get a general indication of the effects induced by a chemical in male germ cells, the spermatid micronucleus assay should include at least two time intervals, focused on the response of premeiotic S-phase and meiotic divisions to the treatment. As for the other approaches based on direct cytogenetic inspection, five animals should be treated per dose. Examples of early mouse spermatids with micronuclei are shown in Fig. 5.

1. Kill the animals according to regulatory guidelines of your country. Dissect and isolate both testes in TIM (33°C) (see Note 15).
2. Isolate seminiferous tubules as described in Subheading 3.1.1 (steps 3–5). The only difference is that the incubation step in the shaking water bath is performed at 33°C (see Note 15) and accordingly the duration is extended: shake for 20 min.
3. At the end of the incubation time, transfer the suspension containing the tubules in a 50 ml centrifuge tube. Here, the tubules which appear only mildly digested are allowed to settle down.
4. Discard the medium and add about 10 ml fresh TIM to the tubules.
5. Repeat twice step 4. Then, by gently pipetting, release germ cells from the seminiferous tubules.
6. Filter the cell suspension through a 90-µm nylon membrane into a 10-ml tube.
7. Centrifuge at 112×*g* for 10 min.

8. Resuspend the pellet in 1 ml TIM and stratify this volume onto 30% Percoll (diluted in TIM).
9. Centrifuge at 450×*g* for 30 min.
10. A layer of cells is visible after centrifugation. By using a Pasteur pipette, pick and transfer these cells into a tube with 5 ml fresh medium.
11. Centrifuge the cell suspension for 10 min at 112×*g*, discard the supernatant, and resuspend the pellet in 5 ml TIM.
12. Repeat twice step 11 and finally resuspend in 2 ml fresh medium. Keep the cell suspension on ice.
13. Slide preparations can be obtained by cytocentrifugation. Recommended volume per slide is 500 μl (see Note 16). Spin cells onto perfectly clean slides at 72×*g* for 1 min. Then, immediately dip the slides into PBS and check for the quality of the preparations under the phase contrast microscope (see Note 17). An alternative protocol allows to make preparations when a cytocentrifuge is not available (see Note 18).
14. Transfer the slides to a Coplin jar containing Helly's fixative and incubate for 30 min (see Notes 19 and 20). Skip this step if slides were prepared manually.
15. Rinse the slides in running tap water for 30 min and then wash them briefly in distilled water. Transfer the preparations to 70% ethanol and store them at +4°C until use (see Notes 20 and 21).
16. PAS reaction: briefly dip the slides into distilled water; incubate horizontally placed slides with 1% periodic acid (5–10 min). Rinse 5 min in tap water, then in distilled water. On horizontally placed slides, add Schiff's reagent and incubate for 20–40 min (see Note 22). Rinse subsequently in: sulphurous solution (5% sodium metabisulfite, 3×2 min each), tap water (5 min), and distilled water (3 min).
17. Let dry the slides and immediately counterstain with Mayer's hemallume (2 min). Dehydrate in 70, 90, and 100% ethanol (3 min each) and mount with permanent mounting medium (see Note 23).
18. Scoring is based on the detection of PAS positive acrosome (a pinkish structure). In the testis populations, early spermatids are small round cells with evident nucleolus and only little cytoplasm. Golgi-phase spermatids show a small globular acrosome (one or two spots); in Cap-phase spermatids the acrosome is shaped onto one pole of the nuclear membrane. At least 2,000 Golgi phase spermatids must be scored per animal. The number of Cap phase spermatids must be recorded in parallel. Standard criteria must be used to define micronucleated cells: micronuclei must have round or oval shape and stain as intensively as the main nucleus (see Note 24). Record

separately Golgi- and Cap-phase spermatids with micronuclei. A Golgi–Cap phase ratio can be calculated as an index of cytotoxicity or meiotic delay.

19. For statistical comparisons, the frequencies of micronucleated spermatids must be calculated independently for the two spermatid phases scored. Individual MN frequencies in treated or control animals must be pooled and compared by chi-square analysis or G statistics (preliminary chi-square analysis should be done to exclude inter-individual variability). A trend test should be applied to verify dose–effect relationships. The Golgi–Cap phase ratios should be compared by applying the Student's *t* test on transformed mean values (e.g. square root transformation) or a non-parametric test (19).

20. If you are interested in molecular characterization of micronucleus content, read carefully Notes 20 and 25. A simple protocol for telomere/centromere detection in mouse spermatid micronuclei can be found in ref. 34.

3.3. Sperm FISH Assay

The sperm FISH assay was originally developed by Wyrobeck and his group (35) to detect aneuploidy in mouse and human sperm. The beauty of the assay is that sperm samples of any species can be analyzed provided chromosome-specific DNA-probes are available. In fact, applications have been published not only for the rat (36, 37) but also for Rhesus monkey (38). A detailed description of PCR-based preparations of DNA-probes for specific chromosomes is given in ref. 39.

For experimental purposes, the mouse sperm-FISH assay is most commonly used and the largest database exists for the mouse. The method is fairly simple. Sperm are collected from the *caudae epididymes* of experimental males. The time of sampling is chosen on the basis of the timing of mouse spermatogenesis (Fig. 1) except when results from BrdU-labelling studies indicate a chemically caused alteration of timing (40). Sperm are spread on slides and decondensed as described below. Hybridizations with fluorochrome-labeled DNA probes for chromosomes 8, X, and Y are performed as previously described (41) with some modifications (42, 43).

Within the frame of an EU-funded R&D project, a number of suspected and known aneugens have been tested with the sperm-FISH assay to validate the method (44). Unique studies were designed to compare the effects on chromosome segregation of diazepam in human and mouse meiosis prior to spermiogenesis. The most unexpected result was that humans seemed more sensitive than mice (45, 46). This result threatens the general paradigm in genetic toxicology that mice or rats are more sensitive than humans so that precautions taken upon these animal data will err on the safe side. However, more comparative data are necessary.

Positive and negative data have been collected on sperm aneuploidy or diploidy of probands exposed to chemotherapy (47–49), pesticides (50–52) or adverse lifestyle (alcohol; smoking) (53, 54).

Molecular technology is progressing and the original method of sperm-FISH is now extended or supplemented. An experimental design was described to score for structural chromosome aberrations by FISH in human and mouse sperm (55) which eventually may be combined with FISH for scoring numerical chromosome aberrations in the same samples. In human infertility studies, three-colour FISH was extended to five-colour FISH (56) and automated scoring of colour signals was compared to manual scoring (57, 58). Another new method to determine aneuploidies in human sperm used sequential primed in situ labelling (PRINS) of three or four chromosomes (59). All these developments will provide perfect tools to perform comparative studies between experimental animals and humans.

1. Mice are sacrificed 22 days after administration of the test compound according to the rules of your country.
2. The two epididymes of each male will be prepared and five to six incisions be made before placing them in an Eppendorf cup containing 300 μl of foetal calf serum. Each cup has to be labelled with the animal number. The cups are incubated for 30 min at 32°C which allows the sperm to actively leave the epididymes. After removal of the tissue, the sperm suspensions can be stored at –80°C. Sperm slides are prepared by spreading 5 μl of sperm suspension onto grease-free slides. The slides are allowed to dry overnight and freeze-stored at –20°C for later use. At least six slides are prepared per animal. It is important to label the slides with the animal number, e.g. with a diamond pen.
3. Remove the slides from the freezer and allow to thaw at room temperature for 30 min before opening the bag. Decondense the sperm nuclei by incubation of the slides in DTT for 30 min followed by 30 min in LIS solution, both on ice. The slides are then dried on a hot plate at 70°C for 5 min.
4. Place the denaturation solution (70% formamide, 2× SSC, pH 7.0) in a water bath, turn on to 78°C.
5. Make an excess of hybridization-mix, then use 20 μl per slide.
   (a) 21 μl MM 2.1 (55% formamide, 10% dextran sulphate, 1× SSC)
   (b) 9 μl probe-mix obtained as follows:
      - 2 μl salmon sperm DNA (10 mg/ml)
      - +20 μl probe (DIG)
      - +20 μl cot-1
        - precipitate with Ethanol and Glycogen

- incubate at −20°C for 30 min.
- centrifuge at 4°C for 30 min.
- speed vox and solve the DNA in 9 μl water.

6. Denature the hybridization-mix at 78°C for 8 min and then place it on ice.
7. Denature the slides for 5 min at 78°C in prewarmed 70% formamide (in 2× SSC).
8. Run the slides through ethanol series for 2 min each: 70%, 90%, and 100% at 4°C.
9. Dry on a 37°C slide warmer for about 3 min and then apply the hybridization-mix. Apply a coverslip onto the slides and seal with rubber cement.
10. Place the slides in a moisture chamber in a 37°C incubator for 24–48 h.
11. A total of five post-hybridization washings are carried out at 45°C in 50% formamide (15 min), 2× SSC at pH 7.0 (3×10 min) and PN buffer at pH 8.0+1% Nonidet P-40 (2×15 min).
12. Immuno-detection at room temperature: incubation with
    (a) 40 μl anti-DIG-FITC (for chromosomes X and Y)
    (b) 40 μl streptavidin-CY3 (for chromosomes 8 and X)
13. Amplification of signals: incubation with
    (a) 40 μl biotinylated anti-streptavidin
    (b) 40 μl Streptavidin-CY3
    (c) 40 μl FITC-anti-sheep made in rabbit (RAS)

    all in PNBR buffer. Wash in PN buffer for 20 min between incubations.
14. Counterstaining with 40 μl of DAPI per slide and incubate for 10 min at room temperature.
15. Apply 20 μl Vectashield to the slide, cover with a coverslip (24×50 mm) and seal with wax. Store the slides at 4°C in the dark until scoring. All slides are coded by a non-involved person before scoring.
16. Fluorescent signals are counted in ≈10,000 sperm per animal. Sperm are designated as normal (X8 or Y8), hyperhaploid (X88, Y88, XY8) or diploid (XY88, XX88, YY88) (Fig. 6) under the criteria previously described (60).
17. For a statistical overall comparison of the frequencies of hyperhaploid or diploid sperm the χ^2 test with Yate's correction can be used and comparisons on an individual animal basis can be carried out with the Mann–Whitney *U*-test (19).

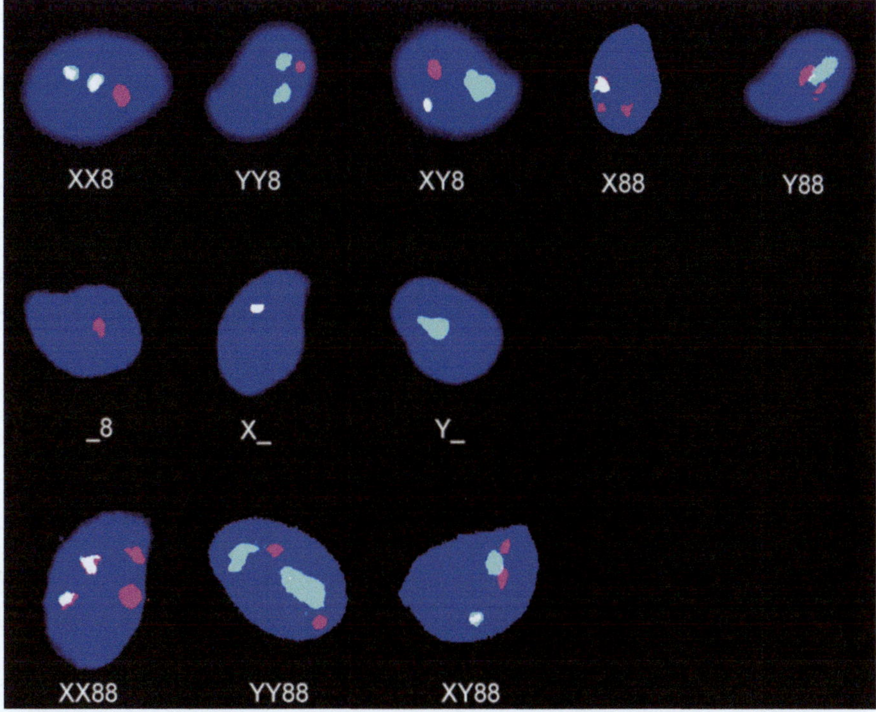

Fig. 6. Aneuploid and diploid mouse sperm with colour domains for chromosomes 8 (*red*), X (*white*) and Y (*green*). By the courtesy of Sabry M. Attia.

3.4. Dominant Lethal Assay

Dominant lethal mutations are due to structural chromosome aberrations in germ cells which are not compatible with embryonic survival and give indirect evidence of clastogenic effects of test substances (61, 62). Chromosomal aberrations which lead to loss of genetic material cause embryonic death after fertilization at the time of implantation into the uterus.

Dominant lethal experiments do not require mouse or rat strains with a specific genetic background, however, the litter size of the strains used should be relatively large (9–12 pups per female) and relatively constant (1, 63). After treatment, male animals (usually mice, sometimes rats) are mated to untreated virgin females in intervals of 4–7 days. The 4-day mating period reflects the oestrus cycle of the female mouse. In mice, successful copulation is indicated by a vaginal plug, in rats, vaginal smears are performed to verify copulation.

Depending on the spontaneous rate of dead implants in a given strain of experimental animals, differing numbers of pregnant females are required to assess statistical differences between treatment and solvent control groups with a certain accuracy (64, 65). Analysis of a large control database showed that variability of dead implantation rates stems predominantly from the females as embryonic or foetal loss can have physiological as well as genetic reasons (66). As a rule of thumb, at least ten males should be treated per experimental group, and each male should be mated to three females. Non-parametric procedures should be applied because the end points do not underlie a normal distribution (66). Using a data-clustering scheme to achieve the desirable distributional properties of variables requires larger sample sizes (67).

The pregnant females are sacrificed at mid-pregnancy and the uterus contents are inspected for live and dead implants. The dominant lethal effect is expressed as the ratio between live implants in the treated group divided by the live implants in the control group (in percent). This ratio includes pre- and post-implantation losses. The actual dominant lethal mutations, however, are the post-implantation losses (resorptions, early deaths, and late deaths), since pre-implantation losses can have physiological reasons such as reduced sperm counts of the treated males. Only at high rates of dead implants and normal pregnancy frequencies will it be likely that pre-implantation loss is due to multiple structural chromosome aberrations per gamete and thus represents a true dominant lethal effect. Therefore, data are often presented as percent dead implants or as dead implants per female.

The fertilization products of different mating intervals represent samples of different stages of male germ cell development (Table 1). Chemical mutagens are characterized by their differential spermatogenic response which means that chemicals affect only certain developmental stages of germ cells which are typical for a specific chemical or chemical class (68). Some chemicals affect mature sperm, some affect only mid-spermatids and late spermatids, and some affect spermatocytes or differentiating spermatogonia depending on their action on physiological processes in these cell types, the accessibility of the DNA, and the repair processes.

Table 1
Mating scheme to sample different stages of mouse spermatogenesis after acute exposure

Mating intervals (days)	Treated spermatogenic stages
1–7	Spermatozoa
8–14	Late spermatids
15–21	Mid and early spermatids
22–28	Spermatocytes
29–35	Spermatocytes
36–42	Differentiating spermatogonia
43 to many months	Stem cell spermatogonia

In a large collaborative study, a standard protocol for dominant lethal tests and their statistical evaluation were validated (63). It is possible to perform a dominant lethal test with chronically treated (over the entire length of the spermatogenic cycle) males. These are then only mated to one set of virgin females for 1 week after the end of the treatment period. This procedure saves females but loses the information on germ cell specificity.

Dominant lethal experiments with treated females are often difficult to judge because embryonic death may result from systemic toxicity rather than true genetic effects. One of the correlations to true dominant lethal effects is seen in the total reproductive performance of females (69). To really prove induction of dominant lethal effects in females, embryo transfer to untreated foster females has been performed (70). Some chemicals could be identified that seemingly only cause dominant lethal effects in female mice (71).

Information on clastogenic effects of chemicals from dominant lethal tests leads to categorization in mutagenicity classes and thereby to regulation (EU category 2). The US EPA requires dominant lethal tests in the second stage of testing pesticides and other toxic substances.

1. After treatment, male animals (usually mice, sometimes rats) are mated to a new set of untreated virgin females in intervals of 4–7 days.
2. Matings are repeated with new females sequentially to sample all stages of spermatogenesis.
3. Female mice are checked for vaginal plugs every day between 6 and 8 am. Vaginal smears of mated female rats are inspected microscopically.
4. Plugged females are removed from the mating cage and housed together until sacrifice (5 mice or 3 rats per cage).
5. The pregnant females are sacrificed at mid-pregnancy (13–14 days after copulation for mice and 15–16 days after copulation for rats) according to the regulatory guidelines of one own country. If copulation was not determined exactly (see step 3) all rat or mouse females are killed 16–18 days after the onset of mating.
6. The abdomen of the pregnant female is opened by incision and the uterus with attached oviducts and ovaries is excised and mounted to a waxen- or wooden board (Fig. 7).
7. The uterus wall is carefully opened by splitting it lengthwise. The numbers of total implantation sites, live foetuses, deciduomata, and dead foetuses are recorded for every female. The latter are scored separately for early and late foetal death. Late foetal death is defined by a visible eye spot ("Augenanlage").

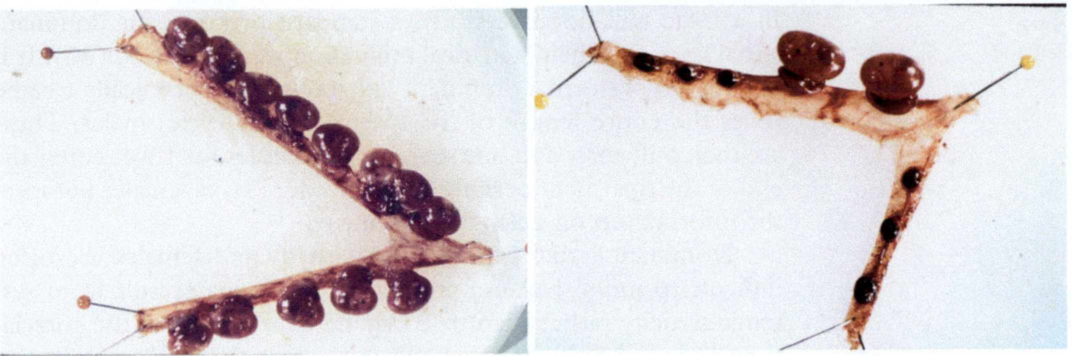

Fig. 7. *Left*: Uterus content of a female mouse with 13 foetuses. *Right*: Mouse uterus with two live foetuses and six deciduomata indicating induced dominant lethal effects.

The data from all females mated to one individual male in each mating interval are averaged because the male is the unit of statistical comparison between treatment and control groups. An example of a score sheet is given in Table 2.

8. If pre-implantation losses are to be calculated by the difference between ovulated and implanted ova, the numbers of yellow bodies (ovulation sites) are counted on each ovary under a dissection microscope (18–20×). Otherwise, preimplantation loss can be calculated by comparing total implant rates between treatment and control groups.

3.5. Female Germ Cells

3.5.1. Cytogenetic Analysis of Oocytes

The cytogenetic analysis of metaphase II oocytes is mainly used to test the possible induction of chromosome segregation errors during the first meiotic division.

For this test, mice are most often used (see Note 26).

Basically, two methods have been described to prepare metaphase II oocytes for chromosome counting. The first method (72, 73) is based on the handling of just one or a few oocytes at a time with capillary pipettes under a dissecting microscope, with final fixation directly on the slide. The second method is based on a so-called mass harvest technique (74, 75) that allows examination of a few hundred oocytes together in custom-made microcentrifuge tubes. Both methods require some skill and exercise before yielding reasonable numbers of good quality preparations. None of them have been standardized and validated by international regulatory bodies. In this chapter, the mass harvest technique will be described in detail, while the readers wishing to try the alternative technique are referred to the relevant literature.

1. Intraperitoneally inject the female mice with 7.5 IU PMS.
2. After 48 h, inject 5 IU HCG.
3. To test a chemical for the possible induction of chromosome segregation errors during the first meiotic division of oocytes,

Table 2
Example of a dominant lethal score sheet. The results from Fig. 7. are entered

Experiment no.: **Test compound:**

Date of treatment: **Mating interval:**

Date of mating: **Dec 2, 2001 to Dec 6, 2001**

Male no.	Female no.	Total implants	Live foetuses	Dead implants	Resorptions + early deaths	Late death	Date of plug
1	1	13	13	0	0	0	12/05/01
1	2	8	2	6	6	0	12/06/01
1	3						
2	1						
2	2						

the chemical can be administered at any time between HCG injection and 8 h later, when the anaphase I occurs (76, 77). Variations in the frequencies of induced aneuploid metaphases have been measured as a function of treatment time, with the peak frequency occurring at different times for different chemicals (24). This finding may result from various factors, including the chemical metabolism and mechanism of action, and may suggest the opportunity to test more than one time interval.

4. Seventeen hours after HCG injection, mice are killed according to regulatory guidelines of one own country. Since, in the absence of fertilization, ovulated oocytes normally arrest at the metaphase II stage, there is no need of an anti-mitotic treatment to collect metaphases for the analysis.
5. Ovaries and oviducts are isolated from the peritoneal cavity into Petri dishes containing HBSS, and after a quick rinse, they are transferred into a watch glass containing only a drop of HBSS.
6. Under a dissecting microscope, ovaries are discarded and the ampulla of each oviduct is pierced with a dissecting needle to free the cumulus mass containing the oocytes surrounded by granulosa cells.
7. Cumulus masses isolated from 20 or more oviducts are transferred together, by a capillary pipette, into a well of a multiwell plate containing 1 ml of 150 IU/ml hyaluronidase solution. After 15–20 min at room temperature, oocytes will appear naked of cumulus cells.
8. The hundreds of oocytes so obtained are washed three times by transferring them in sequence to three wells containing clean HBSS to eliminate the cumulus cells as much as possible.
9. Then, the oocytes are transferred to a well containing the hypotonic solution (0.3% trisodium citrate) where they are kept for 30 min at room temperature. All transfers of oocytes are made by capillary pipettes (see Note 27).
10. During the hypotonic treatment the oocytes are counted under the dissecting microscope. This allows assessment of possible toxic effects of treatment on the ovulation rate. It is also the basis for estimating the success rate of metaphase spreading.
11. The oocytes are then transferred to the microcentrifuge tube in a minimum volume of the hypotonic solution (see Note 28).
12. This and the following four steps describe the gradual fixation procedure (see Note 29). Add one part of fixative to four parts of the oocyte suspension. Mix gently, by an extended capillary pipette, and fix 5 min at room temperature (see Note 30).
13. Centrifuge at $40 \times g$ for 2 min.

14. Remove supernatant but leave 1 cm of volume, then add two parts of fixative before resuspending the pellet. Mix thoroughly by an extended capillary pipette and fix 5 min at room temperature, then repeat the centrifugation step.
15. Remove supernatant up to the 0.5 cm mark. Add fresh fixative in the ratio 7:1; resuspend as above, fix for 5 min, and then centrifuge.
16. Remove supernatant up to the 0.3 cm mark. Add fresh fixative in the ratio 10:1. Mix gently the cell suspension and finally prepare slides by conventional air-drying (see Note 31).
17. Analyze slides by phase contrast microscope at 250× magnification before c-banding. Count oocytes and classify them into metaphase I (MMI) and metaphase II (MMII). Slides must be aged for about a week at room temperature before c-banding (5).
18. For the assessment of chemical effects upon chromosome segregation at MMI and aneuploidy induction, the following scoring criteria are recommended.
    (a) The frequency of metaphase I oocytes must be recorded as an index of meiotic delay/arrest.
    (b) Well spread MMII cells must be classified into euploid ($n=20$), hypoploid ($n<20$), hyperploid ($n>20$) or diploid ($n=40$) based on chromosome counting (Fig. 8). Single unpaired chromatids must be counted as half of a chromosome. Aneuploidy frequency must be calculated as the ratio between the number of hyperploid oocytes to the total number of haploid, hypoploid, hyperploid, and diploid metaphase II oocytes. It is recommended to accumulate about 200 analyzed MMII cells in each experimental group.
    (c) MMII cells with two or more chromosomes split into single chromatids must be separately recorded and their frequency assessed as indicator of premature centromere separation.
    (d) Premature anaphase II cells can be easily recognized and their frequency must be assessed. They can reveal disturbances of cohesion at the centromere. It must be considered that oocyte ageing, i.e. harvesting at increasing intervals after ovulation, may cause premature anaphases II with variable frequencies in different mouse strains.
19. Aneuploidy frequencies in treated and control oocytes should be compared by chi-square analysis or G statistics, or Fisher exact test in the case of small samples. A trend test should be applied to verify dose–effect relationships (19).

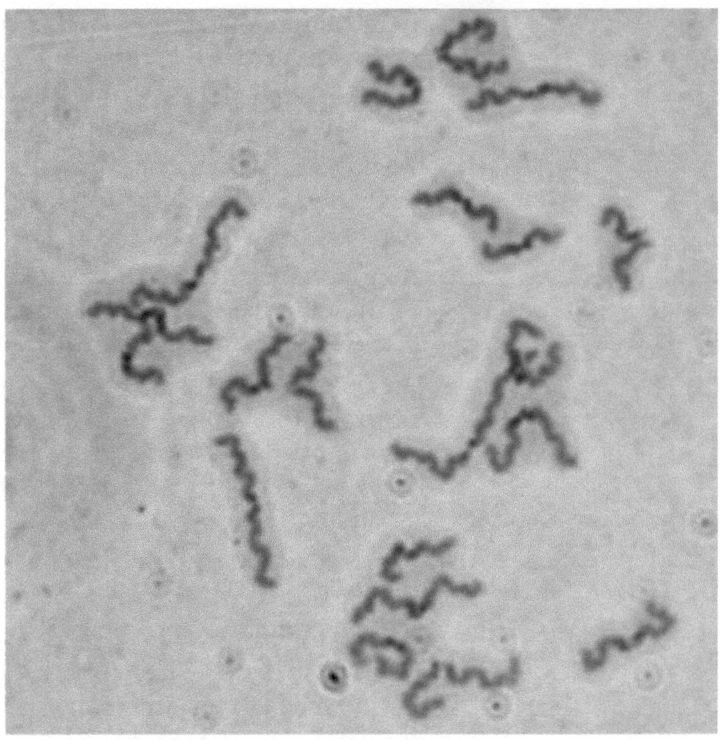

Fig. 8. Second meiotic division chromosomes ($n = 20$) of a mouse oocyte (c-banded).

3.5.2. Cytogenetic Analysis of Zygotes

The cytogenetic analysis of mouse zygotes is mainly carried out to evaluate the transmission to the embryo of structural chromosome damage induced in paternal or maternal germ cells by the exposure to potentially clastogenic chemicals. The induced frequencies of stable and unstable chromosome aberrations detected at the first cleavage metaphase after exposure of paternal germ cells in vivo have shown a high correlation with dominant lethality and heritable translocation data obtained under similar exposure conditions, confirming the suitability of this method to evaluate heritable chromosome damage (62). The cytogenetic analysis of zygotes can be also used to detect effects upon chromosome segregation at the second meiotic division of the oocyte, that takes place after sperm penetration. The transmission of aneuploidies induced in oocytes at the first meiotic division to the embryo can also be assessed.

To evaluate the induction and transmission of structural chromosome aberrations by the cytogenetic analysis of zygotes, treated males are mated with untreated females at post-treatment times chosen on the basis of the timing of mouse spermatogenesis to sample effects on specific meiotic or post-meiotic stages (26). Post-meiotic stages have been shown to be particularly vulnerable to the induction of DNA damage that will ultimately lead to chromosome aberrations in the paternal chromosome complement of the zygote (78). Single acute exposures or repeated daily treatments during the latest stages of sperm maturation, known to be defective in DNA repair, are usually applied. To assess the clastogenic effects potentially induced during oocyte maturation, pre-ovulatory oocytes appear to be more sensitive than immature ones (79). Therefore, females can be optimally treated with a single acute dose shortly before ovulation, usually between −24 and +4 h with respect to HCG injection.

A possible alternative approach to assess the presence of structural chromosome aberrations in the zygote is to score micronuclei in the blastomere's cytoplasm of two-cell embryos. In fact, many unstable aberrations and especially acentric fragments in the zygote, will give rise to micronuclei after the first cleavage division. Two-cell embryos can be fixed and prepared for microscopic analysis by a mass harvest technique similar to that described for the fixation of oocytes and zygotes (as well as by any variation of the single cell fixation technique). Slides will then be stained with DAPI and micronuclei scored by the same criteria used for the analysis of micronuclei in cytochalasin B-blocked mammalian somatic cells (80). The advantage over the analysis of aberrations in zygote metaphases is that much larger numbers of fixed embryos will be scorable for micronuclei. The drawback is that only a proportion of aberrations give rise to micronuclei and no detailed analysis can be carried out of the type of induced aberrations. Theoretically, micronuclei are only formed by unstable chromosomal aberrations, and thus, the induced rate of micronucleated two-cell embryos should represent the induced dominant lethality. An example of a two-cell mouse embryo with micronucleus is shown in Fig. 9.

To evaluate the induction of aneuploidy during the second meiotic division of the oocytes, female mice must be treated 10 h after HCG, after expulsion of the first polar body, but before the onset of anaphase II (81).

1. Female mice are treated with PMS and HCG to induce superovulation as for the cytogenetic preparation of oocyte metaphases.
2. Male mice are mated 1:1 with females immediately after HCG injection.
3. Females are checked for the presence of a vaginal plug 8 h after mating. Plug-negative females can be checked again within 24 h. When males are exposed, record the percentage of plugged females, because a decrease with respect to the untreated matched control group can reflect systemic toxicity induced by the exposure.
4. Twenty-four to twenty-six hours after HCG, vaginal plug positive females are injected with 0.2 ml of the colchicine solution.
5. Females are sacrificed for harvesting zygotes 5 h after colchicine injection. This time can be extended if a treatment-induced delay of the first cleavage cell cycle is suspected (see Note 32).

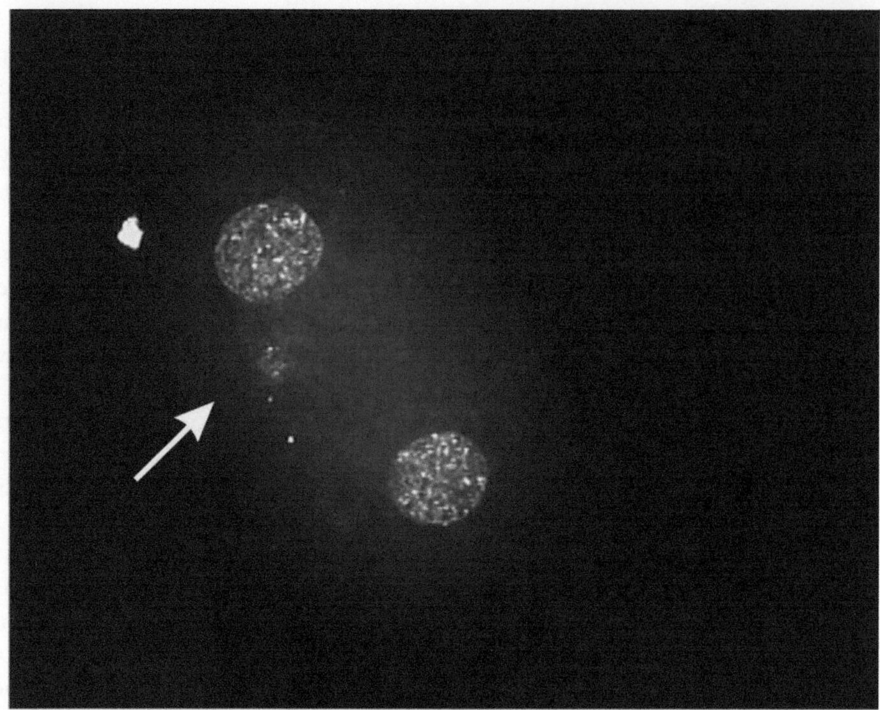

Fig. 9. A two-cell mouse embryo (DAPI-stained) with micronucleus (*arrow*).

6. Oviducts are isolated as for the collection of oocytes, with additional care not to break the infundibulum. Zygotes are no more surrounded by a mass of cumulus cells and located in the ampulla. They must be collected from the distal parts of the oviducts by flushing them with a 1-ml syringe filled with HBSS and mounted with a 30G blunt-end needle inserted into the infundibulum. This is a very demanding process and can be best performed when the syringe is mounted to a board and fluid is released under pressure-controlled conditions. The number of zygotes harvested from each couple of oviducts can be counted at this stage.
7. Zygotes are then transferred for a quick incubation in hyaluronidase (150 IU/ml HBSS) to soften the zona pellucida.
8. Hypotonic treatment, fixation and deposition of zygotes onto microscopic slides are carried out as for the preparation of oocyte metaphases.
9. C-banding of slides is recommended for chromosome counting and is mandatory for the analysis of structural aberrations. Structural aberrations can be alternatively scored after painting with chromosome-specific fluorescent DNA-probes. Chromosome painting is especially appropriate to detect stable aberrations, such as reciprocal translocations, that can be

detected by conventional staining only when they lead to the formation of extremely short and long chromosomes. An optimal protocol combining the use of several chromosome specific probes and centromere staining by DAPI has been applied to score in the same metaphases unstable aberrations over the whole genome and stable aberrations over the painted chromosomes (62).

10. For the assessment of structural aberrations in the first cleavage metaphases, the following scoring criteria are recommended.
    (a) Structural aberrations must be scored only in euploid and hyperploid metaphase cells. A total sample of about 200 analyzed cells is recommended for each experimental group (see Note 33).
    (b) Chromosome-type and chromatid-type aberrations must be separately recorded with breaks and exchanges distinguished in each class. It is noteworthy that chromosome-type aberrations have been almost exclusively observed after exposure of male germ cells, while chromatid-type aberrations can be frequently seen after oocyte exposure.
    (c) If chromosome painting is applied, aberrations involving painted chromosomes must be classified according to the PAINT nomenclature (82).
    (d) In addition to type and frequency of structural aberrations, also the frequency of unfertilized oocytes, as an index of pre-fertilization toxicity on male germ cells, and the frequency of pronuclear stage zygotes, as an index of developmental delay, should be recorded.
11. For the assessment of aneuploidy, the frequencies of hyperploid metaphases ($2n > 40$) must be calculated based on at least 200 analyzed metaphases per experimental group. Polyploid metaphases ($51 < 2n < 60$) must be separately recorded.
12. Chromosome aberration frequencies or aneuploidy frequencies in treated and control zygotes should be compared by chi-square analysis or G statistics, or Fisher exact test in the case of small samples. A trend test should be applied to verify dose–effect relationships (19).

4. Notes

1. The solutions can be prepared in stock and autoclaved for storing. Alternatively, simply use freshly prepared saline solutions.
2. You can use methanol instead of ethanol. Prepare fixative immediately before use and never store it (it undergoes rapid degradation).

3. To prepare the stock solution (0.5%) dissolve 50 mg of colchicine in 10 ml of sterile distilled water. Filter and store in 2 ml vials at −20°C. After thawing, the solution is stable at +4°C for several weeks.
4. Slides must be perfectly degreased to obtain the highest yield of well spread analyzable metaphases. To clean slides several procedures can be followed, for example: keep them in absolute ethanol for at least 1 week, then dry and transfer them to hot water with soap for 2 days. The day before use wipe them with gauze and soap water until they can form a homogeneous water film when put under tap water. Then rinse them with tap water and wash in distilled water. Keep them in distilled water at +4°C until use. Although pre-cleaned slides are commercially available, they should not be considered as immediately ready for chromosome preparations.
5. Cytocentrifuge is optional: you can drop the cells directly onto slides.
6. Helly's fixative is recommended to perform the periodic acid Schiff (PAS) reaction allowing visualization of the developing acrosome. You must consider during manipulation that this salt solution has carcinogenic potential.
7. Different commercial sources of these hormones are available. However, these might change over time and, moreover, not all products are equally effective. Updates and tips on the best source of hormones can be found at: http://www3.imperial.ac.uk/lifesciences/services/research/transgeniclist.
8. A precise relation to the body weight is not necessary. However, note that the accumulation of a suitable number of meiotic (spermatocyte) metaphases is achieved at higher concentrations and longer time intervals with respect to those efficient in the bone marrow compartment (25). This effect reflects the existence of the blood testis barrier.
9. This time interval is recommended in view of the low proliferative rate of spermatogonia. You can try to apply short time intervals but you must increase the colchicine concentrations, up to the levels used for spermatocytes (see Subheadings 2.2 and 3.1.2, step 1).
10. During testis isolation, it is important to avoid fat tissue residues. To obtain high quality preparations, clean as quickly as possible the fat residues from the testis before releasing the seminiferous tubules.
11. Dissolve collagenase immediately before use to avoid predigestion of the enzyme.
12. The cell density is correct when the suspension has an opalescent colour. You will learn quickly how to adjust by eye the first

volume of this solution. A good starting point is to resuspend cells in 0.5 ml of fixative.

13. You can score 200 cells per animal to improve the statistical power of your analysis, if you suspect that the chemical under study is not highly effective and/or inter-individual variability could affect your data.
14. Do not increase the centrifugation speed to avoid sedimentation of the many spermatozoa present in the suspension.
15. The temperature is the physiological one in the testis.
16. Do not vary the volume suggested: until fixation cell morphology must be preserved from rapid dehydration. You can establish empirically the appropriate cell density and further adjust it as necessary. Start for example with a 1:20 dilution from the 2 ml cell suspension. Remember to maintain the cell suspensions on ice.
17. Never let slides dry: cell morphology is well preserved only if slides remain wet at this stage. Without proper cell fixation, cells lose their integrity and cytoplasmatic vacuolization can occur. This can be a confounding element when MN analysis is carried out.
18. Place five to six drops of the cell suspension onto each slide, wait for 15 min (the time necessary for cells to sediment), and then add gradually Helly's fixative. After the whole slide is covered with about 2 ml fixative, incubate for 1 h. Finally, rinse the slides as described in step 15.
19. This step must be started within 10 min from slide preparation to avoid loss of cell integrity.
20. Helly's fixative does not appear appropriate for the subsequent application of molecular cytogenetics (e.g. localization of centromeric or telomeric regions). In this case, slides must be prepared using the cytospin method, rinsed in PBS then transferred in absolute ethanol and maintained at −20°C until use.
21. The slides can be preserved several weeks up to 1 year at these conditions.
22. PAS reaction is a classical technique for the demonstration of carbohydrates in tissue or cell samples. Schiff's reagent can be prepared as follows: dissolve 1 g basic fuchsin in boiling water. Let the temperature decrease to 50°C and filter. Add 20 ml HCl 1 N. Let the temperature decrease to 25°C and add 2 g sodium metabisulfite. Stir overnight at +4°C. Add 6 g activated carbon, incubate 1 min and filter. The solution must be colourless or pale yellow. If brownish, incubate again with activated carbon. Keep in the dark at +4°C and discard if it turns to pink. The activity of Schiff reagent can vary among batches.

The time indicated is indicative, but the reaction can be stopped as soon as the liquid turns pinkish.

23. Mayer's hemallume staining distinguishes the nucleus as well as the micronucleus boundaries.

24. According to conventional criteria, micronuclei which do not appear clearly separated from the main nucleus should be discarded. However because so little cytoplasm is present in spermatids, micronuclei with evident membrane can be accepted even if juxtaposed to the main nucleus.

25. The acrosome is not visible when molecular cytogenetics is applied, and the lack of distinction of the two developmental phases can result in underestimation of the effect. Therefore it is advisable to carry out the conventional analysis in parallel to the molecular cytogenetic assay, which is necessary to characterize the possible mechanism of micronucleus origin.

26. There is no specific strain requirement, but it must be considered that superovulation protocols and subsequent oocyte yields can be largely different for different strains, since oocyte maturation is partially under genetic control. In some strains, such as the Balb/C, 8–12 week-old females provide the highest yield of oocytes after exogenous hormone stimulation, while in the case of C57Bl strain, prepuberal females respond much better than 2-month or older animals. Even under the best superovulation protocol, the average number of oocytes harvested per female may vary between 20 and more than twice this value. In the context of research with transgenic models, various discussion lists were established in the last few years. *See*, for instance, the tg-list at http://www3.imperial.ac.uk/lifesciences/services/research/transgeniclist that might offer a useful forum and source of information also for issues such as superovulation protocols.

27. In the transfer to the hypotonic solution try to keep the HBSS volume containing the oocytes as small as possible not to dilute the hypotonic solution with HBSS.

28. This step is critical: the tube must be filled starting from the bottom and avoiding air bubbles up to the 4 cm notch. Control your pulled pipette before starting to check that it is long and thin enough to reach the bottom of the microcentrifuge tube.

29. Variations of this protocol can be envisaged as a function of the temperature/humidity environmental conditions to improve the metaphase spreading/quality. For instance, in the last one or two fixation steps 2:1 or even 1:1 alcohol–acetic acid can be used instead of the 3:1 fixative. Enrichment of the fixative with acetic acid will enhance metaphase spreading.

30. This and all the following resuspension steps are tricky: if bubbles are formed, at their level oocytes will attach to the wall of the tube or the pipette and be lost for ever.

31. The cell suspension can be dropped from a variable distance (10–40 cm) onto very clean microscope slides (see Subheading 2). It is recommended to handle clean slides with forceps. Depending upon the humidity and temperature conditions of the laboratory, the slides can be used frosted (maintained at –20°C for about 1 h before use), kept in distilled water until use to drop the oocytes onto a thin layer of water, or used perfectly dry. The optimal conditions must be verified in each laboratory. To obtain well-spread chromosome preparations free of cytoplasm the fixative should evaporate quickly, and this is achieved by gently blowing onto the slides after dropping the cell suspension, or by placing slides on a warm (60°C) plate.

32. A long in vivo colchicine exposure might also be applied following a protocol that optimizes the use of a cohort of exposed females to assess, in the same experiment, the induction of aneuploidy in metaphase II oocytes and its transmission to the embryo. This might be the case when, for instance, females are chronically exposed during oocyte maturation for several weeks and it would be demanding to conduct independent exposures (83). In this experiment, females are mated rightly after HCG injection (e.g. at 4 p.m.) and checked for vaginal plugs 17 h later (at 9 a.m.); negative ones are sacrificed immediately for the preparation of oocyte metaphases, while those positives are injected with colchicine 24 h after HCG (at 4 p.m.) and sacrificed the day after (at 9 a.m., 17 h after colchicine injection) for the preparation of zygote metaphases. This protocol is suitable for chromosome counting at both the second meiotic and the first cleavage metaphase, but it is not recommended for the analysis of structural aberrations, because the chromosomes shorten during prolonged colchicine exposure and become less optimal for scoring breaks and exchanges.

33. In principle, it would be ideal to separately score aberrations in the paternal and maternal set of chromosomes. This can be done only when the two chromosome sets did not intermingle. However, this is not always the case, depending upon the duration of colchicine treatment, and even when the two sets are separate, the differential degree of chromosome condensation that could help to distinguish them (with maternal chromosomes less condensed than male ones) is not always a reliable indication. For these reasons, several authors propose to calculate the frequency of chromosome aberrations relatively to the whole diploid complement of the zygote.

References

1. Anderson, D (1984) The dominant lethal test, in *Mutagenicity Testing – A Practical Approach* (Venitt, S., and Parry, J.M., eds.), IRL Press, Oxford, UK, pp. 307–335.
2. Witt, K.L., Hughes, L.A., Burka, L., McFee, A.F., Mathews, J.M., Black, S.L., and Bishop, J.B., (2003) Mouse bone marrow micronucleus results do not predict the germ cell mutagenicity of N-hydroxymethylacrylamide in the mouse dominant lethal assay. *Environ. Mol. Mutagen.* **41**, 111–120.
3. Oakberg, E.F. (1956) Duration of spermatogenesis in the mouse and timing of stages of the cycle of the seminiferous epithelium. *Am. J. Anat.* **99**, 507–516.
4. Dietrich, A.J.J., Scholten, R., Vink, A.C.G., and Oud, J.L. (1983) Testicular cell suspensions of the mouse in vitro. *Andrologia* **15,** 236–246.
5. Salamanca, F. and Armendares, S. (1974) C-bands in human metaphase chromosomes treated by barium hydroxide. *Ann. Genet.* **17,** 135–136.
6. Bonnet-Garnier, A., Lacaze, S., Beckers, J.F., Berland, H.M., Pinton, A., Yerle, M., and Ducos, A. (2008) Meiotic segregation analysis in cows carrying the t(1;29) Robertsonian translocation. *Cytogenet. Genome Res.* **120**, 91–96.
7. Pinton, A., Faraut, T., Yerle, M., Gruand, J., Pellestor, F., and Ducos A. (2005) Comparison of male and female meiotic segregation patterns in translocation heterozygotes: a case study in an animal model (Sus scrofa domestica L.). *Hum. Reprod.* **20**, 2476–2482.
8. Adler, I.-D. (1980) New approaches to mutagenicity studies in animals for carcinogenic and mutagenic agents. I. Modification of the heritable translocation test. *Teratogen. Carcinogen. Mutagen.* **1**, 75–86.
9. Backer, L.C., Dearfield, K.L., Erexson, G.L., Campbell, J.A., Westbrook-Collins, B., and Allen, J.W. (1989) The effects of acrylamide on mouse germ-line and somatic cell chromosomes. *Environ. Mol. Mutagen.* **13**, 218–226.
10. Ciranni, R., and Adler, I.-D. (1991) Clastogenic effects of hydroquinone: induction of chromosomal aberrations in mouse germ cells. *Mutat. Res.* **263,** 223–229.
11. Zhang, H., Zheng, R.L., Wang, R.Y., Wei, Z.Q., Li, W.J., Gao, Q.X., Chen, W.Q., Wang, Z.H., Han, G.W., and Liang, J.P. (1998) Chromosomal aberrations induced by $^{12}C^{6+}$ heavy ion irradiation in spermatogonia and spermatocytes of mice. *Mutat. Res.* **398,** 27–31.
12. Russo, A., Gabbani, G., and Simoncini, B. (1994) Weak genotoxicity of acrylamide on premieiotic and somatic cells of the mouse. *Mutat. Res.* **309,** 263–272.
13. Regulation (EC) No 440/2008 (2008) Official Journal of the European Union, 31/05/08, L 142/1
14. Regulation (EC) No 1272/2008 (2008) Official Journal of the European Union, 31/12/2008, L 353
15. Monesi, V. (1962) Autoradiographic study of DNA synthesis and the cell cycle in spermatogonia and spermatocytes of mouse testis using tritiated thymidine. *J. Cell Biol.* **14**, 1–18.
16. Russo, A., Gabbani, G., and Dorigo, E. (1994) Evaluation of sister-chromatid exchanges in mouse spermatogonia: a comparison between the classical fluorescence plus Giemsa staining and an imunocytochemical approach. *Mutat. Res.* **323,** 143–149.
17. Russo, A. (2000) In vivo cytogenetics: mammalian germ cells. *Mutat. Res.* **455**, 175–198.
18. Adler, I.-D. (1984) Cytogenetic tests in mammals, in *Mutagenicity Testing – A Practical Approach* (Venitt, S. and Parry, J.M., eds), IRL Press, Oxford, UK, pp 275–306.
19. Sokal, R.R. and Rohlf, F.J. (1995) *Biometry: The Principles and Practice of Statistics in Biological Research.* 3rd Edition. W.H. Freeman & Co, New York.
20. Evans, E.P., Breckon, G., and Ford C.E. (1964) An air-drying method for meiotic preparations for mammalian testes. *Cytogenet.* 3, 289–294.
21. Cattanach, M.B. (1957) Induction of translocations in mice by triethylenemelamine. *Nature* **189**, 1364–1365.
22. Allen, J.W., Liang, J.C., Carrano, A.V., and Preston, R.J (1986) Review of literature on chemical-induced aneuploidy in mammalian male germ cells. *Mutat. Res.* **167,** 123–137.
23. Adler, I.-D. (1993) Synopsis of the in vivo results obtained with the 10 known or suspected aneugens tested in the CEC collaborative study. *Mutat. Res.* **287**, 131–137.
24. Pacchierotti, F. and Ranaldi, R. (2006) Mechanisms and risk of chemically induced aneuploidy in mammalian germ cells. *Curr. Pharm. Des.* **12**, 1489–1504.
25. Liang, J.C., Hsu, T.C., and Gay M. (1985) Response of murine spermatocytes to the metaphase-arresting effect of several mitotic arrestants. *Experientia* **41**, 1586–1588.
26. Adler. I.-D. (1996) Comparison of the duration of spermatogenesis between male rodents and humans. *Mutat. Res.* **352**, 169–172.

27. Lähdetie, J. and Parvinen M. (1981) Meiotic micronuclei induced by X-rays in early spermatids of the rat. *Mutat. Res.* **81,** 103–115.
28. Kallio, M. and Lähdetie, J. (1993) Analysis of micronuclei induced in mouse early spermatids by mitomycin C, vinblastine sulfate or etoposide using fluorescence in situ hybridization. *Mutagenesis* **8**, 561–567.
29. Kallio, M. and Lähdetie J. (1997) Effects of the DNA topoisomerase II inhibitor merbarone in male mouse meiotic divisions in vivo: cell cycle arrest and induction of aneuploidy, *Environ. Mol. Mutagen.* **29**, 16–27.
30. Tates, A.D., Dietrich, A.J.J., de Vogel, N., Neuteboom, I., and Bos, A. (1983) A micronucleus method for detection of meiotic micronuclei in male germ cells of mammals. *Mutat. Res.* **121**, 131–138.
31. Adler, I.-D., Anderson, D., Benigni, R., Ehling, U.-H., Lähdetie, J., Pacchierotti, F., Russo, A., and Tates, A.D. (1996) Synthesis report of the Step Project "Detection of germ cell mutagens". *Mutat. Res.* **353,** 65–84.
32. Pacchierotti, F., Adler, I.-D., Anderson, D., Brinkworth, M., Demopoulos, N.A., Lähdetie, J., Osterman-Golkar, S., Peltonen, K., Russo, A., Tates, A., and Waters, R. (1998) Genetic effects of 1,3-butadiene and associated risk for heritable damage. *Mutat. Res.* **397**, 93–115.
33. Hayashi, M., MacGregor, J.T., Gatehouse, D.G., Adler, I.-D., Blakey, D.H., Dertinger, S.D., Krishna, G., Morita, T., Russo, A., and Sutou S. (2000) In vivo rodent erythrocyte micronucleus assay. II. Some aspects of protocol design including repeated treatments, integration with toxicity testing, and automated scoring. *Environ. Mol. Mutagen.* **35**, 234–252.
34. Russo A. (2006) Evaluation of chromosome instability in mammalian cells by PRINS: detection of repetitive DNA sequences in micronuclei. *Methods Mol. Biol.* **334**, 89–104.
35. Wyrobek, A.J. and Adler, I.-D. (1996) Detection of aneuploidy in human and rodent sperm using FISH and applications of sperm assays of genetic damage in heritable risk evaluation. *Mutat. Res.* **352**, 173–179.
36. Lowe, X.R., de Stoppelaar, J.M., Bishop, J., Hoeboe, B., Moore, D., and Wyrobek, A.J. (1998) Epididymal sperm aneuploidies in three strains of rats detected by multicolor fluorescence in situ hybridization. *Environ. Mol. Mutagen.* **31**, 125–132.
37. de Stoppelaar, J.M., van de Kuli, T., Bedaf, M., Verharen, H.W., Stob, W., Mohn, G.R., Hoebee, B., and van Benthem, J. (1999) Increased frequencies of diploid sperm detected by multicolor FISH after treatment of rats with carbendazim without micronucleus induction in peripheral blood erythrocytes. *Mutagenesis* **14**, 621–632.
38. Froenicke, L., Hung, P.H., Van de Voort, C.A., and Lyons, L.A. (2007) Development of a non-human primate sperm aneuploidy assay tested in the rhesus macaque (Macaca mulatta). *Mol. Hum. Reprod.* **13**, 455–460.
39. Baumgartner, A., Weier, J.F., and Weier, H.U. (2006) Chromosome-specific DNA repeat probes. *Histochem. Cytochem.* **54**, 1363–1370.
40. Schmid T.E., Attia S., Baumgartner A., Nuesse M., and Adler, I.-D. (2001) Effect of chemicals on the duration of male meiosis in mice detected with laser scanning cytometry. *Mutagenesis* **16**, 339–343.
41. Schmid, T.E., Wang Xu, and Adler, I.-D. (1999) Detection of aneuploidy by multicolor FISH in mouse sperm after in vivo treatment with acrylamide, colchicine, diazepam or thiabendazole. *Mutagenesis* **14**, 173–179.
42. Attia, S.M., Badary, O.A., Hamada, F.M., Hrabé de Angelis, M., and Adler, I.-D. (2008) The chemotherapeutic agents nocodazole and amsacrine cause meiotic delay and non-disjunction in spermatocytes of mice. *Mutat. Res.* **651**, 105–113.
43. Attia, S.M., Schmid, T.E., Badary, O.A., Hamada, F.M., and Adler. I.-D. (2002) Molecular cytogenetic analysis in mouse sperm of chemically induced aneuploidy: studies with topoisomerase II inhibitors. *Mutat. Res.* **520**, 1–13.
44. Adler, I.-D., Schmid, T.E., and Baumgartner, A. (2002) Induction of aneuploidy in male mouse germ cells detected by the sperm-FISH assay: A review of the present data base. *Mutat. Res.* **504**, 173–182.
45. Baumgartner, A. (1999) Molekularzytogenetische Untersuchungen von Spermaproben des Menschen zum Nachweis von Aneuploidie-Induktion: Ein Vergleich zu experimentellen Daten bei der Maus. Dissertation Ludwig-Maximilians-Universität, München, March 1999.
46. Baumgartner, A., Schmid, T.E., Schutz, C.G., and Adler I.-D. (2001) Detection of aneuploidy in rodent and human sperm by multicolor FISH after chronic exposure to diazepam. *Mutat. Res.* **490**, 11–19.
47. Robbins, W.A., Meistrich, M.L., Moore, D., Hagemeister, F.B., Weier, H.U., Cassel, M.J., Wilson, G., Eskenazi, B., and Wyrobek, A.J. (1997) Chemotherapy induces transient sex chromosomal aneuploidy in human sperm. *Nat. Genet.* **16**, 74–78.
48. Martin, R.H., Ernst, S., Rademaker, A., Barclay, L., and Summers, N. (1999) Analysis of sperm

chromosome complements before, during and after chemotherapy. *Cancer Genet. Cytogenet.* **108**, 133–136.

49. Tempest, H.G., Ko, E., Chan, P., Robaire, A., and Martin, R.H. (2008) Sperm aneuploidy frequencies analysed before and after chemotherapy in testicular cancer and Hodgkin's lymphoma patients. *Hum. Reprod.* **23**, 251–258.
50. Recio, R., Robbins, W.A., Boria-Aburto, V., Morán-Martinez, J., Froines, J.R., Hernández, R.M., and Cebrián, M.E. (2001) Organophosphorous pesticide exposure increases the frequency of sperm sex null aneuploidy. *Environ. Health Perspect.* **109**, 1237–1240.
51. Smith, J.L., Garry, V.F., Rademaker, A.W., and Martin, R.H. (2004) Human sperm aneuploidy after exposure to pesticides. *Mol. Reprod. Dev.* **67**, 353–359.
52. Härkönen, K. (2005) Pesticides and the induction of aneuploidy in human sperm. *Cytogenet. Genome Res.* **111**, 378–383.
53. Shi, Q. and Martin, R.H. (2000) Aneuploidy in human sperm: a review of the frequency and distribution of aneuploidy, effects of donor age and lifestyle factors. *Cytogenet. Cell Genet.* **90**, 219–226.
54. Robbins, W.A., Elashoff, D.A., Xun, L., Jia, J., Wu, G., and Wei, F. (2005) Effect of lifestyle exposures on sperm aneuploidy. *Cytogenet. Genome Res.* **111**, 371–377.
55. Marchetti, F., Cabreros, D., and Wyrobek, A.J. (2008) Laboratory methods for the detection of chromosomal structural aberrations in humans and mouse sperm by fluorescence in situ hybridization. *Methods Mol. Biol.* **410**, 241–271.
56. Durakbasi-Dursun, H.G., Zamani, A.G., Kutiu, R., Görkemli, H., Bahce, M., and Acar, A. (2008) A new approach to chromosomal abnormalities in sperm from patients with oligoasthenoteratozoospermia: detection of double aneuploidy in addition to single aneuploidy and diploidy by five-color fluorescence in situ hybridization using one probe set. *Fertil. Steril.* **89**, 1709–1717.
57. Perry, M.J., Chen, X., and Lu, X. (2007) Automated scoring of multiprobe FISH in human spermatozoa. *Cytometry* **71**, 968–972.
58. Carrell, D.T. and Emery, B.R. (2008) Use of automated imaging and analysis technology for the detection of aneuploidy in human sperm. *Fertil. Steril.* **90**, 434–437.
59. Pellestor, F., Andréo, B., Puechberty, J., Lefort, G., and Sandra, P. (2006) Analysis of sperm aneuploidy by PRINS. *Methods Mol. Biol.* **334**, 49–59.
60. Lowe, X., O'Hogan, S., Moore, D., Bishop, J., and Wyrobek, A. (1996) Aneuploid epididymal sperm detected in chromosomally normal and Robertsonian translocation-bearing mice using a new three-chromosome FISH method. *Chromosoma* **105**, 204–210.
61. Brewen, J.G., Payne, H.S., Jones, K.P., and Preston, R.J. (1975) Studies on chemically induced dominant lethality. I. The cytogenetic basis of MMS-induced dominant lethality in post-meiotic male germ cells. *Mutat. Res.* **33**, 239–246.
62. Marchetti, F., Bishop, J.B., Cosentino, L., Moore, D., and Wyrobek, A.J. (2004) Paternally transmitted chromosomal aberrations in mouse zygotes determine their embryonic fate. *Biol. Reprod.* **70**, 616–624.
63. Ehling, U.H., Machemer, L., Buselmaier, W., Dycka, J., Frohberg, H., Kratochvilova, J., Lang, R., Lorke, D., Müller, D., Peh, J., Röhrborn, G., Roll, R., Schulze-Schenking, M., and Wiemann, H. (1978) Standard protocol for the dominant lethal test on male mice. *Arch. Toxicol.* **39**, 173–185.
64. Haseman, J.K. and Soares, E. R. (1976) The distribution of fetal death in control mice and its implications on statistical test for dominant lethal effects. *Mutat. Res.* **41**, 277–288.
65. Hasemann, J.K. and Kupper, L.L. (1979) Analysis of dichotomous response data from certain toxicological experiments. *Biometrics* **35**, 281–293.
66. Lockhart, A.M., Piegorsch, W.W., and Bishop, J.B. (1992) Assessing overdispersion and dose-response in the dominant lethal assay. *Mutat. Res.* **272**, 35–58.
67. Adler, I.-D., Bootman J., Favor, J., Hook, G., Schriever-Schwemmer, G., Weitzl, G., Whorton, E., Yoshimura, I., and Hayashi, M. (1998) Recommendations for statistical design of in vivo mutagenicity tests with regard to subsequent statistical analysis. *Mutat. Res.* **417**, 19–30.
68. Ehling, U.H. (1974) Differential spermatogenic response of mice to the induction of mutations by antineoplastic drugs. *Mutat. Res.* **26**, 285–295.
69. Bishop, J.B., Morris, R.W., Seely, J.C., Hughes, L.A., Cain, K.T., and Generoso, W.M. (1997) Alterations in the reproductive pattern of female mice exposed to xenobiotics. *Fund. Appl. Toxicol.* **40**, 191–204.
70. Katoh, M.A., Cain, K.T., Hughes, L.A., Bishop, J.B., and Generoso, W.M. (1990) Female-specific dominant lethal effects in mice. *Mutat. Res.* **230**, 205–217.
71. Sudman, P.D. and Generoso, W.M. (1991) Female-specific mutagenic response of mice to hycanthone. *Mutat. Res.* **246**, 31–43.

72. Tarkowski, A.K. (1966) An air-drying method for chromosome preparations from mouse eggs. *Cytogenet.* **5**, 394–400.
73. Hodges, C.A. and Hunt, P.A. (2002) Simultaneous analysis of chromosomes and chromosome-associated proteins in mammalian oocytes and embryos. *Chromosoma* **111**, 165–169.
74. Payne, H.S. and Jones, K.P. (1975) Technique for mass isolation and culture of mouse ova for cytogenetic analysis of the first cleavage mitosis. *Mutat. Res.* **33**, 247–249.
75. Mailhes, J.B., and Yuan, Z.P. (1987) Cytogenetic technique for mouse metaphase II oocytes. *Gamete Res.* **18**, 77–83.
76. Tiveron, C., Marchetti, F., Bassani, B., and Pacchierotti F. (1992) Griseofulvin-induced aneuploidy and meiotic delay in female mouse germ cells. I. Cytogenetic analysis of metaphase II oocytes. *Mutat. Res.* **266**, 143–150.
77. Marchetti, F. and Mailhes, J.B. (1994)Variation of mouse oocyte sensitivity to griseofulvin-induced aneuploidy and meiotic delay during the first meiotic division. *Environ. Mol. Mutagen.* **23,** 179–185.
78. Marchetti, F. and Wyrobek, A.J. (2005) Mechanisms and consequences of paternally-transmitted chromosomal abnormalities. *Birth Defects Res. C Embryo Today* **75**, 112–129
79. Albanese, R. (1987) The induction and transmission of chemically induced chromosome aberrations in female germ cells. *Environ. Mutagen.* **10**, 231–243.
80. Fenech, M. (2008) The micronucleus assay determination of chromosomal level DNA damage. *Methods Mol. Biol.* **410**, 185–216.
81. Marchetti, F., Mailhes, J.B. Bairnsfather, L. Nandy, I., and London S.N. (1996) Dose-response study and threshold estimation of griseofulvin-induced aneuploidy during female mouse meiosis I and II. *Mutagenesis* 11, 195–200.
82. Tucker, J.D., Morgan, W.F. Awa, A.A., Bauchinger, M., Blakey, D., Cornforth, M.N., Littlefield, L.G., Natarajan, A.T., and Shasserre C. (1995) A proposed system for scoring structural aberrations detected by chromosome painting. *Cytogenet. Cell Genet.* **68,** 211–221.
83. Pacchierotti, F., Ranaldi R., Eichenlaub-Ritter, U., Attia, S., and Adler, I.-D. (2008) Evaluation of aneugenic effects of bisphenol A in somatic and germ cells of the mouse. *Mutat. Res.* **651**, 64–70.

Chapter 17

Transgenic Animal Mutation Models: A Review of the Models and How They Function

Steve Dean

Abstract

In regulatory genetic toxicology, the endpoints available for routine study in vivo have been limited to looking at chromosomal damage or unscheduled DNA synthesis in a very limited number of tissues. With the development of transgenic gene mutation systems in rodents came the opportunity to investigate a new endpoint. The better-known λ*LacI* and λ*LacZ* are covered in some detail and the less well established models do receive mention with appropriate references for those wishing more information. Using a recommended experimental design it is now possible to look at the ability of a compound to induce gene mutation following in vivo exposure, in any tissue from which suitable DNA can be isolated.

Key words: Transgenic, Genetic toxicology, Regulatory, In vivo gene mutation, Testing strategy, Mutamouse, Bigblue

1. Introduction

In this chapter, we are going to focus specifically on transgenic rodent models that can be used for the detection of mutations induced in vivo following exposure of the animal itself. We will begin by briefly reviewing the production and general characteristics of the models which are currently available and how they evolved and explore the ways in which they could compliment and enhance existing test systems. We will then look in more detail at perhaps the two best-known reporter gene systems, Mutamouse and BigBlue, with a view to understanding how they work and how they might be used. We will also look at the features of some other interesting though less widespread models. The second part of this chapter will explore some of the data generated and describe much of the protocol-related discussions that took place under the auspices of the International Workshop on Genotoxicity Testing (IWGT) in order to develop the most reliable approach for using

James M. Parry and Elizabeth M. Parry (eds.), *Genetic Toxicology: Principles and Methods*, Methods in Molecular Biology, vol. 817, DOI 10.1007/978-1-61779-421-6_17, © Springer Science+Business Media, LLC 2012

these models in a regulatory setting. There have been two very comprehensive reviews which can provide a great deal of additional information and details and the reader is encouraged to use these as sources of further reading (1, 2).

1.1. What Do We Mean by "Transgenics"

A transgenic animal carries foreign DNA integrated into its chromosomal DNA. This is present in all cells and is stably transmitted to somatic and germinal progeny. In the case of transgenic mutation models, the purpose of the transgene is to express mutations which take place in vivo following treatment of the whole animal, allowing analysis of those mutations in vitro following expression using an appropriate in vitro system. Generally, the animal itself is created using an initial construct which contains the reporter transgene itself and a shuttle vector for recovering the reporter gene from whole DNA following isolation from the tissues of the experimental animal.

1.2. How Are Transgenics Produced?

Pronuclear microinjection is the most widely used method of producing transgenic animals. In this process, the transgene is microinjected into the male or female pronucleus of a fertilised egg and resultant viable embryos are then transferred to the oviduct of a surrogate female. Each resultant animal (founder) is a new hemizygous transgenic strain from which homozygous individuals can be produced by selective breeding (3).

1.3. General Characteristics of Transgenic Mutation Models

Assessment of genotoxicity in vivo is important for several reasons. In simple terms, it assists in the understanding, interpretation, and putting in context of a positive result from an in vitro test, as well as giving additional reassurance that negative in vitro results are a valid reflection of the in vivo situation. Such data also help with the understanding of the absorption, distribution, metabolism, and excretion of a test compound and its metabolites and, in fact, such data will often influence the selection of the in vivo investigations themselves. Finally, in vivo analyses can help understand the mechanism of action of compounds and the impact of, or upon, DNA repair mechanisms which themselves may influence tumorigenic potential. This is one reason why the ability to examine mutations in vivo is so interesting as the demonstration that a mutation has occurred is compelling evidence of genuine mutagenic action which could lead to cancer.

There are examples of the use of endogenous genes for detecting mutation, including assays built around loci such as hprt, aprt, tk, and Dlb-1 (4–6). However, there are serious limitations as mutation analysis can be in very limited tissue types (e.g. Dlb-1 in the small intestine, the only tissue in which it is expressed) or due to the need to isolate and culture viable cells to express mutations in vitro (e.g., hprt, aprt, and tk). Transgenic mutation models allow the in vitro analysis of mutations in any tissue from which intact, high quality DNA can be extracted.

1.4. Use of Transgenes for Detecting Mutation

These systems are built around reporter genes based on bacteriophage or plasmid shuttle vectors which are introduced into the genome as described above. It is important that the reporter genes chosen remain genetically neutral/unexpressed in that the gene product has no function nor inhibitory impact within the tissue or animal in vivo. It should be noted that because the transgenes are non-transcribed in the whole animal, they may be treated or repaired differently to active endogenous genes, particularly as they lack transcription and associated transcription coupled repair (7). It is also worth noting that bacterial DNA in such transgenes does differ from endogenous (mammalian) DNA in that it exhibits a higher GC content, a higher density of dinucleotide CpG, and of associated methylcytosine. Furthermore, multicopy, head-to-tail concatamer structures can result in hypermethylation. However, differences among transgenic loci are considered minor compared to those differences seen between different endogenous genes.

1.5. Models Currently Available

The models in current use and those that are described listed in Table 1. We shall look in particular detail at the Muta™Mouse (λ*LacZ*) and Big Blue® (λ*LacI*) systems as these have been used most frequently. They both feature elements of the lac operon which has been known and used for many years.

1.6. The Lac Operon

One of the most widely used reporter genes in molecular biology. The lac operon, present in wild type *Escherichia coli*, is one of the key elements of lactose metabolism as it allows the cell to produce enzymes responsible for lactose metabolism only when the substrate is present. Thus β-galactosidase, the *lacZ* gene product and the enzyme responsible for the cleavage of lactose into glucose and galactose, is only transcribed in the presence of lactose (8). This is because the presence of lactose interferes with the *lacI* gene product, the lactose repressor protein, which represses β-galactosidase production thus releasing the block and allowing transcription of the *LacZ* gene to proceed. This is illustrated in Fig. 1. When used

Table 1
Transgenic models discussed in this chapter

Big Blue®: λ*LacI*
Muta™Mouse: λ*LacZ*
λ*LacZ* plasmid mouse
λgpt-delta/*spi-*
cII
pKZ1 transrecombinational model
rpsL, supF, and PhiX174

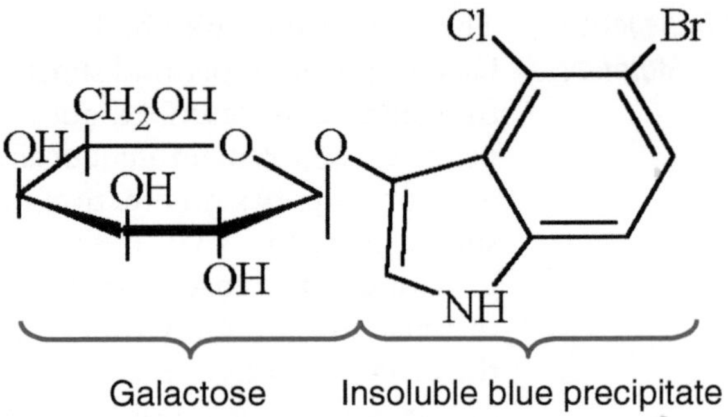

Fig. 1. Structure and cleavage of X-gal (5-bromo-4-chloro-3-indolyl-b-galactopyranoside). β-*Galactosidase* is encoded by the bacterial gene *lacZ*. In bacteria, β-galactosidase cleaves the disaccharide lactose into glucose and galactose. It can also cleave the colourless substrate X-gal (5-bromo-4-chloro-3-indolyl-b-galactopyranoside) into galactose and a blue insoluble product of the cleavage. Because mammalian genomes do not contain *lacZ*, it can be used as a reporter gene.

as a reporter gene, substrates such as X-gal (5-bromo-4-chloro-3-indolyl-b-galactopyranoside) are used as this yields an insoluble blue precipitate when cleaved by β-galactosidase, thus providing a simple colourimetric indication of *LacZ* activity.

A brief time-line of transgenic mutation models is given in Table 2, along with the relevant references describing each milestone publication.

An evaluation of the literature demonstrates the extensive range of tissues which have been examined and are therefore available for mutation analysis using transgenic systems. These are outlined in Table 3.

1.7. Big Blue® and Muta™Mouse: Strain and Species Differences

Big Blue® was created by injecting the lambda construct into fertilised eggs from C57Bl/6 mice. The founder mouse was then crossed with a non-transgenic C57BL/6, and F1 offspring from this cross were those initially used. Another line, the A1, was derived by crossing with the C3H line to give B6C3F1 which is the same strain used for the US NTP bioassay (see http://eden.ceh.uvic.ca). A Big Blue® Fischer F344 Rat has also been developed and is described by Dycaico et al. (25).

In the case of Muta™Mouse, fertilised eggs of CD2 F1 (BALB/c×DBA/2) were used from which procedure four progeny mice were selected, and Muta™Mouse is strain 40.6, carrying 80 head-to-tail concatameric copies per cell, 40 on each chromosome 3 (9).

Both Big Blue and MutaMouse rely on the activity of the β-galactosidase enzyme which is encoded by the *LacZ* gene as described above (Fig. 1). In bacteria, β-galactosidase cleaves the disaccharide lactose into glucose and galactose. It can also cleave the colourless substrate X-gal (5-bromo-4-chloro-3-indolyl-b-

Table 2
Time-line of publications describing the development of Muta™Mouse and Big®Blue

Gossen et al. (9)	Development of Muta™Mouse model in CD2F1 mice using a lambda-based shuttle vector (lgt10) carrying ~80 copies per cell of a *lacZ* reporter gene and using X-gal for mutant analysis
Kohler et al. (10, 11)	Development of the Big®Blue mouse model using C57BL/6 and B6C3F1 mice and a lambda-based shuttle vector (lLIZ) carrying ~40 copies per cell of a *lacI* reporter gene and using X-gal for mutant analysis. Also described the development of a sequencing method
Gossen et al. (12)	Muta™Mouse Model Improvement of mutant analysis using a positive selection system (P-gal)
Douglas et al. (13)	Development of sequencing method for Muta™Mouse
Dycaico et al. (14)	Development of Big Blue rat model F344 rat lambda-based shuttle vector (lLIZ) carrying ~15–20 copies per cell of the *lacI* reporter gene and using X-gal for mutant analysis
Gorelick and Thompson (15)	Statistical analysis of the sources of variability
Piegorsch et al. (16)	Protocol recommendations for Big the Blue model
Gossen et al. (17) and Dollé et al. (18)	Use of pUR288 plasmid as a shuttle vector for the detection of large deletions
Jakubczak et al. (19)	Development of CII positive selection system
Nohmi et al. (20)	Development of Gpt-delta/spi mouse model in the C57BL/6 mouse using a lambda-based shuttle vector (leEG10) carrying ~80 copies per cell. Use of a gpt reporter gene for detecting point mutations with a 6-thioguanine selection system and l spi$^-$ for detecting large deletions using a positive selection system. Sequencing methods for both endpoints
De Boer et al. (21)	Comparison of rat and mouse using Big Blue models
Piegorsch et al. (22)	Statistical analysis of the sources of variability and protocol recommendations for Muta™Mouse
Heddle et al. (23)	Summary of Washington IWGTP Meeting (1999)
Thybaud et al. (24)	Summary of Plymouth IWGT Meeting (2003)
Wahnschaffe et al. (2)	WHO
Lambert et al. (1)	Health Canada

galactopyranoside) into galactose and a blue insoluble product of this cleavage (Fig. 2). Because mammalian genomes do not contain *lacZ*, it can be used as a reporter gene as it has no function within mammalian cells. The genes and the way in which they and their gene products interact are also explained in Fig. 2. The *LacI*

Table 3
The range of tissues which have been examined for mutation analysis using transgenic systems

Germ cells		
Testis	Seminiferous tubules Epididymis Spermatozoa	
Somatic cells		
Site of contact tissues	Skin Nasal mucosa Lung Stomach	Small intestine Colon Urinary bladder
Systemically exposed	Bone marrow Spleen Liver Lung Kidney	Small intestine Colon Urinary bladder Brain Heart

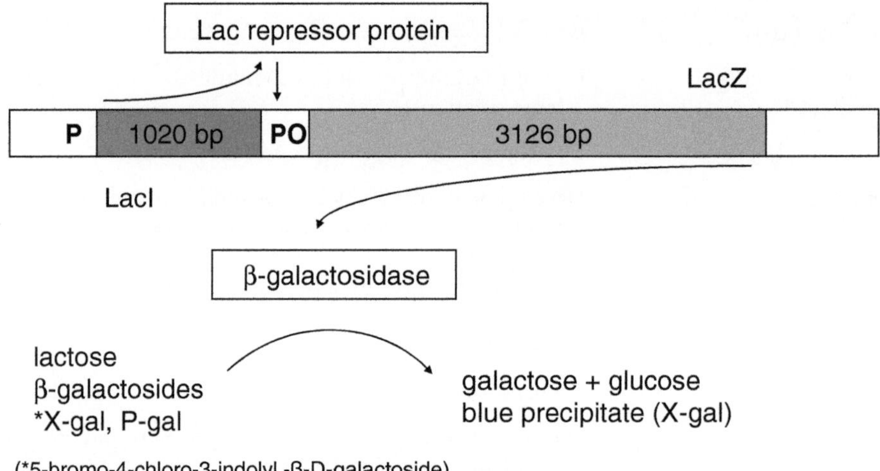

Fig. 2. Diagram showing the interaction of *LacI* and *LacZ* gene products.

gene codes for a protein that represses and prevents the transcription of the *LacZ* gene, which itself codes for β-galactosidase. Thus, in the case of Big Blue, a mutation which inhibits or inactivates the function of the *LacI* gene may reduce or prevent the transcription of an effective *LacZ* repressor protein, allowing full or partial expression of *LacZ* and the subsequent production of β-galactosidase⁻ mutant plaques therefore contain β-galactosidase and would appear blue in the presence of X-gal. In MutaMouse, only the *LacZ* gene is present and a mutation in that gene may reduce or eliminate

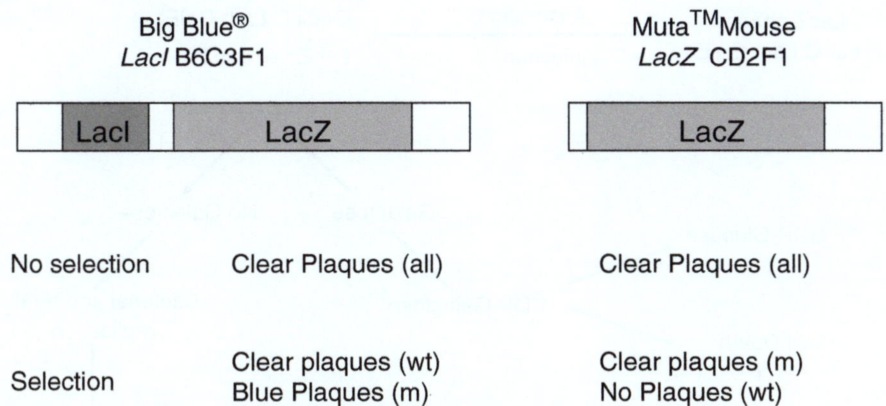

Fig. 3. Figure summarising the appearance pf plaques for both Muta™Mouse and Big Blue®.

the production of β-galactosidase and reduce or prevent the convertion of X-gal, mutant plaques thus appearing colourless, as summarized in Fig. 3.

For both systems, the total population of sequences evaluated is calculated using the titre plates. In the case of Big Blue, the total number of mutants is determined by counting blue (mutant) plaques among a background of clear plaques. For Mutamouse, the total number of mutants is determined by counting clear (mutant) plaques in a background of blue plaques.

1.8. Positive Selection

The idea behind positive selection was to find a mechanism through which non-mutant plaques could be eliminated altogether, leaving only the mutants and, therefore, cutting down on the area required to visualise the mutants and on the necessity to differentiate between coloured and clear plaques. A positive selection system was devised which used engineered host bacteria which would be unable to sustain the replication of the unmutated phage under selection conditions (26). The bacterial strain *E. coli C Lac⁻ GalE⁻* was constructed to carry a plasmid which contains the bacterial galT (galactose-1-phosphate uridyl transferase) and *galK* (galactose kinase) genes. As indicated in Fig. 4, galactose available to these bacteria would be rapidly converted to UDP galactose but cannot further be converted to UDP-glucose due to the presence of *GalE⁻*. This leads to the accumulation of UDP-galactose which is toxic and leads to death of the bacteria before the phage can replicate. Using P-gal (phenylgalactose) as the substrate, the intact phage *lacZ* is needed to convert to UDP-galactose (bacteria are *lac⁻*) to the toxic UDP-glucose whereas in the case of a mutated, *lacI⁻* gene, there is no conversion and the cells survive (12, 27, 28).

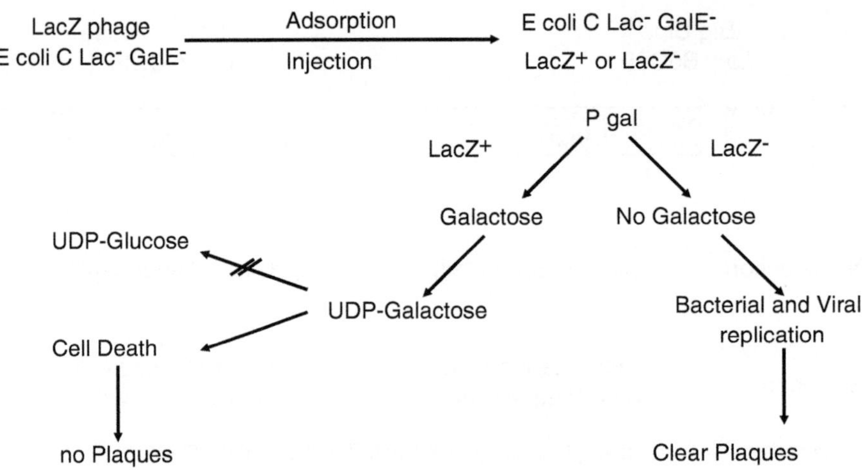

Fig. 4. Figure summarising the principles behind positive selection in Muta™Mouse.

1.9. Other Models Available

1.9.1. The Gpt Delta Rodent Model

This model was produced following microinjection of λEG10 phage DNA into C57Bl/6J mouse oocytes and a Gpt delta rat was established in Sprague–Dawley rats using the same construct. The homozygous mice carry 2 × 80 copies of the construct per cell in a head-to-tail concatamer at a single site on chromosomes 17. The coding region is a relatively short 456 bp, which makes this very convenient for sequencing. For the analysis of mutations, this system employs two different positive selection systems in the same transgene. There is an *E. coli gpt* gene for point mutations and short deletions, selected using 6-thioguanine which the gpt gene product converts into a toxic metabolite (20, 29, 30). Spi⁻ selection is used for larger deletions up to 10,000 bp, and these are mostly intrachromosomal deletions. It is worth noting, however, that as the mice contain concatamers of 80 × 48 kb λEG10 phage DNA, it is possible that deletions as large as 3.8 mbp could theoretically be detected

1.9.2. cII Selection in Lambda Models

Since systems such as Mutamouse and Big Blue are based upon lambda vectors, they lend themselves additionally to cII selection. The gpt delta model is excluded as cII is inactivated in this model. Not only is this useful in itself, as the cII gene is smaller (294 kb) and more convenient for sequencing, it also allows the analysis and comparison of mutations and spectra in two separate genes in the same construct. It also allows differences between mouse genetic background to be explored and, furthermore, the presence of two separate genes provides a good method for detecting jackpot mutations (19, 31).

The cII gene is an essential lambda Phage gene which is involved in controlling lytic/lysogenic regulation. If cII levels are high, then the bacteriophage will lysogenize and become integrated into the DNA of the host bacterium. On the other hand, if cII levels are low, then the phage will enter the lytic pathway, leading to bacteriophage replication and lysis of the host bacterium. Using an *E. coli hfl*⁻ (high frequency of lysogeny) host such as G1250, cII levels remain high and all phage lysogenize immediately, which will be manifest as the presence of a bacterial lawn but no replication nor lysis so, no plaques are formed. Note also that this system is temperature sensitive in that plating at 24°C allows lysis only at low cII levels whereas plating at 37°C will cause lysis even at low cII levels.

Where the cII gene contains a mutation, then reduced levels of the cII gene-product will drive the phage containing the mutated (inactive or reduced activity) cII to enter the lytic pathway and plaques will form if plated at 24°C for 48 h. All phage will form plaques if plated at 37°C overnight to allow calculation of the total population size. Therefore, the frequency of mutations can be derived from selective plating and total titre (mutant frequency = cII^-pfu/total pfu).

1.9.3. pKZ1 Transrecombinational Model

This model is not widely used but offers an interesting approach to the study of somatic intrachromasomal recombination (SICR) which is associated with non-homologous end-joining repair of double-strand breaks. Such lesions can result in chromosomal inversions and deletions. The construct contains the *E. coli* β-galactosidase gene in inverse orientation to a chicken β-actin enhancer/promoter complex. If SICR takes place, the *LacZ* gene becomes inverted and is then in orientation with the enhancer/promoter complex and thus capable of being transcribed. In this model, the gene product, β-galactosidase, can be detected histochemically in frozen tissue using X-gal as a substrate (32, 33).

Data published suggest that pKZ1 is extremely sensitive for low dose radiation and low dose chemical studies and that the inversion responses can be detected over a very wide dose range. Since there is a relatively high endogenous frequency, this also allows detection below endogenous frequency. The technique allows the study of a range of tissues both in vivo (physiologically relevant) and in vitro (amenable to mechanistic studies).

1.9.4. rpsL

This model was created in C56Bl/6J mice which carry the rpsL gene in a pML4 shuttle plasmid containing the *E. coli kanr* (kanamycine-resistance) gene. The target sequence is small at only 375 bp so good it lends itself to sequencing (34).

1.9.5. supF

This carries 80–100 copies of a lambda Phage Vector *supF* and, again, the target is very small at only 85 bp so good for sequencing (35, 36).

1.9.6. PhiX174

This system uses bacteriophage ΦX174am3cs70 as a recovery vector in C57Bl/6J mice. The entire genome is only 5.4 kb and can be recovered by electroporation into host bacteria, following digestion with restriction enzymes and circularization by ligation. The phage ΦX174 has been widely used historically and there already exists a wealth of data on the phage itself. There are advantages in that animals are freely available and vector recovery is inexpensive though laborious. Mutations are identified by Single Burst Analysis of the number of progeny plaques from each bacterium (37–39).

To conclude, this section has described several transgenic mutation systems and some of the principles underlying their function and use – no doubt others models will emerge as the technology matures. Of the models described, Big Blue® (λ*LacI and cII*), Muta™Mouse (λ*LacZ and cII*), the λ*LacZ* plasmid mouse and the λgpt-delta/*spi*$^-$ are those most frequently in use and these are therefore described in more depth as they are more likely to be used for regulatory safety assessment. The less well-established systems do have an important role to play, particularly as research/mechanistic tools. The following sections will discuss how these test systems can be used both in research and in a regulatory testing arena.

2. The Experimental Procedure for Big Blue® and Muta™Mouse Assays

The general procedure for these assays is relatively straightforward and is represented in Fig. 5, although, as we will see later, the details are extremely important. In brief, the experimental transgenic animal is exposed to a potential mutagen for a period of time. It is then sacrificed, tissues removed and frozen. Tissues are selected from which high quality DNA is isolated, transfected into viral vectors and mutations expressed in the appropriate strain of competent bacteria and the appropriate selection conditions. One important development was the technology which permitted the rescue of integrated vector from the animal genome and the subsequent in vitro detection of mutations (Fig. 6). Commercial packaging extracts were developed and can be used routinely to reconstruct viable lambda bacteriophage capable of introducing the transgene into host bacteria. An important factor is the size of the DNA as there is a distance of 38–51 kb between the cos sites necessary for the packaging system to recognise and pack a viable sequence. For this reason, much effort was put into the procedure for isolating high quality DNA. The bacterial strain is also very important and Big Blue® uses *E. coli* strain SCS-8 (*lacZ*MΔ15) whilst Muta™Mouse uses *E. coli* strain C Δ*LacZ*$^-$, *galE*$^-$, *recA*$^-$ and pAA119 (9, 10, 12, 40).

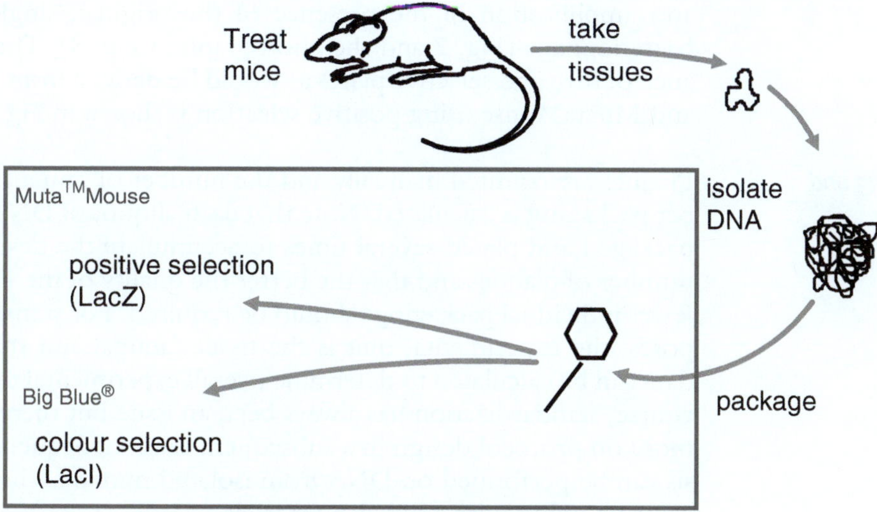

Fig. 5. The experimental procedure for Big Blue® and Muta™Mouse assays.

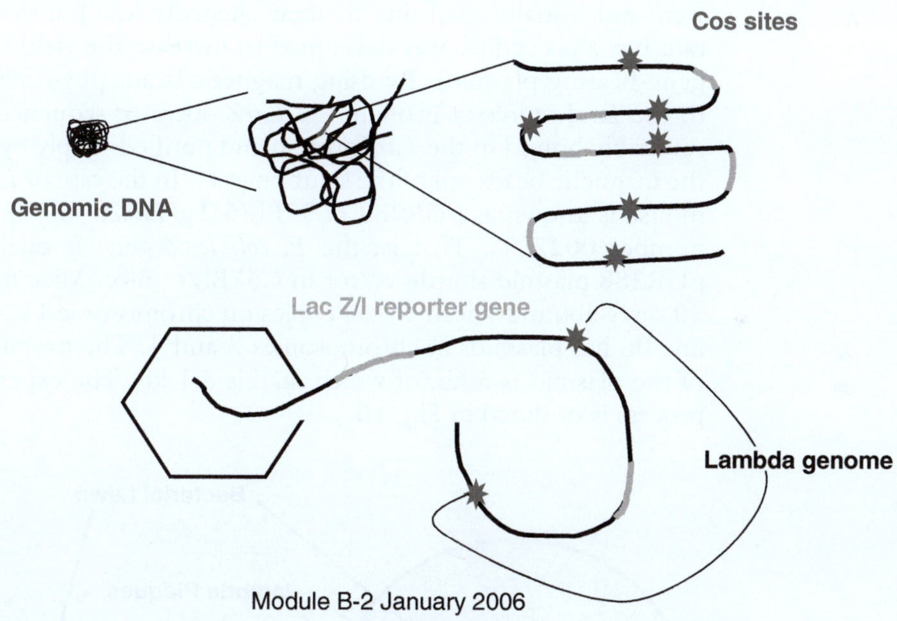

Fig. 6. Packaging of genomic DNA using lambda bactriophage.

The process by which plaques form is a result of the infection, proliferation, and spread of the bacteriophage within the host bacterial lawn. A bacterium infected with a viable phage supports phage reproduction to a point where the phage causes bacterial lysis and the subsequent infection of adject bacteria. This developing circle of infection and lysis results in a clear plaque devoid of intact bacteria and it is these which are counted as a representation and amplification of the presence of the original, single, viable bacteriophage (Fig. 7 and the photograph in Fig. 8). The appearance of titre and selective plates as would be derived from Big Blue and MuataMouse using positive selection is shown in Fig. 9.

2.1. Scoring and Analysis

Plaques are counted manually and the number of mutant colonies per packaging is calculated. Note that each aliquot of DNA may be packaged and plated several times to accumulate the desired total number of plaques and that the better the quality of the DNA, the fewer individual packagings should be required. For statistical purposes, the experimental unit is the tissue/animal and that group data can be calculated to determine overall experimental trends. Of course, standardization has always been an issue but there is much more on protocol design in a subsequent section. Sequence analysis can be performed on DNA from isolated mutants, though this topic would require a full chapter itself.

2.2. The LacZ Plasmid Mouse

Plasmid systems use a plasmid shuttle vector rather than a viral system to recover the transgene from genomic DNA. Plasmids were not initially used due to their allegedly low transformation rate but a procedure was developed to increase the yield of transgene-bearing plasmids. By using magnetic beads physically linked to the *LacI* repressor protein, the *LacZ* operator sequence can be reversibly bound to the *LacI* protein and purified simply by pulling the magnetic beads out of the solution (41). In the case of *LacZ*, the mouse is known as pUR288 (C57Bl/6-Tg(*LacZ*pl)60Vij/J:stock number 002754). That is, the *E. coli lacZ* gene is carried in a pUR288 plasmid shuttle vector in C57Bl/6 mice. Mice from line 30 carry about 20 head-to-tail copies on chromosome 11, whereas line 06 has plasmids at chromosomes 3 and 4. The overall length of the plasmid is 5 kb, of which *lacZ* is 3.1 kb. The experimental process is outlined in Fig. 10.

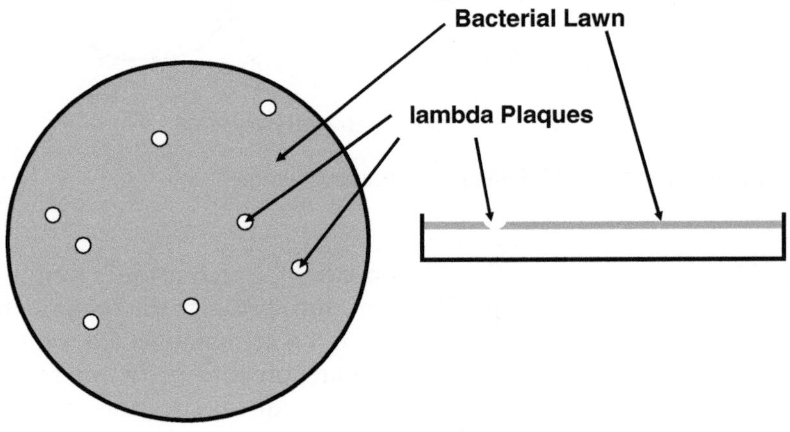

Fig. 7. Diagrammatic explanation of lambda plaques.

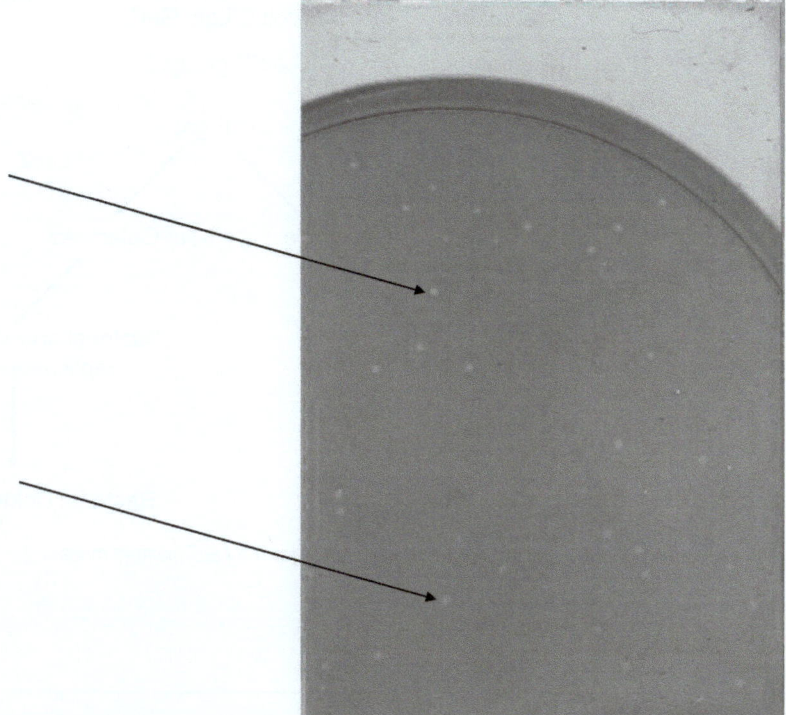

Fig. 8. Photograph showing clear lambda plaques on a bacterial lawn.

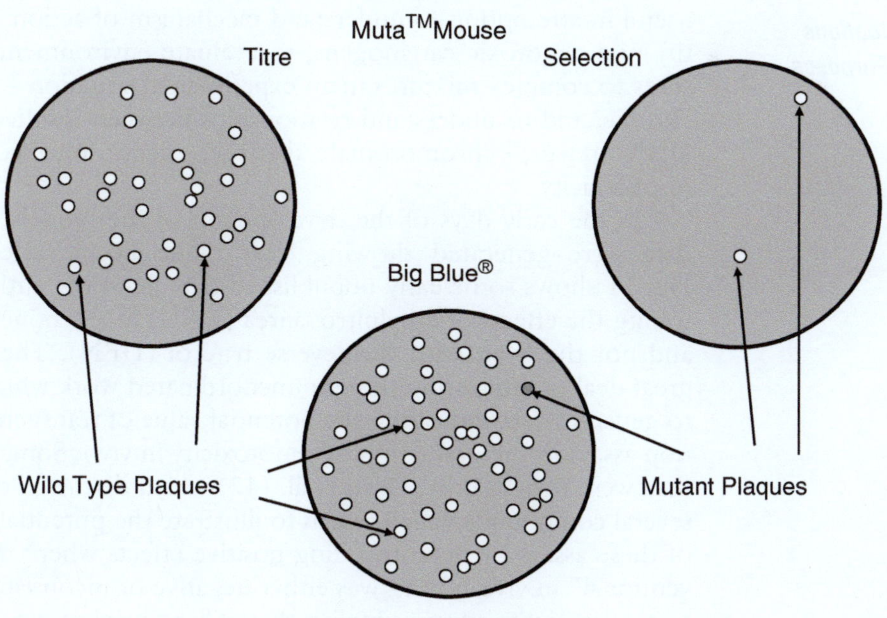

Fig. 9. Diagrammatic representation of mutant and wild-type plaques for Muta™Mouse and Big Blue®.

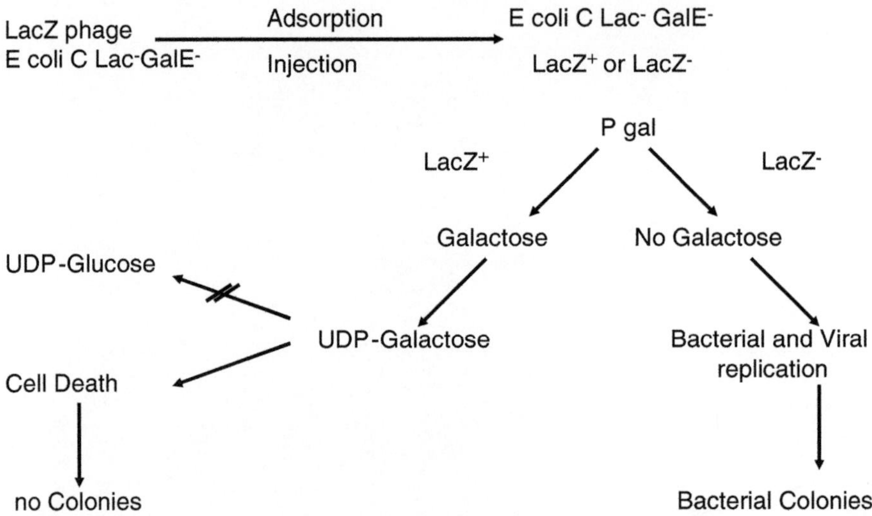

Fig. 10. Figure summarising the principles behind the selection system for the *LacZ* plasmid mouse.

3. Use of Transgenics and the Development of an Accepted Protocol

3.1. Applications for R&D Purposes

There is a blurred boundary between pure R&D and "trouble-shooting" as might be encountered when assessing the safety of a new compound. Transgenic models such as those described can be useful in attempting to understand mechanism of action, to identify non-genotoxic carcinogens, to evaluate environmental exposures to complex mixtures in an experimental situation – acute or chronic, and to understand relationships between insult, adducts, DNA repair, chromosomal damage, gene mutation, and oncogenicity.

In the early days of the development of the models, exciting data were generated showing clear tissue-specific effects, and Fig. 11 shows some early unpublished data from the author illustrating the effects of ethylnitrosourea (ENU) in the bone marrow and not the liver, with the reverse true of (DEN). There was a great deal of interesting though uncoordinated work which began to generate data indicating the potential value of transgenic mutation assays in the evaluation of genotoxicity in vivo. Some of these data were reviewed by, Dean et al. (42) which discussed data with several compounds which began to illustrate the potential benefits of these assays by demonstrating positive effects where the "conventional" in vivo testing was either negative or inconsistent. One particularly interesting example is that of β-propriolactone. A review of the genetic toxicology data for this compound is summarised in Table 4.

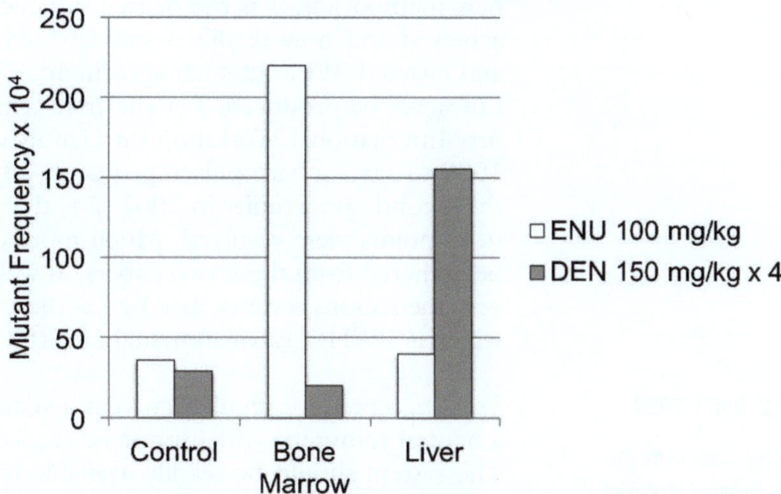

Fig. 11. Early data showing the tissue-specificity of mutation induction by ENU and DEN.

Table 4
β-Propriolactone – summary of genetic toxicology data

Positive in vitro Potent in vitro mutagen (–S9)
Negative in vivo Negative for bone marrow micronucleus (i.p.) Negative for peripheral blood micronucleus (p.o.) Negative In vivo mouse liver UDS (p.o.)
Tumours at site of contact

That such a potent mutagen remained "undetected" in the conventional genetic toxicology was interesting, although it did yield positive results when micronuclei were evaluated in the liver and testis of i.p. dosed rodents. In a study by Brault and Thybaud (43), it was shown that β-propriolactone induced a significant increase in mutations in the stomach of orally dosed rats, peaking 14 days post-treatment but no increase was seen either in the bone marrow or the liver. Furthermore, using the comet assay, DNA damage was seen within hours of dosing in the stomach. This, and similar data, led to the conclusion that these models can be useful for detecting compounds which are thought to be direct-acting or rapidly metabolised and in circumstances where the site-of-contact is more relevant than the tissues conventionally analysed such as bone marrow and liver (42).

One of the most important influencers in the acceptance of new methodologies is the degree to which standardisation can be achieved and how readily a standardised protocol can be agreed and enacted. Without such agreement, a formal OECD guideline can never be produced. For the transgenic mutation systems, the first International Workshop on Genotoxicity Tests took place in 1999 to agree a harmonised protocol (23) though it was not until the second opportunity in 2002 (24) that the outstanding contentious points were resolved. Much more detailed information can be gathered from these two papers, as well as from two recent and very meticulous reviews, one by Lambert et al. (1) and the second which is a WHO Environmental Health Criteria report (2).

3.2. IWGT 1999

3.2.1. Criteria for the Inclusion of an Assay

To be accepted for regulatory work a system should be based upon a neutral transgene which must be recoverable from most tissues. The system should be readily available to laboratories wishing to use it and suitability should be based upon extent of published work, its widespread use in several laboratories and reproducibility. At the time of writing, acceptable systems included *lacI*, *lacZ*, {lambda or plasmid}, cII, and gpt-delta models, though similar criteria should be applied to new systems as they arise.

3.2.2. Treatment Groups

A full set of data must be generated from a minimum of two dose levels, where the top dose should be the MTD and the others should be 2/3 and 1/3 of that MTD. If all three dose groups are complete, the top and second dose levels would be analysed, retaining the low dose for possible future analysis if needed.

3.2.3. Positive Controls

Concurrent positive control animals are not necessary but it is recommended that positive control DNA be included with each plating to confirm the success of the method.

3.2.4. Storage of Tissues

Tissues should be stored at or below –70°C under which conditions they may be kept for several years.

Isolated DNA, stored refrigerated in an appropriate buffer, should be used for mutation analysis within 1 year, but may still generate useful data if stored longer than this.

3.2.5. Methods of Measurement

Standard laboratory or published methods for the detection of mutants have been published and are available for the recommended transgenic models. If modifications are made they should be justified and properly documented. It was agreed that there is no biological justification to set a minimum acceptable number of pfu's from an individual packaging.

3.2.6. Requirements for Reporting

Reporting of a regulatory transgenic mutation study should be as defined for all GLP studies and should include the total number of pfu and MF for each organ and for each animal. Data for individual packagings should be retained but need not be reported.

3.2.7. Statistical Analysis of Data

There should be between five and ten animals per group, assuming 3×10^{-5} MF and 125–300 k pfu/animal. Tissues should be processed and analysed in a block design in order to minimise the impact of process-induced variation. Statistics should involve pairwise analysis and dose–response evaluation if two or more doses are investigated and the concurrent negative controls should be compared with historical data. An appropriate test would be the generalised Cochran-Armitage which allows the analysis of variable data. It is important that the criteria for a positive and negative outcome are defined carefully. For a clear positive result, all treated groups should exhibit a MF twice that of the background or demonstrate a dose–response with one or more treated groups with a MF twice the background. A clear negative result would require statistical non-significance in all treated groups and for all treated groups to lie within the range of the concurrent control ±2 SD. An equivocal result may need some form of repeat.

3.3. IWGT 2002

There were, however, some issues that remained unresolved following the 1999 IWGT workshop and these were addressed subsequently, through discussion and the presentation of data, in 2002 (24). These were duration of treatment, sampling time(s) and whether or not DNA sequencing should be a routine expectation. Let us not forget that these were not simple technical issues but were related to some important scientific points which required discussion and justification. They involved processes which would impact upon the exposure of DNA to induce mutations, the fixation and expression of mutations, the rate of proliferation of the target tissue, tissue and agent specificity and the use of a general versus modified protocol.

3.3.1. Duration of Treatment and Sampling Time(s)

To summarise, the working group recommended that for a general protocol, a treatment schedule of 28 consecutive days plus a single sampling time 3 days later should detect the majority of genotoxic compounds in the majority of tissues. This was based upon the database which the group had reviewed, for 143 agents, published and unpublished, most of which were potent mutagens. It was noted that, for highly proliferating tissues, the optimal sampling time is "early" (~3 days) although for some tissues (e.g. germ cells) sensitivity may be increased by sampling at other time points. Figure 12 (Douglas 2002) shows an example of data generated in the bone marrow following exposure to ENU where the MF peaks very early at 3 days but then falls markedly soon after. Other data did indicate that the longer the treatment, the more mutations are accumulated, but the 28 + 3 protocol was regarded as a suitable compromise when testing multiple tissues. However, other treatment regimens (e.g. weekly) or sampling times may be appropriate under specific circumstances but would need to be scientifically justified, for example, based upon cell proliferation, exposure,

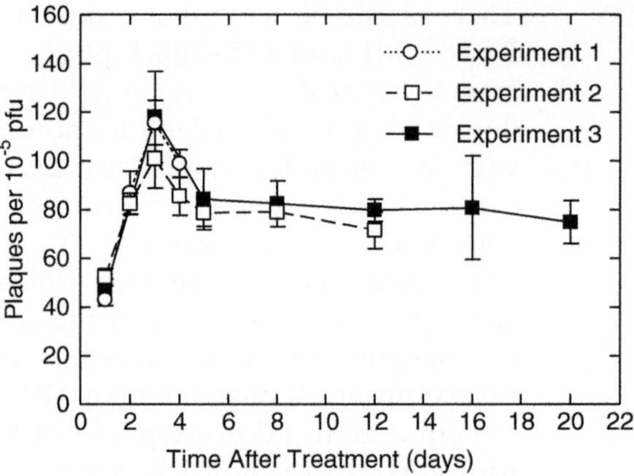

Sampling time of *lacZ* mutations in tansgenic mouse bone marrow cells following a single i.p. dose of 80 mg/kg ENU. Animals received PMSG (5 IU/animal) 48 hr. before sacrifice.

From G Douglas (2002)

Fig. 12. Sampling time of *LacZ* mutations in transgenic mouse bone marrow cells following a single i.p., dose of 80 mg/kg ENU.

TK data, enzyme induction, toxicity (inhibition of cell proliferation), target tissue.

However, a word of warning was issued and advantages and disadvantages of the "28 + 3" protocol were discussed.

Advantages

- Twenty-eight days avoid the confounding effects due to chronic toxicity and carcinogenesis with longer treatments.
- Twenty-eight days allow sufficient accumulation of mutations from weak mutagens.
- Three days avoid a possible decline in MF.
- Other toxicological data are often available for a 28-day treatment (at least in rats).

Disadvantages

- Twenty-eight days might not be necessary for strong mutagens (perhaps 7 or 14 days).
- Longer treatment duration would allow the accumulation of more mutations.
- Three days might not be the optimal sampling time for slowly proliferating tissues (28 days).
- There are several tissues for which there are few or no data.

3.3.2. Selection of Tissues

The criteria used to determine which tissues to sample will depend upon several aspects of experimental design. The route of administration itself might indicate which site-of-contact is likely to be important, such as the lung following inhalation or GI tract following oral administration. Information on the extent of likely systemic exposure, kinetics, and metabolism which can be obtained fro existing.

Information on toxicity/target tissue will also help but, in the absence of any background information, it was recommended that at least one rapidly dividing and one slowly dividing tissue (e.g. bone marrow and liver) should be evaluated.

3.3.3. Sequencing

It was agreed that sequencing is not necessary in the case of either a clearly negative or clearly positive response, but it was acknowledged that sequencing data might be particularly useful to identify and correct for clonal expansion (jackpots). Such jackpot mutations occur when a mutated germ or stem cell leads to the proliferation of cells containing the mutated transgene, and this can lead to abnormally high MF in several tissues from the untreated animal. It is possible to apply appropriate statistical methods for the detection of outliers and removal of outlier values from the data set though it is often advisable to analyse additional samples from the same tissue. Sequencing can also help to investigate high variability, particularly in a control group, when trying to evaluate the relevance of an equivocal result and when investigating molecular mechanisms of mutagenesis.

When sequencing is needed, a minimum of ten mutants per tissue per animal should be sequenced to identify clonal expansion, although more may be necessary to perform accurate clonal corrections. For statistical evaluation of sequence data, there are several relevant publications which describe appropriate methods (REFS). The IWGT group felt there were insufficient data to recommend a minimum number of mutants necessary to evaluate spectral differences.

3.3.4. Other Issues

Some final topics were agreed. It was decided that male animals should normally be used unless there were significant differences between the sexes, in which case then males and females will be required. There may be cases when females alone should be used. All the methods described here should be applicable to the rat as well as the mouse, though very little data exist for the rat, and the choice of rat or mouse as a species will depend upon other toxicological information.

References

1. Lambert I. B., Singer T. M., Boucher S. E. and Douglas G. R. (2008) *OECD Environment, Health and Safety Publications Series on Testing and Assessment (2nd Draft) Detailed review paper on transgenic rodent mutation assays.*
2. Wahnschaffe U., Kielhorn J., Bitcsch A. and Mangelsdorf I. (2006) *WHO Environmental Health Criteria 233, Transgenic Animal Mutation Assays.*
3. Gordon J. W. and Ruddle F. H. (1983) Gene transfer into mouse embryos: production of transgenic mice by pronuclear injection. *Meth Enzymol*, **101**: 411–433.
4. Van Sloun P. P., Wijnhoven S. P, Kool H. J., Slater R., Weeda G., van Zeeland A. A., Lohman P. H., and Vrieling H. (1998) Determination of spontaneous loss of heterozygosity mutations in *Aprt* heterozygous mice. *Nucleic Acids Res.*, **26**(21): 4888–4894.
5. Dobrovolsky, V. N., Casciano D. A. and Heflich R. H. (1999) *Tk+/−* mouse model for detecting *in vivo* mutation in an endogenous, autosomal gene. *Mutat. Res.*, **423**(1–2): 125–136.
6. Winton, D. J., Gooderham N. J., Boobis A. R., Davies D. S. and Ponder B. A. J. (1990) Mutagenesis of mouse intestine *in vivo* using the *Dlb-1* specific locus test: studies with 1,2-dimethylhydrazine, dimethylnitrosamine, and the dietary mutagen 2-amino-3,8-dimethylimidazo(4,5-f)quinoxaline. *Cancer Res.*, **50**(24): 7992–7996.
7. Chen, T., Aidoo A., Manjanatha M. G., Mittelstaedt R. A., Shelton S. D., Lyn-Cook L. E., Casciano D. A., Heflich R. H. (1998) Comparison of mutant frequencies and types of mutations induced by thiotepa in the endogenous *Hprt* gene and transgenic *lacI* gene of Big Blue rats. *Mutat. Res.*, **403**(1–2): 199–214.
8. Jacob, F. and Monod D. (1961) Genetic regulatory mechanisms in the synthesis of proteins. *J Molec Biol*, 3: 318–356
9. Gossen J. A., De Leeuw W. J. F., Tan C. H. T., Zwarthoff E. C., Berends F., Lohman P. H. M., Knook D. L. and Vijg J. (1989) Efficient rescue of integrated shuttle vectors from transgenic mice: a model for studying mutations in vivo. *Proc Natl Acad Sci USA,* **86**: 7971–7975.
10. Kohler S. W., Provost G. S., Fieck A., Kretz P. L., Bullock W. O., Putman D. L., Sorge J. A. and Short J. M. (1991a) Analysis of spontaneous and induced mutations in transgenic mice using a lambda ZAP/*lacI* shuttle vector. *Environ Mol Mutagen*, **18**: 316–321.
11. Kohler S. W., Provost G. S., Fieck A., Kretz P. L., Bullock W. O., Sorge J. A., Putman D. L. and Short J. M. (1991b) Spectra of spontaneous and mutagen-induced mutations in the *lacI* gene in transgenic mice. *Proc Natl Acad Sci USA*, **88**: 7958–7962.
12. Gossen J. A., Molijn A. C., Douglas G. R. and Vijg J. (1992) Application of galactose-sensitive *E. coli* strains as selective hosts for *lacZ*⁻ plasmids. *Nucleic Acids Res,* **20**: 3254.
13. Douglas G. R., Gingerich J. D., Gossen J. A. and Bartlett S. A. (1994) Sequence spectra of spontaneous *lacZ* gene mutations in transgenic mouse somatic and germline tissues. *Mutagenesis*, **9**: 451–458.
14. Dycaico M. J., Provost G. S., Kretz P. L., Ransom S. L., Moores J. C. and Short J. M. (1994) The use of shuttle vectors for mutation analysis in transgenic mice and rats. *Mutat Res,* **307**(2): 461–478.
15. Gorelick N. J. and Thompson E. D. (1994) Overview of the workshops on statistical analysis of mutation data from transgenic mice. *Environ Mol Mutagen*, **23**: 12–16.
16. Piegorsch, W. W., Lockhart A. M., Margolin B. H., Tindall K. R., Gorelick N. J., Short J. M., Carr G. J., Thompson E. D. and Shelby M. D. (1994) Sources of variability in data from a *lacI* transgenic mouse mutation assay. *Environ. Mol. Mutagen.*, 23(1): 17–31.
17. Gossen J. A., Martus H-J., Wei J. Y. and Vijg J. (1995) Spontaneous and X-ray-induced deletion mutations in a *LacZ* plasmid-based transgenic mouse model. *Mutat Res,* **331**: 89–97.
18. Dollé M. E., Martus H. J., Gossen J. A., Boerrigter M. E. and Vijg J. (1996) valuation of a plasmid based transgenic mouse model for detecting in vivo mutations. *Mutagenesis,* **11**: 111– 118.
19. Jakubczak J. L., Merlino G., French J. E., Muller W. J., Paul B., Adhya S. and Garges S. (1996) Analysis of genetic instability during mammary tumor progression using a novel selection-based assay for in vivo mutations in a bacteriophage – transgene target. *Proc Natl Acad Sci USA*, **93**: 9073–9078.
20. Nohmi, T., Katoh M., Suzuki H., Matsui M., Yamada M., Watanabe M., Suzuki M., Horiya N., Ueda O., Shibuya T., Ikeda H., Sofuni T. (1996) A new transgenic mouse mutagenesis test system using Spi− and 6-thioguanine selections. *Environ. Mol. Mutagen.*, **28**(4): 465–470.
21. de Boer J. G., Mirsalis J. C., Provost G. S., Tindall K. R. and Glickman B. W. (1996) Spectrum of mutations in kidney, stomach, and liver from *lacI* transgenic mice recovered after

treatment with tris(2,3-dibromopropyl)phosphate. *Environ Mol Mutagen*, **28**: 418–423.

22. Piegorsch W. W., Lockhart A. C., Carr G. J., Margolin B. H., Brooks T., Douglas G. R., Liegibel U. M., Suzuki T., Thybaud V, van Delft J. H. M. and Gorelick N. J. (1997) Sources of variability in data from a positive selection *lacZ* transgenic mouse mutation assay: an interlaboratory study. *Mutat Res*, **388**: 249–289.
23. Heddle J. A., Dean S., Nohmi T., Boerrigter M., Casciano D., Douglas G. R., Glickman B. W., Gorelick N. J., Mirsalis J. C., Martus H-J., Skopek T. R., Thybaud V., Tindall K. R. and Yajima N. (2000) In vivo transgenic mutation assays. *Environ Mol Mutagen*, **35**: 253–259.
24. Thybaud V., Dean S., Nohmi T., de Boer J., Douglas G. R., Glickman B. W., Gorelick N. J., Heddle J. A., Heflich R. H., Lambert I, Martus H-J., Mirsalis J. C., Suzuki T. and Yajima N. (2003) In vivo transgenic mutation assays. *Mutat Res*, **540**(2): 141–151.
25. Dycaico M. J., Stuart G. R., Tobal G. M., de Boer J. G., Glickman B. W. and Provost G. S. (1996) Species-specific differences in hepatic mutant frequency and mutational spectrum among λ/*lacI* transgenic rats and mice following exposure to aflatoxin B1. *Carcinogenesis*, **17**: 2347–2356.
26. Mientjes EJ, Steenwinkel MJST, van Delft JHM, Lohman PHM & Baan RA (1996) Comparison of the X-gal- and P-gal-based systems for screening of mutant lambda *lacZ* phages originating from the transgenic mouse strain 40.6. *Mutat. Res.*, **360**(2): 101–106
27. Gossen J. and Vijg J. (1993) Transgenic mice as model systems for studying gene mutations in vivo. *Trends Genet*, 9: 27–31.
28. Dean S. W. and Myhr B. (1994) Measurement of gene mutation in vivo using Muta™Mouse and positive selection for *lacZ*– phage. *Mutagenesis*, **9**: 183–185.
29. Nohmi, T., Suzuki M., Masumura K., Yamada M., Matsui K., Ueda O., Suzuki H., Katoh M., Ikeda H., Sofuni T. (1999) Spi – selection: an efficient method to detect gamma-ray-induced deletions in transgenic mice. *Environ. Mol. Mutagen.*, **34**(1): 9–15.
30. Nohmi T., Suzuki T. and Masumura K. (2000) Recent advances in the protocols of trans-genic mouse mutation assays. *Mutat. Res.*, **455**(1–2): 191–215.
31. Swiger RR (2001) Just how does the *cII* selection system work in Muta Mouse? *Environ Mol Mutagen*, **37**(4): 290–296.
32. Matsuoka F., Nagawa F., Okazaki K., Kingsbury L., Yoshida K., Muller U., Larue D. T., Winter J. A. and Sakano H. (1991) Detection of somatic DNA recombination in the transgenic mouse brain. *Science*, **254**: 81–86.
33. Sykes P. J., Hooker A. M., Jacobs A. K., Harrington C. S., Kingsbury L. and Morley A. A. (1998) Induction of somatic intrachromosomal recombination inversion events by cyclophosphamide in a transgenic mouse model. *Mutat Res*, **397**: 209–219.
34. Gondo Y., Shioyama Y., Nakao K. and Katsuki M. (1996) A novel positive detection system of *in vivo* mutations in *rpsL* (*strA*) transgenic mice. *Mutat. Res.*, **360**(1): 1–14.
35. Leach E.G., Narayanan L., Havre P. A., Gunther E. J., Yeasky T. M. and Glazer P. M. (1996a) Frequent spontaneous deletions at a shuttle vector locus in trans-genic mice. *Mutagenesis*, **11**(1): 49–56.
36. Leach E.G., Gunther E. J., Yeasky T. M., Gibson L. H., Yang-Feng T. L. and Glazer P. M. (1996b) Tissue specificity of spontaneous point mutations in lambda *supF* transgenic mice. *Environ. Mol. Mutagen.*, **28**(4): 459–464.
37. Burkhart J. G., Winn R. N., VanBeneden R. and Malling H. V. (1992) The Phi-X174 bacteriophage shuttle vector for the study of mutagenesis. *Environ. Mol. Mutagen.* **19** (Suppl 20) 8
38. Burkhart J. G., Burkhart B. A., Sampson K.S. and Malling HV (1993) ENU-induced mutagenesis at a single A-T base pair in transgenic mice containing Phi-X174. *Mutat Res*, **292**: 69–81.
39. Malling, H.V. and R.R. Delongchamp (2001) Direct separation of *in vivo* and *in vitro am3* revertants in transgenic mice carrying the phiX174 *am3*, *cs70* vector. *Environ. Mol. Mutagen.*, **37**(4): 345–355.
40. Vijg J. and Douglas G. R. (1996*) Bacteriophage lambda and plasmid lacZ transgenic mice for studying mutations in vivo. In: Pfeifer GP ed. Technologies for detection of DNA damage and mutations.* New York, Plenum Press, pp 391–410.
41. Gossen, J.A., de Leeuw W. J. F., Molijn A. C., and Vijg J. (1993) Plasmid rescue from transgenic mouse DNA using *LacI* repressor protein conjugated to magnetic beads. *Biotechniques*, **14**(4): 624–629.
42. Dean S. W., Brooks T. M., Burlinson B., Mirsalis J., Myhr B., Recio L. and Thybaud V. (1999) Transgenic mouse mutation assay systems can play an important role in regulatory mutagenicity testing in vivo for the detection of site-of-contact mutagens. *Mutagenesis*, **14**(1): 141–151.
43. Brault D., Renault D., Tombolan F. and Thybaud V. (1997) Kinetics of detection of β-propiolactone and N-methyl-N -nitro-N-nitrosoguanidine genotoxic effects in mouse gastric mucosa using single cell gel electrophoresis assay and Muta™Mouse assay. *Mutat Res* **379** (Suppl. 1). 150.

Chapter 18

Analysis of Genotoxicity Data in a Regulatory Context

Ian de G. Mitchell and David O.F. Skibinski

Abstract

Analytical methods for regulatory tests usually must be defined before testing. To take this into account and to minimise equivocal interpretations, a sequential strategy is recommended. Assay validity must be verified and results then classed as clearly negative, clearly positive, or uncertain based on historical data. Where there is uncertainty, standard parametric or non-parametric statistical methods should be used with appropriate corrections to assess the significance. The biological importance of statistically significant data should then be evaluated using historical data.

Key words: Assay validity, Statistical significance, Null hypothesis, Parametric, Non-parametric, Multiple comparisons, Biological importance, Historical controls

1. Introduction

Initial rationalisation of statistical analyses for genotoxicity data (1) was followed by UKEMS Guidelines 1989 (2). Data analyses then relied heavily on null hypothesis testing at defined critical (p) values for α error (false positives). Little attention was paid to the problems of multiple comparisons, false negatives (β error), biological importance or quantitative quality control. Here we provide a framework for the analysis of regulatory data which alleviates these problems. A sequential approach is recommended.

Firstly, the assay must be valid in a regulatory context. The treatment data should then be classified clearly negative, clearly positive, or uncertain (based on laboratory historical negative control data). Where there is uncertainty, statistical analyses will be helpful. If treatment-associated increases in response are not significant, the assay is negative. Significant increases should then be assessed for a positive dose–response trend; the absence of such a trend suggests that there is no treatment-relationship and the

James M. Parry and Elizabeth M. Parry (eds.), *Genetic Toxicology: Principles and Methods*, Methods in Molecular Biology, vol. 817, DOI 10.1007/978-1-61779-421-6_18, © Springer Science+Business Media, LLC 2012

assay is negative. Statistically significant increases with a significant positive trend should be assessed for biological importance with important results interpreted as positive.

2. Assay Validity

Regulatory work must be conducted in a GLP (good laboratory practice) accredited laboratory using internationally accepted protocols. However, experimental designs in such guidelines are often sub-optimal for data analyses and can be improved with inexpensive additional work such as duplicating negative controls, reading range-finding tests for the genetic endpoint as well as for toxicity, and repeating in vitro tests. The strategy and methods for data analysis must be defined before testing (a priori) in contrast to the frequently more satisfactory approach of deciding on the analyses after the results have been seen (post hoc or a posteriori). After testing, the data must be checked to determine that they are of adequate quality; that is they should not have excessive variability nor assay-specific artefacts, and there should be consistency of the within-test positive and negative controls with the testing-laboratory's historical control database (which itself must be consistent with the literature). As a guide, for good consistency, negative and positive controls from individual tests should fall within the 99% confidence limits set on the laboratory's historical data (see Subheading 7.1).

3. Where Statistical Analyses Are Necessary and Test Strategy

Clear negative and clear (biologically important) positive results may be defined by use of the laboratory historical negative control data (see Subheading 7.2). For a clear negative treatment, values should be well within the range of historical negative control values, and for a clear positive treatment, values should be extremely rare in the historical negative controls. As a rather arbitrary guide, the 70% confidence limits (CL) and 99.99% CL should be calculated. Clear negatives should fall below the upper of the 70% CL and clear positives would have to exceed the upper of the two 99.99% CL (see Subheading 7.2). This means that 15% [(100 − 70)/2] of the historical negative control data values would be equal to or greater than (≥) the level defined for a clear negative while only 0.005% [(100 − 99.99)/2] would be ≥ the clear positive level (see Subheading 4.1).

Where the outcome is not clear, data should be analysed statistically. The significance of differences between control and treated samples is currently determined by testing the null hypothesis (that there is no difference between treatment and control). The treatment is defined as significant if the calculated probability (p) that control and treatment are derived from the same population is equal to or less than ($\leq$) a defined (critical) probability. Critical probability values are conventionally set at $p = 0.05$ or 0.01 (5.0 or 1.0%) although other critical values are acceptable. Thus if a critical value of $p = 0.05$ is selected, the observed difference between treatment and control is significant only if the chance of finding such a large (or larger) difference in a single population (from which the treated and control samples were derived) is 5.0% or less. A useful addition is to calculate the appropriate confidence interval (CI) by placing confidence limits (CL) on the difference (D) between the control and treatment values. Thus selecting $p = 0.05$ as the critical value, the appropriate 95% CI on D is given by the 95% CL on D. If the CI does not include $D = 0$ then D is significant at $p < 0.05$. For example, for (say) $D = 25$ 95% CI might be 15–35 and this could be written $D = 25$ 95% CI 15–35 $p < 0.05$.

Two types of error can occur; α (type 1) error is the chance of incorrectly rejecting the null hypothesis when it is true (defined by the critical value e.g. with $p = 0.05$, the null hypothesis will be incorrectly rejected in 5.0% of cases), while β (type 2) error is the chance of incorrectly accepting the null hypothesis when it is false ($1 - \beta$, power, being the chance of correct rejection).

4. Inherent Errors and Difficulties in the Null Hypothesis Testing Strategy

4.1. One- or Two-Sided (One- or Two-Tailed) Tests

In a one-sided test, the question is whether treatment values exceed (>) the control values, and in a two-sided test, the question is whether treatment and control differ (< or >). The p value associated with a two-sided test is twice that for a one-sided test for the same data, e.g. $p = 0.05$ two-sided equals $p = 0.025$ one-sided. In genetic toxicology, one-sided tests are nearly always used but there are occasions where two-sided tests may be appropriate (e.g. comet assay where reduced comet size is indicative of DNA cross-linking and increased comet size indicates DNA strand breaks).

Assessment of data is difficult where the experimental p value for α error is very near the selected critical value. For example, with a selected critical value of $p = 0.050$ and an experimental value of 0.051, the data are not significant (negative) but significant (positive) if the calculated p value is 0.050. This is difficult to justify logically.

4.2. Calculation of β Error and Power

Currently, β error is seldom calculated despite the importance of knowing the size of response likely to be missed. To calculate β error it is necessary to know the smallest size of response that it is important to detect, the variance in the data, and the data distribution. Power (rather than β error) is usually stated, e.g. there is an 80% chance of detecting an increase of "*X*" above the control. In the future, β error may need to be calculated before testing (from the laboratory historical data) and after testing.

4.3. Multiple Comparisons

In genetic toxicology, we are concerned with three types of multiple comparisons:

1. Several separate tests may be analysed by meta-analysis (e.g. in biomonitoring). However, in regulatory genotoxicity assays there are usually only one (in vivo) or two (in vitro) separate assays. If the results of two separate assays agree, this is very reassuring; if not a third "decider" test probably will be needed (3) but it is rarely productive to go beyond a total of three separate tests.
2. For multiple endpoints in one test, the Bonferroni correction (4, 5), which is a post-hoc test (see Subheading 6.3), should be used in setting critical values in pre-test protocols. The logic is that in a test with several endpoints each endpoint has the same chance (a priori) of giving an α error (false positive). Thus the Bonferroni corrected critical probability (pc) that one out of n endpoints will be significant by chance is n times the uncorrected critical p value set for each individual endpoint (pi); thus $n \times \text{pi} = \text{pc}$. For example, if only one out of three endpoints is significant at $p = 0.05$, it becomes not significant after the Bonferroni correction as the corrected $p = 0.15$ (3×0.05). Alternatively, the equation above can be re-arranged to calculate the uncorrected p value for an individual endpoint (pi) that is needed to give a Bonferroni corrected critical value set at pc. This is done by dividing the corrected value (pc) by the number of endpoints (n); thus $\text{pi} = \text{pc}/n$. For example, an Ames test with five strains in the presence and absence of S9 has ten endpoints; so for one of these endpoints to be significant at the Bonferroni corrected critical value of $p = 0.05$, one individual endpoint must be significant at an uncorrected critical value of $p = 0.005$ ($0.05/10$).

   Where endpoints are related there may be problems after testing if several give results with p values near the overall critical value. In such cases, the scientist in charge of the study may have to disregard the strategy set out in the protocol and use their scientific judgement to re-assess the data.

   The p values for independent endpoints in a test can be combined to produce a single overall p value (6). This is useful when some or many of the p values for endpoints are low but not individually significant. The test is applied to the a priori p values. With n endpoints the quantity $-2\sum_{i=1}^{n} \log_e p_i$ is distributed as χ^2

with $2n$ degrees of freedom. A significant χ^2 value indicates that the test is significant overall but those endpoints that are individually significant cannot be determined. A useful aspect of this test is that p values from qualitatively different tests and experimental approaches can be combined.

3. Where several treatment levels are compared with control the Bonferroni correction, or, better, one of the modifications (4, 5) or alternatives (see Subheading 6.2) may be used, but these "over-correct". Better methods collectively compare treatment with control, e.g. Dunnett's test (3, 7) or the Williams test (8) or some form of trend analysis, e.g. linear trend. In general, Dunnett's test is preferred because it makes no assumption that the mutation will increase with increasing dose at all dose levels.

4.4. Outliers

An outlier is a data point that clearly does not fit in with the rest of the sample data. Identification of outliers is based on specific tests or on probability generating functions (9, 10). For example, where samples of values that fit the normal distribution, statistical tests, e.g. Grubb's Test (11) are available. Outliers can arise from non-treatment-related confounding factors, resistant or sensitive sub-populations, technical errors or "chance". In statistical analyses they should only be disregarded and removed from the dataset if it is likely that they have arisen from a technical error or a confounding factor. Outliers will have a disproportionate effect on parametric analyses and are best dealt with either by analysing the data with and without the outlier or by using non-parametric (ranking) methodology; the latter is probably the best regulatory compromise.

4.5. Biological Importance

Biological importance does not feature in null hypothesis testing. It is crucial but very difficult to define. Definitions are often entirely arbitrary such as what some expert group thinks is important (e.g. the twofold rule or global evaluation factors). Alternatively, it may be argued that once treatment has shown a statistically significant increase over control with a significant positive trend over a number of dose levels then this demonstrates a hazard and, hence, a biologically important effect.

We suggest basing biological importance on a control value very rare in the laboratory historical control data (12) e.g. for guidance, use the upper of the 99.9% confidence limits above which there will be only 0.05% of the control data (see Subheading 7.2). Results which are not biologically important should normally be classified as negative. However, with a statistically significant increase nearly large enough for biological importance and with a clear dose–response trend, a classification of equivocal may be more realistic.

5. Introduction to Statistical Methods for Testing the Null Hypothesis

5.1. Some Shorthand and Definitions

1. $\sum_{i=1}^{n} X_i$ (often abbreviated to Σx) – add together all n values of a set of numbers designated $x_1, x_2, x_3, \ldots, x_n$.

   x!: multiply x by all preceding whole numbers (e.g. $3! = 3 \times 2 \times 1$).

   df: degrees of freedom; the number of independent comparisons that can be made among a series of observations (normally n observations gives a df of $n-1$).

   m: sample mean ($\Sigma x/n$). Mean is often written *X* bar (with a line over top of the *X*) and is used to estimate μ the global (total) population mean.

   ss: sum of squares (of deviations from sample means). $\Sigma(x-m)^2 = \Sigma x^2 - (\Sigma x)^2/n$.

   s^2 *(V or var)*: variance of the sample (ss/df) and is used to estimate σ^2 the variance for the global population.

   s (SD): standard deviation of a sample $(s^2)^{1/2}$ and is used to estimate σ the standard deviation for the global population.

   CV (VC): coefficient of variation (variance coefficient); (s/m) often given as percentage.

   SE: standard error (SD of the mean) $(s/n^{1/2})$.

   χ^2: chi-squared statistic (ss/m).

   "*m*" statistic (not to be confused with *m* for mean) – (χ^2/df or s^2/mean).

2. *Statistics* are measurements. Statistical analysis is the analysis of measurements either *parametrically* assuming that the sample is derived from a global population with a distribution defined by *parameters* (measurements) that can be estimated from the samples or *non-parametrically* making no distributional assumptions.

3. A *sample* is the part of a population selected for examination to make *estimates* of the population values and *inferences* about population relationships. A *random sample* is where each member of the *global* (total) population has an equal chance of selection for the sample while *bias* occurs where the chances of selection for a sample is not the same for each member of a global population with the result that estimates for the parameters of the global population based on such samples will also be biased (not representative of the global population).

4. Measurements are either *constants* or *variables*; a *dependent variable* depends on the variation of the *independent variable* e.g. in $y = ax + bx^2$; a and b are constants, x is the independent variable, and y is the dependent variable and a *function* of x, $f(x)$.

A *discrete variable* can only take on a restricted set of values while there are no discontinuities in a *continuous variable.* Variab les are *associated* (*correlated*) if a change in one accompanies a change in the other(s). *Monotonic* increases (or decreases) occur when each successive value exceeds (or is less than) the next one.

5. *Frequency distributions* show the frequency of occurrence of differing values for a variable within a population. The proportion of a population in an *interval* between two values defines the probability of an observation occurring in that area. Confidence limits (CL) enclose a defined proportion of the data e.g. 95%, etc.

5.2. Key Measurements for Statistical Analyses

Distributions can differ in shape or central tendency. In genetic toxicology, the question is whether central tendency for treated samples differs significantly from that for the negative controls. This will depend on the size of the difference and on variability. There are three measures of central tendency; mode (most frequent value), median (middle value of a ranked data set) and mean (the average value). The mode is seldom used but the median and mean are both common in statistical methodology, e.g. for a set of values: 1, 3, 4, 4, 4, 4, 5, 5, 5, 7, 12, and 18, mode = 4, median = 4.5 (the average of the two middle values with an even number of data points) and mean = 6. The amount of variation is normally measured as standard deviation; for example:

Data values = 2, 6, 7, 5, 12, 8 and 9.

$n = 7$ (number of units in sample) and $df = (n - 1) = 6$ (degrees of freedom).

$\Sigma x = 2 + 6 + 7 + 5 + 12 + 8 + 9 = 49$ (sum of values).

$m = \Sigma x / n = 49/7 = 7$ (mean value).

$\Sigma x^2 = 4 + 36 + 49 + 25 + 144 + 64 + 81 = 403$.

$(\Sigma x)^2 = 49^2 = 2{,}401$.

$ss = \Sigma x^2 - (\Sigma x)^2 / n = 403 - 2{,}401/7 = 403 - 343 = 60$ (sum of squares).

$s^2 = ss/df = 60/6 = 10$ (variance).

$s = (s^2)^{1/2} = 10^{1/2} = 3.162$ (standard deviation).

$SE = s/n^{1/2} = 3.162/7^{1/2} = 3.162/2.646 = 1.195$ (standard error).

5.3. Expected Distribution of Genotoxicity Data and Appropriate Statistical Strategies

It is crucial to realise that the unit of statistical analysis is the culture or the animal and not the cell, colony, or chromosome (which are the units used for scoring results) (*see* Subheading 6.1). This is because the difference between treatments should be measured against the variation between cultures (or animals) within treatment, not between cells within cultures (or animals). Further, all statistical methods have underlying assumptions which must be checked for applicability to the test.

Most genotoxicity endpoints are discrete variables and would be expected to follow binomial or Poisson distributions. These are based on the assumptions that there are only two possible results per observation (e.g. mutant or non-mutant), the chances of these two results are the same for every observation and successive observations are independent. However, these last two assumptions may well be violated in genotoxicity testing. Also technical variation in the conduct of the tests must be taken into account. Thus the experimental data will have variance in excess of the binomial/Poisson prediction resulting in compound distributions, e.g. negative, correlated or beta binomials. Negative binomials will tend to arise where the chances of two results are not the same for every observation (e.g. clumping of mutant cells or non-random inter-animal variation). Beta binomials will occur where successive observations are not independent (e.g. if a bag contains 50 white balls and 50 black balls removal of a black ball on the first sampling slightly increases the chances of drawing a white ball on the second sampling). Correlated binomials will arise when normally distributed variables (e.g. technical errors) are combined with binomial/Poisson distributed data.

Such distributions are skewed where there are less than five mutants per sample but near-normal with higher numbers (>5–10). However, the variance is unstable being dependent on the mean. Such distributions may be analysed parametrically, e.g. by weighted analysis or after transformation to give, approximately, normal distributions with stable (constant) variance. Alternatively, non-parametric methods that make no distributional assumptions may be applied, e.g. data permutation, rank permutation, and other methods. Non-parametric methods do have some underlying assumptions and are less versatile than parametric methods. They will give more false negatives when the parametric distributional assumptions apply but less false positives where these assumptions are violated. A useful compromise is parametric analysis of ranks (13, 14) which has much of the versatility and power of parametric methods but retains the broader applicability of non-parametric methods.

6. Overview of Statistical Analytical Methods and Strategies

There are a variety of statistical methods (2, 15) which will give reasonably reliable statistical analyses appropriate for genotoxicity data. For most routine work, statistical analyses are too time-consuming to perform by hand and it is customary instead to use dedicated software packages. It remains important for the researcher to understand basic methodology and the assumptions and limitations of the different statistical methods applicable to a specific problem. Commonly used commercial packages include SAS and SPSS and there is much freeware accessible through the internet. For example, the package Poptools (16), which embeds in Excel, is useful for Monte Carlo methods based on re-sampling the data under consideration. Another very useful free resource is the R integrated suite of software facilities that is continually under development by its enthusiasts.

6.1. Approaches to the Analysis of Discrete Variables

Although discrete and continuous variables are analysed using different methods, discrete data can be treated as continuous for analytical purposes. For example, suppose that the ratio of normal to micronucleated cells on a single treatment slide are 993:7 and the ratio for a control slide is 999:1, in each case the 1,000 cells scored being fixed by the experimenter.

Treated as discrete counts, the two ratios can be compared using a wide variety of tests such as the conventional χ^2 contingency test and Fisher's exact test. Assumptions of such tests are that the cells counted are a random sample of those available for counting and the measures on different cells are independent of each other. Lack of independence might occur if there were local areas on the treatment slide where the chemical agent had for some reason higher effective concentration. Cells with or without micronuclei might then occur in clumps. The experimental design described, with only one treatment and one control slide is not a particularly wise one. The reason is that an apparent treatment effect might be due to a factor unrelated to the treatment, e.g. the treated slide might inadvertently have been subject to slightly different temperature. The way around this problem is to have replicate treatment and control slides where extraneous factors might hopefully by randomised across slides.

Normally priority would be given to increasing the number of treatment slides rather than the number of control slides. Suppose, for example, that there were 20 treatment and four control slides. Clearly the total number of cells with and without micronuclei might be summed over all control and all treatment slides and the two ratios compared by a contingency test as above.

An alternative would be to treat the percentage of micronuclei on each slide as a continuous variable. There would be a sample of $n_1 = 20$ treatment percentage values and a group of $n_2 = 4$ control percentage values. The average percentage values in these two groups could be compared by a t-test, usually after transformation, or a nonparametric equivalent. The advantage of such an approach is that the treatment effect can be tested against the slide to slide variation. This would take account of extraneous factors causing variation between slides unrelated to the treatment, something that could not be done with the initial design with only two slides.

Count data such as that described above can also be analysed by a variety of computer intensive Monte-Carlo methods (17). Applied to the two micronuclei count ratios above the method would randomly generate a re-sampled data set in which the column and row totals are the same as in the original data but the internal cell values might be different. The test statistic, e.g. a χ^2, is then calculated for the re-sampled data set. This process is repeated a large number of times, for example to generate 10,000 re-sampled datasets. The original χ^2 value would be deemed significant if only a small proportion (say 5% or less) of the 10,000 re-sampled data sets had χ^2 values greater than that for the original data. Monte-Carlo methods can be useful when there are complex experimental designs which are difficult to analyse using conventional methods. It might also be desirable to have the re-sampled data sets with row totals but not column totals fixed. This would certainly be appropriate for the micronuclei data where the re-sampled data sets should have the total per slide fixed at 1,000 as in the original data.

6.2. Parametric Analyses, Transformations and Normal Distributions

The most powerful statistical tests generally assume that the data follows some underlying theoretical distribution. These are parametric tests and a common assumption is that the data follows the normal distribution (e.g. the *t*-test). Much data are, however, not normally distributed. A common departure from normality is skewness where the distribution is asymmetrical having a long tail to one side; nearly always towards higher values (positive skew) for genotoxicity data.

Such departures from normal distributions can often be corrected by carrying out a mathematical transformation of the data. Log and square root transformations often correct skewness and the arcsine transformation is useful for percentages. Both before and after transformation the distribution of the data can be checked for normality. Visual examination of the data against the normal curve is useful. This can be supplemented with the one-sample Kolmogorov–Smirnov test which tests for significant overall deviations from normality. Other tests such as Grubb's test (11) identify outliers at the extreme ends of the distribution. These can then be removed in certain circumstances (Subheading 4.4) from the data prior to analysis with parametric statistics. As samples become very large, quite small deviations from normality might be detected as significant even though the data fits well visually to the normal distribution. It is generally accepted that parametric tests are quite robust to (tolerant of) small deviations from normality. Parametric tests such as the *t*-test also assume that the two groups being compared have the same variance. A number of tests, e.g. the Levene test, compare the two groups for homogeneity of variance.

6.3. Analysis of Variance and Post-hoc Tests for Control Versus Treatment

A typical experiment looking at a dose response to a potentially toxic chemical might have several values (e.g. colony counts for individual plates) at each dose of the chemical and also for the control. As there are more than two groups, such data might typically be analysed using analysis of variance (ANOVA) which belongs to a family of general linear models called mixed effects models in which both random and fixed factors can be analysed. Good resources for analysing a huge range of models are available in the R suite computer software package, for example for analysing mixed effects models (18). The result of the ANOVA might indicate that overall there are significant differences in average colony counts between the doses (including control). Of interest, however, is to know whether comparisons between particular doses are statistically significant.

A variety of tests carried out alongside the ANOVA provide the answer. These tests, known as post-hoc tests, take account of the fact that when many tests are made, some will be significant just by chance even with random data (1/20 are expected to be significant at the 5% level). Dunnett's test is a widely used post-hoc test for comparing a series of doses (collectively) with the control. The Bonferroni and LSD tests are used making all possible post-hoc comparisons among the doses and controls. Other tests such as the Ryan-Einot-Gabriel-Welsch F test allow the dose means to be sorted into homogenous subsets. Doses within a subset are not significantly different from each other, whereas doses in different subsets are significantly different.

The Bonferroni procedure (Subheading 4.3 steps 2 and 3) has been criticized as being unduly conservative, that is the procedure gives too many false negative results. More recently, other less conservative extensions of the Bonferroni have been devised. One involves the concept of false discovery rate (FDR) (19). In the FDR approach, perhaps 15 of the 100 comparisons might be identified as being significant at the FDR rate of say 20%. What this means is that 20% of the 15 (i.e. about 3) are not significant whereas the remaining 80% (i.e. about 12) are significant. Application of the Bonferroni would not normally result in as many as 12 significant comparisons in this example. The drawback of the FDR approach is that it would not be known which 12 of the 15 comparisons are the significant ones. The FDR approach is widely used in gene expression studies using microarrays where the expression of thousands of genes might be compared between treatment and control conditions.

Tests such as ANOVA permit analysis of complex experimental designs in which the effects of different factors and their interactions on the dose response can be disentangled. A typical example would be where five doses of a chemical are analysed for their effect on SCE counts in mice. Each dose has four mice and in each mouse 20 cells are scored for SCE count. This is a nested design. The total variation in SCE count can be partitioned into that between doses, between mice within dose, and between cells within mice. It could be that there are highly significant differences between mice within dose when assessed against between cells within mice. In this circumstance, the differences between doses should be assessed against the differences between mice, not against the differences between cells.

6.4. Parametric Approaches Other than ANOVA for Treatment Versus Control

Other approaches exist to analyzing both simpler and complex experimental designs. An example is the application of the concept of likelihood. In this approach, a testing of the difference between the means of two samples would proceed by calculating the probability of observing the data, given an assumed distribution such as normality. This would be done assuming no difference between means (the null hypothesis) and then assuming the mean difference is that actually observed (the alternative hypothesis). The ratio of these two probability values is then called the ratio of the likelihood of the two hypotheses, and can be converted into a single probability value as in a *t*-test. Likelihood can be regarded as a measure of confidence in the hypothesis given the data, and is not the same as the probability of the data given the hypothesis.

Another approach that is gaining popularity in biology is that based on the so-called Bayesian methods. These are related to likelihood but allow prior information to be included in probability calculations. For example, the probability values for the experiment under consideration can be modified taking account of prior historical results regarding toxic effects of the chemical under consideration.

6.5. Non-parametric Analyses of Non-normal Data for Treatment Versus Control

In general, if the departures from normality are judged as unacceptable and transformation is not effective in correcting these, parametric tests such as the *t*-test or ANOVA cannot be used. The alternative is to use Monte-Carlo methods (17) or a variety of available non-parametric tests. An example of a Monte-Carlo approach in place of the conventional *t*-test is the procedure of bootstrapping. As example, suppose that percentage micronuclei values are to be compared in a sample of ten control slides and 50 treatment slides. In the first bootstrap re-sampling, ten values are picked at random, with replacement, from the ten control slides, and 50 values are picked at random, with replacement, from the 50 treatment slides. The difference between the means is calculated for this first bootstrap sample. This procedure is then repeated for 10,000 independent bootstrap samples. The difference between means in the original sample is deemed significant if only a small proportion (say 5% or less) of the 10,000 bootstrap samples have difference values which are less than or equal to zero, which represents the null hypothesis of no difference between means. The philosophy underlying bootstrapping is that the re-sampling reproduces that same variation as that estimated by the variance in the original samples when using parametric statistics.

Non-parametric tests have an advantage that they do not assume that the data follow some underlying distribution. Often they operate by analysis of ranking patterns in the data. The cost of fewer assumptions, however, is lower power compared with parametric approaches. Commonly used non-parametric equivalents of the *t*-test are the Mann–Whitney test and the two-sample Kolmogorov–Smirnov test. The former assumes that the two samples being tested have the same distribution and tests for a difference in median between the samples, while the latter tests for a difference between samples in either the shape or location (median) of their distributions. A non-parametric equivalent of a one-way ANOVA, for example, where several doses of the same chemical are being compared, is the Kruskal–Wallis test. Although nonparametric tests are useful they are not so versatile as parametric ANOVAs and not easily extended to more complicated ANOVA designs such as nested or multi-way ANOVAs.

Other non-parametric methods use count data directly and include χ^2 contingency and Fisher exact tests alluded to in Subheading 6.1. Such methods are often used in biomonitoring. However, they are usually not appropriate to routine regulatory toxicology because they are based, incorrectly, on the cell rather than the treated culture or animal as the unit of statistical analysis thereby missing much of the test variance and over-estimating significance. The exception is the use of Fisher exact tests to analyse in vitro metaphase analysis (2) because the data are so sparse that no other methods can be used (and the size of the errors at the level of counting very small numbers, such as 0, 1, 2 or 3 of aberrant cells, dwarf all the other errors).

6.6. Parametric Analyses for Trend

In an experiment where increasing doses of a chemical are tested, it is usually important to determine whether there is a dose-related trend in the response (see Subheading 4.5). That is, does the response, e.g. in terms of number of cells with micronuclei, get progressively greater as the dose increases? Such information does not necessarily come from ANOVA which indicates only whether there are overall significant differences between doses or which specific doses are different from the control or from other doses.

Correlation methods can indicate whether there is, overall, an increase in response with dose, but give no information about the precise form of the dose–response relationship, that is whether it follows a straight line or is curvilinear. The Pearson product-moment correlation coefficient can be used for normally distributed data.

Parametric linear regression is normally used for determining whether the data fit a straight line. A fitted line is determined which minimises the squared deviations of the response values, perpendicular to the line. It is important to note that parametric linear regression should only be used when the X axis variable is measured without error. This would be the case if the X axis values were different fixed doses of a chemical. If the X axis values are measured with error, a special version of linear regression should be used that takes account of this. Alternatively, a method such as principal components analysis should be used. This allows error in both X and Y axes and minimises the squared perpendicular deviations of the data points from the fitted line.

Fitting a linear regression to data is a simple example of model fitting. This model has two parameters, the regression slope and the intercept on the Y axis. Testing for a curvilinear relationship can be done by adding another parameter as in quadratic regression. The two models can be fitted sequentially. If the quadratic explains a significantly greater proportion of the variation in response than the linear model it would be preferred. General methods are now available for comparing a set of models that might explain the data and picking the best model. An example is the use of the Akaike information criterion (AIC) and related approaches (20). These balance the likelihood of the model and the number of parameters. The best models generally have high likelihood with not too many parameters.

6.7. Non-parametric Analyses of Non-normal Data for Trend

Non-parametric equivalents of the parametric Pearson product moment-correlation are Spearman's and Kendall's tests which work by ranking the data and then calculating the correlation of the ranks rather than the original values. Non-parametric, but less powerful, equivalents of parametric linear regression are the Jonckeere-Terpstra test and the median test.

7. Setting Up and Using Historical Databases

The historical negative control database should contain data from at least six or seven separate tests and the data should be consistent with literature data. However, no result should be excluded from the database unless there is good evidence of a technical error or confounding factor. For the positive control database, at least two dose–response experiments should be performed using a well-known mutagen and with the negative control values subtracted from the mutagen-treatment values (to give mutagen-related increases). The positive control dose to be used should be taken from near the bottom of the steepest part of the composite dose–response curve and at least four more tests carried out at that dose. For establishing confidence limits, data should be transformed (usually square root or log) firstly, to give approximately normal distributions (which are needed to enable the calculation of probability from standard deviation e.g. using "*t*" distribution tables) and secondly, to give approximately stable variance (important so that variance and standard deviation remain constant when mean values change).

7.1. Confidence Limits on Historical Data for Assessing Validity from Test Controls

To assess test validity from concurrent negative and positive controls upper and lower confidence limits (CL) are placed on the historical control mean values. First, all the mean data are transformed and CL calculated from the formula: $CL = M \pm (\text{“}t\text{”} \times s)$, where M is the mean of n individual control means, "*t*" is the tabulated value for the degrees of freedom (df) and the desired CL (p value) and s is the calculated standard deviation. As an example, seven mean values from seven sets of historical negative control mouse micronucleus data (individual animal data shown in Subheading 7.2) are used here to set 99% CL using square root data transformation $[(x+½)^{½}]$:

Control means = 2.2 1.2 0.6 1.0 0.4 1.6 2.0

Transformed means = 1.643 1.304 1.049 1.225 0.946 1.449 1.581.

Σx (sum of transformed means) = 9.197, $n = 7$, $df = n - 1 = 6$, $M = \Sigma x / n = 1.314$.

$\Sigma x^2 = 12.495$ $(\Sigma x)^2 = 84.584$ $\Sigma x^2/n = 12.083$.

$ss = \Sigma x^2 - (\Sigma x)^2/n = 12.495 - 12.083 = 0.412$.

$s^2 = ss/df = 0.412/6 = 0.0687$.

$s = (s^2)^{½} = 0.262$.

"*t*" tabulated value = 3.707 for df = 6 and $p = 0.01$ (two-sided) the 99% CL.

99% CL (transformed) = $M \pm (\text{“}t\text{”} \times s) = 1.314 \pm (3.707 \times 0.262) =$ 2.285–0.343.

Back transform (square then subtract ½) 99% CL = 4.721 to −0.382.

These data would indicate that a current test would be valid if its negative control mean was between 4.721 and 0 (values of less than zero such as −0.382 have theoretical but no practical meaning). If the current negative control exceeded 4.721 the test would be invalid and would have to be repeated. However, a current negative control value in excess of 4.721 would still have to be added into the historical database for the future unless that there was clear evidence that it was due to an artefact or confounding factor (see the beginning of Subheading 7). Current test positive control data would be evaluated in the same way with test rejection if the positive control fell outside either of the historical positive control 99% confidence limits.

7.2. Confidence Limits on Historical Negative Controls to Determine Where to Use Statistical Analyses and to Assess Biological Importance

Setting limits for where statistics are needed or for biological importance is a one-sided problem so only the upper of the two confidence limits (CL) is applied and the associated p value is, therefore, half that for where both CL are applied e.g. 99% upper and lower CL enclose 99% of the area under the frequency curve ($p=0.01$) while the upper limit CL encloses 99.5% of the area ($p=0.005$).

CL on historical negative control data are used for setting limits (thresholds) for where statistics are needed or for biological importance. It is crucial to differentiate between tests where all the negative control samples can be considered to belong to a single global population with a constant true mean and where the global population and true mean vary from test to test. Such variation in true mean will occur either where cells or animals are derived from highly heterogeneous populations (e.g. human lymphocytes) or, more seriously, where there is clonal expansion of mutants (21) in the pre-treatment inocula. In practice, the controls for most in vivo and in vitro cytogenetic assays may be considered to have a constant true mean whereas the controls for most bacterial and mammalian cell mutation assays will have true means that vary from test to test. To give an example of how to assess both types of control data the same set of micronucleus data is assessed below firstly assuming (correctly) that the global mean is the same for all tests and secondly assuming (incorrectly) that it varies from test to test.

Where the true mean is approximately constant for all tests, the formula and calculation are the same as in Subheading 7.1. However, here we are only interested in the upper CL. Thus the upper 99% confidence limit will be 4.721. It should be noted that if the number of statistical units differs in the control and treated groups, a correction may be needed. This uses the approximation $(1/n_t+1/n_c)^{½}/(1/n_c+1/n_c)^{½}$ where n_t and n_c are respectively the number of units in the treated and control groups.

Where the true mean varies from test to test, the CL have to be based on the mean for the test under consideration because its true test mean will not be the same as the true mean for the historical data. However, transformation will render variance approximately constant across all tests. Thus historical average standard deviation, s (calculated from historical average variance) can be applied to the current test mean to enable CL calculation from the formulae: $d=\text{“}t\text{”}\times s(1/n_c+1/n_t)^{½}$ and $\text{CL}=d\pm m$, where d is the maximum difference expected between control and treatment at the specified p value if treatment has no effect, "t" is the value appropriate to the CL and degrees of freedom (df), (n_c) and (n_t) are the respective number of concurrent control and treated units, and m is the mean for the test in question e.g. setting the upper of the 99% CL with the data square root $(x+½)^{½}$ transformed and a concurrent test control mean (m) of, say, 1.286:

Individual animal

Historical control values						Mean (m)	Sum of sq. (ss)	Var. (s^2)
Test 1	3	2	2	4	0	(2.2)		
(trans.)	1.87	1.58	1.58	2.12	0.71	1.572	1.132	0.283
Test 2	3	1	0	1	1	(1.2)		
(trans.)	1.87	1.22	0.71	1.22	1.22	1.248	0.679	0.170
Test 3	0	0	2	1	0	(0.6)		
(trans.)	0.71	0.71	1.58	1.22	0.71	0.986	0.636	0.159
Test 4	0	2	0	3	0	(1.0)		
(trans.)	0.71	1.58	0.71	1.87	0.71	1.116	1.278	0.320
Test 5	0	2	0	0	0	(0.4)		
(trans.)	0.71	1.58	0.71	0.71	0.71	0.884	0.606	0.151
Test 6	3	1	2	2	0	(1.6)		
(trans.)	1.87	1.22	1.58	1.58	0.71	1.392	0.794	0.199
Test 7	5	1	0	1	3	(2.0)		
(trans.)	2.35	1.22	0.71	1.22	1.87	1.474	1.637	0.409
								$\Sigma s^2 = 1.691$
								$\text{av.}s^2 = 0.242$

$s\ (s^2)^{1/2} = 0.492$, $n_t = n_c = 5$, df $= 8$ $(n_c - 1 + n_t - 1)$.

"t" for df $= 8$ and $p = 0.01$ (two-sided; equals $p = 0.005$ one-sided) is 3.355 (99.0% CL).

$d = \text{"}t\text{"} \times s(1/n_c + 1/n_t)^{1/2} = 3.355 \times 0.492 \times 0.632 = 1.043$.

$m = 1.286$, $[(x + 1/2)^{1/2}]$ transformed $= 1.336$.

CL 99% (transformed) $= (d \pm m) = 1.336 \pm 1.043 = 2.379\text{–}0.293$.

Back transform (square then subtract ½) CL = 5.160 to −0.414.

The upper of the two 99% CL (the 99.5% limit) = 5.160.

Two points need to be made. First, just as an example of the calculation method, the 99% confidence limits have been determined. It should be noted that these are not what we recommend for determining the limit for a clear negative (70% CL) or for a clear positive (99.99% CL) or for biological importance of statistical positives (99.9% CL). In any case, these recommendations are not immutable but are only for guidance. Second, the two methods of calculation give quite similar values (4.721 and 5.160) for the upper of the 99% CL. Had we used assay data where there was a lot of test to test variation in true mean and, incorrectly, calculated the upper CL as if the true mean were constant, the CL value would have been very much higher than by the correct calculation.

The potential error is much greater if we assume that the true control mean is constant when it is not, than if we assume that the true mean does vary from test to test when it does not.

8. Conclusion

The strategy set out here should enable the reader to achieve a balance between false positive and false negative results while applying methods that are relatively simple, transparent, and acceptable to regulators. However, there are occasions when analytical methods set before testing are not appropriate for the observed results. In such cases, the scientist responsible must use scientific judgement and disregard the pre-set strategy.

The analysis of individual tests is only the start of defining hazard but with very little input into risk. Risk is assessed on weight-of-evidence when a battery of in vitro and in vivo genotoxicity assays are set in the context of other absorption, distribution, metabolism, excretion, toxicology (animal and human), and biomonitoring studies.

References

1. Parry, J., Brooks, T., Mitchell, I., and Wilcox, P. (1984) Genotoxicity studies using yeast cultures. In, *UKEMS Sub-Committee on Guidelines for Mutagenicity Testing (Part 2 A) Supplementary tests* (Ed. B. J. Dean, UKEMS Pub. Swansea, UK) **Ch. 3**. pp. 27–61.
2. *Statistical Evaluation of Mutagenicity Test Data.* (1989) (Ed. David J. Kirkland, Cambridge University Press, Cambridge, UK).
3. Mahon, G.A.T., Middleton, B., Robinson, W.D., Green, M.H.L., Mitchell, I. de G., and Tweats, D.J. (1989) Microbial assays from microbial colony assays. *In, Statistical Evaluation of Mutagenicity Testn Data.* (Ed. David J. Kirkland, Cambridge University Press, Cambridge, UK.) **Ch. 2**. pp. 26–65.
4. Abt, K. (1981) Problems of repeated significance testing. *Controlled Clinical Trials* **1**, 377–381.
5. Hothorn, L. (1994) Multiple comparisons in long-term toxicity studies. *Env. Hlth. Pers.* **102** (Suppl. 1.), 3–38.
6. Sokal, R.R. and Rohlf, F.J. (1995) *Biometry* (3rd edition) (W.H.Freeman and Company, USA).
7. Bechhofer, R. E. (1988) Percentage points of multivariate Student 't' distributions. In, *Selected Tables in Mathematical Statistics Vol. 11.* (Ed. R. E. Oden and J. M. Davenport, Am. J. Math. Soc. Pub. Providence, USA) pp. 1–319.
8. Williams, D.A. (1972) The comparison of several dose levels with a zero dose control. *Biometrics* **28**, 519–531.
9. Barnett, V., and Lewis, T. (1984) *Outliers in Statistical Data* (2nd ed.) (John Wiley & Sons Pub. Chichester, UK).
10. Mitchell, I. de G., Carlton, J. B., and Gilbert, P. J. (1988) The detection and importance of outliers in the *in vivo* micronucleus assay. *Mutagenesis* **3**, 491–495.
11. Grubbs, F. (1969) Procedures for Detecting Outlying Observations in Samples. *Technometrics* **11**, 1–21.
12. Mitchell, I. de G., Rees, R. W., Gilbert, P. J., and Carlton, J. B. (1990). The use of historical data for identifying biologically unimportant but statistically significant results in genotoxicity assays. *Mutagenesis* **5**, 159–164.
13. Conover, W. J., and Iman, R. L. (1981) Rank transformations as a bridge between parametric and non-parametric statistics. *American Statistician* **35**, 124–133.
14. Mitchell, I. de G., Amphlett, N. W., and Rees, R. W. (1994) Parametric analysis of rank transformed data for statistical assessment of genotoxicity data with examples from cultured mammalian cells. *Mutagenesis* **9**, 125–132.

15. Campbell, R.C. (1989) *Statistics for Biologists* (3rd edition) (Cambridge University Press, Cambridge, UK).
16. Hood, G. M. PopTools – Software for the analysis of ecological models. Version 3.0.6, CSIRO, 2000.
17. Manly, B.F.J. (2007) *Randomization, bootstrap and Monte Carlo methods in biology* (3rd edition) Chapman and Hall. Boca Raton.
18. Pinheiro, J.C. and Bates, D.M. (2004) *Mixed-effects models in S and S-plus.* Springer Science, New York.
19. Verhoeven, K.J.F., Simonsen, K.L., McIntyre, L.M. (2005) Implementing false discovery rate control: increasing your power. OIKOS **108**, 643–647.
20. Burnham, K.P., Anderson, D.R. (2001) Kullback-Leibler information as a basis for strong inference in ecological studies. Wildlife Research **28**, 111–119
21. Luria, S. E., and Delbrück, M. (1943) Mutations of bacteria from virus sensitivity to virus resistance. *Genetics* **28**, 491–511.

INDEX

James M. Parry and Elizabeth M. Parry (eds.), *Genetic Toxicology: Principles and Methods*, Methods in Molecular Biology, vol. 817, DOI 10.1007/978-1-61779-421-6, © Springer Science+Business Media, LLC 2012

H

N

T

V

W

X

Z

MIX
Papier aus verantwortungsvollen Quellen
Paper from responsible sources
FSC® C105338

If you have any concerns about our products,
you can contact us on
ProductSafety@springernature.com

In case Publisher is established outside the EU,
the EU authorized representative is:
Springer Nature Customer Service Center GmbH
Europaplatz 3, 69115 Heidelberg, Germany

Printed by Libri Plureos GmbH
in Hamburg, Germany